Database Systems

Database Systems

A Pragmatic Approach

Third Edition

Elvis C. Foster
with Shripad V. Godbole

CRC Press is an imprint of the
Taylor & Francis Group, an **informa** business
AN AUERBACH BOOK

Third Edition published 2023
by CRC Press
6000 Broken Sound Parkway NW, Suite 300, Boca Raton, FL 33487-2742

and by CRC Press
2 Park Square, Milton Park, Abingdon, Oxon, OX14 4RN

First Edition published 2014 by Apress

Second Edition published 2016 by Apress
CRC Press is an imprint of Taylor & Francis Group, LLC

ISBN: 978-1-032-21732-1 (hbk)
ISBN: 978-1-032-20202-0 (pbk)
ISBN: 978-1-003-27572-5 (ebk)

DOI: 10.1201/9781003275725

Typeset in Garamond
by Deanta Global Publishing Services, Chennai, India

Contents

SECTION C THE STRUCTURED QUERY LANGUAGE

Preface

This book has been compiled with three target groups in mind: The book is best suited for undergraduate students of computer science (CS) or a related discipline who are pursuing a course in database systems. Graduate students who are pursuing an introductory course in the subject may also find it useful. Finally, practicing software engineers and/or information technology (IT) professionals who need a quick reference on database design may find it useful.

The motivation that drove this work was a desire to provide a concise but comprehensive guide to the discipline of database design, construction, implementation, and management. Having worked in the software engineering and IT industries for several years before making a career switch to academia, it has been my observation that many IT professionals and software engineers tend to pay little attention to their database design skills; this is often reflected in the proliferation of software applications with inadequately designed underlying databases. In this text, the discipline of database systems design and management is discussed within the context of a bigger picture — that of software engineering. The reader is led to understand from the outset that a database is a critical component of a software system, and that proper database design and management is integral to the success of the software system. Additionally and simultaneously, the reader is led to appreciate the huge value of a properly designed database to the success of a business enterprise.

The book draws from lecture notes that have been compiled and tested over several years, and with outstanding results. The lecture notes draw on personal experiences gained in industry over the years, as well as the suggestions of various professionals and students. The chapters are organized in a manner that reflects my own approach in teaching the course, but each chapter may be read on its own.

The text has been prepared specifically to meet three objectives: comprehensive coverage, brevity, and relevance. Comprehensive coverage and brevity often operate as competing goals. In order to achieve both, I have adopted a pragmatic approach that gets straight to the critical issues for each topic and avoids unnecessary fluff, while using the question of relevance as the balancing force. Additionally, readers should find the following features quite convenient:

- Short paragraphs that express the salient aspects of the subject matter being discussed
- Bullet points or numbers to itemize important things to be remembered
- Diagrams and illustrations to enhance the reader's understanding
- Real-to-life examples
- Introduction of a few original methodologies that are applicable to database design; the main ones are: the Relations–Attributes List (RAL as in Section 3.8); the Object Naming Convention (ONC as in Section 3.9); the Object/Entity Specification Grid (O/ESG as in Section 5.8); the User Interface Topology Chart (UITC as in Section 6.3); the Dynamic

Menu Interface Designer (DMID as in Section 21.3); and the Information Topology Chart (ITC as in Section 21.4.9)

- Step-by-step, reader-friendly guidelines for solving generic database systems problems
- Each chapter begins with an overview and ends with a summary
- A chapter with sample assignment questions (for the student) and case studies (for the student as well as the practitioner)

What Is New in Edition 3

This edition represents an enhancement of the second edition. In each chapter, the content has been revised; figures and illustrations have been revised and new ones added; and additional content has been added. Chapter 26 represents new information not covered in the previous editions. In this chapter, I discuss DBMS alternatives such as the Entity–Attributes–Value model, NoSQL databases, database-supporting frameworks, and other burgeoning database technologies.

Organization of the Text

The text is organized in 31 chapters (26 regular chapters and 5 appendices), placed into six divisions. From start to finish, the text is written as a friendly conversation with students — the way classroom sessions would be organized. The chapters outlined are as follows:

Part A: Preliminary Topics

Chapter 01: Introduction to Database Systems
Chapter 02: The Database System Environment

Part B: The Relational Database Model

Chapter 03: Introducing the Relational Model
Chapter 04: Integrity Rules and Normalization
Chapter 05: Database Modeling and Design
Chapter 06: Database User Interface Design
Chapter 07: Relational Algebra
Chapter 08: Relational Calculus
Chapter 09: Reflective Look at the Relational Model

Part C: Structured Query Language (SQL)

Chapter 10: Overview of SQL
Chapter 11: SQL Definition Statements
Chapter 12: SQL Data Manipulation Statements

Chapter 13: Logical Views and Security
Chapter 14: The System Catalog
Chapter 15: Some Limitations of SQL

Part D: Advanced Topics

Chapter 16: Database Administration
Chapter 17: Distributed Database Systems
Chapter 18: Object Databases
Chapter 19: Data Warehousing and Information Extraction
Chapter 20: Web-Accessible Databases
Chapter 21: Using Relational Databases to Anchor Management support Systems

Part E: Overview of Contemporary DBMS Suites and Technologies

Chapter 22: Overview of Oracle
Chapter 23: Overview of DB2
Chapter 24: Overview of MySQL
Chapter 25: Overview of SQL Server
Chapter 26: Other Emerging Database Methodologies

Part F: Appendices

Appendix 1: Review of Trees
Appendix 2: Review of Hashing
Appendix 3: Review of Information Gathering Techniques
Appendix 4: BNF Syntax for Selected SQL Statements
Appendix 5: Sample Exercises and Examination Questions

Overview of Chapters

Chapter 1 introduces the course in database systems, establishing its importance, scope, and relevance. The chapter proceeds under the following subheadings: Definition and Rationale; Objectives of a Database System; Advantages of a Database System; Approaches to Database Design; Desirable Features of a Database System; Database Development Life Cycle; Summary and Concluding Remarks.

Chapter 2 exposes the student to the environment of a database system. This includes discussion of the following captions: Levels of Architecture; Inter-level Mappings; Database Personnel; Database Management System; Components of the DBMS Suite; Front-end and Back-end Perspectives; Database System Architecture; Database System Classifications; Summary and Concluding Remarks.

Chapter 3 introduces the student to the fundamental principles of the relational model for database design. The chapter discusses the following:

Basic Concepts; Domains; Relations; Relational Database System; Identifying Relationships; Representing Relationships; Implementing Relationships; Relation–Attributes List and Relationship List; Database Naming Convention; Summary and Concluding Remarks.

Chapter 4 discusses various data integrity and normalization rules. The chapter covers captions such as: Fundamental Integrity Rules; Foreign Key Concept; Rationale for Normalization; Functional Dependence and Non-loss Decomposition; First Normal Form; Second Normal Form; Third Normal Form; Boyce–Codd Normal Form; Fourth Normal Form; Fifth Normal Form; Other Normal Forms; Summary and Concluding Remarks.

Chapter 5 applies the principles discussed in Chapters 3 and 4 to the problem of database modeling and design. The chapter discusses various approaches to database modeling and design. Subtopics include Database Model and Database Design; The E–R Model Revisited; Database Design with the E–R Model; The Extended Relational Model; Database Design with the Extended Relational Model; The UML Model; Database Design with the UML Model; Innovation: The Object/Entity Specification Grid; Database Design Using Normalization Theory; Database Model and Design Tools; Summary and Concluding Remarks.

Chapter 6 summarizes fundamental principles of (database) user interface design. This short chapter proceeds under the following captions: Introduction; Deciding on the User Interface; Steps in the User Interface Design; User Interface Development and Implementation; Summary and Concluding Remarks.

Chapter 7 introduces the reader to relational algebra as the foundation for understanding how databases are implemented. Included also is a sample college database, which is used as a reference for subsequent chapters. Subtopics discussed include: Introduction; Basic Operations of Relational Algebra; Syntax of Relational Algebra; Aliases, Renaming and the Relational Assignment; Other Operations; Summary and Concluding Remarks.

Chapter 8 introduces the student to relational calculus as a complement and equivalent of the relational algebra of the previous chapter. After the introduction, the chapter continues discussion via the following captions: Calculus Notations and Illustrations; Quantifiers, Free and Bound Variables; Substitution Rule and Standardization Rules; Query Optimization; Domain Related Calculus; Summary and Concluding Remarks.

Chapter 9 examines the relational model in more detail by discussing the technical requirements of the relational model, as well as Codd's 12 rules for relation database systems. Subheadings are: The Relational Model Summarized; Ramifications of the Relational Model; Summary and Concluding Remarks.

Having laid the foundation in previous chapters, Chapter 10 launches into a focused study of structured query language (SQL), the universal database language of choice. The chapter provides an overview of the language: Important Facts; Advantages of SQL; Summary and Concluding Remarks.

Unlike several texts in the field, the study of SQL begins (at the beginning) with a discussion of the data definition language (DDL) statements in Chapter 11. Captions covered include: Overview of Oracle's SQL Environment; Database Creation; Database Management; Tablespace Creation; Tablespace Management; Oracle Cloud Features; Table Creation Statement; Dropping or Modifying a Table; Working with Indexes; Creating and Managing Sequences; Altering and Dropping Sequences; Creating and Managing Synonyms; Summary and Concluding Remarks.

Chapter 12 discusses the data manipulation language (DML) statements of SQL. This includes: Insertion of Data; Update Operations; Deletion of Data; Commit and Rollback Operations; Basic Syntax for Queries; Simple Queries; Queries Involving Multiple Tables; Queries Involving the Use of Functions; Queries Using LIKE and BETWEEN Operators; Nested Queries; Queries Involving Set Operations; Queries with Runtime Variables; Queries Involving SQL Plus Format Commands; Embedded SQL; Dynamic Queries; Summary and Concluding Remarks.

Chapter 13 discusses SQL logical views and security. Areas covered include: Traditional Logical Views; System Security; Materialized Views; Summary and Concluding Remarks.

Chapter 14 discusses the system catalog (also referred to as the data dictionary) and its importance to a DBMS. The Oracle Data Dictionary is used as a case study. The following captions are covered: Introduction; Three Important Catalog Tables; Other Catalog Tables; Querying the System Catalog; Updating the System Catalog; Summary and Concluding Remarks.

Chapter 15 highlights some of the limitations of SQL. These include: Programming Limitations; Limitations on Views; Stringent Enforcement of Referential Integrity; Limitations on Calculated Columns; If–Then Limitation; Summary and Concluding Remarks.

Chapter 16 addresses the matter of database management. It proceeds under the following captions: Database Installation, Creation, and Configuration; Database Security; Database Management; Database Backup and Recovery; Database Tuning; Database Removal; Summary and Concluding Remarks.

Chapter 17 discusses distributed database systems in the following areas: Introduction; Advantages of Distributed Database Systems; Twelve Rules for Distributed Database Systems; Challenges to Distributed Database Systems; Database Gateways; Future of Distributed Database Systems; Summary and Concluding Remarks.

Chapter 18 discusses object-oriented (OO) databases as an alternative to relational databases. The chapter advances under the following captions: Introduction; Overview of Object-Oriented Database Management Systems; Challenges to Object-Oriented Database Management Systems; Hybrid Approaches; Summary and Concluding Remarks.

Chapter 19 discusses data warehousing in the following areas: Introduction; Rationale for Data Warehousing; Characteristics of a Data Warehouse; Data Warehouse Architectures; Extraction, Transformation and Loading; Summary and Concluding Remarks.

Chapter 20 provides an overview of Web-Accessible Databases in the following areas: Introduction and Rationale; Web-Accessible Database Architecture; Supporting Technologies; Implementation with Oracle; Implementation with DB2; Generic Implementation via Front-end and Back-end Tools; Summary and Concluding Remarks.

Chapter 21 provides insights on using relational databases to anchor management support systems (MSSs). Much of the content covered are based on research previous conducted in this area. The main topics covered are: Overview of Management Support Systems; Building System Security through Database Design; Case Study: Dynamic Menu Interface Designer; Selected MSS Project Ideas; Summary and Concluding Remarks.

Chapters 22–25 are devoted to providing an overview of four prominent DBMS suites, namely Oracle, DB2, MySQL, and SQL Server (one per chapter). Each chapter identifies the main features of the DBMS suite in question, provides an overview of the main components, mentions some shortcomings, and closes with a summary and some concluding remarks.

Finally, Chapter 26 explores other emerging database methodologies such as the Entity–Attributes–Value model, database-supporting frameworks, NoSQL databases, and other burgeoning trends.

Appendices 1 and 2 provide a review of trees and hashing, respectively, as covered in a typical course in Data Structures and Algorithms. Appendix 3 reviews information-gathering techniques typically covered in a course in Software Engineering. Appendix 4 provides the BNF syntax for selected SQL statements. Finally, Appendix 5 provides some sample examination questions and case studies for the student to practice on.

Text Usage

The text could be used as a one-semester or two-semester course in database systems, augmented with material from a specific database management system. However, it must be stated that it is highly unlikely that a one-semester course will cover all 25 chapters. The preferred scenario therefore is a two-semester course. Below are two suggested schedules for using the text: one assumes a one-semester course; the other assumes a two-semester course. The schedule for a one-semester course is a very aggressive one that assumes adequate preparation on the part of the participants; this schedule is shown in Figure P.1. The schedule for a two-semester course gives the student more time to absorb the material and also engage in a meaningful project; this schedule is outlined in Figure P.2.

One-Semester Schedule:	
Week	**Topic**
01	Chapters 01 & 02
02	Chapter 03
03	Chapter 04
04	Chapter 04
05	Chapter 05
06	Chapters 06 & 07
07	Chapter 08
08	Chapters 09 & 10
09	Chapter 11
10	Chapter 12
11	Chapter 12
12	Chapters 13 & 14
13	Chapters 15 & 16
14	Chapters 17 – 19
15	Chapters 20 & 21
16	Review & Final Exam

Figure P.1 Schedule for a one-semester course.

Two-Semester Schedule:	
Week	**Topic**
01	Chapters 01 & 02
02	Chapter 03
03	Chapter 04
04	Chapter 04
05	Chapters 05
06	Chapters 06 & 07
07	Chapter 08
08	Chapters 09 & 10
09	Chapter 11
10	Chapter 12
11	Chapter 12
12	Chapters 13 & 14
13	Chapters 15 & 16
14	Chapters 17 & 18
15	Chapters 19 & 20
16	Chapter 21
16	Review
17	Chapters 22 & 23
18	Chapters 23 & 24
19	Chapters 25 & 26
20-21	Review
22-32	Course Project

Figure P.2 Schedule for a two-semester course.

Approach and Notations

As can be observed, I have employed a principle-then-example approach throughout the course. All principles and theories are first explained and then clarified by examples where deemed necessary. The reason for this approach is that I firmly believe one needs to have a solid grasp of database principles and theories in order to do well as a database designer or administrator. This is also the reason the course covers database theory before delving into practical implementation issues via SQL. Database design is emphasized as a critical component of good software engineering, as well as the key to successful company databases.

Chapters 8 and 9 discuss relational algebra and relational calculus, respectively, as the basis for modern database languages. Then Chapters 10–15 cover the salient features of SQL, the universal database language. In these chapters, the Backus–Naur Form (BNF) notation is extensively used, primarily because of its convenience and brevity, without sacrificing comprehensive coverage.

Feedback and Support

It is hoped that you will have as much fun using this book as I have had in preparing it. You are welcome to access support materials from the CRC Publishing website (https://www.routledge.com/) as well as my personal website (www.elcfos.com). Additionally, your comments will be appreciated.

Acknowledgment

I express profound gratitude to my wife, Jacqueline, as well as children Chris-Ann and K. V. Rhoden, for putting up with me during the extended periods of preparation of this text. There have been countless nights spent in my study at home, and as many forgone opportunities for family time; still my family members understood that this project meant a lot to me, and their support was unwavering.

Also, I must recognize several of my past and current students at Keene State College (KSC), on whom my lecture notes have been repeatedly tested. Of the students taught at the institution, I must single out a few contributions: Georgie Hill committed personal time to assist me in redoing many of the illustrations: Jesse Schmidt, James Dahlen, Joshua Pritchett, Myles Dumas, and Thomas O'Dea participated on various ongoing projects, some of which have been mentioned in Chapter 21.

Prior to my stint at KSC, I have had the good fortune of grooming several students from three different institutions. These earlier experiences have been foundational for the eventual publication of this volume. In this regard, I would like to make special mention of Dionne Jackson, Kerron Hislop, Brigid Winter, Abrams O'Buyonge, Sheldon Kennedy, and Ruth Del Rosario.

Special appreciation is offered to my colleague Shripad Godbole, who has coauthored some of the chapters with me, particularly in Division E of the text. Being a practicing database administrator, Shripad has also served as a valuable resource in these areas.

The editorial and production teams at CRC Press/Taylor & Francis Group deserve mention for their work in facilitating the publication of this volume. I make special mention of John Wyzalek and Todd Perry at CRC Press/Taylor & Francis Group and Vijay Bose and the supporting team at Deanta Global. Earlier editions and drafts were reviewed by Han Reichgelt, Marlon Moncrieffe, and Jacob Mangal; for the current edition, Matthew Taylor has joined this distinguished group; they also deserve mention. Thanks to everyone!

Elvis C. Foster, PhD
Keene State College
Keene, New Hampshire, USA

PRELIMINARY TOPICS

This preliminary division of the course is designed to cover some fundamentals.

The objectives are as follows:

- To define and provide a rationale for database systems
- To identify the many objectives, advantages, and desirable features of a database system
- To discuss the salient features of a database system environment

The division consists of two chapters:

- Chapter 1 — Introduction to Database Systems
- Chapter 2 — The Database System Environment

DOI: 10.1201/9781003275725-1

Chapter 1

Introduction to Database Systems

Welcome and congratulations on your entry to this course in database systems. The fact that you are in this course means that you have covered several fundamental topics in programming, data structures, user interface, and software engineering. Now you want to learn about databases — their significance, the underlying theoretical principles that govern them, how they are constructed, and their management. You are at the right place. This chapter addresses the first issue: the significance of database systems. Topics covered include the following:

- Definition and Rationale
- Objectives of a Database System
- Advantages of a Database System
- Approaches to Database Design
- Desirable Features of a Database System
- Database Development Life Cycle
- Summary and Concluding Remarks

1.1 Definitions and Rationale

A *database system* (DBS) is a computerized record-keeping system with the overall purpose of maintaining information and making it available whenever required. The database typically stores related data in a computer system.

A *database management system* (DBMS) is a set of programs that allow for the management of a database. Starting in chapter 2 and extending to subsequent chapters, we will cover several of the critical functions of a DBMS. Some of the more obvious ones are the following:

- Data definition — creation and management of relations, dependencies, integrity constraints, views, etc.

DOI: 10.1201/9781003275725-2

- Data manipulation — insertion, update, deletion, retrieval, reorganization, and aggregation of data
- System and data security — controlling access to the system, resources, and data
- Programming language support

Components of a DBS include:

- Hardware and operating system
- DBMS
- Database
- Related software systems and/or applications
- Users — including technical users and end users

Database users communicate with the software systems/applications, which in turn communicate (through the programming interface) with the DBMS. The DBMS communicates with the operating system (which in turn communicates with the hardware) to store data in and/or extract data from the database, which is illustrated in Figure 1.1.

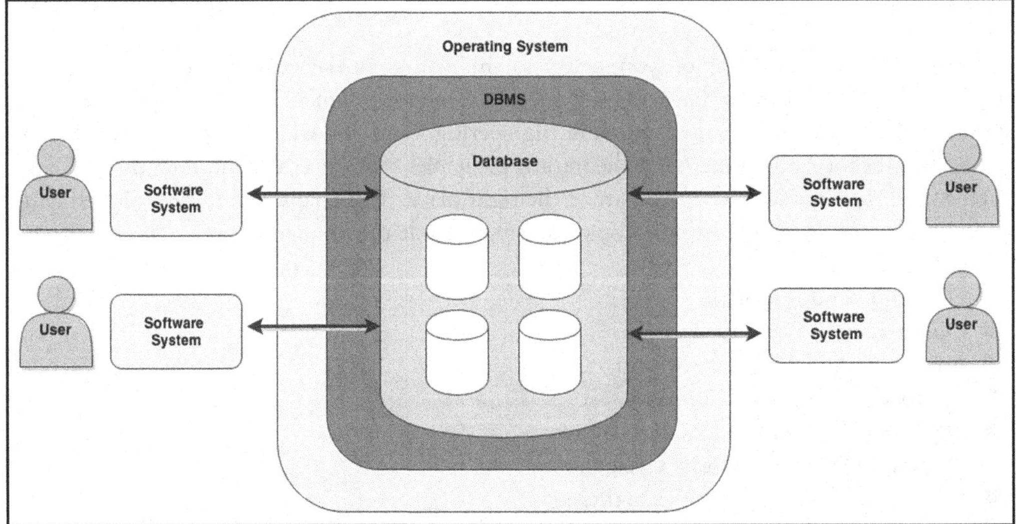

Figure 1.1 Simplified illustration of a DBS.

Databases are essential to software engineering; many software systems have underlying databases that are constantly accessed, though in a manner that is transparent to the end-user. Table 1.1 provides some examples. Companies that compete in the marketplace need databases to store and manage their mission-critical data and other essential data.

In this course, you will learn how to design, implement, and manage databases. In so doing, you will be exposed to various database technologies and methodologies that are common in the software engineering industry.

Before proceeding further, it is important to make a distinction between *data* and *information*. Data refers to the raw materials that software systems act on in order to produce useful information to end-users. Information is processed and assimilated data that conveys meaning to its intended users. A database is configured to store data. Software systems and/or applications

Table 1.1 Illustrations of the Importance of Database

Software Category	*Database Need*
Operating systems	A sophisticated internal database is needed to keep track of various resources of a computer system including external memory locations, internal memory locations, free space management, system files, and user files. These resources are accessed and manipulated by active jobs. A job is created when a user logs on to the system and is related to the user account. This process (also called a job) can in turn create other jobs, thus creating a job hierarchy. When you consider that in a multi-user environment, there may be several users and hundreds to thousands of jobs, as well as other resources, you should appreciate that underlying an operating system is a very complex database that drives the system.
Compilers	Like an operating system, a compiler has to manage and access a complex dynamic database consisting of syntactic components of a program as it is converted from source code to object code.
Information systems	Information systems all rely on and manipulate related databases, in order to provide mission-critical information for organizations. All categories of information systems are included. Common categories include (but are not confined to) decision support systems (DSS), executive information systems (EIS), management information systems (MIS), Web information systems (WIS), enterprise resource planning systems (ERPS), and strategic information systems (SIS).
Expert systems	At the core of an expert system is a knowledge base containing cognitive data, which is accessed and used by an inference engine, to draw conclusions based on input fed to the system.
CAD, CAM, and CIM systems	A computer-aided design (CAD) system, computer-aided manufacturing (CAM) system, or computer-integrated manufacturing (CIM) system typically relies on a centralized database (repository) that stores data that is essential to the successful operation of the system.
Desktop applications	All desktop applications (including hypermedia systems and graphics software) rely on resource databases that provide the facilities that are made available to the user. For example, when you choose to insert a bullet or some other enhancement in an MS Word document, you select this feature from a database containing these features.
CASE and RAD tools	Like desktop applications, computer-aided software engineering (CASE) tools and rapid application development (RAD) tools rely on complex resource databases to service the user requests and provide the features used.
DBMS suites	Like CASE and RAD tools, a DBMS also relies on a complex resource database to service the user requests and provide the features used. Additionally, a DBMS maintains a very sophisticated meta database (called a data dictionary or system catalog) for each user database that is created and managed via the DBMS.

provide the intermediary role of pulling data from the underlying database and synthesizing this data into meaningful information for the end-users.

What would life be like without contemporary database systems? If you know someone who is old enough, ask him/her about such an era of filing cabinets, hand-written records, or typewriter-generated documents. Life was very slow then, but it was the norm. Try to fit that lifestyle into 21st-century life, and you would have a perfect euphemistic definition of misery. Quite simply, it would not work. Databases are here to stay!

1.2 Objectives of a Database System

There are several primary and secondary objectives of a database system that should concern the computer science (CS) professional. Whether you are planning to design, construct, develop, and implement a DBS or you are simply shopping around for a DBMS, these objectives help you to develop an early appreciation for the field; they should also provide useful insight into where the course is heading. As you will soon see, these objectives are lofty, and it is by no means easy to achieve them all.

1.2.1 Primary and Secondary Objectives

The primary objectives of a database system include the following:

- Security and protection — prevention of unauthorized users; protection from inter-process interference
- Reliability — assurance of stable, predictable performance
- Facilitation of multiple users
- Flexibility — the ability to obtain data and effect action via various methods
- Ease of data access and data change
- Accuracy and consistency
- Clarity — standardization of data to avoid ambiguity
- Ability to service unanticipated requests
- Protection of the investment — typically achieved through backup and recovery procedures
- Minimization of data proliferation — new application needs may be met with existing data rather than creating new files and programs
- Availability — data is available to users whenever it is required

In addition to the above, there are some additional objectives that one may argue are just as important. For want of a better term, let us label these as secondary or additional objectives. Included in these additional objectives are the following:

- Physical data independence — storage hardware and storage techniques are insulated from application programs
- Logical data independence — data items can be added or subtracted, or the overall logical structure modified, without affecting existing application programs that access the database
- Control of redundancy — the general rule is to store data minimally and not replicate that storage in multiple places unless this is absolutely necessary
- Integrity controls — range checks and other controls must prevent invalid data from entering the system

- Clear data definition — it is customary to maintain a data dictionary that unambiguously defines each data item stored in the database
- Suitably user-friendly interface — be it graphical, command based, or menu based
- Tunable — easily reorganizing the database to improve performance without changing the application programs
- Automatic reorganization or migration to improve performance

1.2.2 Clarification on Data Independence

Data independence is an important concept that needs further clarification: Data independence is the immunity of application programs to changes in structure and/or access strategy for the related data. It is necessary for the following reasons:

- Different applications and users will need to have different logical views (interpretation) of data.
- The tuning of the system should not affect the application programs.

Physical data independence implies that the user's view is independent of physical file organization, machine, or storage medium. Logical data independence implies that each user (or application program) can have his/her (its) own logical view and does not need a global view of the database.

As an aspiring computer science (CS) professional, you have by now been exposed to various high-level programming languages (HLPLs). These HLPLs have built-in file processing systems that you have no doubt gained mastery in using. You will soon learn that these HLPLs do not support data independence. When you use them to define your data files, there is no separation between the data file you wish to manipulate and the application programs that use them. A database system resolves this dilemma by introducing an additional layer of abstraction between the application programmer and the data files that are manipulated in multiple application programs.

1.3 Advantages of a Database System

A database system brings a number of advantages to its end-users as well as the company that owns it. Some of the advantages are mentioned below:

- Redundancy can be reduced.
- Inconsistencies can be avoided.
- Data can be shared.
- Standards can be enforced.
- Security restrictions can be applied.
- Integrity can be maintained.
- Conflicting requirements can be balanced.
- Improved performance due to speed of processing, reduction in paperwork, etc.
- Maintenance and retrieval of data are very easy — no complicated application program needed.
- Is not solely dependent on the high-level language (HLL) programming for use.

- Logical views of data stored can be easily created.
- Record structures can change without any adverse effect on data retrieval (due to data independence).

1.4 Approaches to Database Design

In examining the management of data via computerized systems, it appears that five broad approaches have been pursued over the past few decades:

- Instant small system — uses one file
- File processing systems — involve many files
- Traditional nonrelational systems, e.g., hierarchical, inverted-list, and network approaches
- Relational databases (the focus of this course) — pioneered by prominent individuals such as Edgar Codd, Ronald Fagin, Christopher Date, among others
- Object databases — an alternate approach (also discussed later in the course)

These five approaches may actually be rearranged into three broad categories as follows:

- Conventional files
- Outdated database approaches such as the hierarchical model, network model, and the inverted-list model
- Dominant contemporary database approaches as in the relational model and the object-oriented model

In addition to the dominant contemporary database approaches, there are three emerging database approaches that are worthy of mention. They are summarized below (for further clarification, see Chapter 26):

- **Hadoop:** This describes a framework for handling distributed processing of large data sets (see Apache 2014)
- **Entity–Attributes–Value (EAV) Model:** This approach reduces a database to three principal storage entities — an entity for defining other entities; an entity for defining properties (attributes) of entities; an EAV entity that connects the other two entities and stores values for entity–attribute combinations (see Wikipedia 2016).
- **NoSQL:** The acronym "NoSQL" is often interpreted as "not only Structured Query Language." This approach refers to a family of non-relational database approaches that are designed for managing large data sets, while providing benefits such as flexibility, scalability, availability, lower costs and special capabilities. Four related methodologies are key value stores, graph stores, column stores, and document stores (see IBM 2015).

1.4.1 Conventional Files

Figure 1.2 illustrates the idea of the conventional file approach. Application programs exist to update files or retrieve information from files. This is a traditional approach to database design that might still abound in very old *legacy systems* (software systems constructed based on old technology and/or methodologies). You have also used this approach in the early stages of your journey as a CS professional: Most high-level programming languages (HLPLs) have built-in file processing systems that you learn to use.

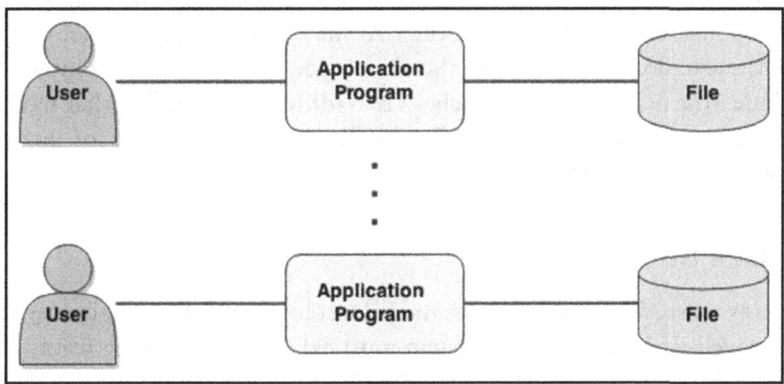

Figure 1.2 Conventional file-based design.

The main problem with the traditional approach is the absence of data independence. To illustrate the problem, consider for a moment, an information system consisting of 30 data files and 150 application programs that manipulate those files. Suppose that each data file impacts 10–15 application programs. Then, whenever it becomes necessary to adjust the structure of a data file in any way, it will be necessary to track down 10–15 application programs and adjust them as well. Certainly, you do realize this is a very inefficient way of managing a complex software system that may be contingent on a far more complex database.

1.4.2 Database Approach

In the database approach, a database is created and managed via a database management system (DBMS) or CASE tool. A user interface, developed with appropriate application development software, is superimposed on the database, so that end users access the system through the user interface. Figure 1.3 (which is a simplification of Figure 1.1) illustrates the basic idea. All the data resides in the database. Various software systems can then access the database.

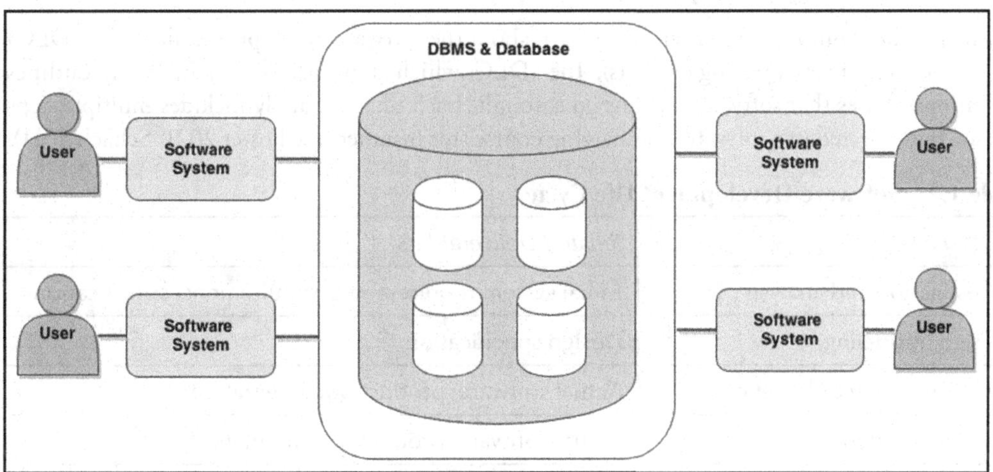

Figure 1.3 Database approach to data management.

Of the five methodologies for database design, the relational model still dominates contemporary software engineering. Since the 1970s, relational databases have dominated the field of database systems. Object databases created some interest for a while, but it appears that they have been replaced by more contemporary approaches such as the EAV model, Hadoop, and NoSQL. Still, relational databases continue to dominate. Later in the course, you will understand why this dominance is likely to continue. The other three approaches are traditional approaches that have been discarded due to their related problems. They will not be discussed any further; for more information on them, see the recommended readings.

1.5 Desirable Features of a DBS

Contemporary database systems must live up to de facto standards set by the software engineering industry. Roughly speaking, a well-designed database system must exhibit the following features (more specific standards will be discussed later in the course):

- Provide most of the advantages mentioned earlier
- Meet most of the objectives mentioned earlier
- Provide for easy communication with other systems
- Be platform independent
- Have a user-friendly interface
- Be thoroughly documented
- Provide comprehensive backup and recovery features
- Provide efficient and effective programming support
- Comprehensive system catalog
- Appropriate transaction management

The importance of these features will become clearer as you proceed through the course. You will also see that for the most part, the choice of the DBMS goes a far way in determining the characteristic features of the database system, and how they are provided.

1.6 Database Development Life Cycle

You are perhaps familiar with or will soon learn about the software development life cycle (SDLC) from your software engineering course(s). The SDLC, which is summarized in Table 1.2, outlines the various phases that software systems go through. Each phase actually includes multiple steps that are best covered in a software engineering course (for instance, see Foster 2021; Schach 2011).

Table 1.2 Software Development Life Cycle

SDLC Phase	*Related Deliverable(s)*
Investigation and analysis	Initial system requirements; requirements specification
Design (modeling)	Design specification
Development (construction)	Actual software; product documentation
Implementation	Actual software; product documentation
Management	Enhanced software; revised documentation

A database system qualifies as a software system. Moreover, the DBMS (which you typically use to create and manage databases) is one of the most complex software systems that you will encounter and work closely with. As you continue through this course and subsequently work with databases, this will become even clearer to you.

As mentioned earlier (Section 1.1), a database does not exist in a vacuum, but is typically part of a software system. A *database development life cycle* (DDLC) may therefore be perceived from two perspectives:

- It may be viewed as being identical and concurrent with the SDLC. At each phase in the SDLC, consideration is given to the database as an integral part of the software product.
- If we consider that in many cases, the database has to be constructed and implemented, and managed as a separate resource that various software products can tap into, then we may construct a similar but also different life cycle for the database as illustrated in Table 1.3.

Table 1.3 Database Development Life Cycle

DDLC Phase	*Related Deliverable(s)*
Database investigation and analysis	Initial database requirements
Database modeling	Database model
Database designing	Database design specification
Database development	Actual database
Implementation	Actual database in use
Management	Enhanced database; revised database design specification

If you compare Tables 1.2 and 1.3, you will clearly see that a database system is really a specialized software system. Here are a few additional points to remember:

1. Applying basic investigation strategies and methodologies that are covered in your software engineering course, you will be able to navigate the database investigation and analysis phase. Appendix 3 provides a summary of these strategies and methodologies. This course primarily concentrates on the other phases.
2. With experience, the database modeling and database designing phases can be merged into one phase. This will be further clarified in Chapters 3–5. However, as a new learner of database systems, you should not rush into this. It is strongly advised that you keep them separate! After a few years of practice, you should be able to look back and smile at the concepts you once struggled with.
3. Once the database is in the implementation phase, management of it becomes an ongoing experience, until the database becomes irrelevant to the organization.

1.7 Summary and Concluding Remarks

Let us summarize what we have covered in this chapter:

- A database system is a computerized record-keeping system with the overall purpose of maintaining information and making it available on demand.

- The DBMS is the software that facilitates creation and administration of the database.
- The DBS is made up of the hardware, the operating system, the DBMS, the actual database, the application programs, and the end users.
- There are several primary and secondary objectives of a DBS, which are of importance to the CS professional.
- Many software systems rely on underlying database systems to provide critical information.
- A DBS brings a number of significant advantages to the business environment.
- There are three traditional approaches to constructing a DBS that are no longer prevalent today. They are the instant small system, the file processing system, and the traditional non-relational approaches.
- There are five contemporary approaches to constructing a DBS. They are the relational approach, the object-oriented approach, the Hadoop framework, the EAV approach, and the NoSQL approach. The relational approach is the most dominant.
- In striving to acquire a DBS, it is advisable to aspire for most of the objectives and advantages. Additionally, one should aim for user-friendliness; thorough documentation; and a DBMS that provides platform independence, comprehensive system catalog, backup and recovery, appropriate transaction management, communication with other systems, and adequate programming support.
- The database development life cycle outlines the main activities in the useful life of a DBS.

Interested? We have just begun to touch the surface. There is a lot more to cover. Most successful software systems are characterized by carefully designed databases. In fact, it is safe to say that the efficacy of the software system is a function of its underlying database. So, stay tuned: the next chapter provides more clarification on the database environment.

1.8 Review Questions

Here are some review questions for you to try answering. You are encouraged to write your responses down; that way you will know whether you need to revisit related sections of the chapter.

1. What is a database system?
2. Why are database systems important?
3. What is a database management system (DBMS)?
4. What are the objectives (primary and secondary) of a DBS?
5. What is data independence, and how important is it?
6. What are the advantages of a DBS?
7. What are the possible approaches to acquiring a DBS?
8. How do database systems relate to software engineering?
9. Compare the software development life cycle to the database development life cycle.

References and/or Recommended Readings

Apace Software Foundation. 2014. "Hadoop." Accessed February 2016. http://hadoop.apache.org/

Connolly, Thomas, & Carolyn Begg. 2015. *Database Systems: A Practical Approach to Design, Implementation and Management* 6th ed. Boston, MA: Pearson. See chapter 1.

Coronel, Carlos, & Steven Morris. 2015. *Database Systems: Design, Implementation & Management* 11th ed. Boston, MA: Cengage Learning. See chapter 1.

Date, Christopher J. 2004. *Introduction to Database Systems* 8th ed. Menlo Park, CA: Addison-Wesley. See chapter 1.

Elmasri, Ramez, & Shamkant B. Navathe. 2011. *Fundamentals of Database Systems* 6th ed. Boston, MA: Pearson. See chapter 1.

Foster, Elvis C. 2021. *Software Engineering: A Methodical Approach* 2nd ed. New York: CRC Press.

Garcia-Molina, Hector, Jeffrey Ullman, & Jennifer Widom. 2009. *Database Systems: The Complete Book* 2nd ed. Boston, MA: Pearson. See chapter 1.

Hoffer, Jeffrey A., Ramesh Venkataraman, & Heikki Topi. 2013. *Modern Database Management* 11th ed. Boston, MA: Pearson. See chapter 1.

IBM Corporation. 2015. "Analytics White Paper." Accessed February 2016. https://cloudant.com/wp-content/uploads/Why_NoSQL_IBM_Cloudant.pdf

Kifer, Michael, Arthur Bernstein, & Phillip M. Lewis. 2006. *Databases and Transaction Processing: An Application Oriented Approach* 2nd ed. Boston: Pearson. See chapters 1 & 2.

Pratt, Phillip J. & Mary Z. Last. 2015. *Concepts of Database Management* 8th ed. Boston: Course Technology. See chapter 1.

Schach, Stephen R. 2011. *Object-Oriented and Classical Software Engineering* 8th ed. Boston: McGraw-Hill.

Ullman, Jeffrey D., & Jennifer Widom. 2008. *A First Course in Database Systems* 3rd ed. Boston: Pearson. See chapter 1.

Wikipedia. 2016. "Entity-Attribute-Value Model." Accessed February 2016. https://en.wikipedia.org/wiki/Entity%E2%80%93attribute%E2%80%93value_model

Chapter 2

The Database System Environment

This chapter discusses the environment of a database system. Topics covered include:

- Levels of Architecture
- Inter-level Mappings
- Database Personnel
- Database Management System
- Components of the DBMS Suite
- Front-end and Back-end Perspectives
- Database System Architecture
- Database System Classifications
- Summary and Concluding Remarks

2.1 Levels of Architecture

Christopher Date (2004) describes three levels of architecture of a database system (DBS), namely, the *external level*, the *conceptual level*, and the *internal level*. These levels are illustrated in Figure 2.1; we will briefly discuss each.

2.1.1 External Level

The *external level* is concerned with individual user views. It therefore varies according to users' perspectives. The external level is defined by the *external schema*.

Typically, the database is accessed through its external schema. The application programmer uses both the *host language* and the *data sublanguage* (DSL) to create a user interface that end users use to access the system:

- The DSL is the language that is concerned specifically with database objects and operations. To illustrate, SQL (*structured query language*) is the industry's standard DSL. Another example of a DSL is knowledge query and management language (KQML). An older example

DOI: 10.1201/9781003275725-3

which is no longer prevalent is QUEL (short for "Query Language"). These languages (SQL and QUEL) will be further discussed later in the course.

- The host language is that which supports the DSL in addition to other non-database facilities such as manipulation of variables, computations, and Boolean logic. Host languages are typically high-level languages (HLL); examples include C++, Java, C#, Python, PHP, and PL/SQL.

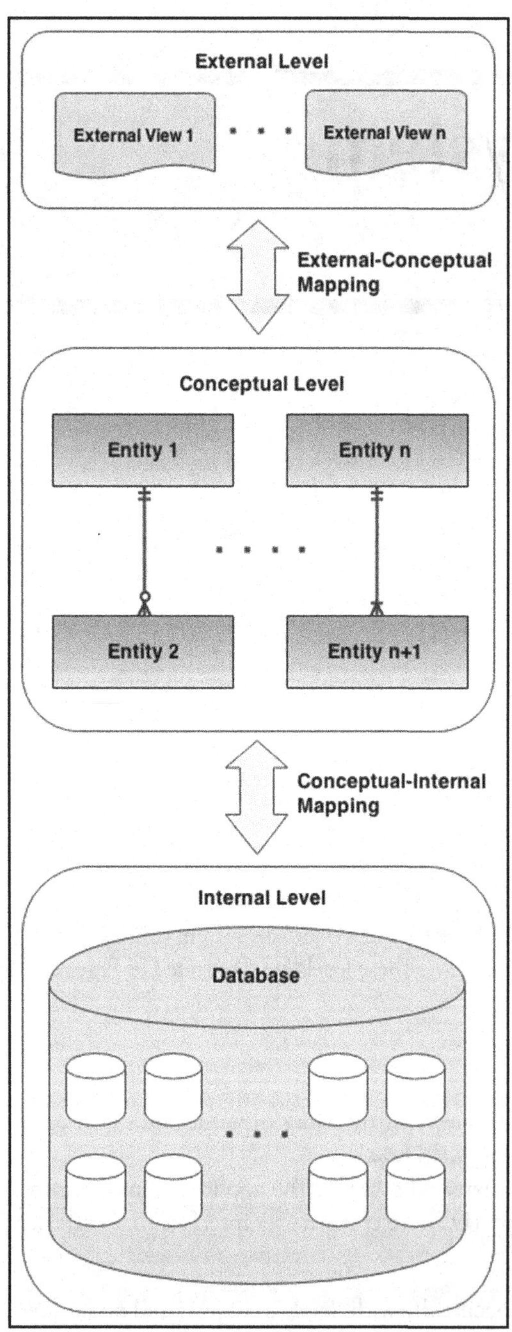

Figure 2.1 Levels of DBS architecture.

Typically, the DSL consists of a *data definition language* (DDL), a *data manipulation language* (DML), and a *data control language* (DCL). These facilities allow users (both technical users and end users) to use the DSL to define assorted logical views of data in the database. In summary, the external schema is the user interpretation of the database but facilitated by the DSL.

2.1.2 Conceptual Level

The *conceptual level* is an abstract representation of the entire information content of the database. It is defined by means of the *conceptual schema* (also called the *logical schema*), which includes the definition of each of the various *persistent objects* comprising the database. By persistent objects, we mean objects that are permanently stored in the database until explicitly deleted. You will learn more about these as the course proceeds.

The conceptual schema includes defining the structure of the database, security constraints, operational constraints, and integrity checks. It represents a closer picture of how data will be actually stored and managed and is the level that most technical users will relate to.

The conceptual schema must adhere to the data independence requirement. Also, it must be comprehensive since it represents the realization of the entire database design.

2.1.3 Internal Level

Also called the *storage view*, the *internal level* is the low-level representation of the database. It is one level above the physical level, which deals with pages, cylinders, and tracks on the storage device.

The internal level is defined by the *internal schema*, which addresses issues such as record types, indexes, field representation, physical storage sequence of records, and data access, and written in the internal DDL. Ultimately, this physical layer has a significant impact on the performance of the database.

2.2 Inter-Level Mappings

Continuing from the previous section, the literature also describes two levels of mappings that connect the three schemas (see Date 2004). Figure 2.1 illustrates the different schemas and their interrelationships with respect to the DBMS. From the figure, it is observed that there are two levels of mapping — the *external-conceptual* mapping and the *conceptual-internal* mapping:

- The conceptual-internal mapping specifies how conceptual records are represented at the internal level. If changes are made at the internal level, this mapping must be updated. Traditionally, the *database administrator* (DBA) maintains this mapping in order to preserve data independence (the DBA is discussed in the next section). In contemporary systems, the DBMS automatically updates and maintains this mapping, transparent to the user.
- The external-conceptual mapping specifies how external views are linked to the conceptual level. In effect, this is achieved by application programs and logical views via the host language and the DSL.

It must be borne in mind that these levels are abstractions that facilitate the understanding of the DBS environment. As an end user, you will most likely not visibly observe these levels of architecture. However, if, as a software engineer, you find yourself in a software engineering team that is constructing or maintaining a DBMS (a huge undertaking), knowledge of these abstractions becomes critical.

2.3 Database Personnel

A database resource team varies on a continuum of a simple one-person team (for small business environments) to a multi-membered team (for more complex business environments). A database systems team may consist of the following role-players: database administrator(s), data architect(s), tools expert(s), application programmer(s), user liaison specialist(s), and network and infrastructure specialist(s). Typically, each team member is tasked with specific responsibilities to ensure the overall success of the database environment. Let us briefly examine each role.

2.3.1 Database Administrator and Data Architect

The database administrator (DBA) has overall responsibility for the control of the system at the technical level. The specific responsibilities will vary across different organizations. Some of the functions of the DBA may include the following:

- Defining the conceptual schema (i.e., logical database design)
- Defining the internal schema (i.e., physical database design)
- Liaising with users and identifying/defining views to facilitate the external schema
- Defining security and integrity checks
- Defining backup and recovery procedures
- Monitoring performance and responding to changing requirements

In smaller organizations, the tendency is to include these functions in the job description of the software engineer. This is quite rational and prudent, since good software engineering includes good database design. However, large corporations that rely on complex company database(s) on an on-going basis usually employ the services of one or more DBAs. Where there are multiple DBAs, one of them is designated the chief DBA.

Database administration is a very coveted specialization that requires advanced training (for example, see Thomas 2014). Because of the importance of having reliable databases, DBAs are among the highest-paid information technology (IT) professionals. Chapter 16 provides a more detailed summary of the profession. However, please note that comprehensive coverage is beyond the scope of this course.

In large and complex database environments, a contemporary trend is to also define a role for a data architect. This makes sense, since in such circumstances, the role of the DBA can be overwhelming. A data architect focuses primarily on database design and integration with assorted data sources. This latter matter of database integration is of particular importance in environments where *data warehousing* and *information extraction* are applicable; these matters are discussed in Chapter 19.

2.3.2 Tools Expert

In very complex database environments, a tools expert is useful in providing expert advice on various technologies and/methodologies required to ensure an efficient and effective database environment. The required number of tools experts depends on the size and complexity of the DBS environment. For instance, special tools may be required in the areas of report generation, data analytics, data preparation, etc. Among the critical functions of the tools expert are the following:

- Developing mastery of the various technologies and methodologies required to enhance the DBS environment
- Providing training and support to the DBS resource team on the relevant technologies and methodologies
- Keeping abreast of current developments in related areas to ensure that the relevance of the DBS is extended for as long as possible
- Developing resource materials that will be useful to the DBS resource team

2.3.3 Application Programmer

Application programmers may be part of the DBS team or the related software engineering team. Whatever the scenario, the functions remain the same. The required number of application programmers varies across different organizations depending on the size and complexity of the DBS environment. Some of the critical functions are listed below:

- Preparation of operation specifications that represent the user needs for the software system(s) that access the database
- Developing application programs corresponding to the operation specifications for the software system(s) that access the database
- Analysis of user requests to determine the best alternative for meeting these requests
- Revision and/or enhancement of existing operation specifications and application programs in response to changing system dynamics and/or user needs
- Identification, diagnosis, and resolution of existing programming problems
- Provision of technical information relating to various application programs so that adequate user documentation can be prepared
- Installation of programming upgrades as required
- Documentation of any changes made to an existing related software system
- Recommendations for related software enhancements as required

2.3.4 User Liaison Specialist

In large, complex DBS environments with very large and diverse end-user populations, it may be necessary to employ the services of one or more user liaison specialist(s) or business analyst(s). In less complex environments where this position is not filled, the responsibilities are spread across the other functional areas of the DBS resource team.

As the title suggests, the user liaison specialist acts as a facilitator who enables effective communication between the end users of the database and the more technical database professionals. The user liaison specialist role includes responsibilities such as:

- Investigation of the user needs from various end-user departments and for the related software system(s)

- Development and management of system change request forms for the end-user departments
- Documentation of the user needs for the related software system(s)
- Reporting the required user needs to the DBS resource team in a timely and organized manner
- Conducting follow-up and/or feedback checks to ensure that user needs are satisfactorily addressed
- Preparation of appropriate system resources for end users
- Planning and execution of training sessions for end users of the related software system(s) as required

2.3.5 Network and Infrastructure Specialist

Network and infrastructure specialist(s) may be part of the DBS team or a complimentary networking team that is already in place. Whatever the scenario, the functions are similar. The required number of network specialists varies across different organizations depending on the size and complexity of the DBS environment. Network/infrastructure design and management is a specialization that is best covered in a course in computer networks. From a DBS perspective, some of the critical functions of the network specialist are listed below:

- Configuration, installation, and maintenance of all related hardware and network servers
- Configuration, installation, and maintenance of all related operating system(s) and other related software systems
- Definition and management of the network topology, spanning all domains, sub-networks, and reinforcement points
- Definition, implementation, and management of the network security constraints, spanning issues such as user permissions, group policies, access control on file servers, intrusion blocking, and other related issues
- Managing software installations on all network servers and client nodes
- Documentation of the network infrastructure
- Ensuring that any network-related problems are resolved in a timely manner
- Overseeing software integration across the network infrastructure to ensure that defined business requirements are met
- Defining and managing network security mechanisms for the environment
- Development and implementation of a preventative maintenance plan for all computers in the organization's network
- Trouble-shooting hardware/software problems in the network, and taking appropriate actions to resolve them
- Management of all related network resources to ensure acceptable operation
- Conducting research into innovative and/or improved ways to enhance network performance
- Keeping abreast of contemporary developments in computer networks to ensure that the organization's network maintains the state-of-the-art status

A full treatment of computer networks is beyond the scope of this course; you would typically cover such information in your computer networking course. Fortunately, there is no shortage of excellent resources on this topic (for instance, see Comer 2015).

2.4 The Database Management System

The database management system (DBMS) is the software that facilitates the creation and management of the database. When a user issues a request via some DSL (typically SQL), it is the DBMS that interprets such a request, executes the appropriate instructions, and responds to the request. Depending on the nature of the initial request, the response may be relayed (by the DBMS) directly to the end user or indirectly to the end user via an executing application program.

Through the DBMS, the objectives of the DBS that were mentioned in Chapter 1 are achieved. The primary functions of this very important software system include the following:

- Data definition (relation, dependencies, integrity constraints, views, etc.)
- Data manipulation (adding, updating, deleting, retrieving, reorganizing, and aggregating data)
- Data security and integrity checks
- Management of data access (including query optimization), archiving, and concurrency
- Maintenance of a user-accessible system catalog (data dictionary)
- Support of miscellaneous non-database functions (e.g., utilities such as copy)
- Programming language support
- Transaction management (either all changes are made or none is made)
- Backup and recovery services
- Communication support (allow the DBMS to integrate with underlying communications software)
- Support for interoperability including open database connectivity (ODBC), Java database connectivity (JDBC), and other related issues

Optimum efficiency and performance are the hallmarks of a good DBMS. To illustrate the critical role of the DBMS, consider the steps involved when an application program accesses the database:

- **Program-A** issues a request to the DBMS (expressed in terms of sub-schema language).
- DBMS looks at **Program-A** sub-schema, schema, and physical description (this information is stored in tables).
- DBMS determines the optimal way to access the data, determining which files must be accessed, which records in the files are needed, and the best method to access DBMS issues instruction(s) (reads or writes) to the operating system.
- Operating system causes data transfer between disk storage and main memory.
- DBMS issues move to transfer required fields.
- DBMS returns control to **Program-A** (possibly with a completion code).

Figure 2.2 provides a graphic representation, but bear in mind that these steps are carried out automatically in a manner that is transparent to the user.

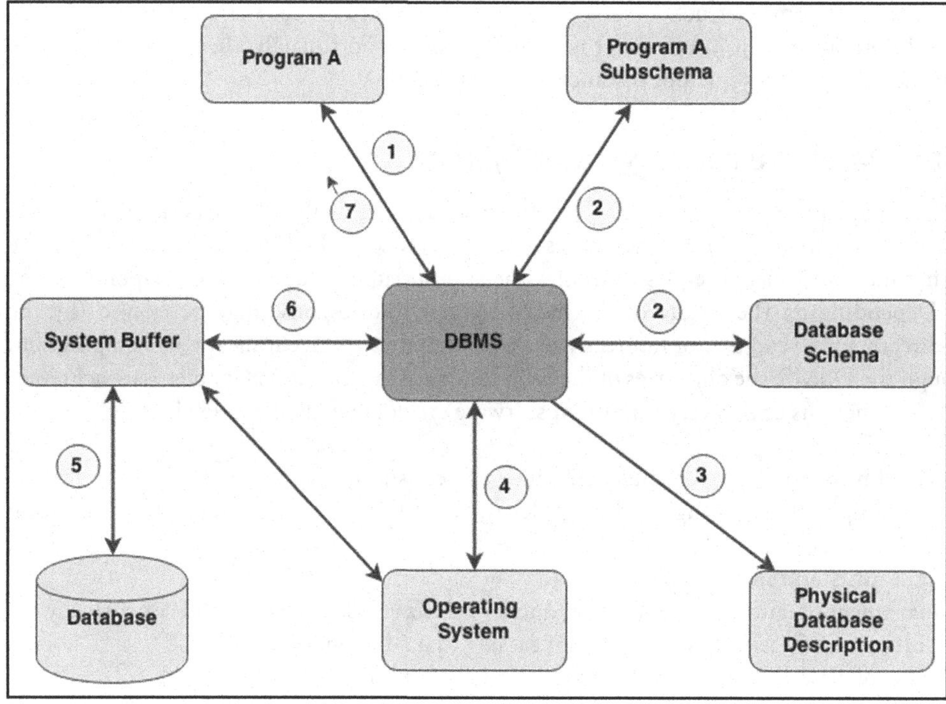

Figure 2.2 Steps involved when application programs access a database.

2.5 Components of DBMS Suite

The DBMS is actually a complex conglomeration of software components working together for a set of common objectives. For the purpose of illustration, we may represent the essential components of the DBMS as the following:

- DBMS Engine
- Data Definition Subsystem
- User Interface Subsystem
- Application Development Subsystem
- Data Administration Subsystem
- Data Dictionary Subsystem
- Data Communications Manager
- Utilities Subsystem

These functional components (illustrated in Figure 2.3) are not necessarily tangibly identifiable, but they exist to ensure the acceptable performance of the DBMS. These functional components are clarified in the upcoming subsections.

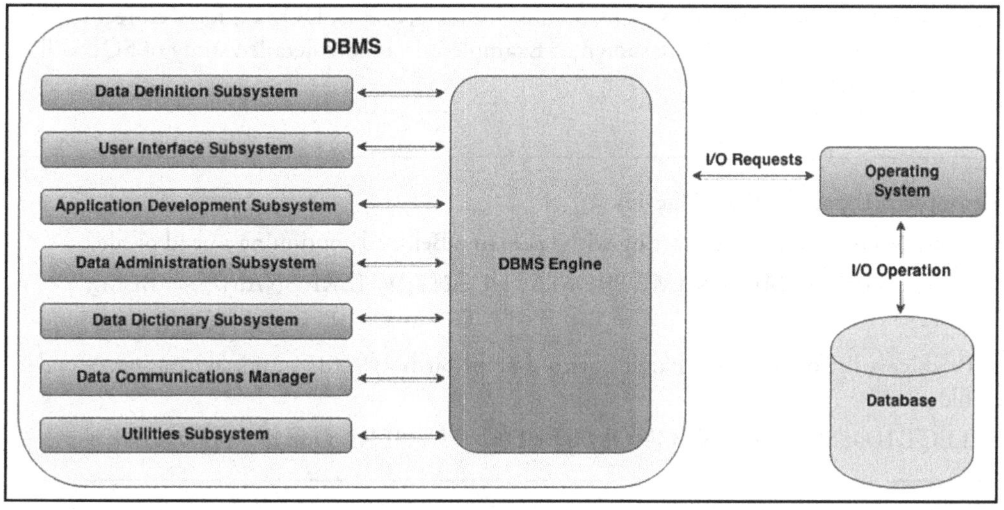

Figure 2.3 Functional components of a DBMS.

2.5.1 The DBMS Engine

The DBMS engine is the link between all other subsystems and the physical device (the computer) via the operating system. Some important functions are as follows:

- Provision of direct access to operating system utilities and programs (e.g., input/output requests, data compaction requests, communication requests, etc.)
- Management of file access (and data management) via the operating system
- Management of data transfer between memory and the system buffer(s) in order to effect user requests
- Maintenance of overhead data and metadata stored in the data dictionary (system catalog)

2.5.2 Data Definition Subsystem

The *data definition subsystem* (DDS or its equivalent) consists of tools and utilities for defining and changing the structure of the database. The structure includes relational tables, relationships, constraints, user profiles, overhead data structures, etc.

The DDL (data definition language) is used to define all database objects that make up the conceptual schema (relations, relationships, constraints, etc.). The DML (data manipulation language) is used to facilitate manipulation and usually includes a query language to insert, remove, update, and find data in the database. The DCL (data control language) is used to set up control environments for data management by the end user. As mentioned earlier, the DDL, DML, and DCL comprise the DSL.

2.5.3 The User Interface Subsystem

The *user interface subsystem* (UIS or its equivalent) allows users and programs to access the database via an interactive query language such as SQL and/or the host language. Suppose, for example, that a file named **Student** has fields {ID#, SName, FName, Status, DOB,...} for each record. Two possible SQL queries on this file are shown in Example 2.1 (a more detailed study of SQL will be covered later in the course):

Example 2.1: Sample SQL Queries

// Produce a list of students, starting with last name Bell and continuing to end of file

SELECT ID#, SNAME, FNAME FROM STUDENT WHERE SNAME >= 'BELL';

// Produce a list of students, starting with date of birth 19960101 and continuing to end of file

SELECT ID#, SNAME, DOB FROM STUDENT WHERE DoB >= 19660101;

The traditional interface was command based; however, the current trend is to provide graphical user interfaces (GUIs). Other more sophisticated DBMS suites may use *natural language* interface.

The user interface may also include DBMS-specific programming language extensions (e.g., Oracle's PL/SQL). These language extensions pertain only to the DBMS in which they are used. Additionally, the DBMS may support multiple high-level languages such as C++ and Java, thus making it more flexible and marketable.

2.5.4 Application Development Subsystem

The *application development subsystem* (ADS or its equivalent) contains tools for developing application components such as forms, reports, and menus. In some cases, it may be merged with the user interface subsystem. Typically, this subsystem provides a graphical user interface (GUI), which is superimposed on an underlying host language. The suite may include an automatic code generator (as in Delphi and Team Developer) or seamless access of the compiler of the host language (as in Oracle).

Much of the application development that takes place in a database environment will involve the use of services of the ADS. Additional facilities that may be provided through the ADS are as follows:

- Quick GUI-based access to data stored in the database
- Report writer for generating formatted reports
- Project manager component for facilitation of the management of database-related projects
- Menu builder for assisting in the quick development of menus for database applications
- Graphic data interpreter

2.5.5 Data Administration Subsystem

The *data administration subsystem* (DAS) consists of a collection of utilities that facilitate the effective management of the database. Included in this subsystem are facilities for backup and recovery, database tuning, and storage management. It is typically used by DBAs as well as software engineers.

2.5.6 Data Dictionary Subsystem

The *data dictionary* (DD) is a traditional term to refer to the *system catalog* (which is the preferred term) in many systems. The system catalog contains information on the database structure, relationships among database objects, system and object privileges, users, integrity constraints, etc. It is automatically created and maintained by the DBMS.

The system catalog contains all metadata for the database. It can be queried using the same commands used to manipulate source data; it is therefore of inestimable value to the DBAs and software engineers. More will be said about the system catalog later in the course.

2.5.7 Data Communications Manager

Traditionally a separate system that is linked to the DBMS, the *data communications manager* (DCM) carries out functions such as:

- Handling communication to remote users in a distributed environment
- Handling messages to and from the DBMS
- Communication with other DBMS suites

Modern systems tend to have this subsystem as an integral part of the DBMS suite. In short, the data communications manager ensures that the database communicates effectively with all client requests in a client-server-based environment. Typically, the server-based portions of the DBMS will be running on machines designated as servers in the network. All other nodes are then deemed as client nodes that can request database services from a server. There may be several database servers in the network; also, a node may act as both a server and a client (provided the essential software components are in place).

2.5.8 Utilities Subsystem

Utilities are programs that perform various administrative tasks. The *utilities subsystem* consists of various utility programs that are applicable to the database environment. Examples of utilities are as follows:

- Load routines to create an initial version of a database from non-database files
- Copy routines for duplicating information
- Reorganization routines to reorganize data in the database
- File deletion routine(s)
- Statistics routines to compute and store file statistics
- Backup and recovery utilities

- Tools for importing data of disparate format(s) to the database and exporting data from the database to different formats
- Other utilities (that might have been) developed by application programmers

2.6 Front-End and Back-End Perspectives

A DBS can be perceived as a simple two-part structure — a *front-end* and a *back-end*. The front-end consists of end users, applications, and a programming interface; the back-end consists of the actual DBMS and the database; it relates to the actual creation and administration of the database, the inner workings of the system, processing requests, memory management, and input/output (I/O) management.

The front-end system may be on a different machine from the back-end system, and the two are connected by a communication network. Typically, an IDE (integrated development environment) fulfills the role of the front-end system. Common examples include the following: Delphi (supporting languages Object Pascal and C++); NetBeans (supporting Java, C++, JavaScript, PHP, XML, HTML, etc.); Visual Studio (supporting C#, C++, JavaScript, etc.); and Qt (supporting C++, C#, Java, Python, Ruby, etc.). The back-end system may be any of the leading DBMS suites in the marketplace — Oracle, DB2, MySQL, SQL Server, or PostGreSQL. Figure 2.4 illustrates this concept of front-end and back-end perspectives.

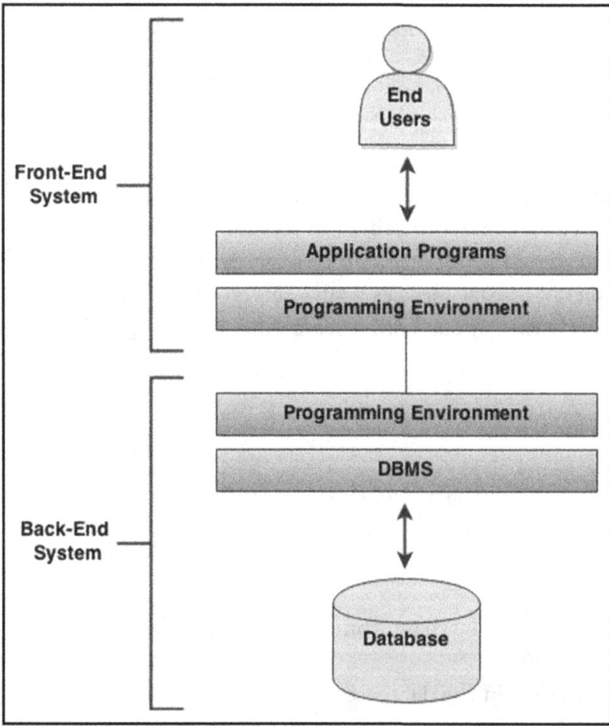

Figure 2.4 Front-end and back-end perspectives.

2.7 Database System Architecture

There may be added benefits of using different machines for the back-end and front-end system. Figures 2.5–2.7 show three possible configurations. Please note also that various network topologies are applicable to any computer network (network topology is outside of the scope of this course; however, it is assumed that you are familiar with such information).

Of the three configurations shown, perhaps the one most commonly used is one requiring a single back-end and multiple front-ends (Figure 2.6). This is so because it provides more flexibility and sophistication than the simpler configuration of Figure 2.5, and less cost and overheads than the more complex configuration of Figure 2.7. The database resides on a single database server (the source) and is accessible from multiple clients.

Discuss: Study Figures 2.5–2.7 and answer the following question: What are some of the advantages of using distributive processing in a DBS?

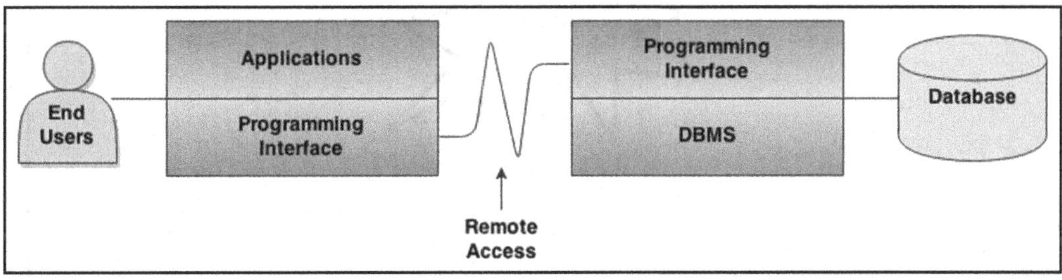

Figure 2.5 Back-end and front-end running on different machines.

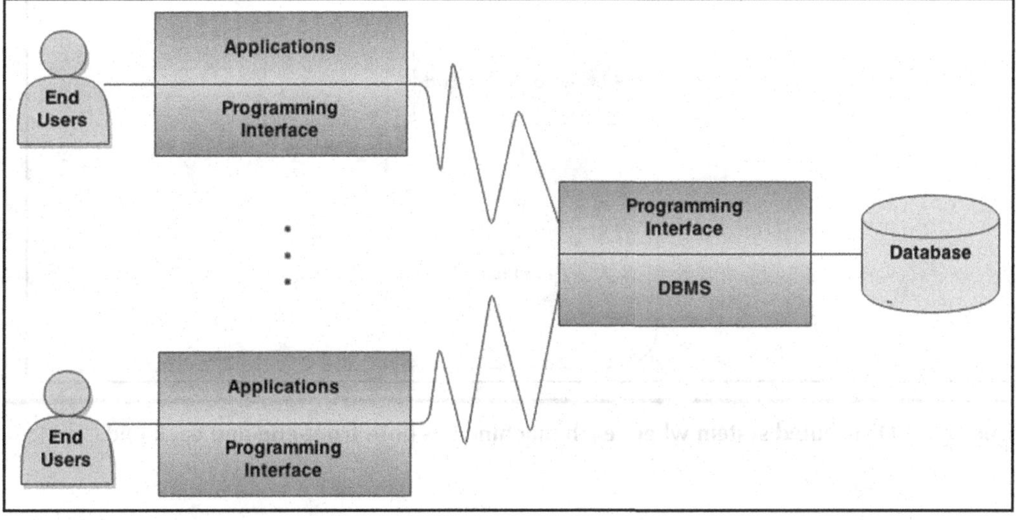

Figure 2.6 One back-end, multiple front-ends.

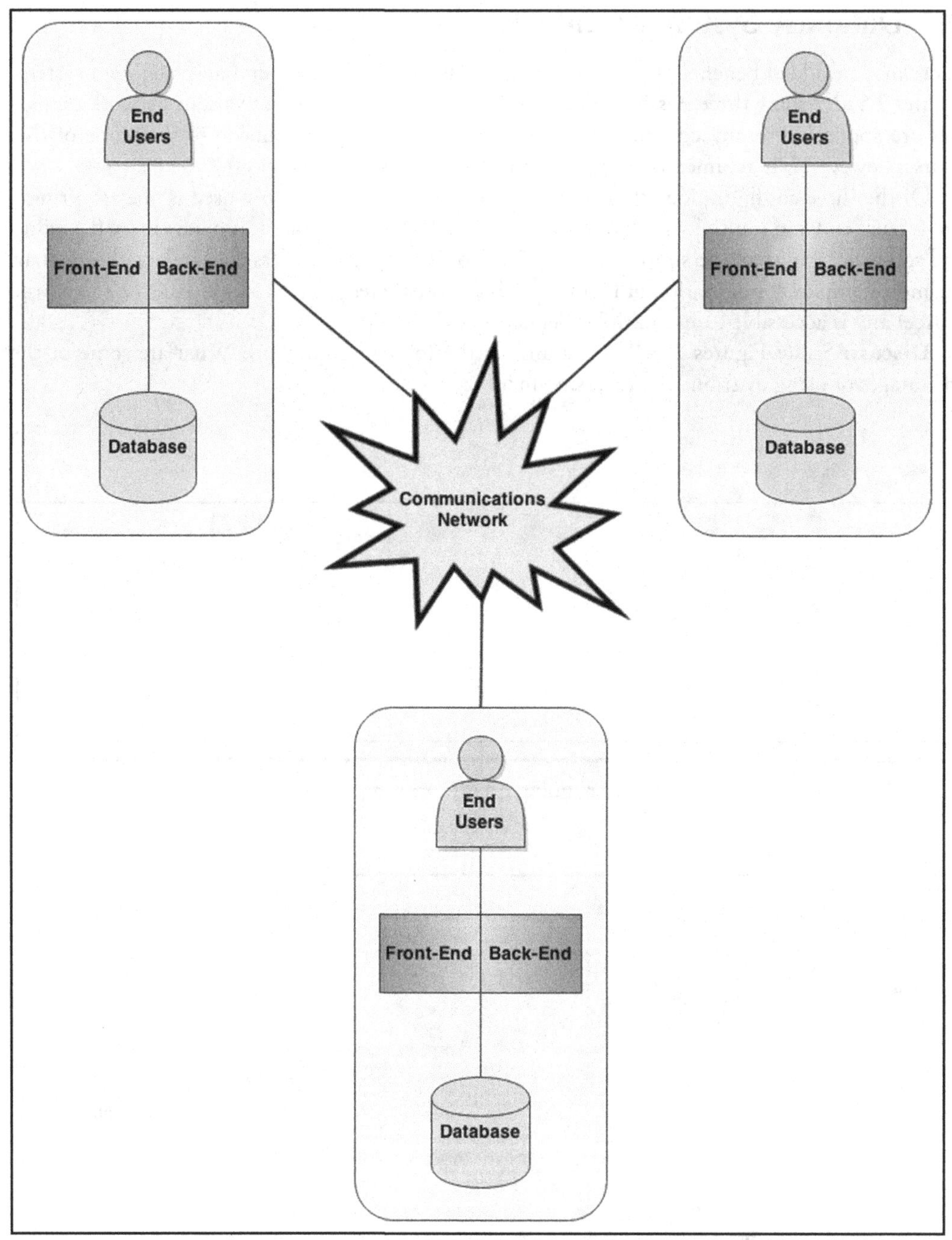

Figure 2.7 Distributed system where each machine has both front-end and back-end.

2.8 Database Management System Classifications

Database systems may be classified based on data model supported, number of users facilitated, site configuration, or purpose. Let us clarify each approach.

2.8.1 Classification Based on Data Models

Classification based on data models can take a number of traditional approaches and some that are more contemporary in nature. The traditional approaches are generally looked at as being obsolete, whereas the contemporary approaches represent those in active growth and development in the software market.

The traditional approaches are the hierarchical model, the network model, and the inverted-list model; these were mentioned in Chapter 1. The hierarchical model was developed by IBM in the 1960s and was epitomized by a product called RAMIS (to be further clarified in Chapter 3). The network model was also developed in the 1960s, inspired by a project that was initiated at General Electric. One of the crowning products of this initiative was the CODASYL system (also clarified in Chapter 3). The inverted-list model prescribed the construction and management of indexes that would map data to their storage locations in database files. This approach has had a significant impact on the modern configuration of DBMSs.

Two dominant contemporary approaches are the relational model and the object-oriented model, with the former occupying most of the market share. Most of the remainder of the course is based on the relational model, except for Chapter 18, which discusses object databases. Prominent relational DBMS (RDBMS) suites include Oracle, DB2, MySQL, SQL Server, Informix, PostGreSQL, and Sybase (recently incorporated into SAP). Object databases seldom appear on their own; they are typically bundled with RDBMS suites and marketed as universal DBMS suites. The leading products such as Oracle, DB2, and Informix fall in this category.

In addition to the relational model and the object-oriented model, other emerging approaches such as the Hadoop framework, the entity–attributes–value (EAV) model, and NoSQL are gaining increased attention (review Section 1.4).

2.8.2 Classification Based on Number of Users

Software engineering firms that market DBMS suites typically provide a single/limited user version for evaluation purposes or limited work. For example, both Oracle and IBM market an Express Edition of their respective DBMS products. In contrast to the limited user edition, the related software engineering firms typically market their complete DBMS suite (allowing multiple users) as an Enterprise Edition. Additionally, there may be different alternatives between these two extremes. Division E of the text provides examples from four of the leading DBMS suites.

2.8.3 Classification Based on Site Configuration

A software engineering firm may market its DBMS suite as either a centralized DBS or a distributed DBS. In a centralized system, the DBMS software and the database reside on a single machine, and all other nodes in the network connect to it. Due to the increase in telecommunication bandwidth, centralized databases have become widespread and will suffice for many small- and medium-sized DBS environments.

In a distributed DBS, the DBMS software and the database are distributed over multiple database servers at different sites in a computer network (review the previous section). The network may be a local area network (LAN), metropolitan area network (MAN), or a wide area network (WAN). This allows for more flexibility and fault tolerance in the system and is applicable for large, complex organizations with multiple branches across the state and/or national borders. If each participating site runs the same DBMS, the system is said to be homogeneous; otherwise, it is heterogeneous. This topic is discussed more fully in Chapter 17.

An emergent trend that has taken off in the marketplace is for the software engineering companies to provide a *cloud implementation* of the DBMS — a facility to configure and store a database in a storage space on the World Wide Web (WWW) in a manner that is seamless and less tedious than previously possible. This alternative provides additional improved benefits such as inexpensive offshore backup, flexibility, and reliability. Indeed, one of the recent versions of the Oracle DBMS — Oracle 12C — is so named to emphasize "C" for the cloud. Other competitors have pursued alternate strategies geared at incorporating this (cloud) technology into their DBMS product offerings.

2.8.4 Classification Based on Database Purpose

The final classification of databases that we will discuss relates to the purpose of the database. A database may be described as *general purpose* if it stores various forms of data as defined by its constituent data files. Alternately, the database may be classified as *specific purpose* if it is configured to store data of a specific type. For instance, databases used in computer-aided design (CAD) and/or computer-aided manufacturing (CAM) are said to be specific-purpose databases.

A database may also be classified based on the nature of processing for which it has been configured. From this perspective, two broad categories are *online transaction processing* (OLTP) databases and *data warehouses.*

- An OLTP database (also called an *operational* database) is a database that facilitates a large volume of concurrent transactions. Such databases are widely used in general business operations (for example, banking, manufacturing, insurance, and marketing).
- A data warehouse is a read-only database (except when it is being updated) that has been configured to store summarized information drawn from various operational databases. Data warehouses are also referred to as *online analytic processing* (OLAP) databases. They are updated by a special process called the *extract–transform–load* (ETL) process; this is clarified in Chapter 19.

The rest of this course will focus on operational databases, except for Chapter 19, which provides an introduction to data warehousing. However, please note that a full treatment of data warehousing is beyond the scope of this course.

2.9 Summary and Concluding Remarks

Here is a summary of what has been covered in this chapter:

- A database system can be construed as having three levels of architecture: the external, the conceptual, and the internal. These levels are seamlessly interlinked by the DBMS.
- The external level constitutes all the external views that end users have of the database.

- The conceptual level relates to the logical structure of the database.
- The internal level relates to the physical structure of the files making up the database.
- In a DBS environment, the personnel may consist of any appropriate combination of the following roles: database administrator, data architect, tools expert, application programmer, user liaison specialist, and network and infrastructure specialist. The DBA is the official responsible for the planning, construction, implementation, and administration of the database. The other roles provide critical supportive responsibilities.
- The DBMS is the software that facilitates the creation and administration of the database.
- A database system can be construed as being comprised of a front-end system and a back-end system. The back-end system relates to the actual creation and administration of the database, the inner workings of the system, processing requests, memory management, and I/O management. The front-end system relates to the creation and administration of the user interface through which end users access the system.
- By applying the principle of separating front-end from back-end, we can conceive of various database architectures.
- Database systems may be classified based on data model supported (traditional, relational, or object-oriented), number of users facilitated (single-user or multi-user), site configuration (centralized or distributed), or purpose (operational or data warehouse).

With this background, you are now ready to move ahead and learn more about the relational database model. You will learn the foundations of the model and why it is so important.

2.10 Review Questions

For review, try answering the following questions:

1. With the use of a diagram, explain the different levels of architecture of a database system.
2. Explain the acronyms DSL, DML, DCL, and DDL. How are they related?
3. Identify the main personnel in a DBS environment. What are the primary functions of each role?
4. What are the main functions of the DBMS?
5. With the aid of an appropriate diagram, explain how the DBMS ensures that requests from end users are satisfactorily addressed.
6. Discuss the functional components of a DBMS. Use an appropriate diagram to illustrate.
7. Explain the concept of front-end and back-end systems, and show how they add flexibility to the implementation of distributed database systems.

References and/or Recommended Readings

Comer, Douglas. 2015. *Computer and Communication Networks* 2nd ed. Boston, MA: Pearson.

Connolly, Thomas, & Carolyn Begg. 2015. *Database Systems: A Practical Approach to Design, Implementation and Management* 6th ed. Boston, MA: Pearson. See chapters 2 & 3.

Coronel, Carlos, & Steven Morris. 2015. *Database Systems: Design, Implementation & Management* 11th ed. Boston, MA: Cengage Learning. See chapters 1 & 2.

Date, Christopher J. 2004. *Introduction to Database Systems* 8th ed. Menlo Park, CA: Addison-Wesley. See chapter 2.

Elmasri, Ramez, & Shamkant B. Navathe. 2011. *Fundamentals of Database Systems* 6th ed. Boston, MA: Pearson. See chapter 2.

Garcia-Molina, Hector, Jeffrey Ullman, & Jennifer Widom. 2009. *Database Systems: The Complete Book* 2nd ed. Boston, MA: Pearson. See chapter 1.

Hoffer, Jeffrey A., Ramesh Venkataraman, & Heikki Topi. 2013. *Modern Database Management* 11th ed. Boston, MA: Pearson. See chapters 1 & 2.

Thomas, Biju, Gavin Powell, Robert Freeman, & Charles Pack. 2014. *Oracle Certified Professional on 12C Certification Kit*. Indianapolis, IN: Wiley.

Ullman, Jeffrey D., & Jennifer Widom. 2008. *A First Course in Database Systems* 3rd ed. Boston, MA: Pearson. See chapter 1.

THE RELATIONAL DATABASE MODEL

B

The next seven chapters will focus on the relational database model. As pointed out in Chapter 1, there are other approaches to database design, but the relational model reigns supreme: it is superior to other traditional approaches; it remains a strong, viable alternative to or complement of (depending on your perspective) other contemporary database models. Even if your choice is to attempt construction of a database via another contemporary approach (such as an object-oriented database, a NoSQL database, or an EAV database), a working knowledge of the relational model will still be required. For these and other reasons, mastery of the relational model is essential to good database administration and software engineering. The objectives of this division are as follows:

- To clearly define, describe, and discuss the relational database model
- To discuss how databases are planned, represented, and implemented
- To discuss the theory, rationale, and practical ramifications of normalization
- To discuss important database integrity rules
- To discuss relational algebra and relational calculus as the foundations to modern database languages
- To discuss the standards to which database management systems ought to attain

Chapters to be covered include:

- Chapter 3 — Introducing the Relational Model
- Chapter 4 — Integrity Rules and Normalization
- Chapter 5 — Database Modeling and Design
- Chapter 6 — Database User Interface Design
- Chapter 7 — Relational Algebra
- Chapter 8 — Relational Calculus
- Chapter 9 — Reflective Look at the Relational Database Model

DOI: 10.1201/9781003275725-4

Chapter 3

Introducing the Relational Model

This chapter introduces you to the fundamental principles and concepts upon which subsequent chapters will be built. Discussion advances under the following captions:

- Basic Concepts
- Domains
- Relations
- Relational Database System
- Identifying Relationships
- Representing Relationships
- Implementing Relationships
- Innovation: Relation–attributes List and Relationship List
- Database Naming Convention
- Summary and Concluding Remarks

3.1 Basic Concepts

The relational model is by far the most widely used model for database design. The model owes its success to the fact that it is firmly founded on mathematical principles (set theory and linear algebra), which have been tested and proven; like the underlying principles, the model itself has been tested and proven over the years.

Before proceeding further, there are some fundamental concepts to be introduced and clarified. These concepts are summarized in Table 3.1. Take some time to review them; in short order, you are required to be comfortable with these concepts, since they form the foundation on which the relational model is built.

As you view Table 3.1, please note that strictly speaking, terms such as *entity* and *entity set* are not considered part of the relational model in some circles, but more correctly belong to the related *entity–relationship model*, which will be discussed later. For the moment, do not obsess yourself with the distinction between the two models; that will come later.

DOI: 10.1201/9781003275725-5

Table 3.1 Basic Concepts of the Relational Database Model

Database Term	*Clarification*
Entity	An object, concept, or thing about which data is stored. Examples include **PurchaseOrder**, **Course**, **Department**, **Program**, and **Student**. Entities are implemented as two-dimensional tables and traditionally perceived as files containing records.
Attributes	Some qualities associated with the entity; e.g., **Order#**, **OrderDate,** and **Item#** of entity **PurchaseOrder**; **Dept#,** and **DeptName** of entity **Department.** Two synonymous terms for attributes are *elements* and *properties*. Attributes correspond to columns of a table and are ultimately implemented as fields of a record.
Entity set	A set of entities of the same type (for example, financial institutions, golfers).
Relationship	An inherent mapping involving two or more entities. Relationships are represented in *relations*. A *binary relationship* involves a mapping between two entities; an *n-ary relationship* involves a mapping among **n** entities.
Relation	A two-dimensional (tabular) representation of entities and/or relationships. A *binary relation* contains two attributes; an *n-ary relation* contains **n** attributes. Other terms used for relation are *relational table* or simply *table*.
Tuples	Correspond to rows of the table or records of a relation. Going forward, we will use the terms *record, row,* and *tuple* as synonyms.
Primary key	An attribute or combination of attributes for which values uniquely identify tuples in the relation. The primary key is chosen from a set of *candidate keys*.
Candidate keys	There may be more than one potential primary keys for a relation. Each is called a candidate key or *super-key*.
Alternate key	An alternate access path to data that is not via the primary key.
Index	A database object that uses predetermined attribute or combination of attributes from a relational table to access records from that table. The index must be created before usage. An index may be designed to retrieve unique records at a time (provided that the attributes used are candidate key attributes) or multiple records at a time.
Composite key	A combination of attributes that act as a candidate key in a relation. Each participating attribute in the composite key (also called *compound key*) is called a *simple key*.
Foreign key	An attribute (or combination of attributes) that is a primary key in another relation.
Domain	A pool of all legal values from which actual attribute values are drawn.
Cardinality	Number of tuples in a relation. The cardinality varies with time.
Degree	Number of attributes in a relation; also called the *arity*. The degree is also used to describe the number of entities participating in a relationship.

Table 3.2 Relational Terms and Their Informal Equivalents

Formal Relational Term	Informal Equivalents
Entity	Object conceptualized as a table, traditionally perceived as a file
Relation	As for entity
Tuple	Conceptualized as a row, traditionally perceived as a record
Attribute	Conceptualized as a column, traditionally perceived as a field
Cardinality	Conceptualized as the number of rows
Degree	Conceptualized as the number of columns
Domain	Conceptualized as a pool of legal values

Table 3.2 provides a list of commonly used relational terms and their informal equivalents. To illustrate, the first row of the list may be interpreted as follows: An entity is an object conceptualized as a relational table and traditionally perceived as a file. Several of the rows that follow could be read in a similar fashion.

Figure 3.1 illustrates how some of the terms of Table 3.1 are applied. Three entities are represented in the figure: **Supplier** stores information on various suppliers to business; **InventoryItem** stores information on the various items used to conduct business; and **SupplierOfItems** stores a schedule of inventory items supplied by each supplier. Observe that as a general rule, even though the data stored in all three entities are related, they are not stored together in one entity; later in the course, you will learn why and how to make similar determinations. Observe further that the ideas of *entity* and *relation* seem to be similar. In fact, the two concepts are similar to some extent, but there is also a subtle difference, which will be clarified shortly. For now, just assume the similarity.

Take some time to carefully go over these concepts. They are foundational to all relational databases; you will not be able to excel at database systems without a solid grasp of them. In short, the database consists of entities, which are implemented as relational tables; each table is defined by a set of attributes; each relational table is also characterized by a primary key, but may also have alternate/candidate keys as well as indexes to facilitate easy access to data; and via foreign key(s), a relational table may be connected to other tables in the database.

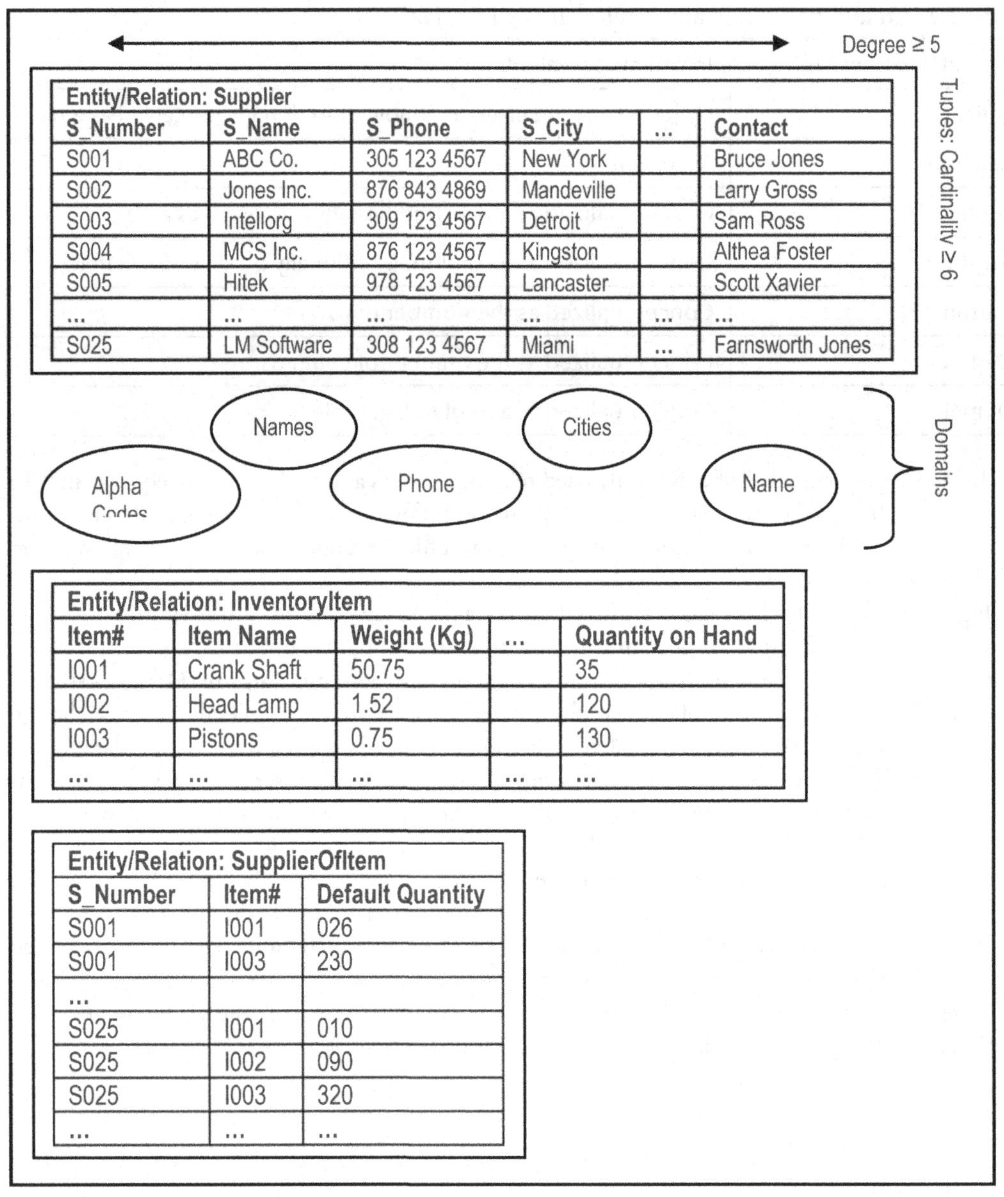

Entity/Relation: Supplier					
S_Number	**S_Name**	**S_Phone**	**S_City**	**...**	**Contact**
S001	ABC Co.	305 123 4567	New York		Bruce Jones
S002	Jones Inc.	876 843 4869	Mandeville		Larry Gross
S003	Intellorg	309 123 4567	Detroit		Sam Ross
S004	MCS Inc.	876 123 4567	Kingston		Althea Foster
S005	Hitek	978 123 4567	Lancaster		Scott Xavier
...	...	...	...	...	...
S025	LM Software	308 123 4567	Miami	...	Farnsworth Jones

Entity/Relation: InventoryItem				
Item#	**Item Name**	**Weight (Kg)**	**...**	**Quantity on Hand**
I001	Crank Shaft	50.75		35
I002	Head Lamp	1.52		120
I003	Pistons	0.75		130
...	...	...	...	...

Entity/Relation: SupplierOfItem		
S_Number	**Item#**	**Default Quantity**
S001	I001	026
S001	I003	230
...		
S025	I001	010
S025	I002	090
S025	I003	320
...	...	...

Figure 3.1 Illustrating some of the basic concepts.

3.2 Domains

A domain is a named set of *scalar values* from which attribute values are drawn. Scalar values are values defined on single domains and are generally perceived to be in their typical formats. For instance, the scalar value for a city's name would be the name that people generally use to refer to that city (e.g., Boston, Philadelphia); similarly, an acceptable scalar value for someone's telephone number would be the valid set of dialed digits that cause his/her telephone to ring.

Each attribute of a relation must be defined on an underlying domain. The attribute values must come from the underlying domain (as illustrated in Figure 3.1).

Domains are conceptual; they may not be (and usually are not) actually stored in the database. Rather, the subsets of the domains containing actual attribute values are stored. Domains are analogous to data types in high-level programming languages such as C++, Java, Python, etc.

A *composite domain* is a combination of simple domains. Whether a composite domain is used or is replaced by its constituent simple domains, is a design decision that should follow thoughtful consideration. Date is an excellent illustration of a composite domain, as explained in Example 3.1. From the example, this Date domain has a total of 12 * 31 * 10,000 possible values, but not all values are valid dates.

EXAMPLE 3.1: ILLUSTRATING DATE AS A COMPOSITE DOMAIN

Date is a combination of
Year which has range 0 .. 9999
Month which has range 1 .. 12
Day which has range 1 .. 31

There can therefore be composite attributes based on composite domains. Composite domains are analogous to Pascal records and C++ structures. However, they do not appear as frequently as simple domains.

Significance of Domains: An understanding of domains is critical for the following reason: If attributes of different relations (entities) come from the same (or a related) domain, then comparisons can be made; otherwise, comparisons are meaningless. Example 3.2 illustrates this point: Refer to the sample database schema of Figure 3.1 and think about how we might retrieve information from relations **InventoryItem** and **SupplierOfItem**. The first retrieval attempt (via structured query language (SQL), which you will soon learn) would produce meaningful results because it makes use of a comparison between attributes defined on the same domain. The second retrieval attempt would not produce meaningful results because it violates this principle.

EXAMPLE 3.2: SAMPLE SQL RETRIEVALS THAT UNDERSCORE THE IMPORTANCE OF DOMAINS

```
// Consider relations InventoryItem and SupplierOfItem of Figure 3.1:
// The following SQL statement would produce meaningful results:
SELECT * FROM SupplierOfItems SI, Supplier S WHERE SI.S_Number = S.S_Number;
/* The following SQL statement would not produce meaningful results since attempt is
   being made to compare attributes (Weight and DefaultQuantity) defined on different
   domains:*/
SELECT * FROM SupplierOfItems SI, InventoryItem I WHERE SI.DefaultQuantity =
   I.Weight;
```

3.3 Relations

A relation R on a collection of domains D1, D2, . . . Dn (not necessarily distinct) consists of two parts — a *heading* and a *body*. The heading consists of a fixed set of attributes or, more precisely, attribute–domain pairs,

{(A1:D1), (A2:D2), . . . (An:Dn)}

such that each attribute corresponds to exactly one domain and n is the degree of the relation. Another term used to describe the heading of a relation is the *predicate* of the relation.

The body consists of a time-varying set of tuples where each tuple consists of a set of attribute–value pairs

{(A1:Vi1), (A2:Vi2), . . . (An:Vin)} (i = 1 . . . m)

where m is the number of tuples (i.e., the cardinality) in the set. The body of the relation is also sometimes referred to as the *proposition* of the relation. The proposition defines a set of tuples whereby for each row in the relation, the respective attributes take on legal values from their respective domains.

Observe that the definition of a relation appears to be similar to that of an entity. There are two subtle differences:

- The term *relation* as used belongs to the field of relational systems. We talk about relations because we are discussing the relational model. Entities, on the other hand, describe identifiable objects and/or situations.
- *Entity*, as defined does not necessarily account for relationships. Relation, on the other hand, accounts for entities as well as relationships. Thus, in the relational model, we represent entities as relations and (M:M) relationships (between entities) as relations. A binary relation, for instance, has two attributes. If both attributes are foreign keys and they both constitute the primary key, this binary relation actually represents a *many-to-many relationship* between two referenced relations; otherwise, it is (a relation that can be construed as) an entity. This point will become clear as we proceed.

The foregoing underscores the point that entities can be construed as special kinds of relations. In designing a database, the software engineer or database designer commences by identifying entities during the requirements specification. After further analysis, these entities are eventually implemented by relations.

Note:

A unary relation differs from a domain in the sense that the former is dynamic and the latter static. For instance, in a database system (DBS) intended for various mathematical maneuverings, we may want to store a readily available value for the universal constant called Pi; we could have a unary relation for storing the Pi value (of course, there are other ways to handle Pi). Or, in DBS intended to exclusively serve a single organization, it may be deemed expedient to store the organization's name in a unary relation that can be referenced from various application programs in order to avoid hard-coding the organization's name in these programs.

3.3.1 Properties of a Relation

Based on the relational model, all relations have the following four properties: They store tuples (records) that are uniquely identified by their respective candidate key values, and therefore do not contain duplicate records; the records are unordered; the attributes are unordered; and the attribute values are atomic (i.e., defined on a single domain).

The first and last properties are constraints that end users should be cognizant of since they have to manage data contained in the database. The second and third properties on the surface are immaterial to end users; they are usually enforced by the database management system (DBMS) in a manner that is transparent to the end user. However, it is imperative that technical users (such as database designers and software engineers) understand the implication of each property. For instance, when the DBMS is written, concern has to be given to access tuples. Further, DBMS suites are typically written to give the illusion that the attributes of a relation are ordered. Finally, remembering that attributes are unordered is important in understanding how SQL works and writing SQL queries.

3.3.2 Kinds of Relations

A database will consist of various types of relations, some of them at different stages of the system. The common categories of relations are mentioned below:

1. **Base Relations** are named and permanently represented in the database. They make up the conceptual schema of the database; they form the foundation of the database. Chapter 11 shows you how to create base relations using SQL.
2. **Logical Views** are virtual relations derived from named (base) relations. A logical view (often referred to as simply a view) stores the definition of the virtual relation, but stores no physical data. It is simply a logical (conceptual/external) interpretation of data stored in base relations. Chapter 13 covers the management of logical views in SQL.
3. **Snapshots** are named, derived relations. They differ from logical views in that they are represented in terms of definition as well as physically stored data. From the perspective of the end user, a snapshot relation is typically (but not necessarily) read-only. To illustrate, consider two systems — System A and System B — which both need to access a database table, Table X. Suppose that System A has update rights to Table X, but System B does not. Table X is therefore stored in System A's database; a duplicate version for read-only purposes is stored in System B and is periodically updated (without user interference) from System A.

   Modern DBMS suites tend to use the term *snapshot* loosely to mean different variations of its original meaning. For instance, in Oracle, snapshots are implemented through *materialized views* (see Chapter 13); in Microsoft SQL Server, the term [snapshot] is used to describe the read-only view of a database.
4. **Query Results:** Queries are typically entered at a command prompt (they may be also embedded in high-level language programs or stored in special query files). Results may be directed to screen, printer, or a named relation. An important principle to note is that a query when executed always results in a new relation. This principle will be elucidated later in the course.
5. **Intermediate Results:** The DBMS may create an intermediate relation to assist in furnishing a final answer to a complex query request. This will also be elucidated later in the course.
6. **Temporary Relations** are named relations that are destroyed in a relatively short space of time (compared to base relations and logical views).

3.4 Relational Database System

A *relational database system* (RDBS) is a collection of time-varying *normalized relations*, managed through an appropriate user interface, and with desirable constraints and features that enhance the effective and efficient management of the database. This working definition will be further clarified as we cover more ground, but for now, it will suffice. The desirable features and constraints will be discussed (see Chapter 9) as we progress through the course. The term *normalized relations* will be fully clarified in Chapter 4; for now, just consider it to mean that the relations are designed to promote efficiency and accessibility.

The relations are conceptualized as tables and ultimately implemented in the underlying operating system as files. Each relation contains one and only one row type, consisting of a specific number of *atomic attributes*. Each relation has a primary key (chosen from a set of candidate keys). In many cases, the primary key is obvious and can be identified intuitively. In situations where this is not the case, the database designer, based on principles to be discussed in the next chapter, typically takes decision about the primary key.

Each record type is made up of *atomic attributes*. This means that each attribute is defined on a single domain and can only have a value from that domain. Moreover, when data is loaded into the database, each tuple from any given relational table has a unique primary key value.

Superimposed on the database is a user interface that facilitates access to the database by end users. The database and the user interface are designed to ensure that certain objectives are met (revisit Section 1.2) and established standards are conformed to.

Steps in Building a Relational Database System: In constructing an RDBS, the following steps may be pursued:

a. Identify entities
b. Identify relationships
c. Eliminate unnecessary relationships
d. Develop *entity–relationship diagram* (ERD), *object–relationship diagram* (ORD), or some equivalent model
e. Normalize the database
f. Revise ERD, ORD, or the equivalent model used
g. Design the user interface
h. Proceed to development phase

Note that these steps are typically pursued within the context of a software engineering project. Accordingly, this course assumes that you are familiar with Steps a–d and h. The rest of this chapter will review Steps b–d, while providing some (additional) insights probably not covered in your (introductory) software engineering course. Chapters 4 and 5 will focus on Steps e and f; Step g is covered in Chapter 6, and Step h is covered in Chapters 10–14.

3.5 Identifying Relationships

As mentioned earlier, a relationship is an inherent mapping involving two or more relations. In planning a relational database, it is very important to know how to identify and represent relationships. Of course, the ultimate objective is successful implementation of the model. Let us take some time to discuss this matter:

To identify relationships, you have to know what a relationship is (review Section 3.1) and what types of relationships there are. There are six types of relationships:

- *One-to-one (1:1) relationship: A tuple in one relation R1 maps to another tuple in a second relation R2.*
- *One-to-many (1:M) relationship: A tuple in one relation R1 maps to several tuples in relation R2.*
- *Many-to-one (M:1) relationship: Several tuples in relation R2 map to a single tuple in relation R1.*
- *Many-to-many (M:M) relationship: Several tuples in relation R1 map to several other tuples in relation R2 and vice versa.*
- *Component relationship: A tuple in relation R1 is comprised of tuple(s) in other relations R1A . . R1Z; moreover, relations R1A . . R1Z are existence-dependent on relation R1.*
- *Subtype relationship: If any tuple of relation R2 contains all the properties of a tuple in relation R1, then R2 is the subtype and R1 is the super-type.*

The first four types of relationships are referred to as *traditional relationships* because up until the object model (for database design) gained preeminence, they were essentially the kinds of relationships that were facilitated by the relational model. Observe also that the only difference between a 1:M relationship and an M:1 relation is a matter of perspective; thus, a 1:M relationship may also be described as an M:1 relationship (so that in practice, there are really three types of traditional relationships). Put another way:

> If R1, R2 are two relations and there is a 1:M relationship between R1 and R2, an alternate way of describing this situation is to say that there is an M:1 relationship between R2 and R1.

For traditional relationships, to determine the type of relationship between two relations (entities) R1 and R2, ask and determine the answer to the following questions:

- How many records of R1 can reference a single record of R2?
- How many records of R2 can reference a single record of R1?

To test for a component relationship between any two relations R1 and R2, ask and determine the answer to the following questions:

- Is (a record of) R1 composed of (a record of) R2?
- Is (a record of) R2 composed of (a record of) R1?

For a subtype relationship, the test is a bit more detailed; for relations R1 and R2, ask and determine the answer to the following questions:

- Is (a record of) R1 also (a record of) an R2?
- Is (a record of) R2 also (a record of) an R1?

Possible answers to the questions are always, sometimes, or never. The possibilities are shown in Table 3.3. Notice that of the nine possibilities, only two outcomes are useful: R1 is a subtype of R2, or R2 is a subtype of R1.

Table 3.3 Testing for Sub-Type Relationship

Possibility	*Implication*
R1 always R2, R2 always R1	R1 and R2 are synonymous
R1 always R2, R2 sometimes R1	R1 is a subtype of R2
R1 always R2, R2 never R1	Makes no sense
R1 sometimes R2, R2 always R1	R2 is a subtype of R1
R1 sometimes R2, R2 sometimes R1	Inconclusive
R1 sometimes R2, R2 never R1	Makes no sense
R1 never R2, R2 always R1	Makes no sense
R1 never R2, R2 sometimes R1	Makes no sense
R1 never R1, R2 never R1	No subtype relationship exists

3.6 Representing Relationships

Having identified the entities and relationships, the next logical question is how do we represent them? Four approaches have been used: *database hierarchies*, *simple networks*, *complex networks*, the *entity–relationship model*, and the *object–relationship model*. The first three approaches are traditional approaches that have made way for the more reputed latter two approaches. Let us therefore start by discussing the latter two approaches.

3.6.1 The Entity–Relationship Model

The popular answer to the challenge of database representation is the *entity–relationship diagram* (ERD or E–R diagram). Figure 3.2 shows the symbols used in an ERD. As the figure conveys, there are two widely used conventions for constructing ERDs: the Crow's Foot notation and the Chen notation. Other notations — for example, the Integration Definition for Information Modeling (IDEF1X) — have been proposed by various software engineering firms, but they will not be discussed here. Figure 3.3 provides an illustrative ERD based on the Crow's Foot notation. In the diagram, the convention to show attributes of each entity has been relaxed, thus avoiding clutter. Note also that relationships are labeled as verbs so that in linking one entity to another, one can read an entity–verb–entity formulation. If the verb is on the right or above the relationship line, the convention is to read from top to bottom or left to right. If the verb is on the left or below the relationship line, the convention is to read from bottom to top or right to left.

As you observe Figure 3.3, you will notice that there are 17 entities and 18 relationships connecting them. Here are the entities: **Customer, PurchaseInvoice, Supplier, PurchaseOrder, Machine, Warehouse, InventoryItem, Machine, Project, Location, Department, Employee, EmployeePersonalInfo, EmployeeEmploymentHistory, EmployeeAcademicLog, EmployeePublications**, and **EmployeeExtraCurricular.** Also note that each relationship is given a name. How you name the relationship depends on your diagramming convention. The relationship-name connecting any two entities is typically a verb. If you are using the Chen notation, the relationship-name goes inside the diamond (review Figure 3.2). If you are using the Crow's Foot notation, the naming abides by the following rule: If you read from left to right, the name goes above the relationship line; if you proceed from right to left, the name goes underneath the relationship line. If you read from top to bottom, the name goes to the right of the relationship line; if you proceed from bottom to top, the name goes to the left of the relationship line.

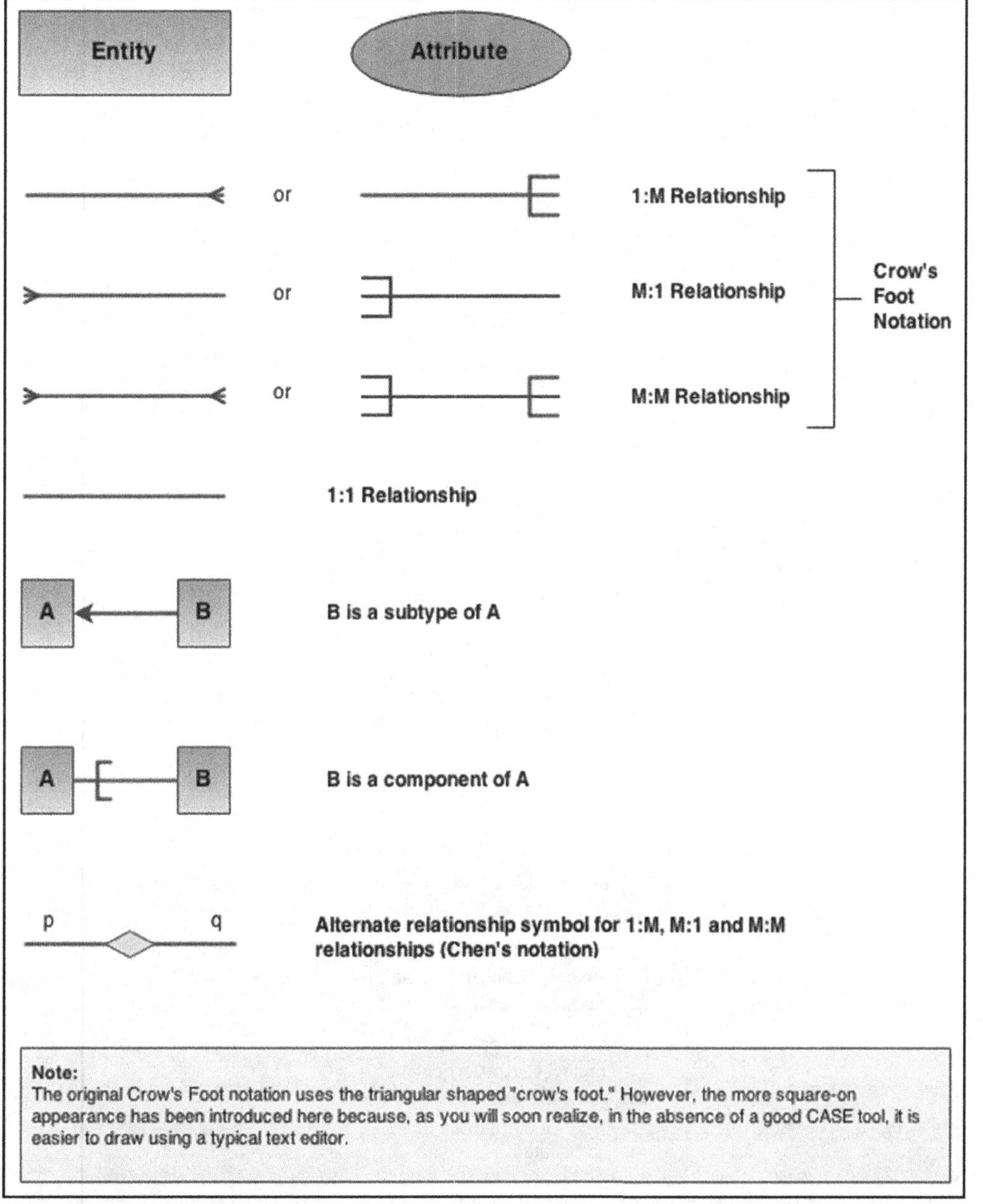

Figure 3.2 Symbols used in E–R diagrams.

Take some time to study the diagram to understand the information it conveys. Here are four examples of how to read the relationships on the ERD: **Employee** belongs to **Department**; **Supplier** sends **PurchaseInvoice(s)**; **PurchaseInvoice** contains **InventoryItem(s)**; **Employee** assigned to **Project(s)**.

Figure 3.4 conveys the same information as Figure 3.3; however, whereas Figure 3.3 depicts the Crow's Foot notation, Figure 3.4 embodies the Chen notation. The Chen notation has been

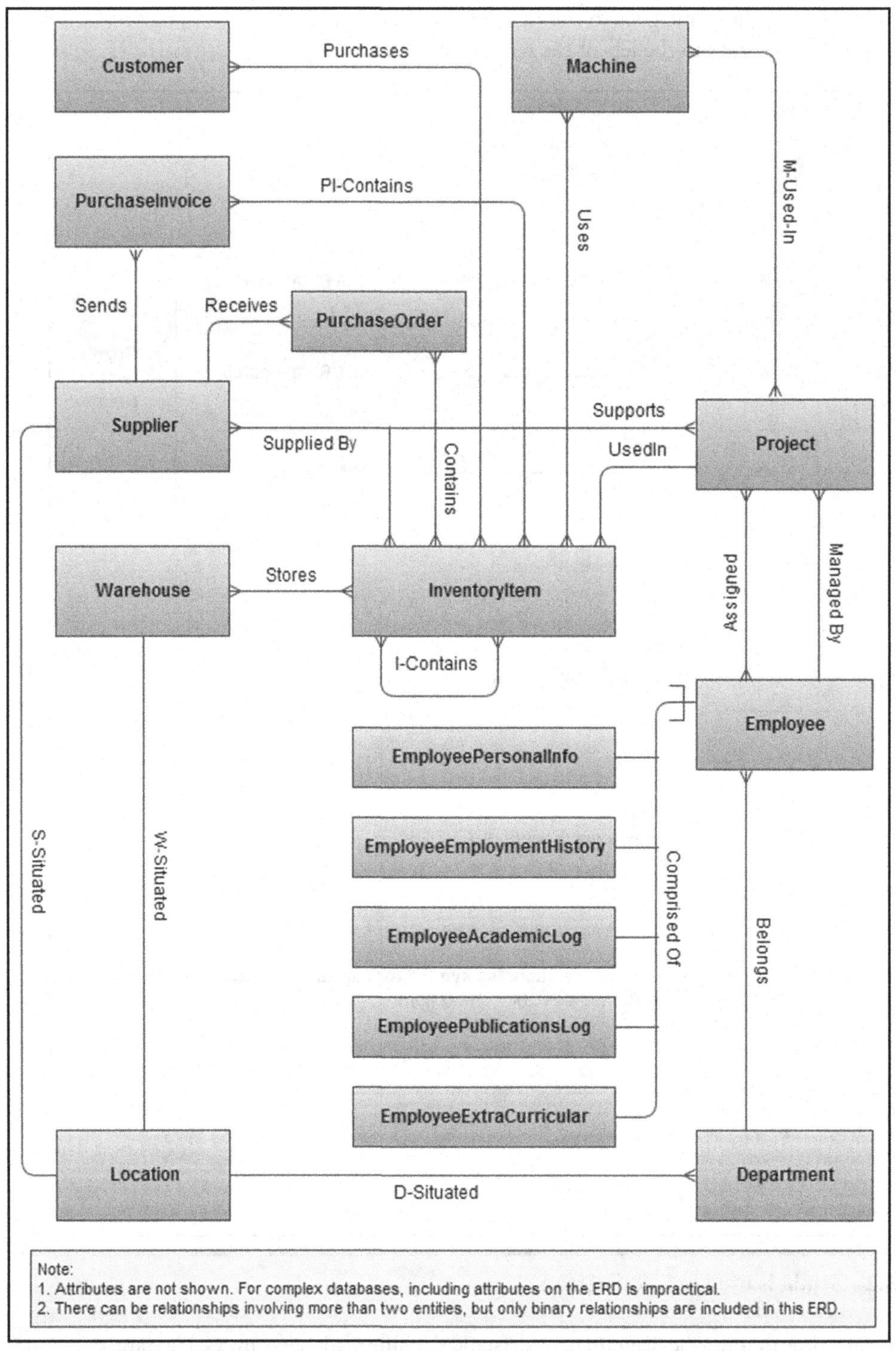

Figure 3.3 Partial ERD for manufacturing environment (Crow's Foot notation).

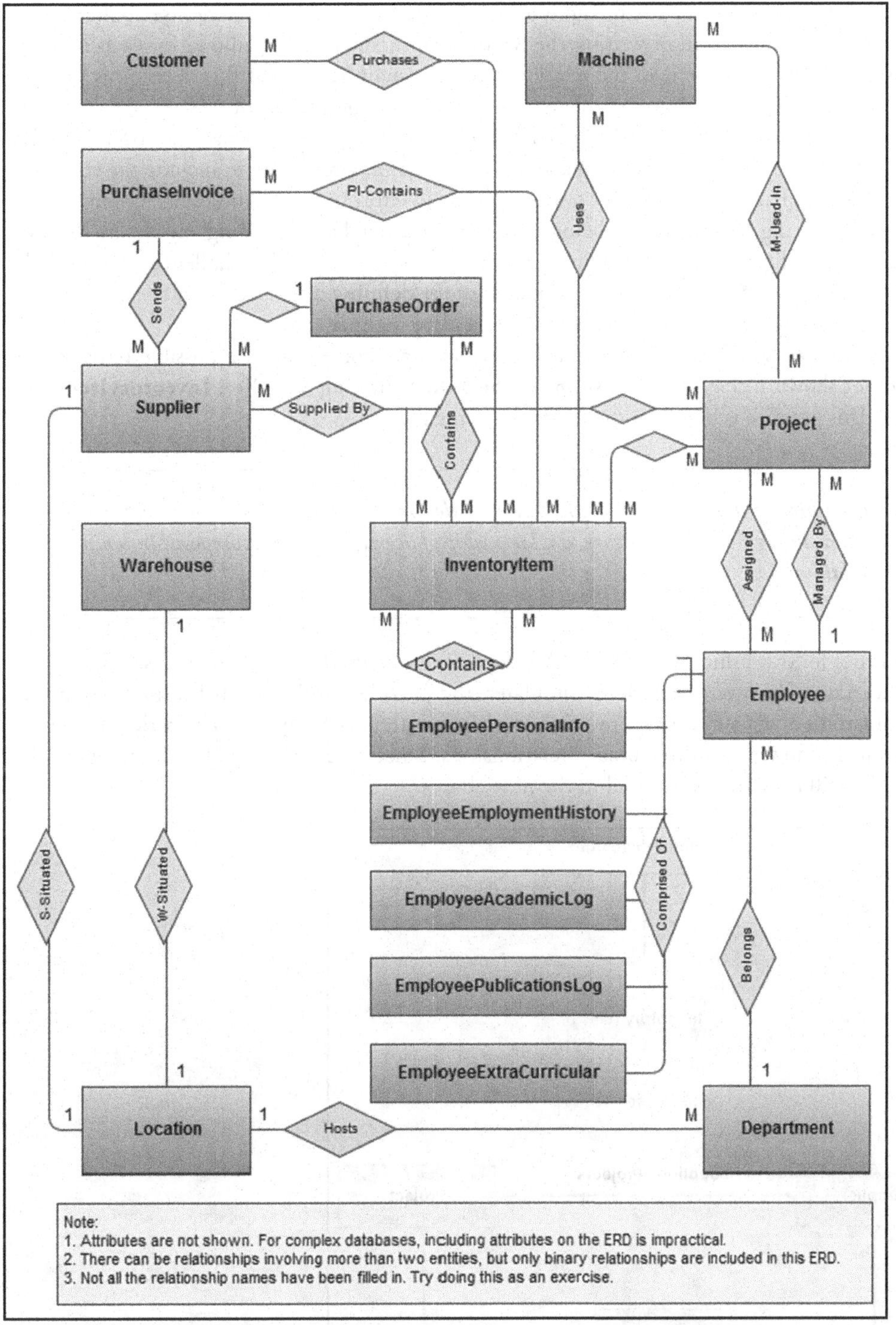

Figure 3.4 Partial ERD for manufacturing environment (Chen notation).

named after its founder, Peter Chen, and is also widely used in academic circles (see [Chen n.d.]). In the figure, the convention of writing the relationship-name in the diamond for each relationship has been relaxed for some of the relationships. The main reason for this decision is that the practice tends to result in a cluttered diagram, especially for complex databases.

Observe that in both Figures 3.3 and 3.4, the attributes for the entities have not been included on the ERD. This is deliberate. For large, complex databases, attempting to include the attributes on the ERD is impractical — virtually tantamount to an effort in futility, unless a computer-aided software engineering (CASE) tool is being used to handle database modeling and design. Section 3.8 discusses a preliminary compensating strategy for circumventing this challenge. Additionally, Chapter 5 (Section 5.8) introduces a methodology for defining and specifying the attributes of each constituent entity of a database in a comprehensive manner.

In many cases, ERDs show only binary relationships. For example, a possible ternary relationship not shown in Figure 3.3 is **Supplier–Schedule** (linking **Supplier**, **InventoryItem**, and **Project**). This practice is based on the following principle:

> *All relationships of degree greater than 2 can be decomposed to a set of binary relationships. Moreover, drawing from graph theory, a relationship of degree n can be decomposed into n(n-1)/2 binary relationships.*

The proof for this principle is beyond the scope of this course (for more insight, see [Dahchour & Pirotte, n.d.]). However, we shall revisit it later, and provide additional clarifications. For now, a simple illustration will suffice: Figure 3.5 shows how the ternary **Supplier–Schedule** relationship may be broken down into three binary relationships. Since care must be taken in applying this principle, it will be further discussed in the next chapter.

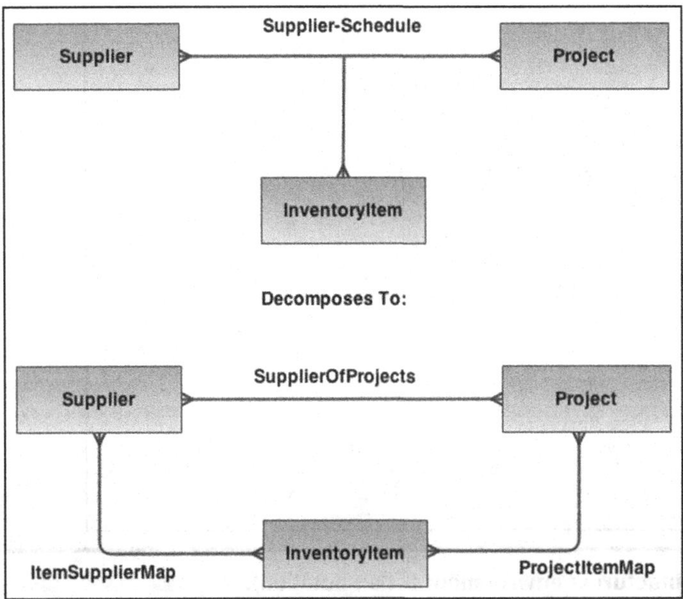

Figure 3.5 Decomposing a ternary relationship.

3.6.2 The Object–Relationship Model

As you are aware, or will soon learn (from your software engineering course), there are, broadly speaking, two alternate paradigms for software construction: the functional approach (which is the traditional approach) and the object-oriented (OO) approach. In an object-oriented environment, the comparative methodology for the E–R diagram is the *object–relationship diagram* (ORD or O–R diagram). The concept of an ORD is similar to that of an ERD, and the diagrams are also similar, but there are a few exceptions:

- In the OO paradigm, the *object type* replaces the entity (type) of the relational model. Like the entity, an object type is a concept or thing about which data is stored. Additionally, the object type defines a set of operations, which will be applicable to all objects (instances) of that type.
- The symbol used to denote an object type is similar to an entity symbol, except that it has two extended areas — one for the attributes of the object type and the other for its operations.
- The preferred diagramming convention is the UML (Unified Modeling Language) notation.
- Depending on the OO development tool, there might be additional notations regarding the cardinality (more precisely, multiplicity) of the relationships represented.

Mastery of the OO approach to database systems is contingent on a good grasp of the relational model. Moreover, a full treatment is beyond the scope of this course. For these reasons, further discussion is deferred until Chapter 5 (Sections 5.6 and 5.7) and Chapter 18. You are no doubt familiar with using UML diagrams in your OO programming courses. However, as you will learn in Chapter 5, and more comprehensively from your software engineering course, there is a lot more to UML than just programming. For a quick review of the fundamentals, please see references Lee (2002) and Schach (2011).

3.6.3 Summary of Traditional Models

Solely for the purpose of historical context, this section provides a brief summary of two database approaches that have be abandoned a long time ago. You do not ever have to master these approaches; however, a brief account is provided so you can have an appreciation of the approaches.

3.6.3.1 Database Tree

A *database tree* (hierarchy) is a traditional alternative that was employed prior to the introduction or the E–R model; it was successfully employed in a system called RAMIS (the original acronym stands for "Random Access Management Information System"). A database tree (hierarchy) is a collection of entities and 1:M relationships arranged such that the following conditions hold:

- The root has no parent
- Each subsequent node has a single parent

Figure 3.6 illustrates a database hierarchy. Observe that it looks like a general tree (review your data structures). Except for the root (node A), each node has a parent node that it references. Note also that all the relationships are 1:M relationships (traditionally referred to as *parent–child relationships*).

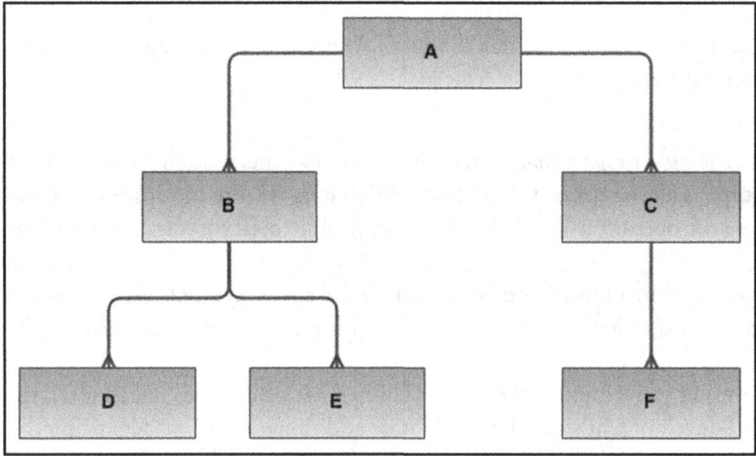

Figure 3.6 Example of a hierarchy (tree).

3.6.3.2 Database Networks

The database network approach is another traditional approach that is no longer employed. A *simple database network* is a collection of entities and 1:M relationships arranged such that any member can have multiple parents, providing that the parents are different entities. Figure 3.7

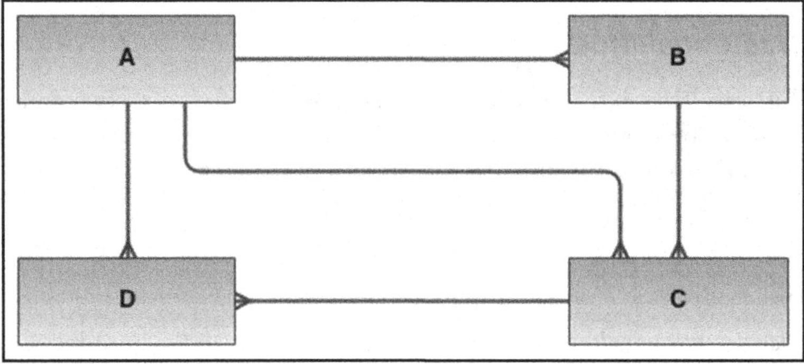

Figure 3.7 A simple database network.

illustrates the approach. It was successfully employed in a database methodology called the CODASYL system (the original CODASYL acronym stands for "Conference on Data Systems Languages").

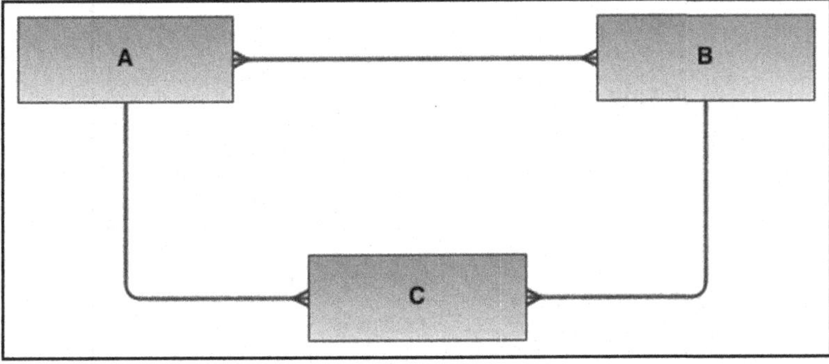

Figure 3.8 Complex database network.

A *complex database network* is a collection of entities and relationships, at least one of the relationships being an M:M relationship. Figure 3.8 illustrates that the complex network can be reduced to a simple network by replacing all M:M relationships with M:1 relationships. The technique for replacing M:M relationships will be discussed in the upcoming section.

3.6.4 Multiplicity of Relationships

It is customary to indicate on the ERD (or ORD), the *multiplicity* (also called the *cardinality*) of each relationship. By this we mean how many occurrences of one entity (or object type) can be associated with one occurrence of the other entity (or object type). This information is particularly useful when the system is being constructed. Moreover, violation of multiplicity constraints could put the integrity of the system in question, which of course is undesirable. DBMS suites do not necessarily facilitate enforcement of all multiplicity constraints at the database level. Nonetheless, it is worthwhile to stipulate these constraints; they are typically enforced at the application level by the software engineering exercise.

Several notations for multiplicity have been proposed, but the Chen notation (first published in 1976 and reiterated in Chen 1994) is particularly clear; it is paraphrased here: Place beside each entity (or object type), two numbers [*x,y*]. The first number (*x*) indicates the minimum participation, while the second (*y*) indicates the maximum participation.

An alternate notation is to use two additional symbols along with the Crow's Foot notation: an open circle to indicate a participation of zero, a stroke (|) to indicate a participation of 1, and the Crow's Foot to indicate participation of many. The maximum participation is always indicated nearest to the entity (or object type) box. With this convention, three combinations are common: a double stroke (||) indicates a participation of one and only one; an open circle followed by a Crow's Foot indicates a participation of zero or many; and a stroke followed by a Crow's Foot indicates a participation of one or many.

For convenience, you could also use the Chen's notation for multiplicity, along with the Crow's Foot notation for representing the relationships. The Chen notation is preferred because of its clarity and the amount of information it conveys. Figure 3.9 provides an illustrative comparison of the two notations.

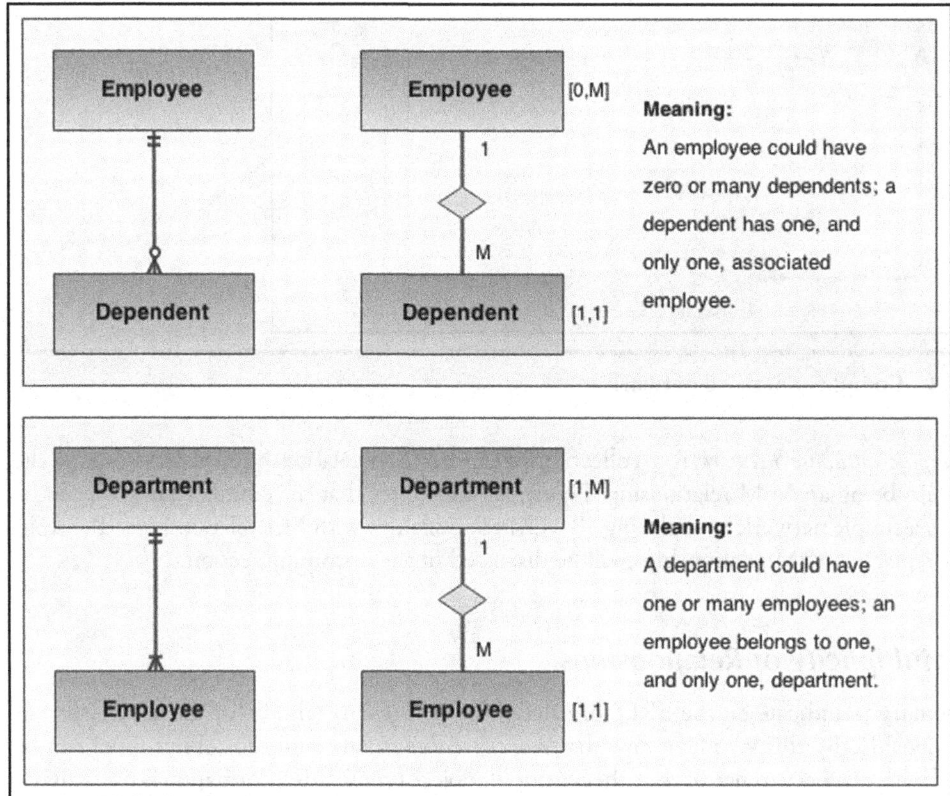

Figure 3.9 Illustrating multiplicity notations.

3.7 Implementing Relationships

Assuming the E–R model, relationships can be implemented by following a set of guidelines as outlined in Table 3.4. Take some time to carefully study these strategies; as you advance through your study of database systems, you will find application of them to be particularly useful during database design (Chapter 5). The strategies in Table 3.4 should underscore forcefully in your mind, the importance of foreign keys in database design. In fact, foreign keys are referred to as the "glue" that holds the database together. We shall revisit this concept in the next chapter.

Note:

In many textbooks and database environments, you will see and/or hear the term *parent–child relationship*. This is a rather lame term, borrowed from preexisting hierarchical database systems, to describe 1:1 and 1:M relationships. In a parent–child relationship, the *parent relation* is the referenced relation; the child relation is the *referencing relation*. Throughout this course, these terms are avoided because they are rather confusing and do not accurately describe several scenarios involving 1:1 and/or 1:M relationships. Alternately, we will use no euphemism for 1:1 and 1:M relationships; instead of parent relation, we'll say the *referenced relation*; instead of child relation, we say *primary relation* or *referencing relation*.

Table 3.4 Guidelines for Implementing Relationships

Relationship	*Recommended Implementation Strategy*
1:M	To implement a 1:M relationship, store the primary key of one as a foreign key of the other (foreign key must be on the "many side"). To illustrate this strategy, carefully compare Figure 3.3 with Figure 3.13. Figure 3.3 includes several 1:M relationships for which there are implementation proposals in Figure 3.13.
M:M	To implement an M:M relationship, introduce a third intersecting relation. The original relations/entities form 1:M relationships with the intersecting relation. The new relation is usually keyed on all the foreign keys (or a *surrogate* #). Figure 3.10 illustrates this principle; to gain additional insights, compare Figure 3.3 with Figure 3.13. Figure 3.3 includes several M:M relationships for which there are implementation proposals in Figure 3.13.
Subtype	To implement a subtype relationship, introduce a foreign key in the subtype, which is the primary key in the referenced super-type. Further, make the foreign key in the subtype, the primary key of that subtype. In the case of *multiple inheritance* (where a subtype has more than one super-types), make the introduced foreign keys in the subtype, candidate keys, one of which will be the primary key. Figure 3.11 includes several subtype relationships for which there are implementation proposals in Figure 3.12.
Component	To implement a component relationship, introduce in the component relation, a foreign key that is the primary key in the summary relation. This foreign key will form part of the primary key (or a candidate key) in the component relation. Figure 3.11 includes several component relationships that were first introduced in Figure 3.3, and for which there are implementation proposals in Figure 3.12.
1:1	To implement a 1:1 relationship, introduce a foreign key in one relation (preferably the primary relation) such that the primary key of one is an attribute in the other. Then enforce a constraint that forbids multiple foreign keys referencing a single primary key. Alternately, treat the 1:1 relationship as a subtype relationship (but ignore enforcing inheritance). Figure 3.3 includes two 1:1 relationships for which there are implementation proposals in Figure 3.13.

Note: A surrogate is an atomic attribute introduced either automatically by the DBMS or manually by the database designer. It is used to uniquely identify each tuple in the relation. This will be further clarified in Chapter 5.

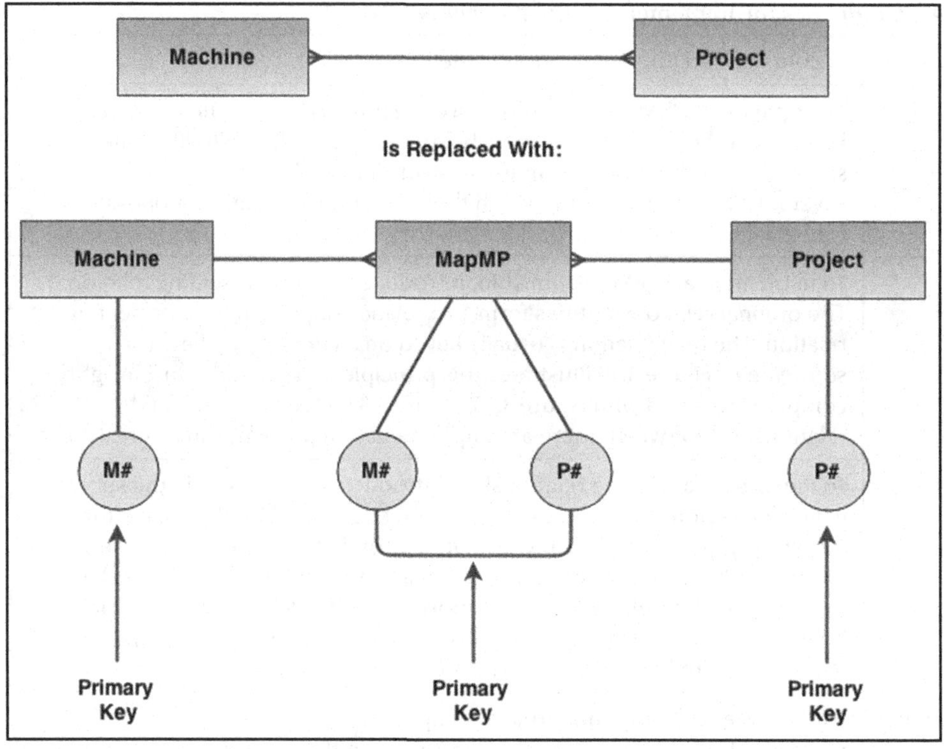

Figure 3.10 Implementing M:M relationships.

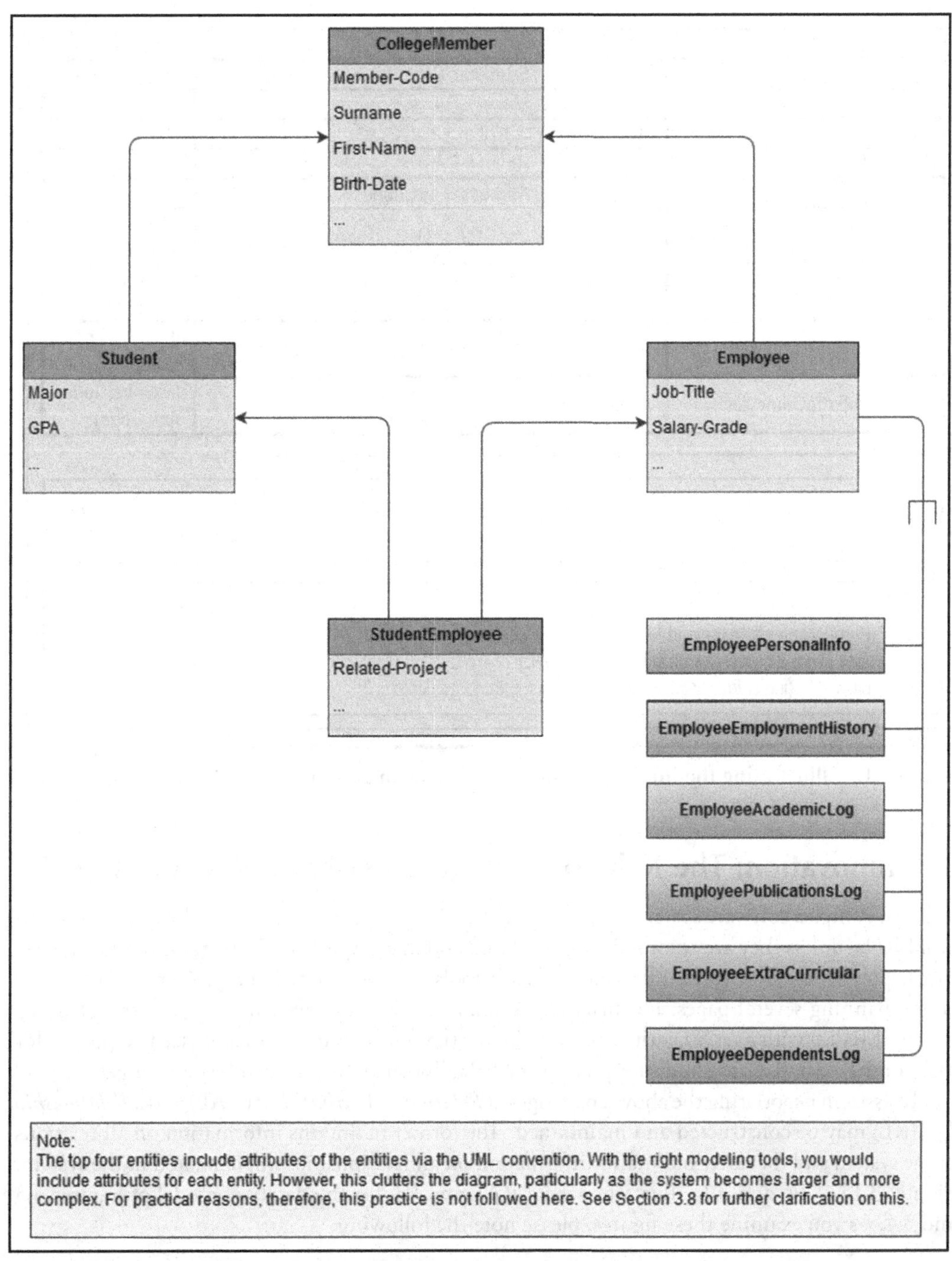

Figure 3.11 Illustrating subtype and component relationships.

Relation	Attributes	Primary Key
College Member	*MemberCode*, Surname, First-Name, BirthDate, ...	MemberCode
Student	*MemberCode*, Major, GPA, ...	MemberCode
Employee	*MemberCode*, JobTitle, SalaryGrade, ...	MemberCode
StudentEmployee	*MemberCode*, Related-Project, ...	MemberCode
EmployeePersonalInfo	*MemberCode*, Address, Telephone, ...	MemberCode
EmployeeEmploymentHistory	*MemberCode, JobSeqNo*, Organization, ...	[MemberCode, JobSeqNo]
EmployeeAcademicLog	*MemberCode, LogSeqNo*, Institution, Period-Attended, Award, ...	[MemberCode, LogSeqNo]
EmployeePublicationsLog	*MemberCode, PubCode*, Title, Book-Journal-Flag, ...	[MemberCode, PubCode]
EmployeeExtraCurricular	*MemberCode, ActvCode*, Activity-Description, ...	[MemberCode, ActvCode]

Each relation and each attribute will need additional clarification prior to database construction and table creation.

Note the following:

1. Primary key attributes and foreign key attributes are in italics.
2. This is not a comprehensive RAL. For several of the relations included, there are additional attributes to be added (indicated by the three periods in the attributes column).

Figure 3.12 Illustrating the implementation of subtype and component relationships.

3.8 Innovation: The Relation–Attributes List and Relationship List

In large, complex software system projects, it is often impractical to attempt to develop and maintain ERDs, unless they are automatically generated and maintained by CASE tools (more on these in Chapter 5). Even when maintained by CASE tools, an ERD for such a project could become large, spanning several pages. For instance, it is not unusual to have *enterprise resource planning systems* (ERPSs) with over 50 entities or even double that number in some cases (see Chapter 21 for more on ERPSs). Reading and interpreting the ERD for such systems can be daunting.

To assist in negotiating the above challenges, a *Relation-–Attribute List* (RAL) and a *Relationship List* (RL) may be constructed and maintained. The former maintains information on all relations of the system and the latter maintains information on all relationships implemented in the system. Figures 3.13, 3.14, and 3.15 illustrate partial RAL and RL for the database model of Figures 3.3 and 3.4. As you examine these figures, please note the following:

- In practice, the format of the RL shown in Figure 3.15 is used as the final list over the format shown in Figure 3.14 (the format used in Figure 3.14 can be deduced by simply identifying all possible relationships among entities; hence, it may contain optional relationships and is therefore useful as a first draft).
- The revised RL of Figure 3.15 has been stripped of all M:M relationships (review Section 3.7 on treating M:M relationships). The relations **PurchaseInvSummary** and **PurchaseInvDetail**

Relation	Attributes	Primary Key
Customer	*Cust#*, CustName, Address, Reference Person, ...	Cust#
Supplier	*Supp#*, SupplName, Address, *SupplLoc#*, ...	Suppl#
Machine	*Mach#*, MachDescription, ...	Mach#
Project	*Proj#*, ProjName, *ProjManagerEmp#*, ...	Proj#,
Warehouse	*Whouse#*, WhouseName, WhouseSize, *WhouseLoc#*, ...	Whouse#
InventoryItem	*Item#*, ItemName, ...	Item#
Location	*Loc#*, LocationName, DistanceFromHQ, ...	Loc#
Department	*Dept#*, DeptName, *DeptLoc#*, ...	Dept#
Employee	*Emp#*, EmpName, *EmpDept#*, DOB,...	Emp#
EmployeeProjAssignments	*EPEmp#*, *EPProj#*, Comment, ...	[EPEmp#, EPProj#]
PurchaseOrdSummary	*OrderRef*, Order#, OrderDate, *OrderSupp#*, OrderStatus, ...	OrderRef
PurchaseOrdDetail	*PODOrderRef*, *OrderItem#*, OrderQuantity, OrderUnitPrice	[PODOrderRef, OrderItem#]
PurchaseInvoiceSummary	*PurchaseRef*, Invoice#, *InvSupp#*, *InvOrderRef*, InvDate, InvAmount, InvStatus, ...	PurchaseRef
PurchaseInvoiceDetail	*PIDPurchaseRef*, *PIDItem#*, PIDItemQuantity, PIDItemUnitPrice	[PIDPurchaseRef, PIDItem#]
SaleInvoiceSummary	*SaleRef*, SIInvoice#, SaleDate, *SaleCust#*, InvoiceStatus, SaleAmount, ...	SaleRef
SaleInvoiceDetail	*SIDSaleRef*, *SaleItem#*, Quantity, UnitPrice	[SIDSaleRef, SaleItem#]
MachineUsageMap	*MUMach#*, *MUItem#*	[MUMach#, MUItem#]
MachProjectsMap	*MPMach#*, *MPProj#*	[MPMach#, MPProj#]
ProjSuppMap	*PSSupp#*, *PSProj#*	[PSSupp#, PSProj#]
ItemProjMap	*IPItem#*, *IPProj#*	[IPItem#, IPProj#]
SuppItemsMap	*SISupp#*, *SIItem#*	[SISupp#, SIItem#]
Stock	*SWhouse#*, *SItem#*	[SWhouse#, SItem#]
ItemStruct	*ISThisItem#*, *ISCompItem#*	[ISThisItem#, ISCompItem#]
EmployeePersonalInfo	*EPIEmp#*, MaritalStatus, Address, Telephone, Email, ...	EPIEmp#
EmployeeEmploymentHistory	*EHEmp#*, *EHJobSeqNo*, Organization, JobTitle, ...	[EHEmp#, EHJobSeqNo]
EmployeeAcademicLog	*ALEmp#*, *ALLogSeqNo*, Institution, PeriodAttended, Award, ...	[ALEmp#, ALLogSeqNo]
EmployeePublicationsLog	*EPEmp#*, *PublCode*, Title, BookJournalFlag, ...	[EPEmp#, PublCode]
EmployeeExtraCurricular	*EXEmp#*, *ExActvCode*, ActivityDescription, ...	[EXEmp#, ExActvCode]

Each relation and each attribute will need additional clarification prior to database construction and table creation.

Note:

1. *Primary key attributes and foreign key attributes are in italics.*
2. Attribute names are unique across the database, even if the attribute is a foreign key. For instance, attribute **EmpDept#** of relation **Employee** is a foreign key that references **Dept#** in the relation **Department**.
3. The attributes **OrderRef** (in **PurchaseOrdSummary** relation), **PurchaseRef** (in **PurchaseInvSummary** relation) and **SaleRef** (in **SaleInvSummary** relation) are examples of surrogates.

Figure 3.13 Partial relation–attributes list for a manufacturing environment database.

of Figure 3.13 are used to replace the M:M relationship between **PurchaseInvoice** and **InventoryItem** in Figure 3.3. Similarly, the relations **PurchaseOrdSummary** and **PurchaseOrdDetail** are used to replace the M:M relationship between **PurchaseOrder** and **InventoryItem** in Figure 3.3. The relation **EmployeeProjAssignments** replaces the M:M relationship between **Employee** and **Project** in Figure 3.3, and so on …

- In constructing the RAL and RL, it is sometimes useful to use the RL to refine the RAL and vice versa. In particular, once you have identified all the (mandatory) relationships, you may use this along with the principles outlined in Section 3.7 to refine the RAL.
- Remember, the ERD of Figure 3.3, RAL of Figure 3.13, and the RL of Figure 3.15 do not represent a comprehensive coverage of the database requirements of a manufacturing environment; neither are they intended to be. Rather, they serve as useful illustrations. For several of the relations included, there are additional attributes to be added (indicated by the three periods in the attributes column). Also, additional relations would be required in order to have a comprehensive model.

Relationship Name	Participating Relations	Type	Comment
SuppliedBy	Supplier, InventoryItem	M:M	Mandatory
Purchases	Customer, InventoryItem	M:M	Mandatory
Uses	Machine, InventoryItem	M:M	Mandatory
Used-In	InventoryItem, Project	M:M	Mandatory
M-Used-In	Machine, Project	M:M	Mandatory
Stores	Warehouse, InventoryItem	M:M	Mandatory
I-Contains	InventoryItem, InventoryItem	M:M	Mandatory
Assigned	Employee, Project	M:M	Mandatory
Belongs	Employee, Department	M:1	Mandatory
D-Situated	Location, Department	1:M	Mandatory
W-Situated	Warehouse, Location	1:1	Mandatory
S-Situated	Supplier, Location	1:1	Mandatory
Sends	Supplier, PurchaseInvoice	M:M	Mandatory
PI-Contains	PurchaseInvoice, InventoryItem	M:M	Mandatory
Contains	PurchaseOrder, InventoryItem	M:M	Mandatory
Receives	Supplier, PurchaseOrder	1:M	Mandatory
Supports	Supplier, Project	M:M	Mandatory
SupplierSchedule	Supplier, InventoryItem, Project	M:M	Optional
ComposedOf	Employee, EmployeePersonalInfo, EmployeeEmploymentHistory, EmployeeAcademicLog, EmployeePublicationsLog, EmployeeExtraCurricular	Comp	Mandatory

These relationships should be represented on the preliminary ERD as depicted in figure 3-3.

Figure 3.14 Relationships list for a manufacturing environment database.

Named Relation	Referenced Relations	Type	Comment
SaleInvoiceSummary	Customer	M:1	Implements relationship Purchases
SaleInvoiceDetail	SaleInvoiceSummary	M:1	Implements relationship Purchases
	InventoryItem	M:1	Implements relationship Purchases
PurchaseInvoiceSummary	Supplier	M:1	Implements relationships Sends and PI-Contains
PurchaeInvoiceDetail	PurchaseInvoiceSummary	M:1	Implements relationship PI-Contains
	InventoryItem	M:1	Implements relationship PI-Contains
MachineUsageMap	Machine	M:1	Implements relationship Uses
	InventoryItem	M:1	Implements relationship Uses
MachProjectsMap	Machine	M:1	Implements relationship M-Used-In
	Project	M:1	Implements relationship M-Used-In
PurchaseOrdSummary	Supplier	M:1	Implements relationships Receives and Contains
PurchaseOrdDetail	PurchaseOrdSummary	M:1	Implements relationship Contains
	InventoryItem	M:1	Implements relationship Contains
Supplier	Location	1:1	Implements relationship S-Situated
SuppItemsMap	Supplier	M:1	Implements relationship SuppliedBy
	InventoryItem	M:1	Implements relationship SuppliedBy
ProjSuppMap	Supplier	M:1	Implements relationship Supports
	Project	M:1	Implements relationship Supports
ItemProjMap	Project	M:1	Implements relationship UsedIn
	InventoryItem	M:1	Implements relationship UsedIn
Stock	Warehouse	M:1	Implements relationship Stores
	InventoryItem	M:1	Implements relationship Stores
ItemStruct	InventoryItem	M:1	Implements relationship I-Contains
	InventoryItem	M:1	Implements relationship I-Contains
EmployeeProjAssignments	Employee	M:1	Implements relationship Assigned
	Project	M:1	Implements relationship Assigned
Employee	Department	M:1	Implements relationship Belongs
Project	Employee	M:1	Implements relationship ManagedBy
Warehouse	Location	1:1	Implements relationship W-Situated
Department	Location	1:1	Implements relationship D-Situated
EmployeePersonalInfo	Employee	M:1	Implements relationship ComposedOf
EmployeeEmploymentHistory	Employee	M:1	Implements relationship ComposedOf
EmployeeAcademicLog	Employee	M:1	Implements relationship ComposedOf
EmployeePublicationsLog	Employee	M:1	Implements relationship ComposedOf
EmployeeExtraCurricular	Employee	M:1	Implements relationship ComposedOf

These relationships should be represented on your refined ERD. Notice that all M:M relationships have been replaced.

Figure 3.15 Refined relationships list for a manufacturing environment database.

Additional Insights:

Both methodologies (RAL and RL) may be automated and also expanded as follows:

1. Notice that both RAL and the RL provide standard information about relations and relationships comprising the database. They could therefore be automated to promote increased efficiency and effectiveness.
2. For each attribute of the RAL, a descriptive narrative and the physical characteristics could also be specified — for instance, the intended data type (numeric, alphanumeric, or otherwise), the number of bytes occupied, etc.
3. In Chapter 5, you will see that both RAL and RL may be integrated into a wider innovative methodology called the *entity specification grid* (ESG).

3.9 Database Naming Convention

The relationship implementation strategies of Section 3.7 are used extensively in database design — a topic that will be revisited in Chapter 5. As you have seen, identification of information entities, their interrelationships, and their constituent attributes is an integral activity that begins with the conceptualization of the DBS. Since the database size (i.e., the number of entities and their related complexities) can grow very quickly, it is a prudent idea to have a carefully designed *database naming convention* from the outset. The naming convention should address issues such as:

- Naming of entities and their related attributes (also called properties);
- Naming of logical views that may be defined on these entities;
- Naming of other database objects such as *stored procedures*, *packages*, and *triggers* (to be clarified later in the course);
- Naming of the application programs (operations) that may access and manipulate the database.

With this in mind, Figure 3.16 provides an example of a comprehensive *object naming convention* (ONC), adopted from Foster (2021). Notice that the proposed convention is fairly comprehensive, covering not just the aforementioned areas, but other aspects typically present in a software system. This ONC will be revisited in Chapters 5 and 6.

When planning a relational database (which this chapter initiates), it is advisable that you apply an ONC for the database. At this preliminary stage, let us revisit the manufacturing environment of earlier discussion (revisit Figures 3.3 and 3.13) and outline a preliminary proposal for the database. Before proceeding, there are two additional principles to be proposed, and they both relate to attributes of entities:

- *It is good design practice to define each attribute so that its name is unique to the database (even if the attribute is a foreign key).*
- *The implementation name for each attribute may be prefixed by an appropriate abbreviation of its entity name.*

Justification for the principle of unique attribute name is provided in Chapter 5 (Section 5.5.2); suffice it to say that easy tracking data throughout the system is facilitated. The second principle is merely a suggestion for application of unique attribute-name principle. For now, let us tentatively accept them both and apply the idea to the manufacturing environment. Applying these ideas, Figure 3.17 shows a more refined RAL for the manufacturing environment. In the figure, you will observe the following:

- The acronym MES (for Manufacturing Environment System) is used to prefix each entity name; this is followed by a descriptive portion and the suffix BR (for base relation).
- Each attribute name is unique.
- There is allocation for a comment on each attribute.
- Primary keys and foreign keys are clearly identified.
- In resolving the M:M relationships, notice the introduction of composite candidate keys. Recall from Section 3.7 and Figure 3.10 that this is the typical outcome. In each case, you have the option of introducing a surrogate and making it the primary key. In Figure 3.17, this strategy was employed for some of the M:M relationships while for others it was not. That was purely arbitrary, simply to expose you to both alternatives. As you become more experienced designing databases, you will instinctively figure out when to introduce surrogates and when not to. The topic will be revisited in Chapter 5.

Object Name: SSSS_XXXXXXX_MMn where interpretations apply: SSSS represents the system or subsystem abbreviation (2 – 4 bytes); MMn represents the object mode or purpose (1-3 bytes); XXXXXXXXX represents the descriptive name of the object (4-15 bytes).
For example, valid subsystem abbreviations for an inventory management system (IMS) may be: AMS: Acquisitions Management FMS: Financial Management SCS: System Controls As another example, a Manufacturing Environment System may be abbreviates as MES.
Valid mode abbreviations include: BR: A base relation (if relational DB model) OT: An object type (if OO DB model) LVn: A logical view (e.g. LV1, LV2, etc.) NXn: An index to a base table or object type (e.g. NX1, NX2, etc.) PK: Primary Key FKn: Foreign Key (e.g. FK1, FK2, etc.) ICn: Integrity Constraint (e.g.IC1, IC2, etc.) AO: An ADD operation MO: A MODIFY operation ZO: A DELETE (Zap) operation IO: An INQUIRE operation FO: A FORECAST operation. RO: A REPORT operation XO: A utility operation DS: A database synonym or alias of a known database table DC: A database constraint DT: A database Trigger DP: A database procedure or function DK: A database package MF: A Message file — a special purpose database table (file) to store the text (and other essential details) for diagnostic error and status messages
The descriptor used for a database base relation or object type is consistently used for other objects that directly relate to that object. For example, the objects related to inventory items may be: MES_InventoryItem_BR: A base relation to store data on inventory items MES_InventoryItem _NX1: An index on the base relation MES_InventoryItem _AO: An operation to ADD inventory items MES_InventoryItem _MO: An operation to MODIFY inventory items MES_InventoryItem _ZO: An operation to DELETE inventory items MES_InventoryItem _IO: An operation to INQUIRE on inventory items MES_InventoryItem _RO: An operation to REPORT on inventory items MES_InventoryItem _XO : A utility operation related to inventory items MES_InventoryItem _LV1: A logical view of the inventory items
Attribute implementation names are merely abbreviations of their more descriptive names, prefixed by an appropriate abbreviation of the entity.

Figure 3.16 Proposed object naming convention.

Relation	Properties (Attributes)	Comment
Core Entities (Including Resolution of 1:M Relationships):		
MES_Customer_BR	*Cust#*	Primary Key (PK)
	CustName . . .	
MES_Department_BR	*Dept#*	PK
	DeptName . . .	
MES_Employee_BR	*Emp#*	PK
	EmpName . . .	
	EmpDept#	Refers to **Department.Dept#**
MES_Supplier_BR	*Supp#*	PK
	SupplName . . .	
	SupplLoc	Refers to **Location.LocCode**
MES_Project_BR	*Proj#*	PK
	ProjName	
	ProjManagerEmp# ...	Refers **Employee.Emp#**
MES_InventoryItem_BR	*Item#*	PK
	ItemName . . .	
MES_Dependent_BR	*DepnEmp#*	Refers to **Employee.Emp#**; K1
	DepnRef	K2
	DepnName	
MES_Location_BR	*LocCode*	PK
	LocName . . .	
MES_Machine_BR	*Mach#*	PK
	MachName . . .	
MES_Warehouse_BR	Whouse#	PK
	WhouseLoc	Refers to **Location.LocCode**
	WhouseName . . .	
Resolution of Component and/or Subtype Relationships		
MES_EmpPersonalInfo_BR	*EPI_Emp#*	Refers to **Employee.Emp#**; PK
	EPI_Specialty . . .	
MES_EmpWorkHistory_BR	*EWH_Emp#*	Refers to **Employee.Emp#**; K1
	EWH_JobSeqNo	K2
	EWH_Organization	
	EWH_JobTitle . . .	
MES_EmpAcademicLog_BR	*EAL_Emp#*	Refers to **Employee.Emp#**; K1
	EAL_SeqNo	K2
	EAL_Institution	
	EAL_StartDate	
	EAL_ExitDate	
	EAL_Award . . .	
Note: Primary key and foreign keys are *italicized*. K1 . . K4 represent composite candidate keys.		

Each of these relations would typically encompass several other attributes; only the essential ones are shown here.

Figure 3.17 Refined RAL for the manufacturing environment. (Continued)

Relation	Properties (Attributes)	Comment
Resolution of Component and/or Subtype Relationships (continued)		
MES_EmpPublicationsLog_BR	*EP_Emp#*	Refers to **Employee.Emp#**; K1
	EP_PubSeqNo	K1
	EP_Title	
	EP_PubType . . .	
MES_EmpExtraC_BR	*EX_Emp#*	Refers to **Employee.Emp#**; K1
	EX_ActivityCode	K2
	EX_ActivityDesc . . .	
Other relations that might have been missed or to be added . . .		
. . .		
Resolution of M:M Relationships		
MES_MachineUsageMap_BR	*MU_Mach#*	Refers to **Machine.Mach#**; K1
	MU_Item# . . .	Refers to **InventoryItem.Item#**; K2
MES_MachProjectMap_BR	*MP_Mach#*	Refers to **Machine.Mach#**; K1
	MP_Proj# . . .	Refers to **Project.Proj#**; K2
MES_SuppItemMap_BR	*SI_Supp#*	Refers to **Supplier.Suppl#**; K!
	SI_Item#	Refers to **InventoryItem.Item#**; K2
	SI_Ref#	Surrogate PK
MES_ItemProjMap_BR	*IP_Item#*	Refers to **InventoryItem.Item#**; K1
	IP_Proj#	Refers to **Project.Proj#**; K2
	IP_Ref#	Surrogate PK
MES_ProjSuppMap_BR	*PS_Proj#*	Refers to **Project.Proj#**; K1
	PS_Supp#	Refers to **Supplier.Suppl#**; K2
	PS_Ref#	Surrogate PK
MES_ProjWorkSchedu;e_BR	*PW_Emp#*	Refers to **Employee.Emp#**; K1
	PW_Proj#	Refers to **Project.Proj#**; K2
	PW_Ref#	Surrogate PK
MES_ItemStruct_BR	*IS_ThisItem#*	Refers to **InventoryItem.Item#**; K1
	IS_CompItem#	Refers to **InventoryItem.Item#**; K2
	IS_Ref#	Surrogate PK
MES_PurchOrdSummary_BR	*OrderRef#*	Surrogate PK
	OrderSupp#	Refers to **Supplier.Suppl#**; K1
	Order#, OrderDate	K2, K3
	OrderStatus . . .	
MES_PurchOrdDetail	*POD_OrderRef#*	Refers to **PurchOrdSummary.OrderRef#**; K1
	POD_Item#	*References* **InventoryItem.Item#**; K2
	POD_Quantity	
	POD_UnitPrice	
Note: Primary key and foreign keys are *italicized*. K1 . . K4 represent composite candidate keys.		

Each associative entity represents the intersecting relation used to implement a M:M relationship.

Figure 3.17 (Continued) Refined RAL for the manufacturing environment. (Continued)

Relation	**Properties (Attributes)**	**Comment**
Resolution of M:M Relationships (continued)		
MES_PurchInvoiceSummary_BR	*PI_Ref#*	Surrogate PK
	PI_Supp#	Refers to **Supplier.Suppl#**; K1
	PI_Invoice#	K2
	PI_OrderRef#	Refers to **PurchOrdSummary.OrderRef#**; K3
	PI_Date	K4
	PI_Amount	
	PI_Status . . .	
MES_PurchInvoiceDetail_BR	*PID_PIRef#*	Refers to **PurchInvoiceSummary.PI_Ref#**; K1
	PID_Item#	Refers to **InventoryItem.Item#**; K2
	PID_Quantity	
	PID_UnitPrice . . .	
MES_SaleInvoiceSummary_BR	*SI_Ref#*	Surrogate PK
	SI_Cust#	Refers to **Customer.CustI#**; K1
	SI_Invoice#	K2
	SI_Date	K3
	SI_Amount	
	SI_Status . . .	
MES_SaleInvoiceDetail_BR	*SID_SIRef#*	Refers to **SaleInvoiceSummary.PI_Ref#**; K1
	SID_Item#	Refers to **InventoryItem.Item#**; K2
	SID_Quantity	
	SID_UnitPrice . . .	
MES_StockPile_BR	*SP_Whouse#*	Refers to **Warehouse.Whouse#**; K1
	SP_Item#	Refers to **InventoryItem.Item#**; K2
	SP_Quantity . . .	
Note: Primary key and foreign keys are *italicized*. K1 . . K4 represent composite candidate keys.		

Each associative entity represents the intersecting relation used to implement a M:M relationship.

Figure 3.17 (Continued) Refined RAL for the manufacturing environment.

3.10 Summary and Concluding Remarks

This very important chapter lays the foundation that will be built upon in subsequent chapters. Let us summarize what we have covered:

- The relational database model is based on a number of fundamental concepts relating to the following: entity, entity set, relation, relationship, tuple, candidate key, primary key, alternate key, index, foreign key, domain, cardinality, degree.
- A domain is a named set of scalar values from which attribute values are drawn.
- A relation consists of a heading and a body. The heading consists of atomic attributes defined on specific domains. The body consists of a set of attribute–value pairs, where each attribute has a value drawn from its domain.

- In a database system, you are likely to find any combination of the following types of relations: base relations, logical views, snapshots, query results, intermediate results, and temporary relations.
- A relational database system (RDBS) is a collection of time-varying normalized relations, managed through an appropriate user interface, and with desirable constraints and features that enhance the effective, efficient management of the database.
- A relationship is an inherent mapping involving two or more relations. There are six types of relationships: one-to-one (1:1) relationship, one-to-many (1:M) relationship, many-to-one (M:1) relationship, many-to-many (M:M) relationship, component relationship, and subtype relationship.
- An E–R diagram (ERD) is a graphical representation of a database model. It is important to know how to represent relations/entities and relationships on the ERD.
- It is important to know how to implement the various types of relationships in the actual database design.
- The relation–attributes list (RAL) and relationship list (RL) are two useful alternatives and/or supplements to the E–R diagram, especially for large, complex systems.
- When planning a database, it is advisable from the outset to use unique attribute names across the database; this allows for easy tracking of data throughout the system.

Database approaches that preexisted the relational approach include the inverted-list approach, the hierarchical approach, and the network approach. Discussions on these have been deliberately omitted because they are no longer relevant. Moreover, the strengths of the relational approach when compared to these approaches are its sound mathematical base, its flexibility, robustness, and simplicity.

Still, there are alternate approaches to database systems. For instance, the *object-oriented* (OO) model has been advanced as an alternative to the relational model on performance and efficiency for certain scenarios. In more recent times, we have seen the advance of other approaches such as Hadoop, the entity–attribute–value (EAV) model, and the more prominent NoSQL approach for large databases in the so-called Big Data Revolution (Thomas & McSharry 2015). Nonetheless, for reasons that will become clear to you by the end of the course, it is reasonable to expect the continued dominance of the relational model. It is also reasonable to expect that these technologies (relational databases, OO databases, and NoSQL models) will continue to peaceably coexist; huge investments have been made in relational database systems, and it is not likely that these will be abandoned. What is more likely to happen is that systems will continue to be built based on relational databases, with superimposed OO user interfaces and NoSQL strategies incorporated where necessary.

Take the time to go over this chapter more than once if necessary and make sure that you are comfortable with the concepts covered. In the upcoming chapter, we will build on the information covered in this chapter, as we discuss integrity rules and normalization. These two topics form the foundation for the rest of the course.

3.11 Review Questions

1. Clarify the following terms:
   - Entity
   - Attributes
   - Entity set
   - Relation
   - Relationship
   - Tuples
   - Primary key
   - Candidate key
   - Foreign key
   - Index
   - Domain
   - Cardinality
   - Multiplicity
   - Degree

   Develop an illustrative data model to explain how these concepts are related.
2. Give the formal definition of a relation. Explain the different kinds of relations that may be found in a database.
3. Outline the steps to be observed in constructing a relational database system.
4. What are the types of relationships that may exist among information entities? For each, explain how it is identified and how it is implemented.
5. Think about a problem that requires a small database system (requiring 6–12 entities or object types). Do the following:
   - Identity all required entities (object types).
   - Identify relationships among the entities (object types).
   - Develop an E–R diagram.
   - Develop a corresponding O–R diagram.
   - Propose a relation–attributes list (RAL) and a relationship list (RL) to represent your model.

References and Recommended Readings

Chen, Peter. 1994. "The Entity-Relationship Model: Toward a Unified View of Data." In *Readings in Database Systems* 2nd ed., pages 741–754. San Francisco, CA: Morgan Kaufmann.

Chen, Peter. n.d. Accessed January 2009. http://bit.csc.lsu.edu/~chen/display.html

Connolly, Thomas, & Carolyn Begg. 2015. *Database Systems: A Practical Approach to Design, Implementation and Management* 6th ed. Boston, MA: Pearson. See chapter 4.

Coronel, Carlos, & Steven Morris. 2015. *Database Systems: Design, Implementation & Management* 11th ed. Boston, MA: Cengage Learning. See chapters 3 & 4.

Dahchour, Mohamed, & Alain Pirotte. n.d. "The Semantics of Reifying N-ary Relationships as Classes." Accessed April 1, 2013. http://www.uclouvain.eu/cps/ucl/doc/iag/documents/WP_35_dahchoursemantics.pdf

Date, Christopher J. 2004. *Introduction to Database Systems* 8th ed. Menlo Park, CA: Addison-Wesley. See chapters 3 and 6.

Elmasri, Ramez, & Shamkant B. Navathe. 2011. *Fundamentals of Database Systems* 6th ed. Boston, MA: Pearson. See chapters 3 and 7.

Foster, Elvis C. 2021. *Software Engineering — A Methodical Approach*, 2nd ed. NY: CRC Publishing See chapters 12 & 13.

Garcia-Molina, Hector, Jeffrey Ullman, & Jennifer Widom. 2009. *Database Systems: The Complete Book* 2nd ed. Boston, MA: Pearson. See chapter 2.

Hoffer, Jeffrey A., Ramesh Venkataraman, & Heikki Topi. 2013. *Modern Database Management* 11th ed. Boston, MA: Pearson. See chapters 2–4.

Kifer, Michael, Arthur Bernstein, & Phillip M. Lewis. 2006. *Databases and Transaction Processing: An Application Oriented Approach* 2nd ed. Boston, MA: Pearson. See chapters 3 & 4.

Kroenke, David M., & David Auer. 2015. *Database Concepts* 7th ed. Boston, MA: Pearson. See chapters 2 & 4.

Lee, Richard C., & William M. Tepfenhart. 2002. *Practical Object-Oriented Development With UML and Java*. Upper Saddle River, NJ: Prentice Hall.

Martin, James, & James Odell. 1989. *Information Engineering Book I: Introduction*. Eaglewood Cliffs, NJ: Prentice Hall.

Martin, James, & James Odell. 1990a. *Information Engineering Book II: Planning and Analysis*. Eaglewood Cliffs, NJ: Prentice Hall.

Martin, James, & James Odell. 1990b. *Information Engineering Book III: Design and Construction*. Eaglewood Cliffs, NJ: Prentice Hall.

Schach, Stephen R. 2011. *Object-Oriented and Classical Software Engineering* 8th ed. Boston, MA: McGraw-Hill.

Silberschatz, Abraham, Henry F. Korth, & S. Sudarshan. 2011. *Database Systems Concepts* 6th ed. Boston, MA: McGraw-Hill. See chapters 2 & 7.

Thomas, Rob, & Patrick McSharry. 2015. *Big Data Revolution: What Farmers, Doctors and Insurance Agents Teach Us about Discovering Big Data Patterns*. Hoboken, NJ: Wiley.

Ullman, Jeffrey D., & Jennifer Widom. 2008. *A First Course in Database Systems* 3rd ed. Boston, MA: Pearson. See chapters 2 & 3.

Chapter 4

Integrity Rules and Normalization

In order to design high-quality databases, you need to be cognizant of the fundamental integrity and normalization rules for the database. We will discuss these rules in this chapter. Subtopics to be discussed include:

- Fundamental Integrity Rules
- Foreign Key Concept
- Rationale for Normalization
- Functional Dependence and Non-loss Decomposition
- First Normal Form
- Second Normal Form
- Third Normal Form
- Boyce–Codd Normal Form
- Fourth Normal Form
- Fifth Normal Form
- An Example
- Other Normal Forms
- Summary and Concluding Remarks

4.1 Fundamental Integrity Rules

Two fundamental integrity rules that the database designer must be cognizant of are the *entity integrity rule* and the *referential integrity rule*.

Entity Integrity Rule: The entity integrity rule states that no component of the primary key in a base relation is allowed to accept nulls. Put another way, in a relational model, we never record information about something that we cannot identify. Three points are worth noting here:

- The rule applies to base relations.
- The rule applies to primary key, not alternate keys.
- The primary key must be wholly non-null.

DOI: 10.1201/9781003275725-6

Recall from the previous chapter that the primary key may be a single attribute or a composite key; either way, a null value is not allowed. If the primary key is a single attribute, it should not be null for any tuple in the relation. If the primary key is a composite key, then no attribute comprising this composite key should be null for any tuple in the relation. Also note that for composite primary keys, the correct attributes must be chosen — choosing fewer attributes than required causes the primary key to lose the uniqueness quality required for distinguishing tuples; choosing more attributes than required makes the primary key redundant and increases the database overhead.

Referential Integrity Rule: The referential integrity rule states that the database must not contain unmatched foreign key values. By unmatched foreign key value, we mean a non-null foreign key value for which there is no match in the referenced (target) relation. Put another way, if B references A then A must exist. The following points should be noted:

- The rule requires that foreign keys must match primary keys, not alternate keys.
- Foreign key and referential integrity are defined in terms of each other. It is not possible to explain one without mentioning the other.

4.2 Foreign Key Concept

The concept of a foreign key was introduced in the previous chapter. Let us revisit this concept by introducing a more formal definition:

> Attribute FK of base relation R2 is a foreign key if and only if (denoted iff from this point) it satisfies the following conditions:
>
> a. Each value of FK is either wholly null or wholly non-null.
> b. There exists a base relation, R1 with primary key PK such that each non-null value of FK is identical to the value of PK in some tuple of R1.

We will use the notation R1 → R2 to mean that relation R1 references relation R2. In this case, R1 is the referencing (primary) relation and R2 is the referenced relation. Since R1 is the referencing relation, it contains a foreign key. We will also use the notion R1{A, B, C, ...} to mean that relation R1 contains attributes A, B, C, and so on. Where specific examples are given, the relation name will be highlighted or placed in the upper case; attribute names of specific examples will not be highlighted when stated with the related relation; however, they will be highlighted when reference is made to them from the body of the text.

This definition of a foreign key is quite loaded, and therefore deserves some clarification. Based on the stated definition, the following consequential points should be noted:

1. The foreign key and the referenced primary key must be defined on the same domain. However, the attribute names can be different (do not be surprised if you see widespread violation of this principle).

2. The foreign key need not be a component of the primary key for the host (holding) relation (in which case nulls may be accepted, but only with the understanding that they will be subsequently updated). To illustrate, in the following two relations, **Employee** references **Department** via an M:1 relationship; you could have an employee who is temporarily not assigned to a department.
   Department {Dept#, DeptName, . . .} PK[Dept#]
   Employee {Emp#, Emp_FName, Emp_LName, Emp_Dept#, . . .} PK[Emp#]
   FK[Emp_Dept#] references **Department.Dept#**
3. If relations Rn, R(n–1), R(n–2) .... R1 are such that Rn → R(n–1) → R(n–2) → .... R2 → R1, then the chain Rn to R1 forms a *referential path.*
4. In a referential path, a relation can be both referenced and referencing. Consider the referential path R3 → R2 → R1. In this case, R2 is both a referenced relation and a referencing relation.
5. The simplest possible referential path is a self-referencing relation. To illustrate, consider a refined version of the aforementioned relation **Employee** where each employee tuple also includes a reference to the employee's manager. In this case, we have a self-referencing relation. Thus,
   Employee {Emp#, Emp_FName, Emp_LName, Emp_Dept#, . . ., Emp_MgrEmp#} PK[Emp#] FK[Emp_Dept#] references **Department.Dept#** and FK [Emp_MgrEmp#] references **Employee.Emp#**
6. More generally, a referential cycle exists when there is a referential path from Rn to itself: Rn → R(n-1) → .... R1 → Rn
7. From this simple discourse, you should note the important role of foreign keys. Foreign keys are said to be the "glue" that holds the database together. As you will later see, relations are joined based on foreign keys.

Now that we have established the importance of foreign keys, we need to address a question: How will we treat deletion of referenced tuples? Three alternatives exist.

- *Restrict* deletion to tuples that are not referenced.
- *Cascade* deletion to all referencing tuples in referencing relation(s).
- Allow the deletion but *nullify* all referencing foreign keys in referencing relations.

The nullification strategy is particularly irresponsible if applied to component relationships, as it could quite quickly plunge the integrity of the database into question, by introducing *orphan records* (i.e., records with unmatched references). However, it could be applied to some M:1 relationships that are not component relationships (more on this in Chapter 5). For instance, referring to the **Employee** and **Department** relations above, we could remove a particular department from the database and then nullify the FK values for employees that belonged to that department; the assumption would be that these employees would be subsequently assigned to different departments.

Traditionally, DBMS suites implement the restriction strategy (and for good reasons). This is the most conservative and safest approach: If you desire to remove a referenced tuple from the database, you must first take appropriate action on all references to that referenced tuple —by either removing those referencing tuples or changing their references.

The cascading strategy has been surfacing in contemporary systems as an optional feature. It must be used with much care, as it is potentially dangerous when used without discretion. This strategy would not be the most prudent for the **Department** and **Employee** relations above. You may want to remove a department without removing all Employee tuples that referenced that department. However, as you will see later in the chapter, an invoice is a perfect example where the cascading strategy would be applicable. An invoice typically consists of several line items (for instance purchasing multiple items from a specific organization). The invoice is typically broken up into a summary portion along with corresponding details. If you desire to remove the invoice summary from the database, the details should also be removed; otherwise, the database would wind up with orphan records.

4.3 Rationale for Normalization

Normalization is the process of ensuring that the database (*conceptual schema*) is defined in such a manner as to ensure efficiency and ease of data access. Normalization ensures and/or enhances the following database features:

- **Entity and Referential Integrity:** The normalization process results in a set of relations that are logically structured such that each relation has its primary key or candidate keys identified. Foreign keys are strategically inserted into specific referencing relations to map back to related primary keys in the target (referenced) relations. The DBMS can then be allowed to enforce appropriate referential integrity constraints when the database is implemented.
- **Control of Redundancy:** Through a process called *non-loss decomposition*, the normalization process is geared toward reducing unnecessary data duplication.
- **Logical Data Independence:** Normalization leaves the logical structure of the database (i.e., the conceptual schema) in a state that easily facilitates the construction of logical integration and aggregation of data from various normalized relations to meet the assorted needs of end-users. These logical maneuvers do not necessarily mandate changes in the application programs that access the database; neither do changes in the application programs necessarily force structural changes in the database.
- **Avoidance of Modification Anomalies:** As you will soon see, normalization resolves undesirable structure of data that causes data redundancy, which in turn causes various modification anomalies as explained below.
- **Efficient Data Storage and Retrieval:** As stated in the definition above, the primary objective of normalization is efficient storage and subsequent retrieval of data.

The contrasting argument is equally important: In the absence of normalization, various problems are likely to occur. Among these problems are the following:

- Data redundancy that leads to the modification anomalies
- Modification anomalies that include are as follows:
  - Insertion anomaly: Data cannot be inserted when it is desirable to do so; one has to wait on some future data due to organization of the data structure
  - Deletion anomaly: Deletion of some undesirable aspect(s) of data necessarily means deletion of some other desirable aspect(s) of data
  - Update anomaly: Update of some aspect(s) of data necessarily means update of other aspect(s) of data
- Inefficient file manipulation; lack of ease of data access

- Inhibition to the achievement of logical data independence
- Compromise on the integrity of data
- Pressure on programming effort to make up for the poor design

Figure 4.1 indicates the six most commonly used *normal forms*. The hierarchy is such that a relation in a given normal form is automatically in all normal forms prior to it. Thus, a relation in the second normal form (2NF) is automatically in the first normal form (1NF); a relation in the third normal form (3NF) is in 2NF, and so on. Edgar Frank Codd defined the first three normal forms in the early 1970s; the Boyce–Codd normal form (BCNF) was subsequently deduced from his work. The fourth and fifth normal forms (4NF and 5NF) were subsequently defined by Ronald Fagin in the late 1970s.

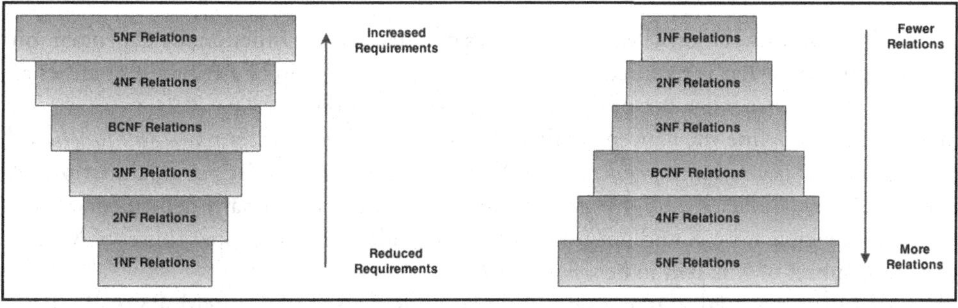

Figure 4.1 Normal forms.

The *normalization procedure* involves decomposing relations into other relations of repeatedly higher normal forms. The process is reversible. Moreover, normalization must be conducted in a manner that ensures that there is no loss of information.

4.4 Functional Dependence and Non-Loss Decomposition

Before discussion of the normal forms, we need to define and clarify two fundamental concepts: *functional dependence* and *non-loss decomposition*.

4.4.1 Functional Dependence

Given a relation, R{A, B, C, ...}, then attribute B is *functionally dependent* on attribute A, written A → B (read as "A determines B") iff each value of A in R has precisely one B value in R at any point in time. Attributes A and B may or may not be composite.

An alternate way to describe functional dependence (FD) is as follows: Given a value of attribute A, one can deduce a value for attribute B since any two tuples that agree on A must necessarily agree on B.

To further clarify the concept of an FD, consider storing information about employees in an organization. Example 4.1 revisits the **Employee** relation of earlier discussion, showing how an FD may be defined on the attribute representing the employee's identification number. In this example, the employee's identification number (**Emp#**) is fulfilling the role of primary key.

EXAMPLE 4.1: AN EXAMPLE ILLUSTRATING THE CONCEPT OF FD

Consider the relation Employee as outlined below:

Employee {Emp#, Emp_FName, Emp_LName, Emp_Dept#, . . ., Emp_MgrEmp#}

The following FD holds:

Emp# → Emp#, Emp_FName, Emp_LName, Emp_Dept#, . . ., Emp_MgrEmp#

What this means is that each employee is uniquely identified by his/her employee number. By knowing the employee number, all other attribute values for that employee can be deduced.

From definition of primary key (PK), all attributes of a relation are functionally dependent on the PK. This is precisely what is required; in fact, an attribute (or group of attributes) qualifies as a candidate key iff all other attributes of the entity are dependent on it.

We need to further refine the concept of FD by introducing two new terms —*full functional dependence* and *partial functional dependence*: Let X and Y represent two distinct combinations of one or more attributes, and let X → Y. In such a scenario, Y is fully functionally dependent on X if it is functionally dependent on X and not functionally dependent on any proper subset of X. Any dependence on a subset of X is said to be a *partial FD*.

To illustrate full and partial FDs, consider the relation R{A, B, C, D,...} with PK[A,B]. Then [A, B] → C, D represents a full FD provided that neither A alone nor B alone determines C, D. However, any of the following FD would represent a partial FD: A → C, A → D, B C, or B → D.

Understanding what FDs are and how to identify them are both essential to designing a database. Based on the aforementioned definition, here are three noteworthy consequences:

1. FD constraints have similarities with referential constraints, except that here, reference is internal to the relation.
2. FDs help us to determine primary keys.
3. Each FD defines a determinant in a relation: the attribute(s) on the right are dependent on the attribute(s) on the left; the attribute(s) on the left constitute(s) a determinant.

Next, we have *trivial functional dependencies*. An FD is trivial if the attribute(s) on the right form(s) a subset of the attribute(s) on the left. For instance, in relation R{A, B, C, D, ...} with PK [A, B], the following FDs are trivial: [A, B] → A; [A, B] → B; and [A, B] → A, B.

4.4.2 Non-Loss Decomposition

We come now to the concept of non-*loss decomposition* (NLD): An NLD of a relation is decomposition of the relation into multiple (at least two) relations in such a way that the new relations can be recombined to yield the original; there is no loss of data from the decomposition of the original relation. Let us start with an example. Suppose that we have a relation **R0** storing basic supplies to a business as follows:

R0{Suppl#, SupplName, Item#, ItemName, Quantity, SupplStatus, Location} PK [Suppl#, Item#]

Functional dependencies of **R0** are represented in the illustrative FD diagram of Figure 4.2; they may also be listed as follows:

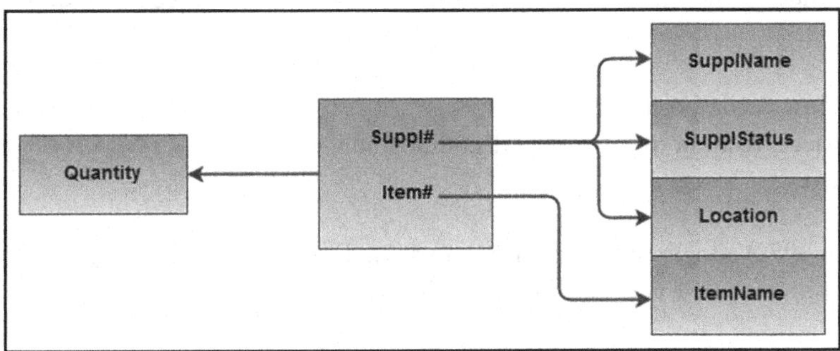

Figure 4.2 FD diagram for relation R0.

- [Suppl#, Item#] → {Quantity, SupplName, SupplStatus, Location, ItemName}
- Suppl# → {SupplName, SupplStatus, Location}
- Item# → ItemName

As you will soon see, storing **R0** as a single entity would introduce a duplication problem that would in turn introduce a number of other problems. As the size of the relation increases (i.e., more data is added over time), these related problems would be exacerbated toward a virtually uncontrollable level. The reason is that **R0** has too many unresolved FDs (i.e., the relation is not sufficiently normalized). As an alternative, we could have the following relations:

R1 {Suppl#, SupplName, Location, SupplStatus} PK[Suppl#]
R2 {Item#, ItemName} PK [Item#]
R3 {Suppl#, Item#, Quantity} PK[Suppl#, Item#]

In this scenario, **R1**, **R2**, and **R3** would constitute an example of a *non-loss decomposition* (NLD) of **R0**. What this means is that we can effectively defer storing **R0** as originally proposed, and instead store **R1**, **R2**, and **R2**. Moreover, in so doing, we would avoid the problems related to storing **R0** in its original form, but still be able to reconstruct it by combining the three replacement relations.

As you will soon see, making NLDs on relations is an integral part of the normalization process. Here is the formal definition of an NLD:

> If R is a relation and P1, P2, ... Pn are projections on R such that
>
> P1 JOIN P2 JOIN ..... JOIN Pn = R,
>
> then P1, P2, ... Pn constitutes a non-loss decomposition of R.

Notice that we have glided into two new terms, *projection* and *join*. These will be formally treated later in the course. Suffice it for now to say that a projection on a relation is an extraction (into a new relation) of some attributes of the original relation; a join requires at least two relations and may be construed as the opposite of a projection: If you can project R into P1 and P2, then you may be able to join P1 and P2 to yield R.

So, has it occurred to you that identifying FDs and managing NLDs are two imperatives for the normalization process? Indeed, they are! Given this definition of an NLD, we need to address the following questions:

1. How do we find non-loss decompositions?
2. When should we replace a relation by a non-loss decomposition?
3. What are the advantages?

Heath's theorem (see Heath 1971) addresses Questions (1) and (2). The answer to the third question has already been stated in Section 4.3. Heath's theorem is stated below:

> If we have a relation R{A, B, C, ...} and if A → B and B → C, then projections P1{A, B} and P2{B, C} constitute a non-loss decomposition of R.

Observe:

Heath's theorem effectively addresses the questions of when to conduct an NLD, and how to do so. Each of the letters [A, B, and C] in the theorem represents a single attribute or a combination of attributes. Example 4.2 provides a proof of the theorem.

EXAMPLE 4.2: PROOF OF HEATH'S THEOREM

We wish to show that for R0 {A, B, C}, if A → B and B → C, then R0 = P1 {A, B} JOIN P2 {B, C}.

Let P1 {A, B} and P2 {B, C} be projections of R0 {A, B, C}. Assume further that A, B, C are single attributes.

Suppose that (a, b, c) is a tuple in R0. Then (a, b) is in P1 and (b, c) is in P2.
So (a, b, c) is in P1 JOIN P2 .... (1)

Suppose that (a, b, c) is in P1 JOIN P2. Then (a, b, c1) is in R0 for some value c1 and (a1, b, c) is in R0 for some value a1. But B → C therefore b → c so that c1 must be c.
Therefore (a, b, c) is in R0 .... (2)

We have shown that any tuple (a, b, c) that is in R0 is also in P1 JOIN P2, and that any tuple (a, b, c) that is in P1 JOIN P2 is in R0. Therefore R0 = P1 JOIN P2.

4.4.2.1 Corollary from Heath's Theorem

An important corollary from Heath's theorem informs us when to discontinue or desist from further decomposition; it is stated below:

> If P1, P2, ... Pn is a non-loss decomposition of R and relations P1, P2, ... Pn all share a candidate key, then there is no reduction in data duplication.

4.4.2.2 Conclusion from Heath's Theorem and Its Corollary

Based on Heath's theorem and its corollary, we can assert with confidence, the following advice:

- Decompose only when there is non-loss decomposition such that the resulting relations do not share a candidate key.
- Do not decompose if each resulting relation does not have a candidate key.

Example 4.3 illustrates the importance of the abovementioned corollary. The example shows two flawed decompositions of the relation **Student** {Stud#, StudName, StudGPA, StudDept} PK [Stud#].

> **EXAMPLE 4.3: ILLUSTRATING USEFULNESS OF THE HEATH'S THEOREM COROLLARY**
>
> Consider the relation **Student** {Stud#, StudName, StudGPA, StudDept#} PK[Stud#].
> Now consider projections S1 {Stud#, StudName} PK[Stud#]
> And S2 {Stud#, StudGPA, StudDept#} PK [Stud#].
> **S1** and **S2** would not represent a useful decomposition because they both share the same primary; such a decomposition would not curtail duplication but worsen it by storing **Stud#** in two different relations.
> Also, decomposition of **Student** into **S3** {Stud#, StudName} and **S4**{StudGPA, StudDept#} makes no sense.

With this established background, let us now proceed to discussing the normal forms. Starting with the first, a section is dedicated to each normal form.

4.5 First Normal Form

We begin with a definition of the first normal form:

> A relation is in the first normal form (1NF) iff all its attributes are atomic (i.e., each attribute is defined on a single domain), and each record in the relation is characterized by a unique primary key value.

In defining atomic attributes, Codd was deliberate about eliminating *repeating groups* — the scenario where a column (attribute) could store an array or list of possible values; this was prevalent in CODASYL and COBOL. Atomicity means each attribute is defined on a single domain and stores a single value for any tuple in the relation. Moreover, implicit in this definition is the enforcement of the entity integrity constraint (Section 4.1). As such, tuples that have null values in their primary key will not be allowed to exist; neither will there be tuples with duplicate primary key values in the relation. [These are problems that persisted and perhaps still do outside of the context of 1NF].

By this definition, all relations are in 1NF. This is by no means coincidental, but by design: We define a relation to consist of atomic attributes, and subject to the entity integrity constraint and the referential integrity constraint. However, as you will soon see, having relations in 1NF only is often not good enough.

Example 4.4 provides an example of a relation that is in 1NF but is poorly designed. It may surprise you to learn that this problem was once very widespread in accounting software systems. As an exercise, try proposing a more efficient design for the relation described in the example.

EXAMPLE 4.4: EXAMPLE OF A POORLY DESIGNED 1NF RELATION

Consider the relation **EndOfMonth** {Acct#, Dept#, Bal1, Bal2, ... Bal13} PK [Acct#, Dept#]

Point to Note:

1. This relation is intended to store monthly balances on various accounts in an organization.
2. **Bal1** ... **Bal13** are defined on the same domain and therefore constitute a vast amount of space wasting (this scenario is often incorrectly described as a repeating group but is different from the original meaning of the term).
3. The only time that **Bal1** ... **Bal13** are all non-null is after **Bal13** is calculated; this is normally done at the end of the financial year.
4. At the end of each accounting period, this relation is cleared and re-initialized for the next accounting period.

Caution:

Having relations in 1NF only is woefully inadequate for ensuring the integrity of a database. Nonetheless, this is an essential first step.

In the previous section, **R0** was presented as a non-normalized relation to be replaced with three other relations. Let us revisit that discussion for further insight. For ease of reference, **R0** is repeated here:

R0{Suppl#, SupplName, Item#, ItemName, Quantity, SupplStatus, Location} PK [Suppl#, Item#]

Relation **R0** is in 1NF only. However, it is undesirable to store it in its current form due to a number of problems. For ease of reference, the functional dependencies of **R0** as illustrated in Figure 4.2 are repeated here:

- FD1: Suppl# → {SupplName, SupplStatus, Location}
- FD2: Item# → ItemName
- FD3: [Suppl#, Item#] → {Quantity, SupplName, SupplStatus, Location, ItemName}

Figure 4.3 shows some sample data depicting various shipments of inventory items for **R0**. A cursory analysis of the data will amplify the following data anomalies:

- **Replication of Data:** Every time we record a supplier–item pair, we also have to record supplier name and item name.
- **Insertion Anomaly:** We cannot insert a new item until it is supplied; neither can we insert a new supplier until that supplier supplies some item.
- **Deletion Anomaly:** We cannot delete an item or a supplier without destroying an entire shipment, as well as information about a supplier's location.
- **Update Anomaly:** If we desire to update a supplier's location or item name, we have to update several records, in fact, an entire shipment, due to the duplication problem.

Insertion, update, and deletion anomalies constitute *modification anomalies,* which are caused by duplication of data due to improper database design. As stated earlier, these problems are exacerbated as more data are added, thus leading to other problems of data access and database performance — a tangled hellish web.

Suppl#	SupplName	Item#	ItemName	Quantity	SupplStatus	Location	
S1	Farnsworth Inc.	I1	Office Chair	12	Experimental	Boston	Duplication
S1	Farnsworth Inc.	I2	Printer Drum Cartridge	12	Experimental	Boston	
S1	Farnsworth Inc.	I3	Printer	12	Experimental	Boston	
S1	Farnsworth Inc.	I4	19" Dell Monitor	12	Experimental	Boston	
S1	Farnsworth Inc.	I5	15" Dell Monitor	10	Experimental	Boston	
...	...						
S2	Bruce Jones Inc.	I1	Office Chair	8	Reliable	Boston	Duplication
S2	Bruce Jones Inc.	I6	Jones HD Encryption	1	Reliable	Boston	
S2	Bruce Jones Inc.	I7	Foster DMID	1	Reliable	Boston	
S2	Bruce Jones Inc.	I8	19" Samsung Monitor	12	Reliable	Boston	
S2	Bruce Jones Inc.	I9	15" Jones Monitor	10	Reliable	Boston	
...	...						
S3	Lambert Ingram Inc.	I10	Ingram CASE Tool	1	Reliable	Seattle	Duplication
S3	Lambert Ingram Inc.	I11	Ingram Security System	1	Reliable	Seattle	
S3	Lambert Ingram Inc.	I12	Ingram Financial Manager	1	Reliable	Seattle	
S3	Lambert Ingram Inc.	I13	Ingram Network Server	1	Reliable	Seattle	
S3	Lambert Ingram Inc.	I14	Ingram Network Sniffer	1	Reliable	Seattle	
...	...	...					

Attention!!! Notice the high level of duplication in this example. Shortly, you will see how we might restructure this file.

Figure 4.3 Sample data representing a shipment of supplies (for relation R0).

4.6 Second Normal Form

The second normal form draws in the concept of functional dependence to shape an elevated benchmark beyond mere 1NF requirement. Here is the definition:

> A relation is in the second normal form (2NF) iff it is in 1NF and every *non-key attribute* is fully functionally dependent on the primary key.

By non-key attribute, we mean that the attribute is not part of the primary key. Relation **R0** (of the previous section), though in 1NF, is not in 2NF, due to FD1 and FD2. Using Heath's theorem, we may decompose relation **R0** as follows (note that the abbreviation PK is used to denote the primary key):

R1 {Suppl#, SupplName, Location, SupplStatus} PK[Suppl#]
R2 {Item#, ItemName} PK[Item#]
R3 {Suppl#, Item#, Quantity} PK[Suppl#, Item#]

We then check to ensure that the resulting relations are in 2NF. Relation **R1** has a single attribute as its primary key, and so does **R2**; there is therefore no possibility of either relation being in violation of 2NF. As for relation **R3**, there is only one non-key attribute and it is dependent on the primary key. We may therefore conclude with confidence that all three relations (**R1**, **R2**, and **R3**) are in 2NF.

So, based on the definition of 2NF, and on the authority of Heath's theorem, we would replace **R0** with **R1**, **R2**, and **R3**. Please note the consequences of our treatment of **R0** so far:

1. The problems with relations in 1NF only have been addressed.
2. By decomposing, we have introduced foreign keys in relation **R3**.
3. JOINing is the opposite of PROJecting. We can rebuild relation **R0** by simply JOINing **R3** with **R1** and **R3** with **R2**, on the respective foreign keys.
4. From the definition of 2NF, two observations should be obvious: Firstly, if you have a relation with a single attribute as the primary key, it is automatically in 2NF. Secondly, if you have a relation with *n* attributes and *n-1* of them form the primary key, the relation may very well be in 2NF, but you must first verify this.

Caution:

Despite the importance of 2NF, it is not adequate for ensuring the integrity of a database. Rather, it is necessary to achieve 2NF and then move beyond that benchmark.

In this example, relations **R2** and **R3** are in 2NF (in fact they are in 3NF), but we still have potential problems with **R1**: What if we have a situation where there may be several suppliers from a given location? Or what if we want to keep track of locations of interest? In either case, we would have modification anomalies as described below:

- Insertion anomaly: We cannot record information about a location until we have at least one supplier from that location.
- Deletion anomaly: We cannot delete a particular location without also deleting supplier(s) from that location.
- Update anomaly: If we wish to update information on a location, we have to update all supplier records from that location.

These problems can be addressed if we take the necessary steps to bring **R1** into the third normal form (3NF). But first, we must define what 3NF is.

4.7 Third Normal Form

The third normal form extends constraints related to functional dependence to *non-key attributes* (i.e., attributes that are not part of the primary key). Following are three alternate definitions:

> A relation is in the third normal form (3NF) iff it is in 2NF and no non-key attribute is fully functionally dependent on other non-key attribute(s).
>
> Put another way, a relation is in 3NF iff non-key attributes are *mutually independent* and fully functionally dependent on the primary key. (Two or more attributes are mutually independent if none of them is functionally dependent on any combination of the others.)
>
> Put another way, a relation is in 3NF iff it is in 2NF and every non-key attribute is *non-transitively* dependent on the primary key. (Non-transitivity implies mutual independence.)

Transitive dependence refers to dependence among non-key attributes. In particular, if A → B and B → C, then C is transitively dependent on A (i.e., A → C transitively).

Let us revisit relations **R1**, **R2**, and **R3** of the previous section (and recall that these are decompositions of **R0**): Relation **R1** is problematic because it is not in 3NF. If it is desirable to store additional information about the locations as indicated in the previous section, then we must be smart enough to discern that location is to be treated as an entity with attributes such as location code, location name (and perhaps others). We may therefore rewrite **R1** as follows:

R1{Suppl#, SupplName, LocationCode, LocationName, SupplStatus} PK[Suppl#]

Now observe that LocationCode → LocationName! **R1** therefore violates 3NF and must be decomposed.

Using Heath's theorem, we may therefore decompose **R1** as follows:

R4{Suppl#, SupplName, LocationCode} PK[Suppl#]
R5{ LocationCode, LocationName} PK[LocationCode]

We now check to ensure that the relations are in 3NF (and they indeed are). Again, please take careful notice of the consequences of our actions to this point:

1. The problems with relations in 2NF only have been addressed.
2. Again, by decomposing, we have introduced a foreign key in relation **R4**.

3. We can obtain the information that relation **R1** represented by simply JOINing **R4** with **R5** on the foreign key.
4. From the definition of 3NF, it should be obvious that if you have a 2NF relation with one candidate key and **n** mutually independent non-key attributes, or only one non-key attribute, that relation is in 3NF.

Caution:

Traditionally speaking, 3NF has been regarded as an acceptable industry benchmark for many databases. However, in many circumstances, this benchmark is inadequate.

Relations **R2**, **R3**, **R4**, and **R5** above are all in 3NF. However, it has been found that 3NF-only relations suffer from certain inadequacies. It is well known that 3NF does not deal satisfactorily with cases where the following circumstances hold:

- There are multiple composite candidate keys in a relation.
- The candidate keys overlap (i.e., have at least one attribute in common).

For these situations, the Boyce–Codd normal form (BCNF) provides the perfect solution. As you shall soon see, the BCNF is really a refinement of 3NF. In fact, where the abovementioned conditions do not hold, BCNF reduces to 3NF.

4.8 Boyce–Codd Normal Form

The Boyce–Codd normal form (BCNF) was developed, thanks to the effort of Raymond F. Boyce and Edgar F. Codd. This normal form is really a refinement of the third normal form. Simply, BCNF requirement states:

> A relation is in BCNF iff every *determinant* in the relation is a candidate key.

Drawing from Section 4.4.1, a determinant is an attribute (or group of attributes) on which some other attribute(s) is (are) fully functionally dependent. Examination of **R2**, **R3**, **R4**, and **R5** above will quickly reveal that they are in BCNF (hence 3NF). We therefore need to find a different example that illustrates the importance of BCNF.

Consider the situation where it is desirous to keep track of animals in various zoos, multiple keepers, and the assigned keepers for these animals. Let us tentatively construct the relation **R6** as shown below:

R6{Zoo, Animal, Keeper}

Assume further that that a keeper works at one and only one zoo. We can therefore identify the following FDs:

- [Zoo, Animal] → Keeper
- Keeper → Zoo

Given the above, we conclude that [Zoo, Animal] is the primary key. Observe that **R6** is in 3NF but not in BCNF, since **Keeper** is not a candidate key but is clearly a determinant. Using Heath's theorem, we may decompose **R6** as follows:

R7{Animal, Keeper} PK[Animal]
R8{Keeper, Zoo} PK[Keeper]

Finally, let us clean up by making the observation that **Zoo**, **Keeper**, and **Animal** are really disguised entities, not mere attributes (similar to **Location** in the previous section), and should be treated as such Minimally, each would require a code and a name. Below is the revised list of normalized relations for the case:

R7 {ZooID, ZooName, . . .} PK[ZooID]
R8 {KeeperID, KeeperName, . . .ZooID} PK[KeeperID]
R9 {AnimalCode, AnimalName, . . .KeeperID} PK[AnimalCode]

Notice that the test for BCNF is quite straightforward — you simply check to ensure that the only the determinant in each relation is a candidate key. As on previous occasions, let us examine the consequences of our action:

1. By achieving BCNF, we benefit from further reduction in data duplication and modification anomalies.
2. A further advantage is that we can now store *dangling tuples*. For instance, in our example, a keeper can be assigned to a zoo even before he/she is assigned an animal.
3. One drawback of this schema is that we cannot store an animal that has not been assigned a keeper or zoo. The schema can be further refined but that requires a discussion of the fourth normal form (4NF); this will be done in the upcoming section.
4. One possible drawback with BCNF is that more relations have to be accessed (joined) in order to obtain useful information. Again, referring to the example, **R9** must be joined with **R8** in order to derive Zoo-Animal pairs; and to obtain the name of the zoo (and no doubt other information), a second join must be made with **R7**. In light of the fact that logical views can facilitate seamless end-user perspectives that obscure these underlying joins (more on this in Chapter 13), and processing power continues on its upswing (according to Moore's law), the benefits of achieving BCNF tend to outweigh the drawbacks by a large margin.

Observe:

The principle of BCNF is very simple but profound. By being guided by it, you can in many circumstances actually bypass obtaining 2NF and 3NF relations and move directly into a set of BCNF relations. Adopting this approach will significantly simplify the analysis process. Moreover, in most practical situations, you will not be required to normalize beyond BCNF. This approach will be further clarified in the next chapter.

4.9 Fourth Normal Form

The fourth normal form (4NF) relates to the situation where mutually independent, but related attributes form a relation and the inefficient arrangement causes duplication and hence modification anomalies. Consider the relation, **CTT–Schedule**, representing course–teacher–text combinations in an educational institution. Assume the following business rules:

a. A course can be taught by several teachers.
b. A course can require any number of texts.
c. Teachers and texts are independent of each other, i.e., the same texts are used irrespective of who teaches the course.
d. A teacher can teach several courses.

Figure 4.4 provides some sample data for the purpose of illustration. Notice the high level of duplication due to the structure of the relation and the prevailing business rules.

Note that the theory so far does not provide a method of treating such a situation, except flattening the structure (by making each attribute part of the primary key) as shown below:

Course	Teacher	Text
MATH151 Calculus I	Prof Lambert Felix	First Principles in Calculus by Kaufman
MATH151 Calculus I	Prof Carolyn Cross	First Principles in Calculus by Kaufman
MATH152 Calculus II	Prof Carolyn Cross	Intermediate Calculus by Cummings
MATH152 Calculus II	Prof Carolyn Cross	Calculus and More Calculus by Rhoden
MATH152 Calculus II	Prof Harold Maitland	Intermediate Calculus by Cummings
MATH152 Calculus II	Prof Harold Maitland	Calculus and More Calculus by Rhoden
...	...	...

Notice the duplication

Identify the duplication points. How might we restructure this relation?

Figure 4.4 CTT–schedule — a relation in BCNF but with a high level of duplication.

R10{Course, Teacher, Text} PK[Course, Teacher, Text]

Since **R10** is keyed on all its attributes, it is in BCNF. Yet, two potential problems are data redundancy and modification anomalies (the former leading to the latter). In our example, in order to record that Calculus II is taught by both Professor B and Professor C, four records are required. In fact, if a course is taught by p professors and requires n texts, the number of records required to represent this situation is p*n. This is extraordinary and could prove to be very demanding on storage space, especially as the number of records increases.

Relation **R10**, though in BCNF, is not in 4NF, because it has a peculiar dependency, called a *multi-valued dependency* (MVD). In order to state the 4NF, we must first define MVD.

4.9.1 Multi-Valued Dependency

A multi-valued dependency (MVD) is defined as follows:

> Given a relation R(A, B, C), the MVD A -» B (read "A multi-determines B") holds iff every B-value matching a given (A-value, C-value) pair in R depends only on the A-value and is independent of the C-value.
>
> Further, given R(A B C), A -» B holds iff A -» C also holds. MVDs always go together in pairs like this. We may therefore write, A -» B/C.

MVDs do not occur as frequently as FDs. Moreover, as a consequence of the definition of MVD, it is important to note the following points:

1. For MVD, at least three attributes must exist.
2. All FDs are MVDs but all MVDs are not necessarily FDs.
3. A -» B reads "A multi-determines B" or "B is multi-dependent on A."

Let us get back to **R10**: Course -» Text/Teacher. Note that **Course** is the pivot of the MVD. **Course -» Teacher** since **Teacher** depends on **Course**, independent of **Text**. Course -» Text since **Text** depends on **Course**, independent of **Teacher**. So how do we resolve this? Fagin's theorem provides the answer.

4.9.2 Fagin's Theorem

Fagin's theorem (named after Ronald Fagin who proposed it) may be stated as follows:

> Relation R{A, B, C} can be non-loss decomposed into projections R1{A, B} and R2{A, C} iff the MVDs A -» B/C both hold.

Note that like Heath's theorem, which prescribes how to treat FDs, Fagin's theorem states exactly how to treat MVDs. With this background, we can proceed to defining the 4NF:

> A relation is in 4NF iff whenever there exists an MVD, say A -» B/C, then all attributes of R are also functionally dependent on A.
>
> Put another way, R{A, B, C…} is in 4NF iff every MVD satisfied by R is implied by the candidate key of R.
>
> Put another way, R{A, B, C…} is in 4NF iff the only dependencies are of the form [candidate key] [other non-key attribute(s)].
>
> Put another way, R{A, B, C ...}is in 4NF iff it is in BCNF and there are no MVD's that are not FDs.

The last formulation is particularly instructive: To paraphrase, whenever you encounter a relation that has an unresolved MVD, use Fagin's theorem resolve by replacing the MVD with equivalent FDs. In the current example, **R10** is not in 4NF. This is so because although it is in BCNF, an MVD exists. Using Fagin's theorem, we may decompose it as follows:

R11{Course, Text} PK[Course, Text]
R12{Course, Teacher} PK[Course, Teacher]

Note:

Fagin's theorem prescribes a method of decomposing a relation containing an MVD that is slightly different from the decomposition of an FD as prescribed by Heath's theorem. Figure 4.5 clarifies this:

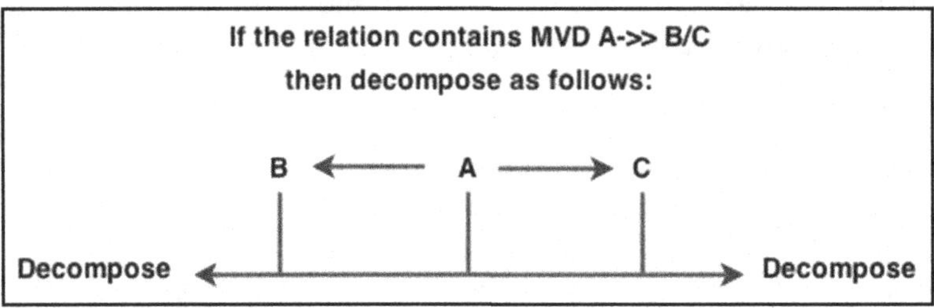

Figure 4.5 Treating MVDs.

As an additional step to resolving the **CTT-Schedule** problem, it is a good idea to refine the solution by applying the following strategies: replace euphemistic relation names (**R11** and **R12**) with more meaningful names; recognize that **Course, Text**, and **Teacher** are actually entities, not mere attributes; return to the use of unique attribute names by using the prefixing strategy that was introduced toward the end of Chapter 3. A more refined solution is shown below:

Course {Crs#, CrsName, CrsCredits, . . .} PK[Crs#]
Teacher {TeacherEmp#, Emp_FName, Emp_LName, . . .} PK[TeacherEmp#]
Text {TextCode, TextTitle, . . .} PK[TextCode]
CourseTextMap{CT_Crs#, CT_TextCode} PK[CT_Crs#, CT_TextCode]
CourseTeacherMap{CR_Crs#, CR_TeacherEmp#} PK[CR_Crs#, CR_TeacherEmp#]

From this proposed solution, it should be obvious to you that the attributes of **CourseTextMap** and **CourseTeacherMap** are foreign keys. You could then construct an RAL (review the latter part of Chapter 3) that clearly outlines the database conceptual schema for the stated problem; this is left as an exercise for you.

4.9.3 The Zoo Revisited

Let us revisit the zoo problem of the previous section. It was mentioned that further refinement was needed to allow for storing animals not yet assigned to keepers or zoos. To facilitate this, we need to recognize the presence of an MVD:

Animal -» Keeper/Zoo

The partial solution given in Section 4.8 is repeated here, with meaningful relation names introduced to replace the more cryptic (euphemistic) names used up to this point in the chapter.

Zoo {ZooID, ZooName, . . .} PK[ZooID]
Keeper {KeeperID, KeeperName, . . .ZooID} PK[KeeperID]
Animal {AnimalCode, AnimalName, . . .KeeperID} PK[AnimalCode]

Given the presence of the MVD Animal -» Keeper/Zoo, we need to refine the three relations by removing prematurely inserted FK from **Animal** and then apply Fagin's theorem to introduce two new relations. The revised set of normalized relations follows (note the use of continued use of meaning relation names and unique attribute names):

Zoo {ZooID, ZooName, . . .} PK[ZooID]
Keeper {KeeperID, KeeperName, . . .KeeperZooID} PK[KeeperID]
Animal {AnimalCode, AnimalName, . . .} PK[AnimalCode]
AnimalKeeperMap {AK_AnimalCode, AK_KeeperID} PK [AK_AnimalCode, AK_KeeperID]
AnimalZooMap {AZ_AnimalCode, AZ_ZooID} PK [AZ_AnimalCode, AZ_ZooID]

As in the previous section (with the **CTT–Schedule** problem), it should be obvious to you what the foreign keys are in entities **Keeper**, **AnimalKeeperMap**, and **AnimalZooMap**; construction of a detailed RAL for the stated problem is left as an exercise for you.

4.10 Fifth Normal Form

So far, we have been treating relations that are decomposable into two other relations. In fact, there are relations that cannot be so decomposed, but can be decomposed into n other relations where n > 2. They are said to be *n-decomposable* relations (n > 2). The fifth normal form (5NF) is also commonly referred to as the *projection-join normal form* (PJNF) because it relates to these (n > 2) projections (of a relation not in 5NF) into decompositions that can be rejoined to yield the original relation.

Specifically, drawing from Chapter 3 (Section 3.6.1), a relationship of degree n (also called an n-ary relationship) can be decomposed into *n(n-1)/2* relationships. If you have a relation consisting of attributes that are all foreign keys and the primary key includes those FKs, then that relation may be the consequence of an n-ary relationship; and if it is, consideration should be given to decomposing it according to the aforementioned formula.

Recall the **SupplierSchedule** relationship (linking suppliers, inventory items, and projects) mentioned in Chapter 3 (Figures 3-3 3.5, 3.13, and 3.17); it is represented here as outlined below:

SupplierSchedule{Suppl#, Item#, Proj#} PK[Suppl#, Item#, Proj#]

Let us assume that specific suppliers can supply specific items for specific projects. The relation represents an M:M relationship involving **Suppliers, InventoryItems**, and **Projects.** Observe the following features about the relation:

1. **SupplierSchedule** is keyed on all attributes; therefore by definition it is in BCNF. By inspection, it is also in 4NF. There is no unresolved MVD as were the cases for the CTT–Schedule

problem and the zoo case of earlier discussions. Here, the attributes are dependent on each other — suppliers supply inventory items for various projects.
2. It is not possible to decompose this relation into two other relations without losing critical information.
3. If there are S suppliers, N items, and J projects, then theoretically there may be up to S*N*J records. Not all of these may be valid: a supplier may supply specific item(s) for specific project(s); not every item may be applicable to every project; and a supplier does not necessarily support every project.
4. If we consider S suppliers, each supplying N items to J projects, then it does not take much imagination to see that a fair amount of duplication will take place despite the fact that the relation is in 4NF.

Let us examine a possible decomposition of **SupplierSchedule** as shown in Figure 4.6. If we employ the first two decompositions only, this will not result in a situation that will guarantee us the original **SupplierSchedule**. In fact, if we were to join these two decompositions (**SI** and **IP**), we would obtain a false representation of the original relation. The third projection (**PS**) is absolutely necessary, if we are to have any guarantee of obtaining the original relation after joining the projections.

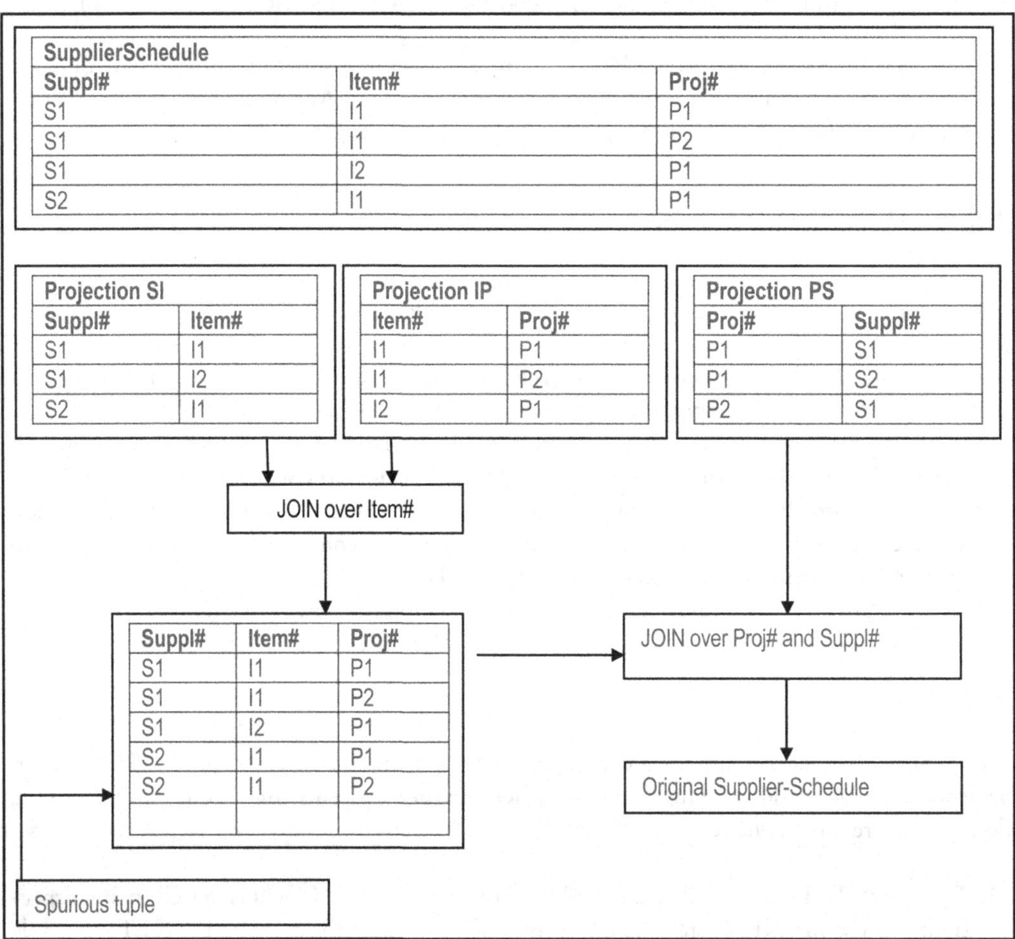

SupplierSchedule

Suppl#	Item#	Proj#
S1	I1	P1
S1	I1	P2
S1	I2	P1
S2	I1	P1

Projection SI

Suppl#	Item#
S1	I1
S1	I2
S2	I1

Projection IP

Item#	Proj#
I1	P1
I1	P2
I2	P1

Projection PS

Proj#	Suppl#
P1	S1
P1	S2
P2	S1

Suppl#	Item#	Proj#
S1	I1	P1
S1	I1	P2
S1	I2	P1
S2	I1	P1
S2	I1	P2

Figure 4.6 Illustrating possible decompositions of SupplierSchedule.

As you examine the figure, observe that the first join produces **SupplierSchedule** plus additional *spurious* tuples. The effect of the second join is to eliminate the spurious tuples. To put it into perspective, **SupplierSchedule** is subject to a (time independent) *3-decomposable (3D) constraint*, namely:

> If(s, i) is in SI
> and (i, p) is in IP
> and (p, s) is in PS
> then (s, i, p) is in **SupplierSchedule**

This is an example of a *join dependency (JD) constraint.* This constraint exists because of the assumption made at the outset. If the prevailing business rules dictate otherwise, then the constraint would likely be different.

4.10.1 Definition of Join Dependency

A join dependency (JD) constraint may be defined as follows:

> Relation R satisfies the JD P1, P2, ... Pn iff R = P1 JOIN P2 JOIN ... JOIN Pn
> where the attributes of P1 ... Pn are subsets of the attributes of R.

Relations that are in 4NF, but not in 5NF (such as **SupplierSchedule**) suffer from duplication, which in turn leads to modification anomalies. These problems are directly related to the presence of the JD constraint(s) in such relations. Fagin's theorem for 5NF relations provides the solution.

4.10.2 Fagin's Theorem

Here now is Fagin's theorem for the fifth normal form (5NF):

> A relation R is in 5NF (also called PJNF) iff every JD in R is a consequence of the candidate keys of R.
>
> In layman's terms, if a relation R is in 4NF and it contains a JD that renders it n-decomposable into P1, P2 .. Pn, such that R = P1 JOIN P2 ... JOIN Pn where n > 2, then such relation is not in 5NF. It may therefore be decomposed to achieve 5NF relations.
>
> Put another way, a relation R is in 5NF iff it is in 4NF and it is not decomposable, except the decompositions are based on a candidate key of R, and the number of projections is n(n-1)/2.

Now examine relation **SupplierSchedule. SupplierSchedule** is not in 5NF because it has a JD (i.e., the JD constraint) that is not a consequence of its candidate key. In other words, **SupplierSchedule** can be decomposed, but this is not implied by its candidate key [Suppl#, Item#, Proj#]. We should therefore proceed with the decomposition represented in Figure 4.6 in order to achieve 5NF. Finally, we can refine the solution by observing the following strategies: assigning meaningful relation names; treating suppliers, projects, and items as entities instead of mere attributes; observing the principle of unique attribute names. Accordingly, a revised list of relations is provided below:

Supplier {Suppl#, SupplName, . . .} PK [Suppl#]
InventoryItem {Item#, ItemName, . . .} PK [Item#]
Project {Proj#, ProjName, . . .} PK [Proj#]
SuppItemMap {SI_Supp#, SI_Item#, SI_Ref#} PK [SI_Ref#]
ItemProjMap {IP_Item#, IP_Proj#, IP_Ref#} PK[IP_Ref#]
ProjSuppMap {PS_Supp#, PS_Proj#, PS_Ref#} PK [PS_Ref#]

As for the previous cases (for instance Sections 4.9.2 and 4.9.3), you should be able to easily identify the foreign keys; you should also be able to construct a detailed RAL for the case (review Figure 3.17 of Chapter 3). Also note that for relations **SuppItemMap**, **ItemProjMap**, and **ProjSuppMap**, a surrogate primary key has been introduced in each case as an alternative to use a composite candidate key consisting of the foreign keys.

Observe:

For most practical purposes, you only have to worry about 5NF if you are trying to implement an M:M relationship involving more than two relations. Once in 5NF, further decompositions would share candidate keys and are therefore to no avail (recall corollary of Heath's theorem). Notwithstanding this, other normal forms have been proposed, as will be discussed in Section 4.12.

4.11 An Example

Now consider the case where it is desirous to keep track of students enrolled in various academic programs in a college environment. Assume further that an academic program is comprised of various courses, and that multiple students may enroll in any given program; also, students may enroll in various courses for each college term (semester or quarter). At minimum, it would be necessary to store for each program, its code and name; similarly for each course (typically, it would be desirable to store additional information about each, but let us keep the illustration simple for the time being). It would be desirable to define several attributes about each student record. However, for the purpose of this illustration, let us keep these to the minimum, namely, student number, name, and grade point average (GPA which is normally calculated from other data, but we will sidestep that complication for this discussion). Let us attempt to derive a set of normalized relations from this scenario.

One way to start the normalization process is to begin with the proposal of a 1NF relation, followed by identification of FDs and subsequent NLDs based on those FDs; this is the classical approach. As you become more experienced, you will be able to employ more expeditious pragmatic approaches (some of which are covered in Chapter 5), but for now, let us take it slowly.

Step 1: Let us propose an initial 1NF relation, **InitialProgram**, with four FDs from the narrative as shown below:

InitialProgram {Pgm#, PgmName, Crs#, CrsName, Stud#, StudFName, StudLName, StudGPA, Term#} PK[Pgm#, Crs#, Stud#]
FD0: [Pgm#, Crs#, Stud#] → PgmName, CrsName, StudFName, StudLName, StudGPA, Term#
FD1: Pgm# → PgmName
FD2: Crs# → CrsName
FD3: Stud# → StudFName, StudLName, StudGPA

Step 2: The second step is to derive 2NF relations from Step 1. To do so, you must recall the definition of 2NF, and then examine the stated FDs to see if there is a violation of 2NF. Recall that 2NF requires full functional dependence on the primary key. Observe that the presence of FD1, FD2, and FD3 renders **InitialProgram** to be in 1NF only, and in clear violation of 2NF. To resolve this dilemma, you must decompose via Heath's theorem to obtain the following relations:

Program {Pgm#, PgmName} PK[Pgm#]
Course {Crs#, CrsName} PK[Crs#]
Student {Stud#, StudFName, StudLName, StudGPA} PK[Stud#]
ProgramStudCourseMap {Pgm#, Crs#, Stud#, Term#} PK[Pgm#, Crs#, Stud#]

Note the use of meaningful relation names to improve readability. Also observe that the relation **ProgramStudCourseMap** is what is left of the original relation **InitialProgram** after it has been decomposed.

Step 3: The third step is to obtain 3NF relations from Step 2. To do so, you must examine the relations derived from Step 2 to see if there is any violation of 3NF. Remember, 3NF requires mutual independence among non-key attributes. A quick check shows that there is no violation, so you can assert with confidence that for this particular scenario, the relations of Step 2 are also in 3NF.

Step 4: This step requires that you obtain BCNF relations. Recall that BCNF requires that every determinant must serve as a candidate within the relation that it appears. A quick eye inspection of the relations from Step 2 reveals no violation of this principle, so you can assert that for this particular case, the relations of Step 2 are also in BCNF.

Step 5: This step requires obtaining higher normal forms where necessary. In most cases, you would stop the normalization process after acquiring BCNF. However, this particular example reveals a peculiar case of data duplication that needs to be resolved. Take a close look at the proposal for relation **ProgramStudCourseMap**. It contains three attributes, each of which forms part of the primary key. Thus, the relation is in BCNF. Why is it necessary to key on all three attributes? It is necessary to do so because it is desirable to represent all program–student–course combinations at the college where this database is implemented. Similar to the previous section, we have a case here where if there are P programs, C courses, and S students, then the total number of combinations possible will be P * C * S. However, it is highly unlikely and illogical that the database would require all the possibilities. The key point here is to observe that there is a JD among **Program**, **Course**, and **Student**. That is what is reflected in relation **ProgramStudCourseMap**. Storing the relation as proposed would not be efficient for the following reasons: It should be possible to easily obtain program–course combinations, program–student combinations, and student–course combinations without having to read triplets as suggested by **ProgramStudCourseMap**. We therefore need to apply Fagin's theorem to resolve the JD into the following projections:

ProgramStructure {Pgm#, Crs#, CrsSeqn} PK[Pgm#, Crs#] or PK[Pgm#, CrsSeqn]
ProgramStudentMap {Pgm#, Stud#} PK[Pgm#, Stud#]
StudentSchedule {Stud#, Crs#, Term#, …} PK[Stud#, Crs#, Term#]

Again, note the use of meaningful relation names. As you view these proposed relations, here are a few additional points worth noting:

1. These three relations would replace relation **ProgramStudCourseMap**. Notice that each new relation implements a binary relationship. Thus, a ternary relationship (represented by **ProgramStudCourseMap**) has been replaced by three binary relationships (represented by relations ProgramStructure, **ProgramStudentMap**, and **StudentSchedule**).
2. In the **ProgramStructure** relation, the attribute **CrsSeqn** has been introduced to facilitate sequencing of courses in a particular major. This is optional; technically speaking, it is not part of the normalization process; the attribute has been introduced merely for convenience.
3. In the **StudentSchedule** relation, the attribute **Term#** has been promoted to be part of the primary key; this is done for increased flexibility. For instance, a student may repeat a course in order to obtain a more favorable grade.

Final Step: The final step is to list the normalized relations that will form the conceptual schema of the database. Sticking with the convention of using unique attribute names, the finalized relations for this case are listed below:

Program {Pgm#, PgmName, . . .} PK[Pgm#]
Course {Crs#, CrsName, . . .} PK[Crs#]
Student {Stud#, StudFName, StudLName, StudGPA, . . .} PK[Stud#]
ProgramStructure {PS_Pgm#, PS_Crs#, PS_CrsSeqn} PK[PS_Pgm#, PS_Crs#]
    or PK[PS_Pgm#, PS_CrsSeqn]
ProgramStudentMap {SP_Pgm#, SP_Stud#} PK[SP_Pgm#, SP_Stud#]
StudentSchedule {SS_Stud#, SS_Crs#, SS_Term#, …} PK[SS_Stud#,SS_ Crs#, SS_Term#]

As for the previous cases (Sections 4.9.2, 4.9.3, and 4.10.2), you should be able to easily identify the foreign keys in the latter three relations; you should also be able to construct a detailed RAL for the case (review Figure 3.17 of Chapter 3). Finally, note that this case covers only a small cross-section of a college database; it is not meant to provide comprehensive coverage of such a problem domain; rather, it is meant to help you develop mastery of the normalization process.

4.12 Other Normal Forms

The field of database systems is potentially a contemptuous one. Indeed, there are accounts of former friends or colleagues becoming foes over database quibbles. Various individuals have proposed several database theorems and methodologies, but they have not all gained universal acceptance as have the normal forms of the previous sections. Two additional normal forms that have been and will no doubt continue to be the subject of debate are the *domain-key normal form* (DKNF) and the *sixth normal form* (6NF). Without picking sides of the debate on these two normal forms, this section will summarize each.

4.12.1 The Domain-Key Normal Form

The domain-key normal form (DKNF) was proposed by Ronald Fagin in 1981. Unlike the other normal forms that all relate to FDs, MVDs, and JDs, this normal form is defined in terms of domains and keys (hence its name). In his paper, Fagin showed that a relation DKNF has no modification anomalies, and that a relation without modification anomalies must be in DKNF. He therefore argued that a relation in DKNF needed no further normalization (at least, not for the purpose of reducing modification anomalies). The definition of DKNF is as follows:

> A relation is in DKNF if every constraint on the relation is a logical consequence of the definition of its keys and domains.

This loaded definition contains three important terms that need clarification; they are *constraint*, *key*, and *domain*:

- A constraint is used to mean any rule relating to static values of attributes. Constraints therefore include integrity rules, editing rules, foreign keys, intra-relation references, FDs, and MVDs, but exclude time-dependent constraints, cardinality constraints, and constraints relating to changes in data values.
- A key is a unique identifier of a row (as defined in Chapter 3).
- A domain is a pool of legal attribute values (as defined in Chapter 3).

The implication of the DKNF is clear: If we have a relation that contains constraint(s) that is (are) not a logical consequence of its (candidate) key and domains, then that relation is not in DKNF and should therefore be further normalized. The DKNF as proposed by Fagin, therefore, represents an ideal situation to strive for.

Unfortunately, a number of problems have arisen from consideration of DKNF; these are summarized below:

- Any constraint that restricts the cardinality of a relation (i.e., the number of tuples in the relation) will render it in violation of DKNF. (It was perhaps for this reason that Fagin excluded from his definition of constraints, time-dependent constraints, or constraints relating to data values.) However, there are many relations for which such constraints are required.
- There is no known algorithm for converting a relation to DKNF. The conversion is intuitive and for this reason described as artistic rather than scientific.
- Not all relations can be reduced to DKNF (relations with cardinality constraints fall in this category).
- It is not precisely clear as to when a relation can be reduced to DKNF.

For these reasons, the DKNF has been compared by Date (see Date 2006) to a "straw man … of some considerable theoretical interest but not yet of much practical ditto."

4.12.2 The Sixth Normal Form

A sixth normal form (6NF) has been proposed by Date (2003), after several years of exploitation, expounding, and research in the field of database systems. It relates to so-called *temporal databases*. Date wrote a whole book on the subject; a summary of the essence is presented in this subsection. Date defines a temporal database as a database that contains historical data as well as, or instead of, current data. Temporal databases are often related with read-only databases, or update-once databases, but they could be used otherwise. In this sense, a temporal database may be considered as a precursor to a data warehouse (discussed in Chapter 19).

For the purpose of illustration, assume that we are in a college or university setting and desire to store the relation **Course** as defined below:

Course {CourseNo, CourseName, CourseCred}

Suppose further that we desire to show different courses at the time they existed in the database. To make our analysis more realistic, let us also make the following additional assumptions:

- The primary key is **CourseNo**; for any given course, the attribute **CourseNo** cannot be changed.
- For any given course, the attribute **CourseName** may be changed any point in time.
- For any given course, the attribute **CourseCred** may be changed any point in time.

We may be tempted to introduce a timestamp on each tuple, and therefore modify the definition of **Course** as follows:

Course {CourseNo, CourseName, CourseCred, EffectiveDate}

Figure 4.7 provides some sample data for the **Course** relation. By introducing the attribute **EffectiveDate**, we have actually introduced a new set of concerns as summarized below:

1. If we assume that the FD CourseNo → {CourseName, CourseCred, EffectiveDate} is (still) applicable, then **Course** is in 5NF. However, in this case, if the **CourseName** or **CourseCred** of a given **Course** tuple changes at a given effective date, there is no way of showing what it was before, unless we create a new course and assign a new **CourseNo**. In either case, this is clearly undesirable.
2. Assume for the moment, the presence of FDs
   CourseNo → CourseName and [CourseNo, EffectiveDate] → CourseCred.

   Then, the relation would be in violation of 2NF, and would therefore need to be decomposed into two decompositions:
   CourseDef {CourseNo, CourseName} PK [CourseNo] and
   CourseTimeStamp {CourseNo, EffectiveDate, CourseCred} PK [CourseNo, EffectiveDate]

   Both of these relations would now be in 5NF. However, if we now desire to change the **CourseName** of a course for a given effective date, we would not be able to represent this in the current schema.
3. We could introduce a surrogate (say **CourseRef**) into relation **Course**, and key on the surrogate, while ignoring the FDs stated in (1) and (2) above. In this case, **Course** would be in violation of 3NF, and if we attempt to decompose, we would revert to the situation in Case (2) above.

CourseNo	CourseName	CourseCred	EffectiveDate
CS120	Introduction to Computer Science	3	1990
CS120	Introduction to Computer Science	4	2005
CS140	Computer Programming I	3	1990
CS140	Computer Programming I	4	2005
CS145	Computer Programming II	3	1990
CS145	Computer Programming II	4	2005
CS130	Pascal Programming	3	1990
...	...	...	...

Impossible with PK [CourseNo]

If we key on [CourseNo, EffectiveDate], the relation would be in violation of 2NF. How do we resolve this?

Figure 4.7 Sample data for the course relation.

The reason for these problems can be explained as follows: The relation **Course** as described defines the following predicate:

- Each course is to be accounted for (we say **Course** is under contract).
- Each course has a **CourseName** which is under contract.
- Each course has a **CourseCred** which is under contract.

The predicate involves three distinct propositions. We are attempting to use the timestamp attribute (**EffectiveDate**) to represent more than one proposition about the attribute values. This, according to Date, is undesirable and violates the sixth normal form.

Here now is Date's theorem for the sixth normal form (6NF):

> A relation R is in 6NF iff it satisfies no non-trivial JDs at all. (A JD is trivial iff at least one of its projections is over all of the attributes of the relation.)
>
> Put another way, a relation R is in 6NF iff the only JDs that it satisfies are trivial ones.

Observe that 6NF as defined, essentially refines 5NF. It is therefore obvious from the definition that a relation in 6NF is necessarily in 5NF also.

Let us now revisit the **Course** relation: With the introduction of the timestamp attribute (**EffectiveDate**), and given the requirements of the relation, there is a non-trivial JD that leads to the following projections:

CourseInTime {CourseNo, EffectiveDate} PK [CourseNo, EffectiveDate]
CourseNameInTime {CourseNo, CourseName, EffectiveDate} PK [CourseNo, EffectiveDate]
CourseCredInTime {CourseNo, CourseCred, EffectiveDate} PK [CourseNo, EffectiveDate]

Observe that the projection **CourseInTime** is strictly speaking, redundant, since it can be obtained by a projection from either **CourseNameInTime** or **CourseCredInTime**. However, in the interest of clarity and completeness, it has been included.

This work by C. J. Date represents a significant contribution to the field of database systems and will no doubt be a topical point of discussion in the future.

4.13 Summary and Concluding Remarks

This brings us near the conclusion of one of the most important topics in your database systems course. Take some time to go over the concepts. Figure 4.8 should help you to remember the salient points.

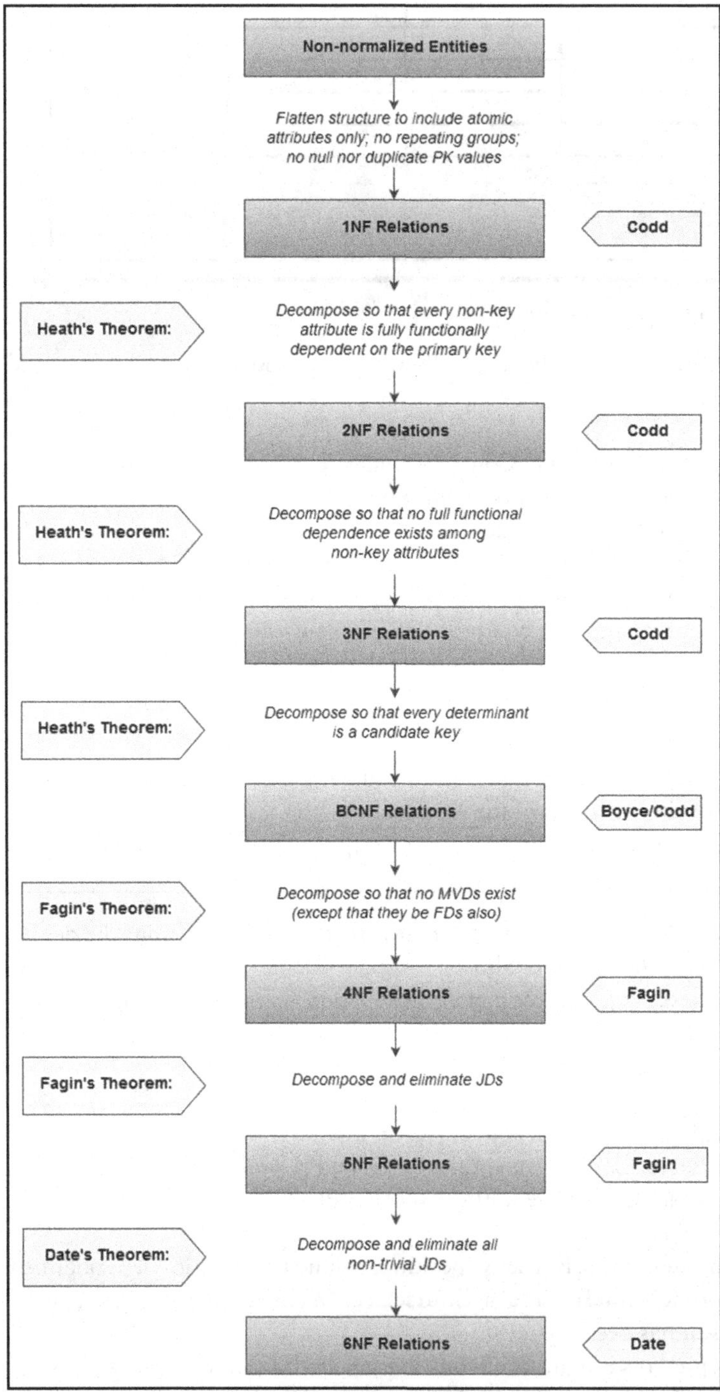

Figure 4.8 Summary of normalization.

Traditionally, it has been widely accepted that for most databases, attainment of 3NF is acceptable. This course recommends a minimum attainment of BCNF, but that database designers strive for 5NF at every reasonable and feasible opportunity. Recall that as stated earlier (in Section 4.8), BCNF is really a refinement of 3NF, and the normalization process can bypass 2NF and 3NF and go straight to BCNF. In some circumstances, it may even be best to proceed all the way to 6NF.

Despite all the virtues of normalization, the methodology has its limitations. Here are three main limitations:

- The normalization process can be quite tedious and challenging for inexperienced database designers. Because of this, the occurrence of inappropriately designed databases is quite high.
- As the size of the database increases over time, the performance on queries involving logical joins tends to deteriorate. However, with the significant advances in processing power (Moore's law), this is not as critical a concern as it was historically.
- In scenarios such as data warehousing (discussed in Chapter 19), where data volume is very high and accessibility is a primary objective, normalization is not always as desirable as it is for operational databases.

Normalization is a technique that must be mastered by database designers. It improves with practice and experience and ultimately becomes almost intuitive. As your mastery of normalization improves, you will find that there is a corresponding improvement in your ability to design and/or propose database systems for various software systems. However, be aware that the converse is also true: Failure to master fundamental principles of database design will significantly impair one's ability to design quality software systems involving underlying databases. Notwithstanding, be careful not to be antagonistic about your views, as informed as they may be. Database feuds are common and passions can sometimes run high.

If after reading and reviewing this chapter, you arrive at the conclusion that normalization is integral to sound database design, you would be correct. Indeed, it is! With time and practice, you should get quite good at applying the technique. The upcoming chapter builds on this topic as you learn about different approaches and methodologies related to database design.

4.14 Review Questions

1. State the two fundamental integrity rules and explain their significance.
2. What is a foreign key? Illustrate using an appropriate example. How should foreign keys be treated?
3. Clarify the following: functional dependence; non-loss decomposition; Heath's theorem. Provide examples that will illustrate the significance of these terms.
4. State the normal forms covered in this chapter. For each normal form, state it, explain in your own words what it means, and provide an example of a relation that conforms to it and one that does not.
5. What are the main advantages and challenges related to database normalization?
6. Review the case presented in Section 4.9.2 and propose a refined RAL for the CTT–Schedule problem.
7. Review the case presented in Section 4.9.3 and propose a refined RAL for the Zoo problem.

8. Review the case presented in Section 4.11 and propose a refined RAL for the partial college database problem.
9. Practice working with simple non-normalized database ideas and refining them to obtain normalized relations. Start off by identifying the main entity/entities, followed by FDs; then apply the appropriate normalization rules.

References and Recommended Readings

Codd, Edgar F. 1972. "Further Normalization of the Database Relational Model." *Database Systems, Courant Computer Science Symposia Series 6*. Eaglewood Cliffs, NJ: Prentice Hall.

Codd, Edgar F. 1974. "Recent Investigations into Relational Data Base Systems." In Proceedings of the IFIP Congress. Stockholm, Sweden.

Connolly, Thomas, & Carolyn Begg. 2015. *Database Systems: A Practical Approach to Design, Implementation and Management* 6th ed. Boston, MA: Pearson. See chapters 14 & 15.

Coronel, Carlos, & Steven Morris. 2015. *Database Systems: Design, Implementation & Management* 11th ed. Boston, MA: Cengage Learning. See chapter 6.

Date, Christopher J. 2004. *Introduction to Database Systems* 8th ed. Menlo Park, CA: Addison-Wesley. See chapters 11–13, 23.

Date, Christopher J. 2006. "Database Debunking." Accessed June 2006. http://www.dbdebunk.com/page/page/621935.htm

Date, Christopher J., Hugh Darwen, & Nikos A. Lorentzos. 2003. *Temporal Databases and the Relational Model.* San Francisco, CA: Morgan-Kaufmann.

Elmasri, Ramez, & Shamkant B. Navathe. 2011. *Fundamentals of Database Systems* 6th ed. Boston, MA: Pearson. See chapters 15 & 16.

Fagin, Ronald. 1977. "Multi-valued Dependencies and a New Normal Form for Relational Databases." *ACM Transactions on Database Systems* vol. 2, no. 3, September 1977.

Fagin, Ronald. 1979. "Normal Forms and Relational Database Operations." In Proceedings of the 1979 ACM SIGMOD International Conference on Management of Data. Boston, MA, May-June 1979.

Fagin, Ronald. 1981. "A Normal Form for Relational Databases that is Based on Domains and Keys." *ACM Transactions on Database Systems* vol. 6, no. 3, September 1981, pp. 387–415.

Garcia-Molina, Hector, Jeffrey Ullman, & Jennifer Widom. 2009. *Database Systems: The Complete Book* 2nd ed. Boston, MA: Pearson. See chapter 3.

Heath, I. J. 1971. "Unacceptable File Operations in a Relational Database." In Proceedings of the 1971 ACM SIGFIDET Workshop on Data Description, Access, and Control. San Diego, CA, November 1971.

Hoffer, Jeffrey A., Ramesh Venkataraman, & Heikki Topi. 2013. *Modern Database Management* 11th ed. Boston, MA: Pearson. See chapters 4.

Kifer, Michael, Arthur Bernstein, & Phillip M. Lewis. 2006. *Databases and Transaction Processing: An Application Oriented Approach* 2nd ed. Boston, MA: Pearson. See chapter 6.

Kroenke, David M., & David Auer. 2014. *Database Processing: Fundamentals, Design, and Implementation* 13th ed. Boston, MA: Pearson. See chapters 3 & 4.

Pratt, Phillip J., & Mary Z. Last. 2015. *Concepts of Database Management* 8th ed. Boston, MA: Cengage Learning. See chapter 5.

Silberschatz, Abraham, Henry F. Korth, & S. Sudarshan. 2011. *Database Systems Concepts* 6th ed. Boston, MA: McGraw-Hill. See chapter 7.

Ullman, Jeffrey D., & Jennifer Widom. 2008. *A First Course in Database Systems* 3rd ed. Boston, MA: Pearson. See chapters 3 & 4.

Chapter 5

Database Modeling and Design

Some textbooks on database systems make much ado about having separate chapters for database modeling and database design. In this course, we will not, and the reason will become clear shortly. All discussions in Chapters 3 and 4 contribute to database modeling and design. In particular, entity–relationship diagrams (ERDs) and normalization are paramount to a logical database model and design. In this chapter, we will clarify the roles of database modeling and database design and then take a more detailed look at some approaches in these areas.

The chapter proceeds under the following captions:

- Database Model and Database Design
- The E–R Model Revisited
- Database Design with the E–R Model
- The Extended Relational Model
- Database Design with the Extended Relational Model
- The UML Model
- Database Design with the UML Model
- Innovation: The Object/Entity Specification Grid
- Database Design Using Normalization Theory
- Database Model and Design Tools
- Summary and Concluding Remarks

5.1 Database Model and Database Design

Database modeling and database design are closely related; in fact, the former leads to the latter. However, it is incorrect to assume that the path from database modeling to database design is an irreversible one. To the contrary, changes in your database model will affect your database design and vice versa. This course therefore purports that you can work on your database model and your database design in parallel, and with experience, you can merge both into one phase. For the purpose of discussion, let us look at each phase.

DOI: 10.1201/9781003275725-7

5.1.1 Database Model

The database model is the blueprint for the database design. Database modeling is therefore the preparation of that blueprint. In database modeling, we construct a representation of the database that will be useful toward the design and construction of the database. Various approaches to database modeling have been proposed by different authors; the prominent ones are as follows:

- The Entity–Relationship (E–R) Model
- The Object–Relationship (O–R) Model
- The Extended Relational Model

The E–R and O–R models were introduced in Chapter 3 (Sections 3.6.1 and 3.6.2). Chapter 3 also introduced the Relation–Attributes List (RAL) and the Relationship List (RL, Section 3.8) as an alternative to the E–R model in situations where E–R modeling is impractical. We will revisit the E–R model later in this chapter and then introduce the extended relational model.

5.1.2 Database Design

The database design is the (final) specification that will be used to construct the actual database. Database designing is therefore the preparation of this specification. In preparing the database specification, the database model is used as input. As such, the guidelines given in Chapter 3 (Section 3.7) on implementing relationships are applicable. Five approaches to database design that will be discussed later in this chapter are as follows:

- Database Design via the E–R Model
- Database Design via the Extended Relational Model
- Database Design via the UML Model
- Database Design via the Entity/Object Specification Grid
- Database Design via Normalization Theory

5.2 The E–R Model Revisited

Recall that in Chapter 3, the similarity and differences between an entity and a relation were also noted. If we assume the similarity, then the E–R model can be construed as merely a specific interpretation of the relational model.

In order to unify the informal E–R model with the formal relational model, Codd introduced a number of conventions specific to the E–R model. These are summarized in Table 5.1 and illustrated in Figures 5.1. Note that the model displayed in the figure represents only a section of the (partial) database model introduced in Chapter 3 (review Figure 3.3). Figure 5.1a employs the Chen notation, while Figure 5.1b employs the Crow's Foot notation. In both cases, the attributes of entities have been omitted, except for primary key attributes (as mentioned in Chapter 3 and emphasized later in this chapter, there are other more creative ways to represent attributes).

For super-type and subtype relationships, we will employ the Unified Modeling Language (UML) convention of having an arrowhead pointing from the subtype to the super-type (as explained in Chapter 3). Figure 5.2 provides an illustration.

Table 5.1 E–R Model Conventions

Concept	*Clarification*
Fundamental concepts	See Table 3.1 of Chapter 3.
Weak entity	A *weak entity* (also called a *dependent entity*) is an entity that cannot exist by itself. For instance, if employees have dependents, then **Dependents** is a weak entity and **Employee** is a *regular entity* (also called a *strong entity*). On the ERD, weak entities are represented by double lines.
Relationship notation	Relationships may be represented by either Chen's notation or the Crow's Foot notation (the latter is preferred).
Chen notation clarifications	If Chen's notation is used to represent relationships, then the following apply: • The double diamond indicates a relationship between weak and regular entities. • The name of the relationship is written inside the diamond. • A double relationship line represents *total participation*; a single relationship line represents *partial participation*. For instance, if all dependents must have a reference employee, participation is total on the part of dependents; however, all employees need not have dependents, so that participation is partial on the part of employees.
Subtype and super-type	An entity can be a *subtype* of another entity that is a *super-type*. For instance, in an organization, the entity **Programmer** would be a subtype of the entity **Employee**, the super-type. All properties of a super-type apply to its subtype; the converse does not hold. Figure 5.2 illustrates how subtypes and super-types are represented.
Relationship implementation	Implementation of relationships can be realized as discussed in Section 3.7. Furthermore, the fundamental integrity rules (Section 4.1) must also be upheld.

One additional matter remains — that of naming of the entities in the ERD. Consistent with a longstanding convention, it is recommended that you give entities a singular name; you may use plural entity names but singular names are more commonly used. Whether you use singular or plural names, you should be consistent for the entire ERD. Also, if you use plural entity names, be sure to adjust your relationship names so that the entity–verb–entity convention is maintained when transitioning from an entity through a relationship to another entity on the ERD. Notice that with very few exceptions, the singular entity naming convention has been used in the illustrations provided in this course (review Sections 3.6–3.8 as well as Chapter 4). Finally, when you get to the database design stage, be sure to employ a carefully constructed database naming convention such as the one presented in Section 3.9.

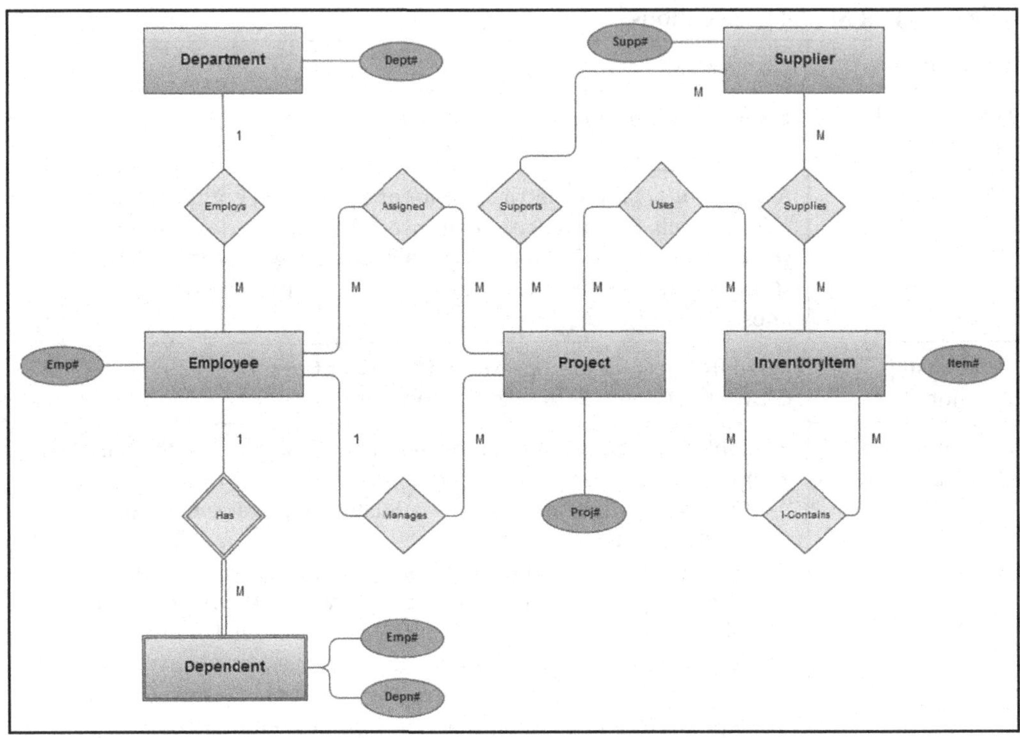

Figure 5.1a Partial ERD for manufacturing firm (Chen notation).

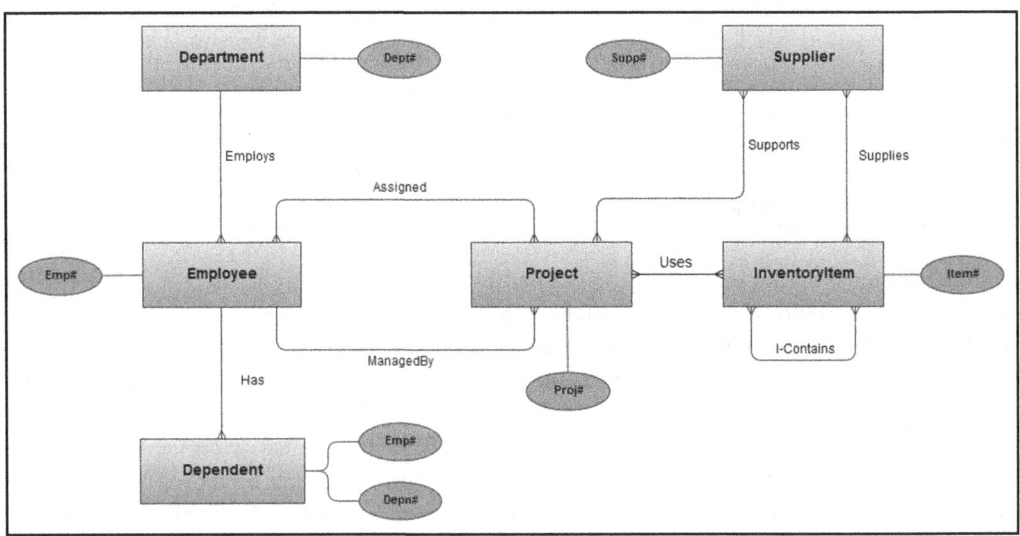

Figure 5.1b Alternate partial ERD for manufacturing firm (Crow's Foot notation).

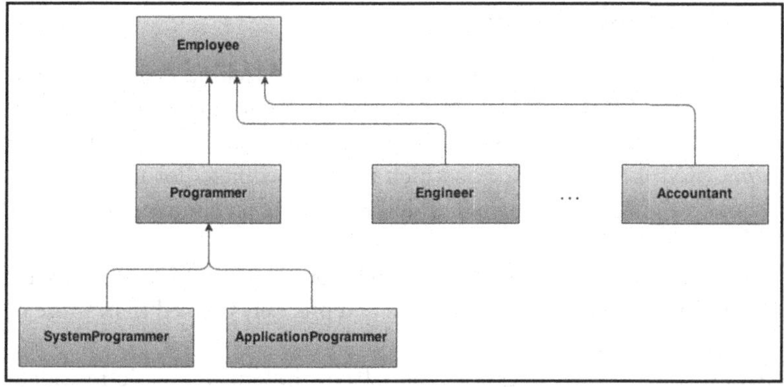

Figure 5.2 Example of type hierarchy.

5.3 Database Design via the E–R Model

Database design with the E–R model simply involves following the rules established in Chapter 3 on implementing relationships (Section 3.7). These rules tell you exactly how to treat the various kinds of relationships and take some time to review them. We may therefore construct a procedure for database design via the E–R model as shown in Table 5.2.

Table 5.2 Database Design Procedure Using the E–R Model

Step	*Activity*
1	Identify all entities and a preliminary set of attributes for each entity.
2	Classify the entities (weak vs strong).
3	Identify all relationships among the entities.
4	Classify the relationships (mandatory vs. optional); decide on which optional relationships will be retained and which ones will be eliminated.
5	Construct an ERD or the equivalent (review Chapter 3).
6	Refine the model.
7	Using the guidelines for implementing relationships (Section 3.7), construct a final set of relations, clearly indicating for each relation, its attributes, candidate key(s), and primary key. The RAL and RL of Chapter 3 (Section 3.8) may be employed.
8	By consistently following this procedure, you will obtain a set of relations that will be normalized to at least the 3NF and in many cases BCNF. You can then apply normalization theory to refine the model and achieve the desired level of normalization.

5.4 The Extended Relational Model

Even with the abovementioned conventions, the E–R model was found lacking in its treatment of certain scenarios. Recognizing this, Codd and Date introduced an alternate *extended relational model* (which for convenience will be abbreviated as the XR model); the original work was called the *Relational Model/Tasmania* (RM/T), the essence of which is described here [for more details, see Date (1990, 2004)]. Although the XR model is not widely referenced in contemporary database literature, the work itself remains quite insightful, hence its inclusion.

The XR model makes no distinction between entities and relations; an entity is a special kind of relation. Structural and integrity aspects are more extensive and precisely defined than those of the E–R model.

The XR model introduces its own special operators apart from those of the basic relational model. Additionally, entities (and relationships) are represented as a set of E-relations and P-relations. The model includes a formal catalog structure by which relationships can be made known to the system, thus facilitating the enforcement of integrity constraints implied by such relationships.

As you will see later in the course, it turns out that the XR model forms the basis of how the system catalog is handled in most contemporary DBMS suites. We shall therefore spend a few moments to look at the main features of the model.

5.4.1 Entity Classifications

Under the XR model, the following entity classifications hold: *kernel entities*; *characteristic entities*; *designative entities*; *associative entities*; and *subtype/super-type entities*. In the interest of comprehensiveness and flexibility, this course recommends a sixth category, namely, *component entities*.

Kernel Entities: *Kernel entities* are the core entities that have independent existence in the database; they are what the database is really about. For example, in an inventory system, kernel entities might be **Purchase Order**, **Purchase Invoice**, **Inventory Item**, **Department**, and **Issuance of Goods** (to various departments). Referring to the example used in the previous section (Figure 5.1b), kernel entities would be **Supplier**, **InventoryItem**, **Project, Employee**, and **Department**.

Characteristic Entities: *Characteristic entities* describe other entities. For instance (referring to Figure 5.1b), **Dependent** is a characteristic of **Employee.** Characteristics are existence dependent on the entity they describe.

Designative Entities: An entity, regardless of its classification, can have a property (attribute) whose function is to *designate* (reference) some other entity, thus implementing a 1:M relationship. For instance (referring to Figure 5.1b), **Employee** is *designative* of **Department** and **Project** is designative of **Employee** (due to the relationship labeled **ManagedBy**). Put another way, a designation is the implementation of an M:1 relationship. The designating entity is the entity on the "many-side" of a 1:M relationship. Note that a characteristic entity is necessarily designative since it designates the entity on which it is existence dependent. Note, however, that a designative entity is not necessarily characteristic. Entities **Project, Employee**, and **Dependent** amplify these points (see Figure 5.1).

Associative Entities: *Associative entities* represent M:M relationships among two or more entities. For example, from Figure 5.1b, the relationships **Assigned**, **Supports**, **Supplies**, **Uses**, and **I-Contains** would be implemented as associative entities. In a college database with kernel entities **Program** and **Course** (among others of course) and an M:M relationship between them, the associative entity representing the relationship could be **ProgramStructure**, which would include foreign keys referencing **Program** and **Course**, respectively. Note that the associative entity is the intersecting relation in the implementation of an M:M relationship (review Section 3.7).

Subtype/Super-type Entities: If we have two entities E1 and E2, such that a record of E1 is always a record of E2, and a record of E2 is sometimes a record of E1, then E1 is said to be a subtype of E2. The existence of a subtype implies the existence of a super-type: To say that E1 is a subtype of E2, it is equivalent to saying that E2 is the super-type of E1. For illustration, review Figure 5.2.

Component Entities: Component entities are featured in the construction of more complex entities. A good example of this comes from Chapter 3 (Figeures 3.3 and 3.11), where **Employee** comprises components such as **EmployeePersonalInfo**, **EmployeeEmploymentHistory**, **EmployeeAcademicLog**, **EmployeePublicationsLog**, and **EmployeeExtraCurricular**.

5.4.2 Surrogates

Recall that the concept of a surrogate was first introduced in Chapter 3 (Section 3.7). In understanding the X–R model, the role of surrogates is very important; we therefore revisit the concept here. Surrogates are system-controlled primary keys, which are defined to avoid identification of records (tuples) by user-controlled primary keys. They are also often used to replace composite primary keys. Two consequences of surrogates arise (both of which can be relaxed with a slight deviation from the XR model, which does not enforce surrogates, *E-relations* and *P-relations* as mandatory):

- Primary and foreign keys can be made to always be non-composite.
- Foreign keys always reference the corresponding E-relations (more on this shortly).

The main benefit of using surrogates is simplification of the design for complex relations involving composite primary keys. For these scenarios, composite primary keys are sometimes cumbersome; surrogates are useful replacements in these circumstances.

To demonstrate the usefulness of surrogates in simplifying database model and ultimate design (with respect to avoiding cumbersome composite primary keys), let us suppose that we want to track purchase orders and their related invoices. The related entities that we would need to track are **Supplier, InventoryItem, PurchaseOrder**, and **PurchaseInvoice**. These are not all included in Figure 5.1; however, they are represented in Figure 3.3 of Chapter 3 (please review). By following through on the E–R model, or by applying normalization principles of Chapter 4, we may construct a tentative set of normalized relations as illustrated in Figure 5.3a. Notice how potentially cumbersome the composite keys would be, particularly on relations **PurchaseOrdDetail** and **PurchaseInvDetail**. However, by introducing surrogates as illustrated in Figure 5.3b, we minimize the need to use complex composite keys.

Relation	Attributes	Primary Key
Supplier	Suppl#, SuppName, Address, E-mail, ContactPerson, Telephone, ...	Suppl#
InventoryItem	Item#, ItemName, QuantityOnHand, LastPrice, AveragePrice, ...	Item#
PurchaseOrdSummary	Order#, OrderDate, *OrderSupp#*, OrderStatus, OrderEstimate, ... // See note 3 below	[Order#, OrderDate, OrderSuppl#]
PurchaseOrdDetail	*PODOrder#, PODOrderDate*, POD*OrderSupp#, PODItem#*, OrderQuantity, ...	[PODOrder#, PODOrderDate, PODOrderSuppl#, PODItem#]
PurchaseInvSummary	{Invoice#, *InvOrder#, InvOrderDate, InvSupp#*, InvDate, InvAmount, InvStatus, InvDiscount, InvTax, InvAmountDue, ... // See note 4 below	[Invoice#, InvoiceDate, InvOrder#, InvOrderDate, InvSuppl#]
PurchaseInvDetail	*PIDInvoice#, PIDInvoiceDate, PIDInvOrder#, PIDInvOrderDate, PIDInvSupp#, PIDItem#*, PIDItemQuantity, PIDItemUnitPrice	[PIDInvoice#, PIDInvoiceDate, PIDInvOrder#, PIDInvOrderDate, PIDInvSuppl#, PIDItem#]

Each relation and each attribute will need additional clarification prior to database construction and table creation.

Note:
1. Only essential attributes are shown in this representation.
2. Foreign keys are *italicized.*
3. Assume that order numbers may be repeated after a cycle of several years. Keying on **Order#** alone would not be prudent, since this strategy would not guarantee identification of unique records.
4. Keying on **Invoice#** alone would not be prudent. Doing so would not guarantee identification of unique records. Moreover, the host organization has no control over invoice numbers since this is assigned by respective suppliers.

Figure 5.3a Relation–attributes list for purchase orders and invoices (without surrogates).

Relation	Attributes	Primary Key
Supplier	Suppl#, SuppName, Address, E-mail, ContactPerson, Telephone, ...	Suppl#
InventoryItem	Item#, ItemName, QuantityOnHand, LastPrice, AveragePrice, ...	Item#
PurchaseOrdSummary	OrderRef, Order#, OrderDate, *OrderSupp#*, OrderStatus, OrderEstimate, ...	OrderRef // See note 3 below
PurchaseOrdDetail	*PODOrderRef, PODItem#*, OrderQuantity	[PODOrderRef, PODItem#] // See note 4 below
PurchaseInvSummary	PurchaseRef, Invoice#, *InvOrderRef*, InvDate, InvAmount, InvStatus, InvDiscount, InvTax, InvAmountDue, ...	PurchaseRef // See note 5 below
PurchaseInvDetail	*PIDPurchaseRef, PIDItem#*, PIDItemQuantity, PIDItemPrice	[PIDPurchaseRef, PIDItem#] // See note 6 below

Each relation and each attribute will need additional clarification prior to database construction and table creation.

Note:
1. Only essential attributes are shown in this representation.
2. Foreign keys are *italicized.*
3. **OrderRef** is a surrogate that is guaranteed to be unique; therefore it is designated as the primary key.
4. Optionally, we may introduce another surrogate **PODCode**, and make it the primary key.
5. **PurchaseRef** is a surrogate that is guaranteed to be unique; therefore it is designated as the primary key.
6. Optionally, we may introduce another surrogate **PIDCode**, and make it the primary key.
7. In each relation containing a surrogate, it is recommended that you create indexes for the alternate keys that the surrogates are substituted for.

Figure 5.3b Alternate relation–attributes list for purchase orders and invoices (with surrogates).

5.4.3 E-Relations and P-Relations

The original XR model specification prescribes the use of *E-relations* and *P-relations*. The database would contain one *E-relation* for each entity type — a unary relation that lists surrogates for all tuples of that entity (type). To illustrate, let us revisit the manufacturing firm's partial database (Figure 5.1) of previous discussions: Suppose for a moment that the **Supplier** relation contains two tuples, the **InventoryItem** relation contains three tuples, and the **Department** relation contains three tuples. A possible internal representation of the E-relations for this scenario is illustrated in Figure 5.4 (Part 5-4a). An "E" is inserted in front of the original relation name (for example, **E-Supplier**) to denote the fact that this is an E-relation being represented. The percent sign (%) next to the attributes (for example, **Suppl%**) is used to denote the fact that these attributes are really surrogates.

In addition to the E-relations concerned with tuples, a special binary E-relation would be required to link the so-called E-relations to the original relations. We will call this the *host E-relation*. It is illustrated in Figure 5.4 (Part 5-4b). This would allow users to relate to the database using relation names that they are familiar with; translation would be transparent to them.

Properties (attributes) for a given entity type are represented by a set of *P-relations*. The P-relation stores all property characteristics and values of all tuples listed in the corresponding E-relation. Properties can be grouped together in a single n-ary relation, or each property can be represented by P-relation, or there can be a convenient number of P-relations used; the choice depends on the designer. In the interest of simplicity, let us assume the third approach; let us assume further that there is a P-relation for each E-relation. A convenient possible representation for the E-relations of Figure 5.4 (Part 5-4a) is illustrated in Part 5-4c of the figure. A "P" is inserted in front of the original relation name (for example, **P-Supplier**) to denote the fact that this is a P-relation being represented. Notice also that each P-relation contains a foreign key that ensures that each tuple is referenced back to its correspondent in the associated E-relation.

Carrying on with the assumption that there is a P-relation for each E-relation: In addition to the basic P-relations, a special P-relation would be required to store the characteristics of each property (to be) defined in the database. Let us call this the *host P-relation*. It is represented in Figure 5.4 (Part 5-4d). By including this relation, we allow users the flexibility of adding new properties to a relation, modifying existing properties in a relation, or deleting pre-existent properties from a relation. These changes are referred to as *structural changes* to a relation; they will be amplified later in the course (Chapter 11).

As you will later see (in Chapter 14), it is application of this methodology that assists in the implementation of sophisticated system catalogs that characterize contemporary DBMS suites. However, with this knowledge, you can actually model and design databases to mirror E-relations and P-relations as described. One obvious advantage is that if you used one P-relation instead of one for each E-relation, then in accessing the database for actual data, you would be accessing fewer relations (in fact just one relation) than if you had used another approach (such as the E–R model). The flip side to this advantage is that this relation would be extremely large for a medium-sized or large database; this could potentially offset at least some of the efficiency gained from just having to access one relation for data values.

Part 5-4a: Illustrating E-relations

E-Supplier: Suppl%	E-InventoryItem: Item%	E-Department: Dept%
SE1	IE1	DE1
SE2	IE2	DE2
SE2	IE3	DE3
...		

Part 5-4b: Illustrating the Host E-relation

E-Host:

Relation	E-relation
Department	E-Department
InventoryItem	E-InventoryItem
Supplier	E-Supplier
...	

There is an E-entry for each entity comprising the database. Entities will eventually be implemented as relational tables.

Part 5-4c: Illustrating P-relations

P-Supplier:

Suppl%	Suppl#	SuppName	Address
SE1	S1	Smithsonian	11 Sydney Way ...
SE2	S2	Bruce Jones Inc.	14 Maple Street ...
...			

P-InventoryItem:

Item%	Item#	ItemName	
IE1	I1	HP 500 Printer	
IE2	I2	Epson 1070 Printer	
IE3	I3	Xerox Laser Printer	
...			

P-Department:

Dept%	Dept#	DeptName	
DE1	D1	Design	
DE2	D2	Research	
DE3	D3	Synthesis	
...			

There is a P-relation for each E-relation (at minimum, there is the equivalence of this). P-relations store data for the properties associated with various entities comprising the database.

Part 5-4d: Illustrating the Host P-relation

P-Host:

Property	E-relation	Type	Length
Dept#	E-Department	Number	04
DeptName	E-Department	Character	40
...			
Item#	E-InventoryItem	Character	08
DeptName	E-InventoryItem	Character	30
...			
Suppl#	E-Supplier	Character	08
SupplName	E-Supplier	Character	40
Address	E-Supplier	Character	45
...			

There is an entry for each property (i.e. attribute) defined in the database. The physical characteristics for each property are specified here.

Figure 5.4 Illustrating E-relations and P-relations.

5.4.4 Integrity Rules

With this set-up, accessing and manipulating data in the database is accomplished by the DBMS through the E- and P-relations. For this reason, additional integrity rules must be imposed. The complete list of integrity rules is as follows:

1. Entity Integrity: The primary key (simple or composite) is not allowed to contain null value(s) (review Section 4.1).
2. Referential Integrity: The database must not have unmatched foreign key values (review Section 4.1).
3. XR Model Entity Integrity: E-relations accept insertions and deletions but no updates (surrogates don't change).
4. Property Integrity: A property cannot exist in the database unless the tuple (entity) it describes exists.
5. Characteristic Integrity: A characteristic entity (tuple) cannot exist unless the entity (tuple) it describes (i.e., references) exists.
6. Association Integrity: An association entity (tuple) cannot exist unless each participating entity (tuple) also exists.
7. Designation Integrity: A designative entity (tuple) cannot exist unless the entity (tuple) it designates also exists.
8. Subtype Integrity: A subtype entity (tuple) cannot exist except there is a corresponding super-type entity (tuple) that it references.
9. Component Integrity: A component tuple cannot exist except there is a corresponding tuple that it references.

Additionally, integrity rules are required for subtypes and super-types. The following rules fill that void:

1. All characteristics of a super-type are automatically characteristics of the corresponding subtype(s), but the converse does not hold.
2. All associations in which a super-type participates are automatically associations in which the corresponding subtype(s) participate, but the converse does not hold.
3. All properties of a given super-type apply to the corresponding subtype(s), but the converse does not hold.
4. A subtype of a kernel is still a kernel; a subtype of a characteristic is still a characteristic; a subtype of an association is still an association.

Discussion:

Here is a nice database research exercise that would take considerable time to implement: Model and design a database using the E–R model. Repeat the exercise using the X-R model. Implement both databases and compare their performance.

5.5 Database Design via the XR Model

The following approach, developed by Date and outlined in Date (1990), employs the basic XR method, but relaxes the requirement that surrogates, E-relations, and P-relations are mandatory. The (partial) database model represented in Figure 5.1 will be used for the purpose of illustration.

Before proceeding, it has become necessary to clarify a notation that was sneaked in toward the end of Chapter 3 (Section 3.9) and used sparingly throughout Chapter 4: The notation R.A will be used to mean attribute A in relation (or entity) R. For instance, **Department.Dept#** denotes attribute **Dept#** in the relation **Department.** The database design approach involves nine steps as summarized in Table 5.3 and clarified in the upcoming sections.

Table 5.3 Database Design Procedure Using the XR Model

Step	*Activity*
1	Determine kernel entities.
2	Determine characteristic entities.
3	Determine designative entities.
4	Determine associations.
5	Determine subtypes and super-types.
6	Determine component entities.
7	Determine properties of each entity.
8	Construct a relation–attribute list (RAL) for each relation.
9	By consistently following this procedure, you will obtain a set of relations that will be normalized to at least the 3NF and in many cases BCNF. You can then apply normalization theory to refine the model and achieve the desired level of normalization.

5.5.1 Determining the Kernel Entities

The first step involves determining the kernel entities. As mentioned earlier, kernels are the core relations. In the example, the kernels are **Department**, **Employee**, **Supplier**, **Project**, and **InventoryItem**. Each kernel translates to a base relation. The primary key of each could be the user-controlled ones, or surrogates may be introduced.

5.5.2 Determining the Characteristic Entities

The second step involves determining and properly structuring the characteristic entities. As mentioned above, a characteristic entity is existence dependent on the entity it describes. One characteristic exists in the example, namely **Dependent**. Characteristics also translate to base relations. The foreign key in **Dependent** would be **DepnEmp#**, which would reference **Employee.Emp#**.

Notice that we did not use the attribute name **Emp#** as the foreign key in **Dependent**, but **DepnEmp#**. This decision is based on the unique attribute-name principle that was introduced in Section 3.9. You are likely to come across various scenarios where this principle has been violated (so that you may observe a foreign key having an identical name as the attribute it references). Moreover, in Chapter 7, you will observe that the discussion on *natural joins* presumes violation of this principle. The reason for this rookie mistake is not clear; it may very well have to do with how traditional databases were construed and implemented. Be that as it may, you should avoid this mistake and uphold the stated principle. Here are two strong arguments that can be given in defense of the unique attribute-name principle:

- Simply, the principle makes good sense and leads to a cleaner, more elegant database design.
- Maintaining unique attribute names in a database avoids confusion when queries involving the joining of multiple relations are constructed and executed on the database. This will become clearer in division C (particularly Chapter 12) of the text.

Moving on to data integrity, we would require the following integrity constraints on the **Dependent** relation:

- Null FKs should not be allowed; the FK must always reference the tuple on which the characteristic is existence dependent.
- Deletion is cascaded from the referenced to the referencing records; if the referenced tuple is removed, so should all the characteristic (referencing) tuples.
- Update cascaded from the referenced to the referencing records; any change made to the referenced tuple necessarily applies to the related referencing tuples.

Next, we must address the matter of data accessibility. Two alternatives exist for the choice of primary key:

a. The foreign key combined with the attribute that distinguishes different characteristics within the host entity. For example, the Dependent relation could be structured as follows:
   Dependent {DepnEmp#, DepnName,...} PK [DepnEmp#, DepnName]
   This approach is neither elegant nor reliable, since we have no control over what values **DepnName** may have; we are hoping for unique combinations but what if two dependents have the same name? That would be unlikely but not impossible.
b. Introduce a surrogate and manage it in such a way as to allow you to key solely on it. Or you could use a simple sequence number that recycles for each value of the foreign key and combine this with the foreign key. For example, below are two possible ways the **Dependent** relation could be structured:
   Dependent {DepnRef, DepnEmp#, DepnName...} PK [DepnRef]
   Dependent {DepnEmp#, DepnSeqn, DepnName,...} PK [DepnEmp#, DepnSeqn]

5.5.3 Determining the Designative Entities

This third step involves identifying and properly structuring the designative entities. As mentioned earlier, a designation is a 1:M or 1:1 relationship between two entities. In the example, designative entities are **Employee**, **Project**, and **Dependent**. However, **Dependent** is also a characteristic that has already been identified above (note that all characteristics are designations, but not vice versa).

From the theory established in Chapter 3 (Section 3.7), a designation is implemented by the introduction of a foreign key in the relation for the designating entity. Following this principle, we would introduce foreign key, **EmpDept#** in relation **Employee** (where **EmpDept#**

references **Department.Dept#**), and foreign key **ProjManagerEmp#** in relation **Project** (where **ProjManagerEmp#** references **Employee.Emp#**).

Specification of integrity constraints for designations will depend on the prevailing circumstances. Below are four options to be considered:

- Null FKs are allowed in the designating entity if the participation is partial.
- Null FKs not allowed in the designating entity if the participation is full.
- Deletion of referenced tuples should be restricted. This is consistent with the referential integrity rule of Section 4.1; deletion of referenced tuples would plunge the integrity of the database into an abyss.
- Update of referenced tuples may be restricted or cascaded, depending on the circumstance. Generally speaking, surrogates should not be used for designations, but in the event that they are, update of referenced tuples should be restricted; you do not want to covertly change something that is being referenced. If as recommended, surrogates are not used for designations, then update of referenced tuples can (and should) be cascaded.

Typically, the foreign key in a designative entity is a non-key attribute. Consequently, there are normally no keying issues. The PK is normally chosen from other attributes in the designative entity.

5.5.4 Determining the Associations

Step 4 involves identification and implementation of all associations. As mentioned earlier, associations are the implementation of M:M relationships. They translate to base relations. In the example, associations are **Assigned**, **Supports**, **Supplies**, **Uses**, and **I-Contains**. Again, relying on the theory established in the previous two chapters (Sections 3.7 and 4.10) on the implementation of M:M relationships, we would introduce five base relations (called intersecting relations in Section 3.7) for the five associations — **ProjWorkSchedule**, **SuppItemMap**, **ProjSuppMap**, **ItemProjMap**, and **ItemStruct**. Additionally, and in keeping with the principle of having unique attribute names for the entire database, we will change the foreign key attribute names to names that are unique but easily traceable to the attributes they reference. The proposed associative entities are outlined below:

SuppItemMap {SI_Supp#, SI_Item#, SI_Ref#}; PK [SI_Ref#]; AK [SI_Supp#, SI_Item#]
ItemProjMap {IP_Item#, IP_Proj#, IP_Ref#}; PK [IP_Ref#]; AK [IP_Item#, IP_Proj#]
ProjSuppMap (PS_Supp#, PS_Proj#, PS_Ref#} PK [PS_Ref#]; AK [PS_Supp#, PS_Proj#]

Note the introduction of an alternate key (AK) for each of the associative entities. While in each case, use of the surrogate is convenient, it is often important to preserve the original composite candidate key. Later, when it becomes time to actually create these relations, the AKs can be implemented as unique indexes (Chapter 11 discusses creation and management of indexes).

Turning to data integrity, the following strategies would be necessary for preserving the association relationship:

- Null FKs are not allowed; otherwise, the referential integrity rule (Section 4.1) would be violated.
- Deletion of referenced is restricted, and also in keeping with the referential integrity rule.

- Update of referenced tuples is allowed so long as the update does not change the PK or AK values.

With respect to the matter of data accessibility, two alternatives exist for the choice of primary key:

a. Use the aggregation of the foreign keys to form a composite PK.
b. Introduce a surrogate and key on it; preserve the AK by creating a unique index for the aggregated composite key.

5.5.5 Determining Entity Subtypes and Super-Types

The fifth step relates to identifying and properly implementing subtype–super-type relationships among the entities. Care should be taken here in not introducing subtype–super-type relationships where traditional relationships would suffice. Each entity type translates to a base relation. Each base relation will contain attributes corresponding to the properties of the entities that apply within the type hierarchy. Again, being guided by principles established in Chapter 3 (Section 3.7), each subtype will share the primary key of its super-type. Further, the primary key of a subtype is also the foreign key of the said subtype. The illustrations provided in Chapter 3 (Figures 3.11 and 3.12) are also applicable here.

No subtype–super-type relationship appears in the model of Figure 5.1. However, in Figure 5.2, there are a few: **Programmer**, **Engineer**, and **Accountant** are subtypes of **Employee**; **SystemProgrammer** and **ApplicationProgrammer** are subtypes of **Programmer**. Note also that in a subtype, except for the primary key (which is also a foreign key), no additional attributes of the super-type need to be repeated, since they are inherited. However, additional attributes may be specified (in the subtype). For example (still referring to Figure 5.2), the **Programmer** entity may contain the attribute **MainLanguage** to store the programmer's main programming language; this would not apply to **Employee.**

In the model shown in Figure 5.2, support for *multiple inheritances* (the case where a subtype inherits from multiple super-types) would be difficult. For instance, how would we represent the possibility of having a programmer who has credentials for both **System Programmer** and **Application programmer**? One strategy could be to promote that case to the level above, namely, **Programmer**. But what about a **System Programmer** and an **Engineer** — how would we represent that? That would be problematic. One possible treatment would be to resort to the use of surrogates and introduce a subtype that inherits (i.e., references) both **SystemProgrammer** and **Engineer** via references to their respective surrogate PKs.

Let us now address the matter of data integrity. For subtype–super-type relationships, the following integrity constraints would be applicable:

- Null FKs are not allowed; this would violate the referential integrity rule.
- Deletion of referenced tuples is restricted, again in keeping with the referential integrity rule. Alternately, one may consider a cascaded deletion that removes a referenced tuple and all referencing tuples to (from subtypes below in the hierarchy). As a general rule, it would be prudent to associate a warning with this latter approach.
- Deletion or update of referencing tuples in a subtype is allowed (assuming those tuples are not also fulfilling the role of a super-type).
- Update of referenced tuples is cascaded to the referencing (subtype) tuples, assuming no surrogates are being used; if surrogates are being used, update should be restricted.

5.5.6 Determining Component Entities

Component entities were not discussed in the original work on the database design via the XR model. However, in the interest of comprehensive coverage, this section has been added. As mentioned in the previous section, care should be taken in not introducing component relationships where traditional relationships would suffice. Each entity type translates to a base relation. Each base relation will contain attributes corresponding to the properties of the entities that apply within the type hierarchy. Again, being guided by principles established in Chapter 3 (Section 3.7), each component will include a foreign key, which is the primary key in the *summary relation* (i.e., the referenced relation). Further, this foreign key will form part of the primary key (or a candidate key) in the component relation. For examples, please refer to Figures 3.11 and 3.12 of Chapter 3.

Turning to the matter of integrity constraints based on the foreign and primary keys: for components, the following strategies would be applicable:

- Null FKs are not allowed; this would violate the referential integrity rule.
- Deletion of referenced tuples is restricted in the summary entity, again in defense of referential integrity. Alternately, one could consider a cascaded deletion that ripples to the referencing details, but at best, this should be accompanied with a warning.
- Deletion of referencing tuples is allowed in the component entity only (but not in the summary).
- Update is cascaded from the summary to the detail; however, if surrogate(s) appear(s) in the summary, update should be restricted.

5.5.7 Determining Additional Properties

The final step in this (modified) XR approach is to carefully determine the properties in each relation (entity). This is actually easy, but in order to avoid mistakes, you must be diligent:

- Except for associations, in your initial system investigation (which would be part of the required software engineering or systems analysis), you would have identified the basic properties for each identified entity. That is your starting point.
- Next, go through Steps 1–3, 5, and 6 and observe the guidelines for dealing with characteristics, designations, subtypes, and components. These steps tell you when and where to introduce foreign keys.
- Next, observe Step 4 above for treating associations.

By following this procedure, you will be able to confidently determine the properties for each relation in the database; in fact, you will end up with a list that is similar to the one provided in Figure 5.3. This finalized list is illustrated in Figure 5.5. As you examine the figure, please note the following:

1. All primary key and foreign key attributes are italicized.
2. The principle of having each attribute with a unique attribute name (including foreign keys) has been followed.
3. The database specification is presented via a Relation–Attributes List (RAL) — a technique introduced in Chapter 3 (Section 3.8).

Relation	Properties (Attributes)	Comment
Kernels and Designations:		
Department	*Dept#*	The primary key
	DeptName ...	
Employee	*Emp#*	The primary key
	EmpName ...	
	EmpDept#	References **Department.Dept#**
Supplier	*Supp#*	The primary key
	SupplName ...	
Project	*Proj#*	The primary key
	ProjName	
	ProjManagerEmp# ...	References **Employee.Emp#**
InventoryItem	*Item#*	The primary key
	ItemName ...	
Characteristics:		
Dependent	*DepnRef*	Surrogate and primary key
	DepnName	
	DepnEmp#	References **Employee.Emp#**
Associations:		
SuppItemMap	*SI_Supp#*	References **Supplier.Suppl#**
	SI_Item#	References **InventoryItem.Item#**
	SI_Ref#	Surrogate and primary key
ItemProjMap	*IP_Item#*	References **InventoryItem.Item#**
	IP_Proj#	References **Project.Proj#**
	IP_Ref#	Surrogate and primary key
ProjSuppMap	*PS_Proj#*	References **Project.Proj#**
	PS_Supp#	References **Supplier.Suppl#**
	PS_Ref#	Surrogate and primary key
ProjWorkSchedule	*PW_Emp#*	References **Employee.Emp#**
	PW_Proj#	References **Project.Proj#**
	PW_Ref#	Surrogate and primary key
ItemStruct	*IS_ThisItem#*	References **InventoryItem.Item#**
	IS_CompItem#	References **InventoryItem.Item#**
	IS_Ref#	Surrogate and primary key
Note: Primary key attributes and foreign keys are *italicized.*		

Each of these relations would typically encompass several other attributes; only the essential ones are shown here.

Each associative entity represents the intersecting relation used to implement a M:M relationship (review section 3.7 of chapter 3).

Figure 5.5 Partial database specification for section of manufacturing firm's database.

With this additional information, you can now revisit the database specification of Chapter 3 (Figures 3.3, 3.13, and 3.17) and revise it accordingly (left as an exercise for you). In so doing, please observe that while the ERD of Figure 5.1 is similar to that of Figure 3.3, they are not identical; they highlight different aspects of a manufacturing environment, with various areas of overlap. By examining both figures, you should come away with a better sense of what a database model and specification for such an environment would likely entail. The key is to apply sound information gathering techniques (as learned in your software engineering course and summarized in Appendix 3), coupled with your database knowledge.

5.5.8 Additional Application of the XR Model

Since its introduction, the XR model has apparently influenced several developments in contemporary database systems in at least two significant ways — the system catalog and the *big data* movement.

System Catalog: Each of the major contemporary DBMS suites is characterized by a comprehensive system catalog. As you will learn later in the course (Chapter 14), the system catalog is a meta-database (i.e., database of the database) that stores critical information about the database in various relational tables. These tables are organized in a manner that is similar to the concepts of E-relations and P-relations. Because of this, many of the desirable features (identified in Chapter 1 and further discussed in Chapter 9) are realized.

Big Data Movement: The principles of the XR model have influenced the design of large, complex databases, as designers search for ways to reduce the required number of relational tables as well as increase the flexibility to store assorted data including very large data objects, hence the term *big data* which is used to describe extremely large data sets. As mentioned in Section 1.4, two such developments are the NoSQL movement and the Entity–Attributes–Value (EAV) model. Indeed, the EAV model embodies the idea of reducing a database to three principal types of storage entities — an entity for defining other entities; an entity for defining properties (attributes) of entities; and a set of EAV entities each of which connects the other two entities to stored values for entity–attribute combinations.

These two developments have already made positive impacts on the arena of database systems as we know it, and will no doubt continue to do so for the foreseeable future.

5.6 The UML Model

An alternate methodology for database modeling is the *Unified Modeling Language* (UML) notation. UML was developed by three contemporary software engineering paragons — Grady Booch, James Rumbaugh, and Ivar Jacobson (see Booch 2005). These three professionals founded Rational Software during the 1990s and, among several other outstanding achievements, developed UML for the expressed purpose of being a universal modeling language. Although UML was developed, primarily for *object-oriented software engineering* (OOSE), it is quite suitable for database modeling. Figure 5.6 provides a description of the main symbols used in UML. Note that with UML comes a slight change in the database jargon ("entity" is replaced with "object type"), consistent with the fact that UML is primarily for OOSE.

Note that the UML notation makes a distinction between a component relationship and an *aggregation relationship* — terms that are well known in OOSE. In the case of the component relationship, the constituent object types are existence dependent on a main object type. In the case of an aggregation relationship, the constituent object types are existence independent of the aggregation object type. Figure 5.7 illustrates the UML diagram for the (partial) college database model that was introduced in Chapter 3 (Figure 3.11). Notice that except for the **StudentEmployee** object type, which has been omitted from Figure 3.11 and replaced with the **Department** object type, the information represented is essentially similar to what was represented in Figure 3.11. The only difference here is that the appropriate UML symbols have been used.

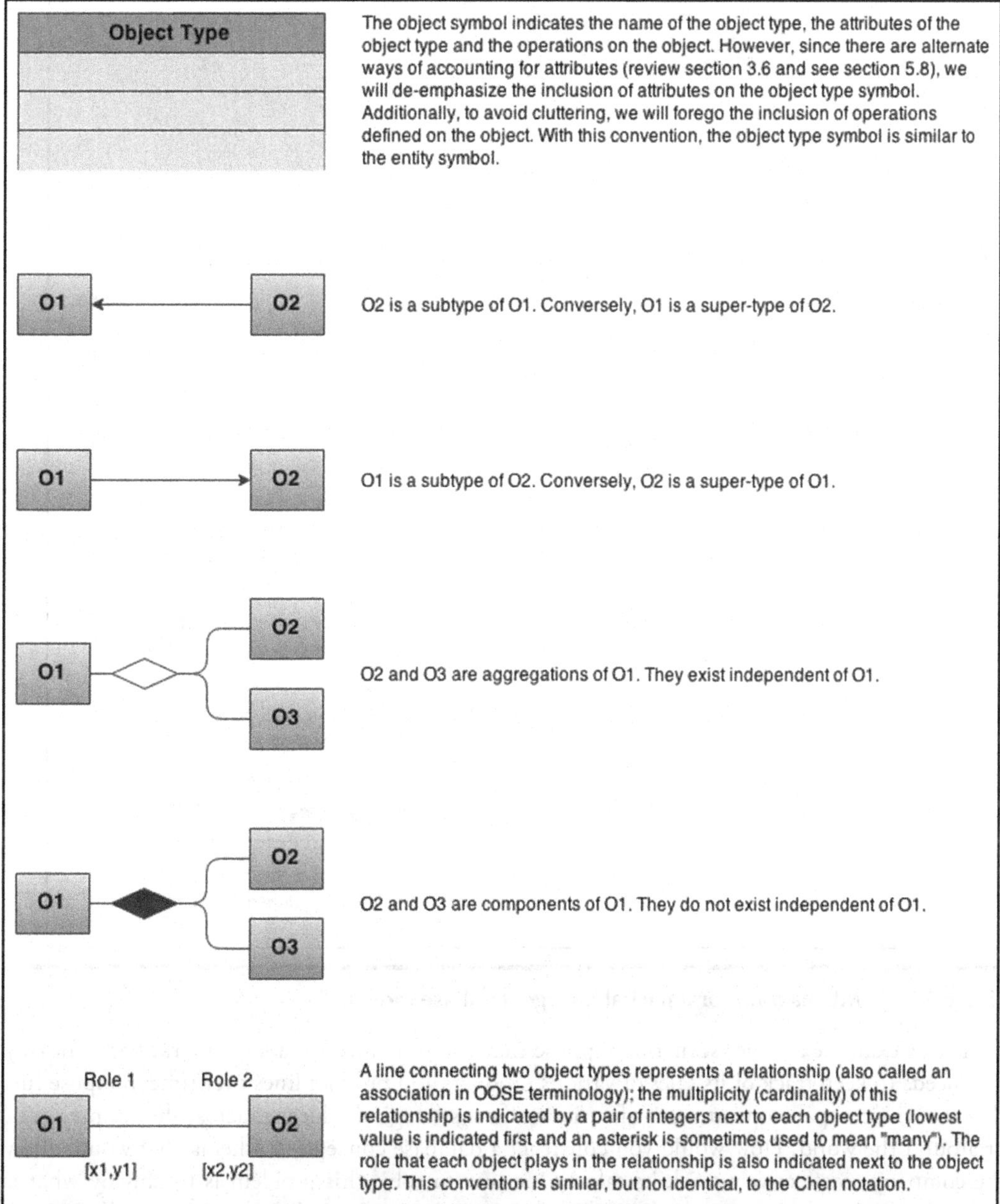

Figure 5.6 Symbols used in UML notation for database modeling.

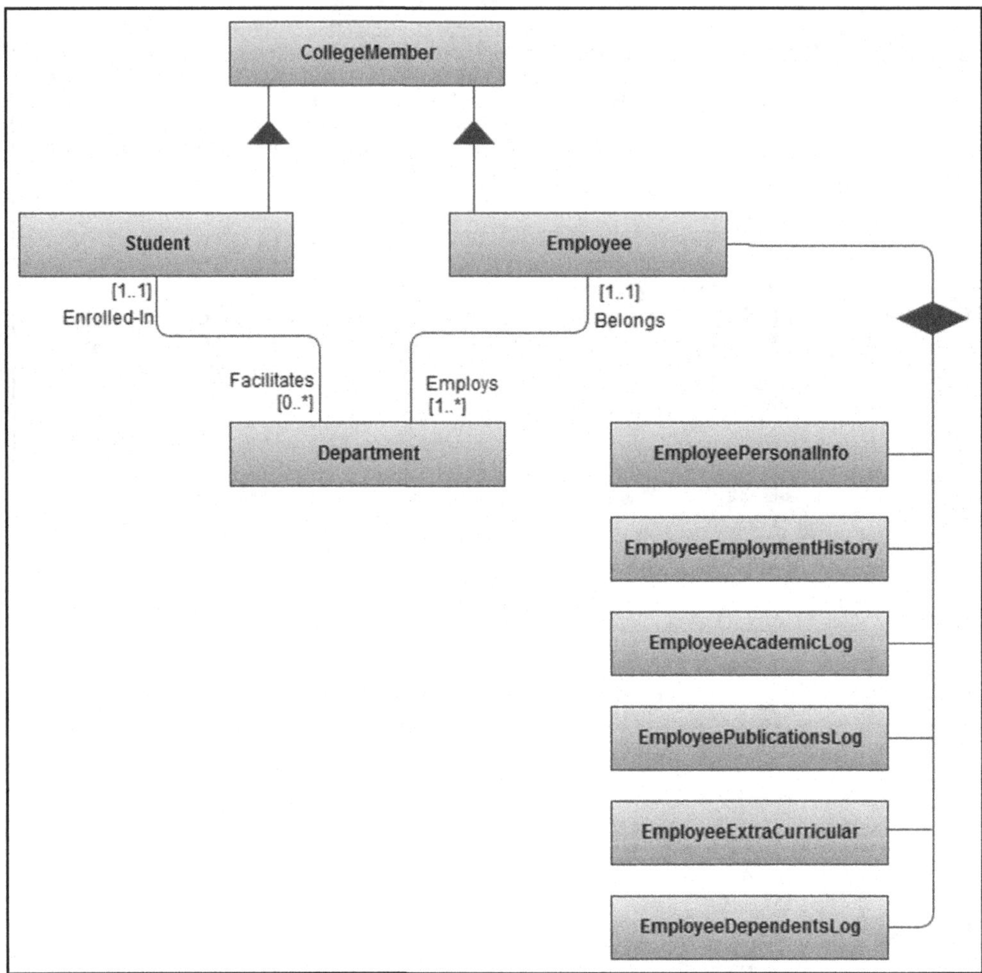

Figure 5.7 UML diagram for a partial college database model.

Let us examine another scenario: Suppose that you were hired at a large marketing company that needs to keep track of its sales of various products and product lines over time. Suppose further that the company operates out of various offices strategically located across the country and/or around the world. How would you construct a database conceptual schema that would allow the company to effectively track its sales? One way to solve this problem is to employ what is called a *star schema* — a central relation (or object type) is connected to two or more relations (or object types) by forming an M:1 relationship with each. Figure 5.8 illustrates such a schema for the marketing company. The central relation (often referred to as the *fact table*) is **SalesSummary**. The surrounding relations (often referred to as *dimensional tables*) are **Location**, **TimePeriod**, **ProductLine**, and **Product**. Each forms a 1:M relationship with **SaleSummary,** the central relation. Notice that consistent with the theory, **SalesSummary** has a foreign key that references each of the referenced relations (object types). Finally, observe also that in this illustration, the attributes for each relation (object type) have been included in the diagram.

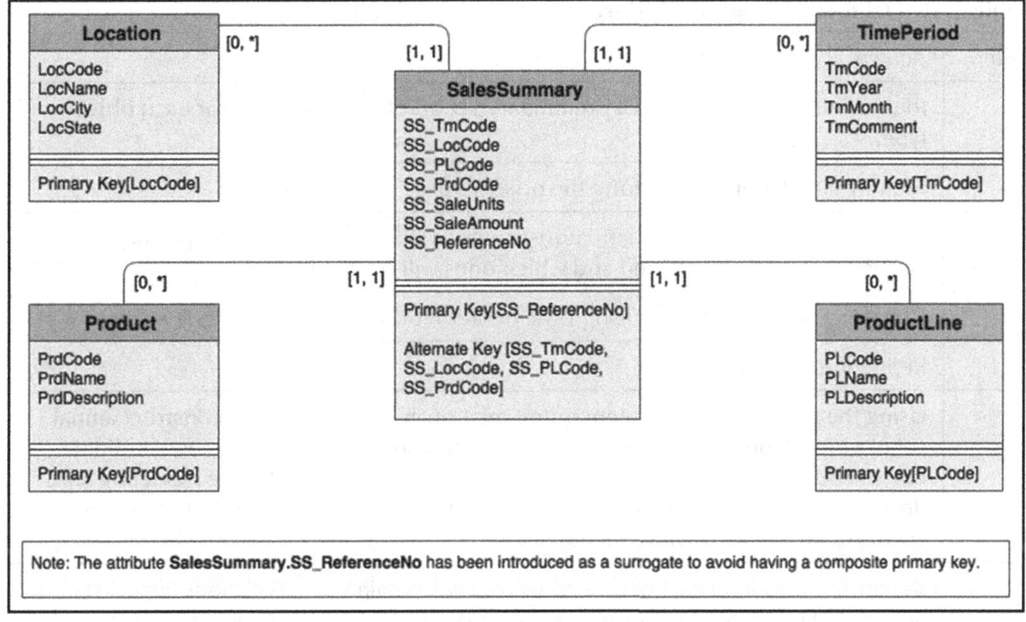

Figure 5.8 UML diagram for tracking sales summary for a large marketing company.

5.7 Database Design via the UML Model

Database design with the UML model is somewhat similar to database design with the E–R model. The points of divergence relate to the differences in notation between the two approaches and the semantic jargon used. The rules that prescribe how to treat various types of relationships (Section 3.7) are still applicable. However, in order to be consistent with object-oriented (OO) terminology, you would replace the term relation (or entity) with the term object type. With this in mind, we may construct a procedure for database design via the UML model as portrayed in Table 5.4.

The prospect of an OO database has its own challenges; some of these are addressed in Chapter 18. However, there is nothing wrong with using the UML standards in modeling and designing a database. To the contrary, it is actually a good idea to do so, especially for complex database systems.

Table 5.4 Database Design Procedure Using the UML Model

Step	*Activity*
1	Identify all object types and a preliminary set of related attributes for each object type.
2	Identify all relationships among the object types.
3	Classify the relationships (mandatory vs optional); decide on which optional relationships will be retained and which ones will be eliminated.
4	Construct an O–R diagram using UML notation or the equivalent (review Chapter 3).
5	Refine the model.
6	Using the guidelines for implementing relationships (Section 3.7), construct a final set of object types, clearly indicating for each object type, its attributes, candidate key(s), and primary key. If you do not have the appropriate database modeling and design tools, construct an object–type–attributes list (OAL) similar to the RAL of Lecture 3 and an RL (review Section 3.8).
7	By consistently following this procedure, you will obtain a set of relations that will be normalized to at least the 3NF and in many cases BCNF. You can then apply normalization theory to refine the model and achieve the desired level of normalization.

5.8 Innovation: The Object/Entity Specification Grid

This section introduces an innovative approach to database design specification that has been successfully used by the author on a number of major projects. The approach may be construed as an extension of the UML model, but is applicable to any database model. As mentioned in Chapter 3 (Section 3.8), for large complex projects (involving huge databases with tens of entities or object types), unless a computer-aided software engineering (CASE) or RAD tool which automatically generates the ORD or ERD is readily available, manually drawing and maintaining this important aspect of the project becomes virtually futile. Even with a CASE tool, perusing several pages of O–R (E–R) diagram may not be much fun. In such cases, an *object/entity specification grid* (O/ESG) is particularly useful. In a relational database environment, the term *entity specification grid* (ESG) is recommended; in an object-oriented environment, the term *object specification grid* (OSG) is recommended.

Figure 5.9 shows the template used for the O/ESG. As the figure portrays, the grid contains the following components:

- Descriptive name of the entity (object type)
- Implementation name of the entity (object type) — typically indicated in square brackets
- Reference identification for each entity to facilitate easy referencing
- Descriptive name, implementation name (in square brackets), and characteristics (in square brackets) for each attribute
- References (implying relationships) to other entities in the system (indicated in curly braces)
- Comments on the entity and selected attributes
- Indexes (including primary key or candidate keys) to be defined on the entity
- Operations to be defined on each entity (object type)
- Optionally, implementation names of operations are be indicated in square brackets next to respective operations

Reference Number — Descriptive Name [Implementation Name]
Attributes: // Itemized specification of all attributes of this object type (or entity)
Comments: // A brief description of the storage purpose of this object type (or entity)
Indexes: // Itemized specification of anticipated indexes, starting with the primary key
Valid Operations: // Itemized list of operations to be defined on this object type (or entity)

Figure 5.9 Template showing main O/ESG components.

The O/ESG presents the specification for each object type (or entity) as it will be implemented in a database consisting of normalized relations. The template is completed by observing a set of conventions as summarized below:

- Each object type (or entity) is identified by a reference code, a descriptive name, and an implementation name (indicated in square brackets).
- For each object type (or entity), the attributes (data elements) to be stored are identified.
- Each attribute is specified by its descriptive name, the implementation name indicated in square brackets, a physical description of the attribute (as described below), and whether the attribute is a foreign key.
- For physical description, the following letters will be used to denote the type of data that will be stored in that attribute, followed by a number that indicates the maximum length of the field: (A) alphanumeric, (N) numeric, or (M) memo. This is specified within square braces. For instance, the notation [Dept#] [N4] denotes a numeric attribute of a maximum length of 4 bytes. In the case of an attribute that stores a memo (M), no length is indicated because a memo field can store as much information as needed. If a real number value is being stored with a decimal value, two numbers will be used: The first number will indicate the length for the whole number part and the second number will indicate the field length of the decimal part (e.g., [N(9,2)]).
- An attribute that is a foreign key is identified by a comment specifying what object type (or entity) is being referenced. The comment appears in curly braces.
- For each object type (or entity), a comment describing the data to be stored is provided.
- An itemized specification of indexes to be defined (starting with the primary key) is provided for each object type (or entity). It may very well be that except for primary keys, the definition of indexes is deferred until the database is in use and these indexes are deemed necessary; nonetheless, anticipating them is a good idea.
- Each operation to be defined on an object type (or entity) will be given a descriptive name and an implementation name, which is indicated in square brackets. Implicit in this principle is a rejection of the commonly applied thinking of constructing a single operation (application program) per entity. The presumption here is that actions such as data insertion, data modification, data deletion, and information inquiry are separated into distinct operations (application programs) as explained in (Foster 2021).

Figure 5.10 provides an illustration of O/ESG for four of the object types (or entities) that would comprise the manufacturing firm's database of earlier discussion (review Figure 5.1). As mentioned earlier, there would be one for each object type (or entity) comprising the system. You will also notice from the figure that the object names specified are in observation of the proposed object naming convention of Section 3.9.

E1 – Department [RM_Department_BR]
Attributes: 01. Department Number [Dept#] [N4] 02. Department Name [DeptName] [A35] 03. Department Head Employee Number [DeptHeadEmp#] [N7] **{Refers to E2.Emp#}** …
Comments: This table stores definitions of all departments in the organization.
Indexes: 1. Primary Key Index: RMDepartment_NX1 on [01]; constraint RM_Department_PK. 2. RM_Department_NX2 on [02]
Valid Operations: 1. Add Departments [RM_Department_AO] 2. Update Departments [RM_Department_UO] 3. Delete Department [RM_Department_ZO] 4. Inquire on Departments [RM_Department_IQ]

E2 – Employee [RM_Employee_BR]
Attributes: 01. Employee Identification Number [Emp#] [N7] 02. Employee Last Name [EmpLName] [A20] 03. Employee First Name [EmpFName] [A20] 04. Employee Middle Initials [EmpMInitl] [A4] 05. Employee Date of Birth [EmpDOB] [N8] 06. Employee's Department [EmpDept#] [N4] **{Refers to E1.Dept#}** 07. Employee Gender [EmpGender] [A1] 08. Employee Marital Status [EmpMStatus] [A1] 09. Employee Social Security Number [EmpSSN] [N10] 10. Employee Home Telephone Number [EmpHomeTel] [A14] 11. Employee Work Telephone Number [EmpWorkTel] [A14] …
Comments: This table stores standard information about all employees in the organization.
Indexes: 1. Primary Key Index: RMEmployee_NX1 on [01]; constraint RM_Employee_PK. 2. RM_Employee_NX2 on [02, 03, 04] 3. RM_Employee_NX3 on [09] 4. RM_Employee_NX4 on [10] or [11]
Valid Operations: 1. Add Employees [RM_Employee_AO] 2. Update Employees [RM_Employee_UO] 3. Delete Employees [RM_Employee_ZO 4. Inquire on Employees [RM_Employee_IO] 5. Report on Employees [RM_Employee_RO]

Figure 5.10 Partial O/ESG for manufacturing environment. (Continued)

E3 – Supplier [RM_Supplier_BR]

Attributes:
01. Supplier Number [Suppl#] [N4]
02. Supplier Name [SupplName] [A35]
03. Supplier Contact Name [SupplContact] [A35]
04. Supplier Telephone Numbers [SupplPhone] [A30]
05. Supplier E-mail Address [SuppEmail] [A30]
...

Comments:
This table stores definitions of all employee classifications.

Indexes:
1. Primary Key Index: RMSupplier_NX1 on [01]; constraint RM_Supplier_PK.
2. RM_Supplier_NX2 on [02]
3. RM_Supplier_NX3 on [04]

Valid Operations:
1. Add Suppliers [RM_Supplier_AO]
2. Update Suppliers [RM_Supplier_UO]
3. Delete Suppliers [RM_Supplier_ZO]
4. Inquire on Suppliers [RM_Supplier_IO]
5. Report on Suppliers [RM_Supplier_RO]

E4 – Project [RM_Project_BR]

Attributes:
01. Project Number [Proj#] [N4]
02. Project Name [ProjName] [A15]
03. Project Summary [ProjSumm] [M]
04. Project's Manager [ProjManagerEmp#] [N7] **{References E2.Emp#}**
...

Comments:
This table stores definitions of all company projects.

Indexes:
1. Primary Key Index: RMProject_NX1 on [01]; constraint RM_Project_PK.
2. RM_Project_NX2 on [02]

Valid Operations:
1. Add Projects [RM_Project_AO]
2. Update Projects [RM_Project_UO]
3. Delete Projects [RM_Project_ZO]
4. Inquire on Projects [RM_Project_IO]
5. Report on Projects [RM_Project_RO]

Figure 5.10 (Continued) Partial O/ESG for manufacturing environment.

The O/ESG is one of several innovative software engineering methodologies that have been proposed and discussed in (Foster 2021; Foster et al. 2015). Among the benefits of this approach are the following:

- **Comprehensive Coverage:** Each entity/object type comprising the system is thoroughly specified. This leads to the next two benefits.
- **Improved Database Documentation:** The database will be thoroughly documented. This facilitates better understanding among technical users as well as end-users.
- **Improved Maintainability:** It will be easier to revisit the design of any relational table comprising the database and modify or redesign it where necessary.
- **More efficient Database Construction:** By using the O/ESG as a guide, it will be easier to create the related relational tables, indexes, and any other related objects for the database.
- **User Interface Design Aid:** If the operations section of the grid is completed, then this information will be useful toward the design and subsequent construction of the user interface to be superimposed on the database.

5.9 Database Design via Normalization Theory

Although this is seldom done, you can actually use the normalization theory as discussed in Chapter 4 to design the basic conceptual schema (involving the structure of the base relations) of a relational database. In practice, normalization is used as a check-and-balance mechanism to refine the acceptability of a database's conceptual schema. As such, normalization can (and should) be applied to each of the database design approaches discussed.

This section looks at two sample database design problems and shows how the normalization theory can be used to solve them. The discussion advances by using two problem scenarios that you will hopefully identify with.

5.9.1 Example: Mountaineering Problem

Suppose that we wish to record information about the activities of mountaineers in a relational database. Let us make the assumption that a climber can only begin one climb per day. Figure 5.11 illustrates an initial set of attributes (with suggested implementation names in square brackets) for the database.

Beginning Date of the Climb	[ClimbStartDate]
Ending Date of the Climb	[ClimbEndDate]
Climber Name	[ClimberName]
Climber Address	[ClimberAddr]
Name of Mountain	[MtnName]
Height of Mountain	[MtnHeight]
Country of Mountain	[MtnCountryName]
District of Mountain	[MtnDistrict]

Figure 5.11 Attributes for the mountaineering problem.

How may we obtain an appropriate conceptual schema for the mountaineering problem? We may approach this problem in one of two ways:

- *The Pragmatic Approach*: Identify related attributes that form data entities, normalize these entities, then identify and rationalize relationships among the entities.
- *The Classical (Theoretical) Approach*: Start out by creating one large 1NF relation involving all attributes, progressively decompose into relations of higher normal forms until given requirements are met.

In the interest of illustrating application of the theory of normalization, let us pursue the second approach. However, please bear in mind that in most real-life situations, you will be advised to employ the pragmatic approach. Several of the cases in Appendix 5 (for instance, Assignments 1 and 2) provide an excellent opportunity for you to do this.

Step 1 — Create a large 1NF Relation

Drawing from the experiences of Chapter 4 (particularly Sections 4.5–4.10), recognize that we are going to need to recognize **Climber**, **Mountain**, and **Country** as disguised entities. Accordingly, let us introduce at minimum, three new attributes: Climber Identification [**Climber#**], Mountain Identification [**Mtn#**], and Country Code [**CountryCD**]; and since we must achieve 1NF requirement, let us assign [**Climber#, ClimbStartDate**] as the PK. Let us tentatively store all attributes as relation **M** as shown in Figure 5.12. The figure also states the observed functional dependencies.

Relation M:

Beginning Date of the Climb	[ClimbStartDate]
Ending Date of the Climb	[ClimbEndDate]
Climber Identification	[Climber#]
Climber Name	[ClimberName]
Climber Address	[ClimberAddr]
Mountain Identification	[Mtn#]
Name of Mountain	[MtnName]
Height of Mountain	[MtnHeight]
Country Code	[CountryCD]
Country of Mountain	[CountryName]
District of Mountain	[MtnDistrict]

Tentative PK [Climber#, ClimbStartDate]

Functional Dependencies:

FD1: [Climber#, ClimbStartDate] → ClimbEndDate, ClimberName, ClimberAddr, Mtn#, MtnName, MtnHeight, CountryCD, CountryName, MtnDistrict

FD2: Climber# → ClimberName, ClimberAddr

FD3: CountryCD → CountryName

FD4: Mtn# → MtnName, MtnHeight, MtnDistrict, CountryCD, CountryName

Figure 5.12 Revised initial 1NF relation for the mountaineering problem.

Step 2 — Obtain 2NF Relations

The second step is to obtain a set of 2NF relations. From Figure 5.12, it is evident that relation **M** is in violation of 2NF due to the presence of FD2. Because of FD2, the relation can be non-loss decomposed via Heath's theorem to obtain the following two relations:

M1: {Climber#, ClimberName, ClimberAddr} PK [Climber#]

M2: {ClimbStartDate, Climber#, ClimbEndDate, Mtn#, MtnName, MtnHeight, CountryCD, CountryName,
MtnDistrict} PK [**Climber#, ClimbStartDate**]

Step 3 — Obtain 3NF Relations

Next, we seek to obtain 3NF relations; refer again to Figure 5.12. Based on FD3 and FD4, relation **M2** is not in 3NF. Again applying Heath's theorem for non-loss decomposition, we obtain the following relations:

M3: {CountryCD, CountryName} PK [CountryCD]

M4: {Mtn#, MtnName, MtnHeight, CountryCD, MtnDistrict} PK [Mtn#]

M5: {ClimbStartDate, Climber#, ClimbEndDate, Mtn#} PK [ClimbStartDate, Climber#]

Step 4 — Obtain BCNF (and Higher Order) Relations

Next, we seek to obtain relations of higher order normal forms. Observe that relations **M1**, **M3**, **M4**, and **M5** are in BCNF, 4NF and 5NF. Additional decomposition would be pointless, since such decomposition would result in either data loss or increased data duplication, neither of which is desirable. We would therefore conclude that these four relations are normalized, and stop the normalization process.

Step 5 — Refinement

Finally, in keeping with the discussions of Chapter 4 (Sections 4.9–4.11), let us refine the solution by replacing euphemistic relation names with meaningful ones, and apply the unique attribute-name principle. The refined set of relations is shown below (you should be able to easily identify the foreign keys):

Climber: {Climber#, ClimberName, ClimberAddr} PK [Climber#]

Country: {CountryCD, CountryName} PK [CountryCD]

Mountain: {Mtn#, MtnName, MtnHeight, MtnCountryCD, MtnDistrict} PK [Mtn#]

Climb: {ClimbStartDate, ClimbClimber#, ClimbEndDate, ClimbMtn#} PK [ClimbStartDate, ClimbClimber#]

Note:

We could have forgone Steps 1–3 above and gone straight for BCNF relations by simply observing the FDs shown in Figure 5.12, and decomposing via Heath's theorem. As your confidence in database design grows, you will (hopefully) be able to do this. Otherwise, a DBA or Data Architect job is probably not for you.

5.9.2 Determining Candidate Keys and Then Normalizing

In some cases, the database designer may be faced with the problem where basic knowledge of data to be stored is available, but it is not immediately clear how this partial knowledge will translate into a set of normalized relations. For instance, you may be able to identify an entity (or set of entities), but are not sure what the candidate key(s) to this (these) entity (entities) will be. With experience, you will be able to resolve these challenges intuitively. However, what do you do in the absence of that invaluable experience? The relational model provides a theoretical approach for dealing with this problem, as explained in the following example.

Suppose that it is desirable to record the information about the performances of students in certain courses in an educational institution environment. Assume further that a set of functional dependencies have been identified, but it is not clear what the final set of normalized relations will be and how they will be keyed. Figure 5.13 illustrates a summary of the information (assumed to be) known in the case. As usual, we start off by assuming that the relation shown — **StudPerfDraft** is in 1NF.

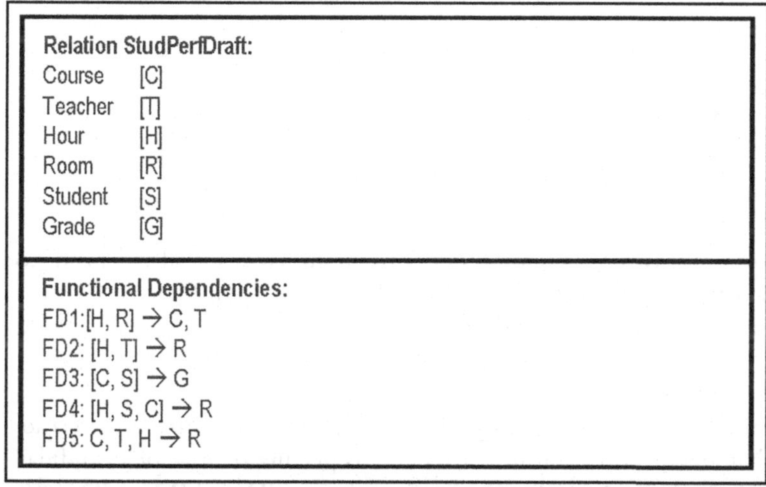

Figure 5.13 Initial 1NF relation for the student performance problem.

Step 1 — Determine the Candidate Key

Having assumed that **StudPerfDraft** is in 1NF, our next step is to determine a candidate key for the relation. We do this by chasing the explicit and implicit dependencies. Any FD that ends up determining all attributes (directly or indirectly) constitutes a candidate key. The technique is referred to as *computing closures.*

> The closure of an FD, denoted FD^+, is the set of all implied dependencies.

To complete the closure for FDs on the **StudPerfDraft** relation, we must examine each explicit FD in turn and identify all the attributes that it determines (explicitly or implicitly), bearing in mind that any attribute or combination of attributes that is a determinant, necessarily determines itself. Hence, we construct the following dependency threads:

- HR → HRCT
- HT → HTR → HTRC
- CS → CSG
- HSC → HSCR → HSCRG → HSCRGT
- CTH → CTHR

From this exercise, observe that [H, S, C] is the only candidate key; it is therefore designated as the primary key (PK). From this point, we would then conduct the normalization procedure as usual.

Step 2 — Obtain 2NF Relations
The next step is to obtain 2NF relations. We may rewrite the initial relation as follows:

StudPerfDraft {H, S, C, T, R, G} PK [H,S, C]

Observe that based on Figure 5.13, the 2NF benchmark for **StudPerfDraft** is violated by FD3 ([C, S] G). We therefore decompose via Heath's theorem to obtain the following relations:

R1 {C, S, G} PK [C, S}
R2 {H, S, C, T, R} PK [H, S, C]

Step 3 — Obtain 3NF Relations
Step 3 is to obtain 3NF relations. Remember, for 3NF, we desire to have mutual independence among non-key attributes. Again, referring to Figure 5.13, observe that both **R1** and **R2** are in 3NF; there is no dependency among the non-key attributes.

Step 4 — Obtain BCNF Relations
Our fourth step is to obtain a set of BCNF relations. Remember, to achieve this, each relation must be constructed so that any determinant in that relation is playing the role of a candidate key. We have already replaced relation **StudPerfDraft** with relations **R1** and **R2**, so our focus needs to be on these latter two relations. Again, being mindful of the FDs of Figure 5.13, observe that **R1** is in BCNF but **R2** is not, due to FD5 ([C, T, H] R), FD4 ([H, S, C] R), FD2 ([H, T] R), and FD1 ([H, R] C, T). To resolve these, decompose via Heath's theorem to obtain the following relations:

R3 {C, T, H, R} PK [C, T, H]
R4 {H, S, C, R} PK [H, S, C]
R5 {H, T, R} PK [H, T]
R6 {H, R, C, T} PK [H, R]

We now have **R1**, **R3**, **R4**, **R5**, and **R6** all in BCNF. Additionally, observe that **R6** is really redundant on **R3** (you could have deduced that from FD1 and FD5), so we can eliminate **R6** from the final solution. Also notice that **R5** is redundant on **R3**, so we can eliminate **R5** from the final solution. We would therefore advance **R1**, **R3**, and **R4** to the next stage of the analysis.

Step 5 — Obtain Higher Order Normalized Relations

There are neither MVDs nor JDs, therefore relations **R1**, **R3**, and **R4** are in 4NF and 5NF. Moreover, we can confidently assert that additional decomposition would be counterproductive since it would result in either data loss or increased data duplication. We would therefore accept these five relations and end the normalization process.

Step 6 — Refinement

As usual, let us now refine the solution by replacing euphemistic relation names, introducing meaningful attribute names, and applying the unique attribute-name principle. Additionally, we will recognize that **Course**, **Teacheer**, **Student**, **Room**, and **Hour** (i.e., Period) are actually disguised entities and should be treated as such. Drawing for discussion of Chapter 4 (Sections 4.9–4.11), following is a proposed set of relations that meet these criteria:

Course {Crs#, CrsName,...} PK[Crs#]
Student {Stud#, StudFName, StudLName, StudGPA,...} PK[Stud#]
Teacher {TeacherEmp#, Emp_FName, Emp_LName,...} PK[TeacherEmp#]
Room {Room#, RoomLocation,...} PK [Room#]
Period {Period#, PeriodStart, PeriodEnd} PK [Period#]
CourseSchedule {CS_Crs#, CS_TeacherEmp#, CS_Room#, CS_Period#,...}
PK [CS_Crs#, CS_TeacherEmp#, CS_Period#]
StudentPerformanceLog {SP_Crs#, SP_Stud#, SP_Grade,...} PK [SP_Crs#, SP_Stud#]
StudentSchedule {SS_Stud#, SS_Crs#, SS_Period#, SS_Room#...} PK [SS_Stud#, SS_Period#, SS_Crs#]

In the solution presented, **R1** is replaced by the **StudentPerformanceLog** relation, **R3** is replaced by the **CourseSchedule** relation, and **R4** is replaced by the **StudentSchedule** relation. Relation **R5** could be implemented as an optional **TeacherAvailabilityLog**, but that information can be deduced from the **Course Schedule** relation. Finally, note that this is an incomprehensive but useful example; a college database is a much more complex system than the representation presented here. However, the representation succeeds in showcasing the application of the normalization theory, and that was the sole intent.

The final activity that would be required prior to creating the tables in a relational DBS is to prepare the database specification for each relation. This can be done via a detailed RAL (as described in Sections 3.8 and 3.9) of an ESG (as described in Section 5.8).

5.10 Database Model and Design Tools

At this point, you must be wondering how in the world are you supposed to model and design a complex database and keep track of all the entities and relationships? The good news is, there are various tools that are readily available, so there's no need to panic. The standard general-purpose word processors (such as MS Office, Word Perfect, Open Office, etc.) are all fortified with graphics capabilities so that if you spend a little time with any of these products, you will figure out how to design fairly impressive database model/design diagrams. Better yet, there is a wide range of CASE tools and/or modeling tools that you can use. Table 5.5 provides an alphabetic list of some of the commonly used products. Some of the products in the list are quite impressive, providing support for the entire database life cycle (including automatic generation of SQL code). The list is by no means comprehensive, so you do not have to be constrained by it. Some of the products are available free of charge; for others, the parent company offers free evaluation copies.

Table 5.5 Some Commonly Used Database Planning Tools

Product	*Parent Company*	*Comment*
ConceptDraw	CS Odessa	Supports UML diagrams, GUI designs, flowcharts, ERD, and project planning charts.
DataArchitect	theKompany.com	Supports logical and physical data modeling. Interfaces with ODBC and DBMSs such as MySQL, PostgreSQL, DB2, MS SQL Server, Gupta SQLBase, and Oracle. Runs on Linux, Windows, Mac OS X, HP-UX, and Sparc Solaris platforms.
Database Design Tool (DDT)	Open Source	A basic tool that allows database modeling that can import or export SQL.
Database Design Studio	Chilli Source	Allows modeling via ERD, data structure diagrams, and data definition language (DDL) scripts. Three products are marketed: DDS-Pro is ideal for large databases; DDD-Lite is recommended for small- and medium-sized databases; SQL-Console is a GUI-based tool that connects with any database that supports ODBC.
DBDesigner 4 and MySQL Workbench	fabFORCE.net	This original product was developed for the MySQL database. The replacement version, MySQL Workspace, is targeted for any database environment and is currently available for the Windows and Linux platforms.
DeZign	Datanamic	Facilitates easy development of ERDs and generation of corresponding SQL code. Supports DBMSs including Oracle, MS SQL Server, MySQL, IBM DB2, Firebird, InterBase, MS Access, PostgreSQL, Paradox, dBase, Pervasive, Informix, Clipper, FoxPro, Sybase, SQLite, ElevateDB, NexusDB, DBISAM.
Enterprise Architect	Sparx Systems	Facilitates UML diagrams that support the entire software development life cycle (SDLC). Includes support of business modeling, systems engineering, and enterprise architecture. Supports reverse engineering as well.
ER Creator	Model Creator Software	Allows for the creation of ERDs, and the generation of SQL and the generation of corresponding DDL scripts. Also facilitates reverse engineering from databases that support ODBC.
ER Studio	Embarcadero	Similar to ER Creator.

(Continued)

Table 5.5 (Continued) Some Commonly Used Database Planning Tools

Product	*Parent Company*	*Comment*
ERWin Data Modeler	Computer Associates	Facilitates creation and maintenance of data structures for databases, data warehouses, and enterprise data resources. Runs on the Windows platform. Compatible with heterogeneous DBMSs.
MagicDraw	No Magic	A relatively new product that has just been introduced to the market. Appears to be similar to Enterprise Architect.
Oracle Designer	Oracle	Supports design for Oracle databases.
Oracle JDeveloper	Oracle	Supports UML diagramming.
Power Designer	Sybase (now owned by SAP)	Supports UML, business process modeling, and data modeling. Integrates with development tools such as .NET, Power Builder (a Sybase Product), Java, and Eclipse. Also integrates with the major DBMSs.
SmartDraw	SmartDraw	A graphics software that facilitates modeling in the related disciplines of business enterprise planning, software engineering, database modeling, and information engineering (IE). Provides over 100 different templates (based on different methodologies) that you can choose from. Supported methodologies include UML, Chen Notation, IE Notation, etc.
TogetherSoft	Borland	Provides UML-based visual modeling for various aspects of the software development life cycle (SDLC). Allows generation of DDL scripts from the data model. Also supports forward and reverse engineering for Java and C++ code.
Toolkit for Conceptual Modeling (TCM)	University of Twente, Holland	Includes various resources for traditional software engineering methodologies as well as object-oriented methodologies based on the UML standards.
Visio	Microsoft	Facilitates modeling in support of business enterprise planning, software engineering, and database management.
Visual Thought	CERN	Similar to Visio but is free.

5.11 Summary and Concluding Remarks

It is now time to summarize what we have covered in this chapter:

- A database model is a blueprint for subsequent database design. It is a representation of the database. We have looked at three database models: the E–R model, the XR model, and the UML model.
- Database design involves preparation of a database specification, which will be used to construct the database. We have discussed database design via the E–R model, the XR model, the UML model, the E/OSG, and normalization theory.
- The E–R model is the oldest model for relational databases that have been discussed. It involves the use of certain predefined symbols to construct a graphical representation of the database.
- Database design via the E–R model involves the following eight steps that lead to a normalized database specification.
- The XR model (conveniently called the XR model in this course) is an alternate model that compensates for the weaknesses in the E–R model. It involves grouping information entities into different predefined categories that will assist in the design phase.
- Database design via the XR model involves the following nine steps that lead to a normalized database specification.
- The UML model is similar to the E–R model. However, it requires taking an object-oriented approach and employing different notations from the E–R model.
- Database design via the UML model involves the following seven steps that lead to a normalized database specification.
- The O/ESG methodology describes an efficient way of developing a comprehensive, normalized database specification.
- Database design via normalization theory describes a rudimentary approach to database design that relies on mastery of the principles of normalization.

Still, other database modeling/design methodologies have been proposed. One of the more interesting proposals is the *object role modeling* (ORM) methodology. This ORM methodology is a fact-oriented modeling approach in which all facts are treated as relationships. One consequence of this is that attributes are also treated as facts and are therefore represented differently than those in the E–R model. One ardent proponent ORM is Terry Halpin; he maintains a website with various resource materials for the methodology (see Halpin 2015).

You are now armed with all the requisite knowledge needed to design quality databases. However, you will likely find that a review of this and the previous two chapters, along with practice, may be necessary until you have gained mastery and confidence with the concepts, principles, and methodologies. The next chapter discusses the design of the user interface for a database system.

5.12 Review Questions

1. Describe in your own words (with appropriate illustrations) the approach to database design based on the following (in each case, describe the approach, then outline the advantages and disadvantages from your perspective):

   - The E–R Model
   - The XR Model
   - The UML Model
   - The O/ESG Methodology
   - Normalization Theory
2. The following are three software systems for which a database is to be designed:
   - An Inventory Management System
   - A Lab Scheduling System for users of a computer lab
   - A Bookshop Manager System for an educational institution

   For each system, develop a database specification based on at least three of the five approaches above. Compare the database specifications derived. Compare the methodologies. Which one(s) are you most comfortable with and why?
3. How important are surrogates in the design of a database? By considering an appropriate example, demonstrate how surrogates may be used in database design.
4. Compare the UML database model with the E–R model. What are the similarities? What are the differences? You may use appropriate illustrations in your response.

References and/or Recommended Readings

Booch, Grady, James Rumbaugh, & Ivar Jacobson. 2005. *The Unified Modeling Language User Guide* 2nd ed. New York: Addison-Wesley.

Connolly, Thomas, & Carolyn Begg. 2015. *Database Systems: A Practical Approach to Design, Implementation and Management* 6th ed. Boston, MA: Pearson. See chapters 12 & 13.

Coronel, Carlos, & Steven Morris. 2015. *Database Systems: Design, Implementation & Management* 11th ed. Boston, MA: Cengage Learning. See chapters 4 & 5.

Date, Christopher J. 1986. *Relational Database: Selected Writings*. Reading, MA: Addison-Wesley.

Date, Christopher J. 1990. *Introduction to Database Systems* 5th ed. Menlo Park, CA: Addison-Wesley. See chapter 22.

Date, Christopher J. 2004. *Introduction to Database Systems* 8th ed. Menlo Park, CA: Addison-Wesley. See chapters 12–14.

Elmasri, Ramez, & Shamkant B. Navathe. 2011. *Fundamentals of Database Systems* 6th ed. Boston, MA: Pearson. See chapters 7–10.

Foster, Elvis C. 2021. *Software Engineering: A Methodical Approach* 2nd ed. New York: CRC Publishing. See chapters 7, 12, and 13.

Foster, Elvis C., Thomas O'Dea, & Myles Dumas. 2015. "Three Innovative Software Engineering Methodologies." In Global Online Conference on Information and Computer Technology. Sullivan University, November 4–6, 2015. http://www.elcfos.com/papers-in-cs

Garcia-Molina, Hector, Jeffrey Ullman, & Jennifer Widom. 2009. *Database Systems: The Complete Book* 2nd ed. Boston, MA: Pearson. See chapter 3.

Halpin, Terry. 2015. "Object Role Modeling." Accessed June 14, 2015. www.orm.net

Hoffer, Jeffrey A., Ramesh Venkataraman, & Heikki Topi. 2013. *Modern Database Management* 11th ed. Boston, MA: Pearson. See chapters 4.

Jensen, C. S., & L. Mark. 1992. "Queries on Change in Extended Relational Model." *IEEE Transactions on Knowledge and Data Engineering* vol. 4, no. 2, April 1992, pp. 192–200.

Kifer, Michael, Arthur Bernstein, & Phillip M. Lewis. 2006. *Databases and Transaction Processing: An Application Oriented Approach* 2nd ed. Boston, MA: Pearson. See chapter 4.

Kroenke, David M., & David Auer. 2014. *Database Processing: Fundamentals, Design, and Implementation* 13th ed. Boston, MA: Pearson. See chapters 4–6.

Lee, Richard C., & William M. Tepfenhart. 2002. *Practical Object-Oriented Development With UML and Java*. Upper Saddle River, NJ: Prentice Hall. See chapter 8.

Lewis, Phillip M., Arthur Bernstein, & Michael Kifer. 2002. *Databases and Transaction Processing: An Application Oriented Approach*. New York: Addison-Wesley. See chapter 8.

Pratt, Phillip J., & Mary Z. Last. 2015. *Concepts of Database Management* 8th ed. Boston, MA: Cengage Learning. See chapters 3–6.

Silberschatz, Abraham, Henry F. Korth, & S. Sudarshan. 2011. *Database Systems Concepts* 6th ed. Boston, MA: McGraw-Hill. See chapters 7 & 8.

Chapter 6

Database User Interface Design

This chapter covers the essentials of good database user interface design. A properly designed database system typically includes a user interface that facilitates end users accessing the system. The chapter presumes that you are familiar with basic software engineering principles as well as user interface design principles as covered in other courses in your undergraduate degree. The chapter is therefore a summary that includes:

- Overview
- Deciding on the User Interface
- Steps in the User Interface Design
- User Interface Development and Implementation
- Summary and Concluding Remarks

6.1 Overview

Recall from Section 1.6 that a database system is a software system with a life cycle of its own. At the stage of the database development life cycle (DDLC) where user interface becomes the focus, we would have already settled on the relations (and related attributes), relationships, constraints, and other related objects comprising the database. Now it's time to design the user interface that will allow end users to access the database.

The user interface should facilitate at least the following basic functions: data insertion, data update, data deletion, and information retrieval (query and print). This is important, as it is not acceptable to give end users direct, unfettered access to the database; were this to be done, the integrity of the system would be compromised in very short order. What is more desirable is to provide the end users with a user-friendly, controlled environment that provides all the privileges and functionalities that they need and nothing more.

The user interface will consist of menus from which user operations can be accessed. Depending on the software development tool used, it will be constructed from various building blocks, and there will be various categories of user interface objects (review your software engineering notes).

DOI: 10.1201/9781003275725-8

The system must also facilitate various user (external) views through logical interpretation of objects. This must be developed using the database management system (DBMS) and/or whatever software development tool is being used. Note that if the Object/Entity Specification Grid (O/ESG) methodology (discussed in Section 5.8) is employed, you will be well on your way with the user interface specification.

By the way of illustration, let us revisit the O/ESG for the partial database specification of the manufacturing environment, discussed in the previous chapters (Sections 3.6 and 5.8). Figure 6.1 shows a repeat of the O/ESG for the **Employee** entity. Applying the object-naming convention of Section 3.9 and drawing from the discussion of Section 5.8, the figure conveys that this entity could be implemented as a relational table named **RM_Employee_BR**, with the constituents shown. The lower part of the grid shows the suggested operations to be defined in a user interface that facilitates access of the entity. Each operation would typically translate to an application program. The user interface should also anticipate and support various logical perspectives on this relational table; these perspectives would typically be provided through one or more inquiry operations (denoted by **RM_Employee_IO**). Following are some examples of different perspectives of **Employee** information:

- Employees arranged by Employee Number
- Employees arranged by Name
- Employees arranged by Telephone Numbers
- Employees arranged by Social Security Number

E2 – Employee [RM_Employee_BR]
Attributes: 01. Employee Identification Number [Emp#] [N7] 02. Employee Last Name [EmpLName] [A20] 03. Employee First Name [EmpFName] [A20] 04. Employee Middle Initials [EmpMInitl] [A4] 05. Employee Date of Birth [EmpDOB] [N8] 06. Employee's Department [EmpDept#] [N4] {**Refers to E1.Dept#**} 07. Employee Gender [EmpGender] [A1] 08. Employee Marital Status [EmpMStatus] [A1] 09. Employee Social Security Number [EmpSSN] [N10] 10. Employee Home Telephone Number [EmpHomeTel] [A14] 11. Employee Work Telephone Number [EmpWorkTel] [A14] ...
Comments: This table stores standard information about all employees in the organization.
Indexes: 1. Primary Key Index: RMEmployee_NX1 on [01]; constraint RM_Employee_PK. 2. RM_Employee_NX2 on [02, 03, 04] 3. RM_Employee_NX3 on [09] 4. RM_Employee_NX4 on [10] or [11]
Valid Operations: 1. Add Employees [RM_Employee_AO] 2. Update Employees [RM_Employee_UO] 3. Delete Employees [RM_Employee_ZO] 4. Inquire on Employees [RM_Employee_IO] 5. Report on Employees [RM_Employee_RO]

Figure 6.1 Excerpt from the partial O/ESG for manufacturing environment.

6.2 Deciding on User Interface

User interfaces can be put into three broad categories — *menu-driven interface, command interface,* and *graphical user* interface (GUI). Figure 6.2 shows a comparison of the approaches in terms of relative complexity of design (COD), response time (RT), and ease of use (EOU). The figure introduces the commonly used acronyms CASE and RAD. As you may know, CASE stands for computer-aided software engineering and RAD stands for rapid application development.

A command interface is characterized by multiple commands (each with a specific syntax) that users have to follow. Command interfaces are the oldest type; they typify traditional operating systems, compilers, and other software development tools. A menu interface is characterized by a menu of options from which the user chooses appropriate actions to be taken. Up until the mid-1990s, menu-driven interfaces were the most frequently used, dominating the arena of business information and application systems. Since the late 1980s, graphical interfaces have become very popular and clearly dominate user interfaces of the current era. They typify features such as direct manipulation and drag-and-drop. Of course, the approaches can be combined.

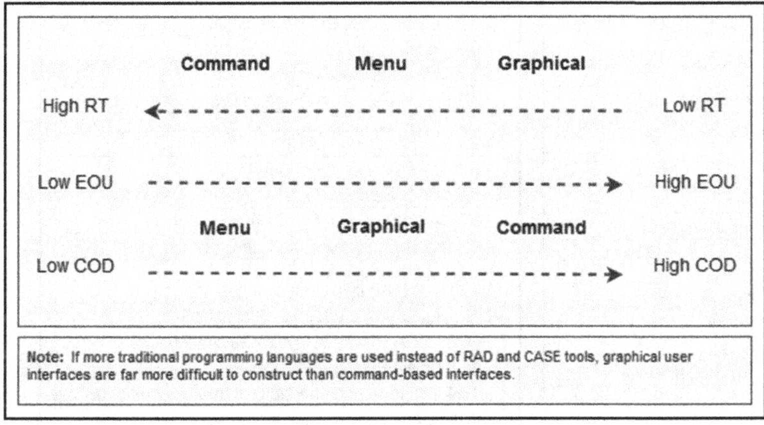

Figure 6.2 Comparison of user interface categories.

6.3 Steps in User Interface Design

How you design the user interface will depend to a large extent on the type of user interface your software requires, as well as the intended users of the software (experts, knowledgeable intermittent, or novices). This section examines two possible approaches.

6.3.1 Menu or Graphical User Interface

If the user interface is to be menu driven or graphical, the following steps are recommended (assuming object-oriented design):

1. Put system objects (structures and operations) into logical groups. At the highest level, the menu will contain options pointing to summarized logical groups.
2. For each summarized logical group, determine the component subgroups where applicable, until all logical groups have been identified.
3. Let each logical group represent a component menu.
4. For each menu, determine options using an object-oriented strategy to structure the menu hierarchy (object first, operation last).

5. Design the menus to link the various options. Develop a menu hierarchy tree or a *user interface topology chart* (UITC).
6. Program the implementation.

Figure 6.3 illustrates a partial *user interface topology chart* (UITC) for a *college/university administrative information system* (CUAIS). The UITC displays the main user interface structure for the system; it is fully discussed in Foster (2021); however, it is intuitive enough for you to understand it. The CUAIS project is also described in Foster (2021); like the UITC, a full discussion is not necessary here

CUAIS Main Menu
1. Infrastructure & Resource Management System
2. Curriculum & Academic Management System
3. Financial Information Management System
4. Student Affairs Management System
5. Public Relations & Alumni Management System
6. Human Relations Management System
7. Cafeteria Services Management System
8. Library Information Management System
9. Other Ad Hoc Services

1. Infrastructure & Resource Management System
1.1 Department Definitions
1.1.1 Add Department Definitions
1.1.2 Modify Department Definitions
1.1.3 Delete Department Definitions
1.1.4 Inquire/Report on Department Definitions
1.2 School/Division Definitions
1.2.1 Add School/Division Definitions
1.2.2 Modify School/Division Definitions
1.2.3 Delete School/Division Definitions
1.2.4 Inquire/Report on School/Division Definitions
1.3 Building Definitions
1.3.1 Add Building Definitions
1.3.2 Modify Building Definitions
1.3.3 Delete Building Definitions
1.3.4 Inquire/Report on Building Definitions
1.4 Lecture Room Definitions
1.4.1 Add Lecture Room Definitions
1.4.2 Modify Lecture Room Definitions
1.4.3 Delete Lecture Room Definitions
1.4.4 Inquire/Report on Lecture Room Definitions
1.5 Fixed Asset Logs
1.5.1 Add Fixed Assets
1.5.2 Modify Fixed Assets
1.5.3 Delete Fixed Assets
1.5.4 Inquire/Report on Fixed Assets
1.6 Inventory/Resource Items
1.6.1 Add Inventory/Resource Items
1.6.2 Modify Inventory/Resource Items
1.6.3 Delete Inventory/Resource Items
1.6.4 Inquire/Report on Inventory/Resource Items

1. Infrastructure & Resource Management System
1.7 Supplier Definitions
1.7.1 Add Supplier Definitions
1.7.2 Modify Supplier Definitions
1.7.3 Delete Supplier Definitions
1.7.4 Inquire/Report on Supplier Definitions
1.8 Purchase Order Summaries
1.8.1 Add Purchase Order Summaries
1.8.2 Modify Purchase Order Summaries
1.8.3 Delete Purchase Orders
1.8.4 Inquire/Report on Purchase Order Summaries
1.9 Purchase Order Details
1.9.1 Add Purchase Order Details
1.9.2 Modify Purchase Order Details
1.9.3 Delete Purchase Order Details
1.9.4 Inquire/Report on Purchase Order Details
1.10 Purchase Invoice Summaries
1.10.1 Add Purchase Invoice Summaries
1.10.2 Modify Purchase Invoice Summaries
1.10.3 Delete Purchase Invoices
1.10.4 Inquire/Report on Purchase Invoice Summaries
1.11 Purchase Invoice Details
1.11.1 Add Purchase Invoice Details
1.11.2 Modify Purchase Invoice Details
1.11.3 Delete Purchase Invoice Details
1.11.4 Inquire/Report on Purchase Invoice Details
1.12 Purchase Return Summaries
1.12.1 Add Purchase Return Summaries
1.12.2 Delete Purchase Returns
1.12.3 Inquire/Report on Purchase Return Summaries
1.13 Purchase Return Details
1.13.1 Add Purchase Return Details
1.13.2 Modify Purchase Return Details
1.13.3 Delete Purchase Return Details
1.13.4 Inquire/Report on Purchase Return Details
1.14 Supplier-Resource Mappings
1.14.1 Add Supplier-Resource Mappings
1.14.3 Delete Supplier-Resource Mappings
1.14.4 Inquire/Report on Supplier-Resource Mappings
. . . .

Figure 6.3 Partial user interface topology chart for the CUAIS project. (Continued)

2. Curriculum & Academic Management System
2.1 Academic Program Definitions
2.1.1 Add Academic Program Definitions
2.1.2 Modify Academic Program Definitions
2.1.3 Delete Academic Program Definitions
2.1.4 Inquire/Report on Academic Program Definitions
2.2 Course Definitions
2.2.1 Add Course Definitions
2.2.2 Modify Course Definitions
2.2.3 Delete Course Definitions
2.2.4 Inquire/Report on Course Definitions
2.3 Academic Department Definitions
2.3.1 Add Academic Department Definitions
2.3.2 Modify Academic Department Definitions
2.3.3 Delete Academic Department Definitions
2.3.4 Inquire/Report on Academic Department Definitions
2.4 Dormitory Definitions
2.4.1 Add Dormitory Definitions
2.4.2 Modify Dormitory Definitions
2.4.3 Delete Dormitory Definitions
2.4.4 Inquire/Report on Dormitory Definitions
2.5 Course Schedules
2.5.1 Add Course Schedules
2.5.2 Modify Course Schedules
2.5.3 Delete Course Schedules
2.5.4 Inquire/Report on Course Schedules
2.6 Examination Schedules
2.6.1 Add Examination Schedules
2.6.2 Modify Examination Schedules
2.6.3 Delete Examination Schedules
2.6.4 Inquire/Report on Examination Schedules
2.7 Course Evaluations
2.7.1 Add Course Evaluations
2.7.2 Modify Course Evaluations
2.7.3 Delete Course Evaluations
2.7.4 Inquire/Report on Course Evaluations
2.8 Academic Program Evaluations
2.8.1 Add Academic Program
2.8.2 Modify Academic Program
2.8.3 Delete Academic Program
2.8.4 Inquire/Report on Academic Program
....

3. Financial Information Management System
3.1 Chart of Accounts
3.1.1 Add Account Definitions
3.1.2 Modify Account Definitions
3.1.3 Delete Account Definitions
3.1.4 Inquire/Report on Account Definitions
3.2 Financial Institutions
3.2.1 Add Financial Institution Definitions
3.2.2 Modify Financial Institution Definitions
3.2.3 Delete Financial Institution Definitions
3.2.4 Inquire/Report on Financial Institution Definitions
3.3 Financial Transactions
3.3.1 Add Financial Transactions
3.3.2 Modify Financial Transactions
3.3.3 Delete Financial Transactions
3.3.4 Inquire/Report on Financial Transactions
3.4 Purchase Orders — Summaries & Details
3.4.1 Add Purchase Orders
3.4.2 Modify Purchase Orders
3.4.3 Delete Purchase Orders
3.4.4 Inquire/Report on Purchase Orders
3.5 Purchase Invoices — Summaries & Details
3.5.1 Add Purchase Invoices
3.5.2 Modify Purchase Invoices
3.5.3 Delete Purchase Invoices
3.5.4 Inquire/Report on Purchase Invoices
3.6 Sale Orders — Summaries & Details
3.6.1 Add Sale Orders
3.6.2 Modify Sale Orders
3.6.3 Delete Sale Orders
3.6.4 Inquire/Report on Sale Orders
3.7 Sale Invoices — Summaries & Details
3.7.1 Add Sale Invoices
3.7.2 Modify Sale Invoices
3.7.3 Delete Sale Invoices
3.7.4 Inquire/Report on Sale Invoices
3.8 Investments
3.8.1 Add Investment Entries
3.8.2 Modify Investment Entries
3.8.3 Delete Investment Entries
3.8.4 Inquire/Report on Investments
....

Figure 6.3 (Continued) Partial user interface topology chart for the CUAIS project. (Continued)

4. Student Affairs Management System
4.1 Student Personal Records
4.1.1 Add Student Personal Records
4.1.2 Modify Student Personal Records
4.1.3 Delete Student Personal Records
4.1.4 Inquire/Report on Student Personal Records
4.2 Student Educational History
4.2.1 Add Student Educational History
4.2.2 Modify Student Educational History
4.2.3 Delete Student Educational History
4.2.4 Inquire/Report on Student Educational History
4.3 Student Academic Qualification
4.3.1 Add Student Academic Qualification
4.3.2 Modify Student Academic Qualification
4.3.3 Delete Student Academic Qualification
4.3.4 Inquire/Report on Student Academic Qualification
4.4 Student Next of Kin Contacts
4.4.1 Add Student Next of Kin Contacts
4.4.2 Modify Student Next of Kin Contacts
4.4.3 Delete Student Next of Kin Contacts
4.4.4 Inquire/Report on Student Next of Kin Contacts
4.5 Student Extracurricular Activities
4.5.1 Add Student Extracurricular Activities
4.5.2 Modify Student Extracurricular Activities
4.5.3 Delete Student Extracurricular Activities
4.5.4 Inquire/Report on Student Extracurricular Activities
4.6 Student Registration Logs
4.6.1 Add Student Registration Logs
4.6.2 Modify Student Registration Logs
4.6.3 Delete Student Registration Logs
4.6.4 Inquire/Report on Student Registration Logs
4.7 Student Academic Performance Logs
4.7.1 Add Student Academic Performance Logs
4.7.2 Modify Student Academic Performance Logs
4.7.3 Delete Student Academic Performance Logs
4.7.4 Inquire/Report on Student Academic Performance
4.8 Student Financial Logs
4.8.1 Add Student Financial Logs
4.8.2 Modify Student Financial Logs
4.8.3 Delete Student Financial Logs
4.8.4 Inquire/Report on Student Financial Logs
....

5. Public Relations & Alumni Management System
5.1 Alumni Information
5.1.1 Add Alumni Information
5.1.2 Modify Alumni Information
5.1.3 Delete Alumni Information
5.1.4 Inquire/Report on Alumni Information
5.2 Alumni Chapters
5.2.1 Add Alumni Chapter Definitions
5.2.2 Modify Alumni Chapter Definitions
5.2.3 Delete Alumni Chapter Definitions
5.2.4 Inquire/Report on Alumni Chapters
5.3 Institutional Sponsors
5.3.1 Add Institutional Sponsors
5.3.2 Modify Institutional Sponsors
5.3.3 Delete Institutional Sponsors
5.3.4 Inquire/Report on Institutional Sponsors
5.4 Alumni Funds Raised
5.4.1 Add Alumni Funds Raised
5.4.2 Modify Alumni Funds Raised
5.4.3 Delete Alumni Funds Raised
5.4.4 Inquire/Report on Alumni Funds Raised
5.5 Other Funds Raised
5.5.1 Add Other Funds Raised
5.5.2 Modify Other Funds Raised
5.5.3 Delete Other Funds Raised
5.5.4 Inquire/Report on Other Funds Raised
5.6 Special Project Definitions
5.6.1 Add Special Project Definitions
5.6.2 Modify Special Project Definitions
5.6.3 Delete Special Project Definitions
5.6.4 Inquire/Report on Special Project Definitions
5.7 Special Project Details
5.7.1 Add Special Project Details
5.7.2 Modify Special Project Details
5.7.3 Delete Special Project Details
5.7.4 Inquire/Report on Special Project Details
5.8 Promotional Initiatives/Activities
5.8.1 Add Promotional Initiatives
5.8.2 Modify Promotional Initiatives
5.8.3 Delete Promotional Initiatives
5.8.4 Inquire/Report on Promotional Initiatives
....

Figure 6.3 (Continued) Partial user interface topology chart for the CUAIS project. (Continued)

6. Human Resource Management System
6.1 Employee Personal Records
6.1.1 Add Employee Personal Records
6.1.2 Modify Employee Personal Records
6.1.3 Delete Employee Personal Records
6.1.4 Inquire/Report on Employee Personal Records
6.2 Employee Academic History
6.2.1 Add Employee Academic History
6.2.2 Modify Employee Academic History
6.2.3 Delete Employee Academic History
6.2.4 Inquire/Report on Employee Academic History
6.3 Employee Work History
6.3.1 Add Employee Work History
6.3.2 Modify Employee Work History
6.3.3 Delete Employee Work History
6.3.4 Inquire/Report on Work History
6.4 Employee Beneficiaries
6.4.1 Add Employee Beneficiary Information
6.4.2 Modify Employee Beneficiary Information
6.4.3 Delete Employee Beneficiary Information
6.4.4 Inquire/Report on Employee Beneficiaries
6.5 Employee Extracurricular Activities
6.5.1 Add Student Extracurricular Activities
6.5.2 Modify Student Extracurricular Activities
6.5.3 Delete Student Extracurricular Activities
6.5.4 Inquire/Report Employee Extracurricular Activities
6.6 Employee Job Definitions
6.6.1 Add Employee Job Definitions
6.6.2 Modify Employee Job Definitions
6.6.3 Delete Employee Job Definitions
6.6.4 Inquire/Report on Employee Job Definitions
6.7 Employee Compensation Packages
6.7.1 Add Employee Compensation Packages
6.7.2 Modify Employee Compensation Packages
6.7.3 Delete Employee Compensation Packages
6.7.4 Inquire/Report Employee Compensation Packages
6.8 Employee Payroll Logs
6.8.1 Add Employee Payroll Logs
6.8.2 Modify Employee Payroll Logs
6.8.3 Delete Employee Payroll Logs
6.8.4 Inquire/Report on Employee Payroll Logs
....

7. Cafeteria Services Management System
7.1 Cafeteria Inventory Items
7.1.1 Add Cafeteria Inventory Items
7.1.2 Modify Cafeteria Inventory Items
7.1.3 Delete Cafeteria Inventory Items
7.1.4 Inquire/Report on Cafeteria Inventory Items
7.2 Cafeteria Meal Plans
7.2.1 Add Cafeteria Meal Plans
7.2.2 Modify Cafeteria Meal Plans
7.2.3 Delete Cafeteria Meal Plans
7.2.4 Inquire/Report on Cafeteria Meal Plans
7.3 Cafeteria Recipes
7.3.1 Add Cafeteria Recipes
7.3.2 Modify Cafeteria Recipes
7.3.3 Delete Cafeteria Recipes
7.3.4 Inquire/Report on Cafeteria Recipes
7.4 Cafeteria Purchase Order Summaries
7.4.1 Add Purchase Order Summaries
7.4.2 Modify Purchase Order Summaries
7.4.3 Delete Purchase Order Summaries
7.4.4 Inquire/Report on Purchase Order Summaries
7.5 Cafeteria Purchase Order Details
7.5.1 Add Purchase Order Details
7.5.2 Modify Purchase Order Details
7.5.3 Delete Purchase Order Details
7.5.4 Inquire/Report on Purchase Order Details
7.6 Cafeteria Sale Order Summaries
7.6.1 Add Sale Order Summaries
7.6.2 Modify Sale Order Summaries
7.6.3 Delete Sale Order Summaries
7.6.4 Inquire/Report on Sale Order Summaries
7.7 Cafeteria Sale Order Details
7.7.1 Add Sale Order Details
7.7.2 Modify Sale Order Details
7.7.3 Delete Sale Order Details
7.7.4 Inquire/Report on Sale Order Details
7.8 Special Catering Projects
7.8.1 Add Special Projects
7.8.2 Modify Special Projects
7.8.3 Delete Special Projects
7.8.4 Inquire/Report on Special Projects
....

Figure 6.3 (Continued) Partial user interface topology chart for the CUAIS project.

6.3.2 Command-Based User Interface

Building a command-driven user interface can be quite involved. Following are some recommended steps to be pursued:

1. Develop an *operations set*, i.e., a list of operations that will be required.
2. Categorize the operations — user operations as opposed to system operations.

3. Develop a mapping of operations with underlying database objects.
4. Determine required parameters for each operation.
5. Develop a list of commands (may be identical to operations set). If this is different from the operations set, each command must link to its corresponding system operations.
6. Define a syntax for each command.
7. Develop a user interface support for each command (and by extension each operation). This interface support must be consistent with the defined command syntax.
8. Program the implementation of each operation.

You are not likely to see a proliferation of command-based user interfaces for database systems; as mentioned earlier, such interfaces are more prevalent in traditional operating systems. Nonetheless, the approach described here would be applicable if you ever encounter a scenario where such an interface is desired. Also bear in mind that while command interfaces are more efficient, they require more [software engineering] effort to design, construct, and implement.

6.4 User Interface Development and Implementation

Designing, constructing, and implementing the user interface really belong to the realm of software engineering, not database systems. However, as you are aware (and as has been emphasized throughout this course), the two fields are closely related. In order to construct the user interface, you will need to have an appropriate set of software development tools. Even if you are not directly involved in the user interface development, having a working knowledge of the process will serve you well. Of course, development and testing of the user interface must proceed according to established software development standards.

The software development tool used to develop the user interface will depend to a large extent on the user requirements (another software engineering matter). Table 6.1 provides some possible scenarios.

It must be emphasized that user interface design and development can and often occur independent of database design and development. This is one of the potent results of data independence (review Sections 1.2 and 2.6): The user interface applications are immune to structural and/or physical changes in the database. In fact, as mentioned in Section 2.6, the user interface (which is part of the front-end system) may reside on a different machine (with a different operating system) from the actual database (which is part of the back-end system). So in addition to data independence, we can have *platform independence.*

Two prominent protocols that facilitate platform independence between database and user interface are *open database connectivity* (ODBC) and *Java database connectivity* (JDBC).

Open Database Connectivity: ODBC is an open standard *application programming interface* (API) for accessing a database. A software product that desires to access an external database must include in its suite, an ODBC driver for that database. The ODBC driver converts the database objects into a generic format that is understood by the software. The target DBMS must also support ODBC. This facilitates communication and transfer of data among heterogeneous databases, irrespective of the platform that they reside on. Microsoft is a strong proponent of ODBC; in fact, the ODBC software is typically bundled with the Windows operating system. ODBC is also supported by other leading operating systems (Unix, Linux, Windows, IBM i, Mac OS, etc.).

Java Database Connectivity: JDBC is a Java-based software component that allows Java programs to access heterogeneous databases, irrespective of their platforms. This API is included in

Table 6.1 Possible Scenarios for User Interface

Scenario	*Solution Alternative*
It is desirable to have front-end and back-end based on the same machine.	Use a DBMS suite that provides facilities for both front-end and back-end systems. Oracle, DB2, and Informix are excellent examples.
Front-end and back-end can be based on different software development tools.	The alternative of an object-oriented RAD tool or Integrated Development Environment (IDE) superimposed on a relational or object database has become the norm. Products such as DB2, Oracle, Informix, Sybase, and MS SQL Server are good candidates for back-end systems. DB2, Oracle, and Informix are regarded as universal databases, supporting both relational and object databases. With respect to the front-end system, your choice of development tool will depend on whether or not Web access is critical.
Front-end must support Web access.	Some popular alternatives are Adobe Cold Fusion, Embarcadero Application Development Product Line (Delphi, C++ Builder, or RAD Studio); Java Development Tools for Web Applications (from Oracle); PHP Development; WebSphere Software Product Line (from IBM); Rational Software Product Line (from IBM); etc.
Front-end need not support Web access.	Any popular IDE will suffice. The software developer would simply forego Web accessibility; however, the contemporary trend is to support Web accessibility.
A purely object-oriented environment is desired.	Most contemporary IDEs are object-oriented. Front-end product such as Rational, Delphi, NetBeans, and Visual Studio (from Microsoft) come readily to mind.

J2SE and J2EE releases. JDBC cooperates with the ODBC protocol; as such, a program running JDBC can reach ODBC-accessible databases.

ODBC and JDBC may be considered as subsets of the wider set of protocols described as Common Object Request Broker Architecture (CORBA). CORBA will be further clarified in Chapter 17 (also see OMG 2015). For the purpose of illustration, Table 6.2 provides a summary of the steps you would take in order to configure an Oracle database server to be accessed from a Delphi RAD tool through ODBC (assuming a Windows environment). Please note that these steps are specific to a particular product and operating system, and are likely to change vary across different platforms. Along this vein, Figure 6.4 exhibits a simple Delphi application that accesses an underlying database table. The top half of the figure provides a picture of the GUI that the end user interacts with (from the front-end environment); the bottom half shows a snippet of Object Pascal code (in a Delphi environment) to facilitate the user interaction with the database (in the back-end). You will notice some embedded SQL in the code. No need to panic about this; you will get a chance to gain mastery of SQL later in the course. Also, the code presented in the figure is not optimized; we'll talk about that later. The intent here is that you have an appreciation of the process of connecting to a database (back-end) from a front-end environment.

Table 6.2 Accessing an Oracle Database from Delphi via ODBC

Step	*Activity*
1	Install Oracle DBMS Server on the database server. This installation will automatically include installation of Oracle's ODBC driver.
2	Install Oracle DBMS Client on client machine(s). This typically includes components such as Oracle Net Manager, Network Transport, and Oracle ODBC Driver.
3	Install front-end system (e.g., Delphi) on client machine(s).
4	On each client machine: a. Configure the Network Client Access file to include a service that connects to the database on the DB Server. This file is called **tnsnames** and may be located in the path **<OracleHome>\network\admin\tnsnames**, where **<OracleHome>** represents the default folder where the Oracle software has been installed. b. Configure ODBC (via the Control Panel) to connect to the DB server through the client service established in Step 4a.
5	In your front-end system (Delphi), select the **Database** option from the main menu; then select from the list, the DB service name established in Step 4, and log on the foreign DB server.
6	From this point, you can now create front-end (Delphi) data sets and data sources that connect to the foreign DB server through the (local) service name established above.

In recent years, other alternatives to ODBC have been proposed. For instance, in the .NET environment (courtesy of Microsoft), Object Linking and Embedding Databases (OLEDB) is preferred. Finally, in many PHP environments, the PHP Data Objects (PDO) methodology is preferred. Figure 6.5 features a PHP rendition of the example provided in the previous illustration. As in Figure 6.4, the first block of Figure 6.5 shows the GUI that end users would interact with, while the second block shows the PHP code to support that GUI. Observe that the PHP code applies the PDO methodology and produces a Web-accessible interface.

Notwithstanding the aforementioned alternatives, there seems to be a strong consensus among leading software engineering companies that the objectives of ODBC are enduring (even if the implementation vehicles change). Additionally, as you observe the illustrations in Figures 6.4 and 6.5, please note the following:

1. PHP is a popular scripting language that is often used in website construction. The actual acronym stands for Personal Home Page — a rather lame name for a language that apparently no one has bothered to change.
2. You have no doubt observed that the Delphi code is significantly shorter than the equivalent PHP code for the same functionality. This is due to the fact that Delphi automatically generates a high level of hidden code for the developer. Remember, Delphi is a RAD tool; PHP is a programming language.
3. The SQL script as presented is vulnerable to an intrusion called SQL injection. This matter will be revisited in Chapter 12; for now, the illustration will suffice.

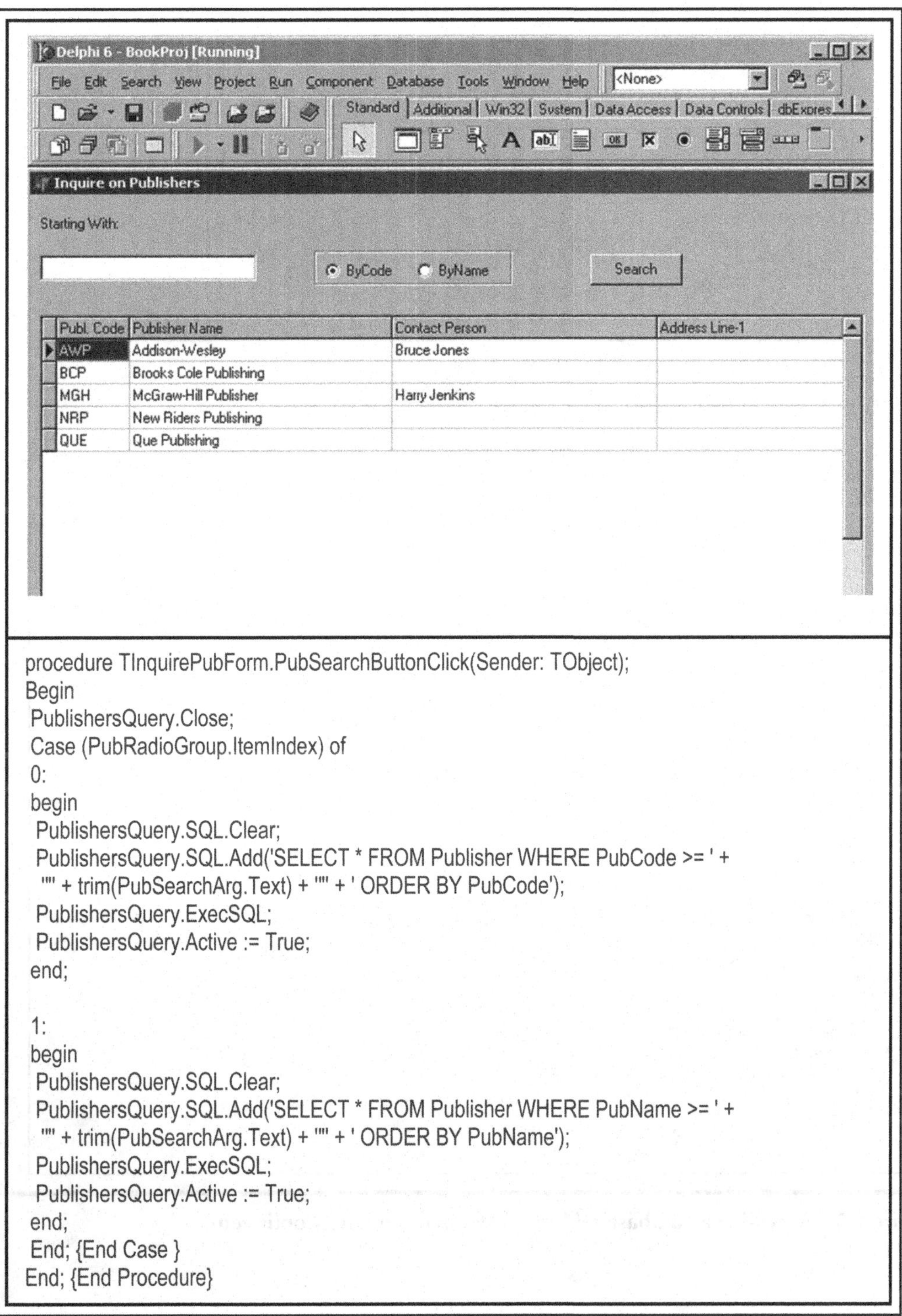


```
procedure TInquirePubForm.PubSearchButtonClick(Sender: TObject);
Begin
 PublishersQuery.Close;
 Case (PubRadioGroup.ItemIndex) of
 0:
 begin
  PublishersQuery.SQL.Clear;
  PublishersQuery.SQL.Add('SELECT * FROM Publisher WHERE PubCode >= ' +
   '''' + trim(PubSearchArg.Text) + '''' + ' ORDER BY PubCode');
  PublishersQuery.ExecSQL;
  PublishersQuery.Active := True;
 end;

 1:
 begin
  PublishersQuery.SQL.Clear;
  PublishersQuery.SQL.Add('SELECT * FROM Publisher WHERE PubName >= ' +
   '''' + trim(PubSearchArg.Text) + '''' + ' ORDER BY PubName');
  PublishersQuery.ExecSQL;
  PublishersQuery.Active := True;
 end;
 End; {End Case }
End; {End Procedure}
```

Figure 6.4 Accessing a database table — Delphi environment.

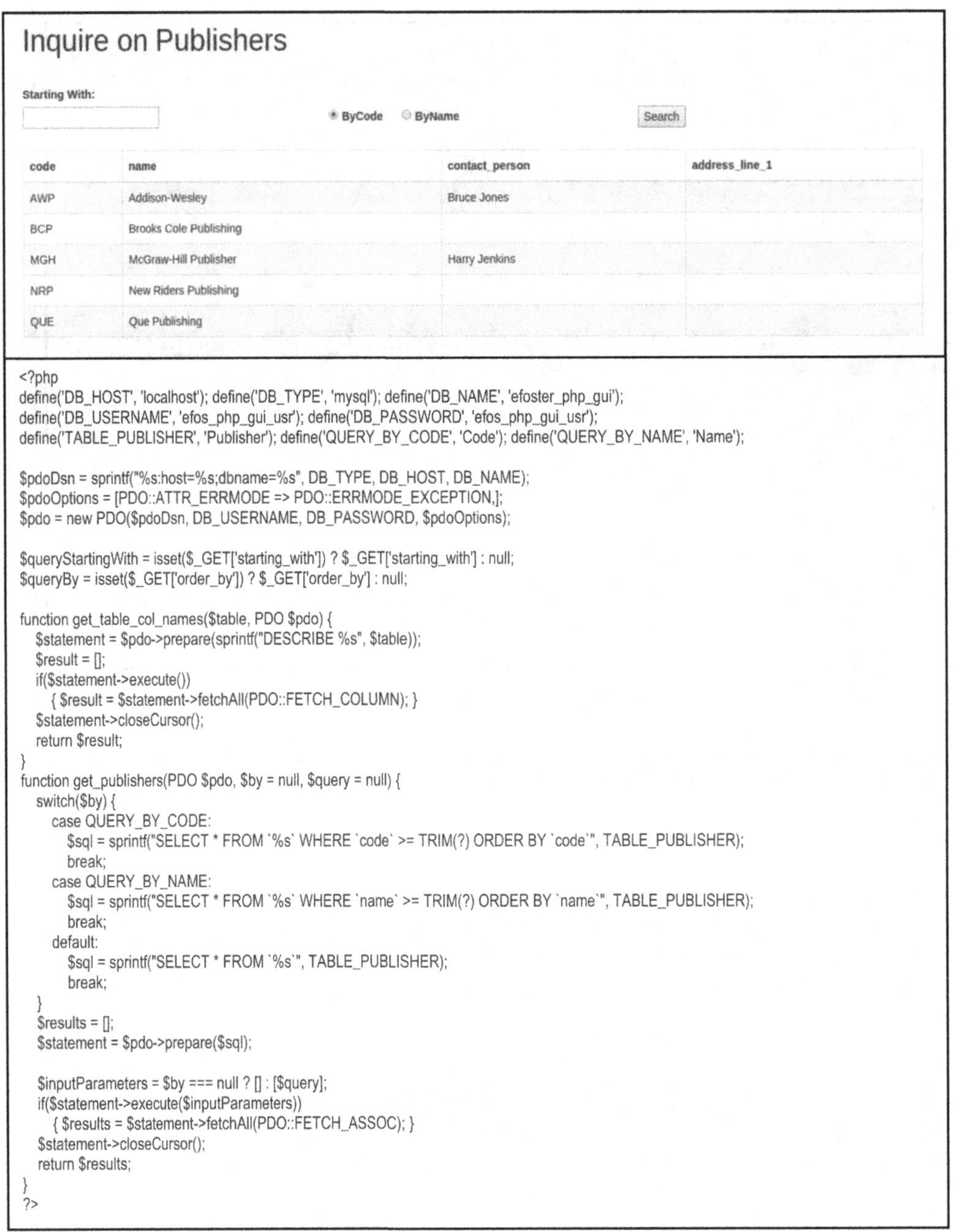

code	name	contact_person	address_line_1
AWP	Addison-Wesley	Bruce Jones	
BCP	Brooks Cole Publishing		
MGH	McGraw-Hill Publisher	Harry Jenkins	
NRP	New Riders Publishing		
QUE	Que Publishing		

```
<?php
define('DB_HOST', 'localhost'); define('DB_TYPE', 'mysql'); define('DB_NAME', 'efoster_php_gui');
define('DB_USERNAME', 'efos_php_gui_usr'); define('DB_PASSWORD', 'efos_php_gui_usr');
define('TABLE_PUBLISHER', 'Publisher'); define('QUERY_BY_CODE', 'Code'); define('QUERY_BY_NAME', 'Name');

$pdoDsn = sprintf("%s:host=%s;dbname=%s", DB_TYPE, DB_HOST, DB_NAME);
$pdoOptions = [PDO::ATTR_ERRMODE => PDO::ERRMODE_EXCEPTION,];
$pdo = new PDO($pdoDsn, DB_USERNAME, DB_PASSWORD, $pdoOptions);

$queryStartingWith = isset($_GET['starting_with']) ? $_GET['starting_with'] : null;
$queryBy = isset($_GET['order_by']) ? $_GET['order_by'] : null;

function get_table_col_names($table, PDO $pdo) {
  $statement = $pdo->prepare(sprintf("DESCRIBE %s", $table));
  $result = [];
  if($statement->execute())
    { $result = $statement->fetchAll(PDO::FETCH_COLUMN); }
  $statement->closeCursor();
  return $result;
}
function get_publishers(PDO $pdo, $by = null, $query = null) {
  switch($by) {
    case QUERY_BY_CODE:
      $sql = sprintf("SELECT * FROM `%s` WHERE `code` >= TRIM(?) ORDER BY `code`", TABLE_PUBLISHER);
      break;
    case QUERY_BY_NAME:
      $sql = sprintf("SELECT * FROM `%s` WHERE `name` >= TRIM(?) ORDER BY `name`", TABLE_PUBLISHER);
      break;
    default:
      $sql = sprintf("SELECT * FROM `%s`", TABLE_PUBLISHER);
      break;
  }
  $results = [];
  $statement = $pdo->prepare($sql);

  $inputParameters = $by === null ? [] : [$query];
  if($statement->execute($inputParameters))
    { $results = $statement->fetchAll(PDO::FETCH_ASSOC); }
  $statement->closeCursor();
  return $results;
}
?>
```

Figure 6.5 Accessing a database table — PHP environment. (Continued)

```
<!DOCTYPE html> <html>  <head>
      <title>Inquire on Publishers</title>
      <link rel="stylesheet" href="https://maxcdn.bootstrapcdn.com/bootstrap/3.3.5/css/bootstrap.min.css" />
      <style type="text/css"> row.top-buffer { margin-top:25px; }
        label[for=order_by-Code] { margin-right:12px;}
        label[for=order_by-Name], label[for=order_by-Code] { margin-left:4px;}
      </style>
   </head>
   <body>
      <div class="container">
         <div class="row top-buffer">
            <div class="col-md-12">
               <h1>Inquire on Publishers</h1>
            </div>
         </div>
         <div class="row top-buffer">
            <div class="col-md-12">
               <label for="starting_with">Starting With:</label>
            </div>
         </div>
         <div class="row">
            <form action="" enctype="application/x-www-form-urlencoded" method="get">
               <div class="col-md-4">
                  <input type="text" name="starting_with" id="starting_with" value="<?= htmlentities($queryStartingWith); ?>" />
               </div>
               <div class="col-md-4">
                  <input type="radio" name="order_by" id="order_by-Code" value="<?= QUERY_BY_CODE; ?>" <?= $queryBy ==
                              QUERY_BY_CODE ?
                              'checked=checked' : ''; ?> /><label for="order_by-Code">ByCode</label>
                  <input type="radio" name="order_by" id="order_by-Name" value="<?= QUERY_BY_NAME; ?>" <?= $queryBy ==
                              QUERY_BY_NAME ? 'checked=checked' : ''; ?> /><label for="order_by-Name">ByName</label>
               </div>
               <div class="col-md-4">
                  <input type="submit" name="submit" id="submit" value="Search" />
               </div>
            </form>
         </div>
         <div class="row top-buffer">
          <div class="col-md-12">
            <table class="table table-bordered table-hover">
              <thead>
                <tr>
                  <?php foreach(get_table_col_names(TABLE_PUBLISHER, $pdo) as $tableColumnName): ?>
                    <th><?= $tableColumnName; ?></th>
                  <?php endforeach; ?>
                </tr>
              </thead>
              <tbody>
                <?php foreach(get_publishers($pdo, $queryBy, $queryStartingWith) as $publisherRow): ?>
                  <tr>
                    <td><?= $publisherRow['code']; ?></td>
                    <td><?= $publisherRow['name']; ?></td>
                    <td><?= $publisherRow['contact_person']; ?></td>
                    <td><?= $publisherRow['address_line_1']; ?></td>
                  </tr>
                <?php endforeach; ?>
              </tbody>
            </table>
          </div>
        </div>
      </div>
   </body>
</html>
```

Figure 6.5 (Continued) Accessing a database table — PHP environment.

6.5 Summary and Concluding Remarks

Here is a summary of what we have covered in this chapter:

- The user interface for a database system should provide the end users with a user-friendly, controlled environment that gives them all the privileges and functionalities that they need and nothing more.
- In planning the user interface, you must first decide what type of interface will be provided. The interface may be command-based, menu-driven, or GUI-based.
- Next, you should design the user interface using established principles of user interface design.
- The final step is the development and implementation of the user interface. Features that the user interface should provide will influence the tools used in developing the user interface. For instance, if the database is to be accessible from the WWW, then the tools used must facilitate such capability.

So now you know how to design a database and a user interface for that database. Later in the course, you will learn Structured Query Language (SQL), the universal database language. In preparation for this, the next two chapters discuss the relational algebra and relational calculus, respectively — two subject areas that will enhance your appreciation of SQL.

6.6 Review Questions

1. What are the types of user interfaces that may be constructed for end users of a database system? How do they compare in terms of efficiency and convenience?
2. Outline the steps to be taken in the construction of a user interface for end users of a database system.
3. Referring to revision Question 2 of the previous chapter, propose a user interface model for any (or all) of the following:
   - An Inventory Management System
   - A Lab Scheduling System for users of a computer lab
   - A Bookshop Manager System for an educational institution
4. Identify four different scenarios for development of a user interface for a database system. Recommend an appropriate software development tool for each scenario.

References and/or Recommend Readings

Carroll, John. 2002. *Human-Computer Interaction in the New Millennium*. Reading, MA: Addison-Wesley.

Foster, Elvis C. 2021. *Software Engineering: A Methodical Approach* 2nd ed. New York: CRC Publishing. See chapters 8, 14, and 15.

Object Management Group. 2015. "OMG® Specifications." Accessed August 14, 2015. http://www.corba.org/

Oracle Corporation. 2015. "Java." Accessed June 19, 2015. http://www.oracle.com/technetwork/java/index.html

Raskin, Jef. 2000. *The Human Interface: New Directions for Designing Interactive Systems*. Reading, MA: Addison-Wesley.

Schach, Stephen R. 2011. *Object-Oriented and Classical Software Engineering* 8th ed. Boston, MA: McGraw-Hill. See chapters 3 & 14.

Shneiderman, Ben, Catherine Plaisant, Maxine Cohen, & Steven Jacobs. 2009. *Designing the User Interface: Strategies for Effective Human-Computer Interaction* 5th ed. Reading, MA: Addison-Wesley.

Sommerville, Ian. 2007. *Software Engineering* 8th ed. Reading, MA, Addison-Wesley. See chapter 16.

Chapter 7

Relational Algebra

One of the reasons for the success and longevity of the relational model is that it is firmly grounded on mathematical principles (of linear algebra and set theory), which are well known and well documented. The (normalized) relational tables are by design organized in such a way as to promote and facilitate manipulation of data to yield meaningful information. As mentioned earlier in the course (Chapter 2), data manipulation relates to the addition, update, deletion, retrieval, reorganization, and aggregation of data. It turns out that the first three aspects are far more straightforward than the latter three aspects.

In this and the next chapter, the focus is on the latter three aspects of data manipulation — retrieval, reorganization, and aggregation. The intent is to expose you to the underlying database theory that had to be developed in order to have DBMS suites that support these data manipulation requirements. As you will soon come to appreciate, mastery of SQL (which is mandatory, if you intend to do well in this field) or any database language is greatly enhanced by an understanding of the fundamental underlying theory. The chapter proceeds under the following captions:

- Overview
- Basic Operations of Relational Algebra
- Syntax of Relational Algebra
- Aliases, Renaming, and the Relational Assignment
- Other Operations
- Summary and Concluding Rema

7.1 Overview

Relational algebra consists of a collection of operations on relations. Each operation produces new relation(s) from one or more already existing relation(s). Through relational algebra, we can achieve the following objectives:

- Defining the scope for data retrieval
- Defining the scope for data update

DOI: 10.1201/9781003275725-9

- Defining virtual information
- Defining constraints related to data access and/or integrity

Earlier in the course, it was mentioned that every DBMS has a data sublanguage (DSL). Typically, the DSL is based on relational algebra, relational calculus (to be discussed in the next chapter), or a combination of both. It is therefore imperative for you to have a good grasp of this topic.

The relational algebra is said to be prescriptive — you specify precisely how an activity is to be carried out. In so doing, there are certain operations that you will learn to use. These operations will be discussed in the upcoming sections.

7.2 Basic Operations of Relational Algebra

There are eight basic relational algebra operations with which you need to become familiar. Each operation acts on one or more data sets (relation(s)) to produce a new resultant data set (relation). The operations are informally described below:

1. **Restriction (RESTRICT):** A restriction (also called selection) extracts specified tuple(s) from a specified relation R1 by imposing some condition on the relation. The resulting relation R2 has only the specified tuples. RESTRICT is replaced by SELECT in modern systems, the latter usually being more powerful than the original. From here on, we will stick to the term *selection* instead of restriction.
2. **Projection (PROJECT):** A projection extracts specified attributes from a specified relation R1 into a new relation R2.
3. **Join (JOIN):** The join of two relations R1 and R2 results in a new relation R3, such that the tuple(s) from R1 and R2 satisfy some specified conditions. However, in the unqualified form, it is generally used to mean *natural join*. A more generic form of the JOIN operation is discussed in Section 7.3.5.

   The *natural join* (implied by the notation R1 JOIN R2) assumes that there is an attribute or a combination of attributes that are common to both R1 and R2 (same attribute name and characteristics). Let us denote the common attribute(s) in R1 and R2 as Z. The natural join creates a relation R3, where every tuple from R1 is concatenated with every tuple from R2, provided that R1.Z = R2.Z. Finally, a projection is performed on the result in the previous step to yield a single copy of each attribute (thus removing either R1.Z or R2.Z) in the result set.
4. **Product (PRODUCT):** The Cartesian product of two relations R1 and R2 is a third relation R3, consisting of the concatenation of every tuple in R1 with every tuple R2.
5. **Intersection (INTERSECT):** The intersection yields a relation R3, from two specified relations R1 and R2, where R3 has tuple(s) that exist in both R1 and R2. Corresponding attributes of R1 and R2 must be defined on the same domain.
6. **Union (UNION):** The union of relations R1 and R2 results in a relation R3 consisting of tuple(s) in either R1 or R2 inclusive. The corresponding attributes of R1 and R2 must be defined on the same domain.
7. **Difference (MINUS):** The difference of relation R1 and relation R2 is a new relation R3, with tuple(s) appearing in R1, but not in R2. The corresponding attributes from R1 and R2 must be defined on the same domain.

8. **Division (DIVIDE BY):** Division takes a relation R1 of degree $m+n$ and a second relation R2 of degree n and produces a third relation R3 of degree m, where all the n attribute values appearing in R2 also appear in R1, concatenated with the other m attribute values.

It should be clear from the foregoing definitions that each relational algebra operation results in the creation of a virtual relation. A virtual relation means that no physical database object is created; rather, a logical interpretation of existing data is created. As you will see later in the course when we discuss SQL, this logical interpretation can be put to good use in servicing the needs of end-users.

7.2.1 Primary and Secondary Operations

The operations may be classified into *primary operations* and *secondary operations*: Primary operations include UNION, DIFFERENCE, SELECT, PROJECT, and PRODUCT. Secondary operations include JOIN, INTERSECT, and DIVIDE BY. Secondary operations are derivable from the primary operations.

7.2.2 Codd's Original Classification of Operations

Codd's original classifications placed the operations in two broad categories as follows: *Traditional operations* included UNION, INTERSECT, DIFFERENCE, and PRODUCT. *Special operations* included RESTRICT, PROJECT, JOIN, and DIVIDEBY.

7.2.3 Nested Operations

The output of each operation is another relation; likewise, the input to each operation is a set of one or more relations. We say that relational algebra is *closed on relational database model.* Since each operation produces a new relation as output, we can nest operations so that the output of one operation forms the input to another. This will become clear in the ensuing examples.

7.3 Syntax of Relational Algebra

Let us take a closer look at a workable syntax for relational algebra operations. In order to do so, we will set up a database scenario that will serve for the remainder of the chapter, as well as through sections of upcoming chapters. First, we'll set up the database scenario, followed by a convention for specifying syntax.

7.3.1 Database Scenario

We shall use as a frame of reference for this and the next few chapters, a section of a college database, as defined in Figure 7.1, which provides an ERD of the database cross section. Figure 7.2 provides a corresponding relations–attribute list for the said database cross section. Notice that Figure 7.2 covers everything represented in Figure 7.1 and also includes additional information about the database cross section.

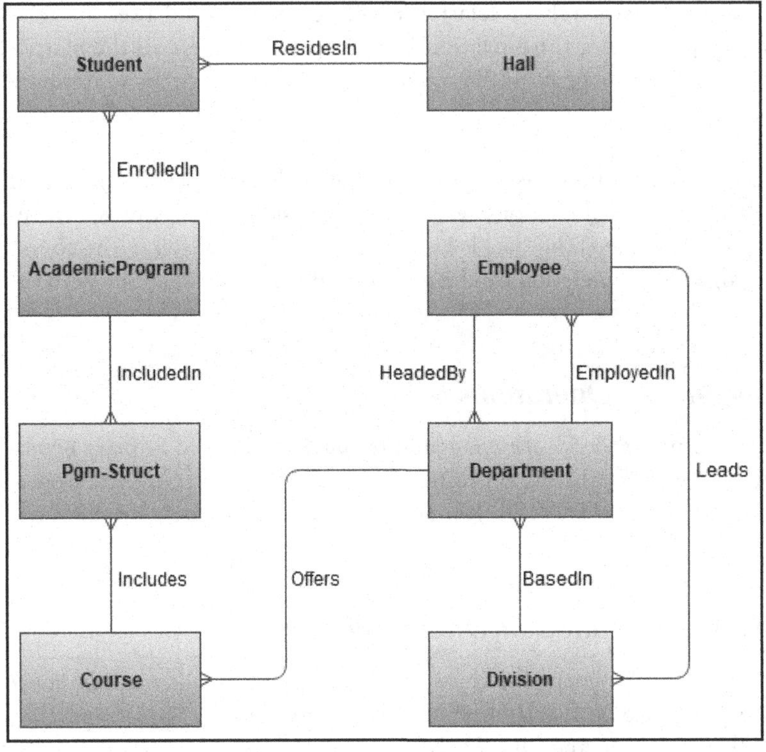

Figure 7.1 ERD for cross section of a college database.

Relation	Attributes	Primary Key	Foreign Key
Student	{Stud#, StudFName, StudLName, StudSex, StudAddr, StudPgm#, StudHall#, StudDoB ...}	[Stud#]	StudPgm# references **AcademicProgram.Pgm#** StudHall# references **Hall.Hall#**
AcademicProgram	{Pgm#, PgmName ...}	[Pgm#]	None
Hall	{Hall#, HallName ...}	[Hall#]	None
Department	{Dept#, DeptName, DeptHead#, DeptDiv#}	[Dept#]	DeptHead# references **Employee.Emp#** DeptDiv# references **Division.Div#**
Employee	{Emp#, EmpName, EmpDept# ...}	[Emp#]	EmpDept# references **Department.Dept#**
Course	{Crs#, CrsName, CrsDept# ...}	[Crs#]	None
Pgm_Struct	{PSPgm#, PSCrs#, PSCrsSeqn}	[PSPgm#, PSCrs#]	PSPgm# references **AcademicProgram.Pgm#** PSCrs# references **Course.Crs#**
Division	{Div#, DivName, DivHead# ...}	[Div#]	DivHead# references **Employee.Emp#**

Each relation and each attribute would need additional clarification prior to database construction and table creation.

Figure 7.2 Relations–attributes list for cross section of a college database.

Some sample data is provided in Figure 7.3 (showing only attributes relevant to the examples discussed). While the data sample is very small, upon close examination, you should note a number of important features of the sample:

1. Although contrived for the purpose of illustration, the data underscores the importance of primary keys, foreign keys, and data integrity.

Student:

Stud#	StudLName	StudFName	StudPgm#	StudHall#	StudDoB
100	Foster	Bruce	BSC1	Chan	1988
105	Jones	Bruce	BSC2	Chan	1982
110	James	Enos	BSC1	Urv	1992
115	James	Yvonne	BSC1	Mars	1994
120	Douglas	Henry	BSC2	Urv	1996
125	Henry	Suzanne	BSC2	Urv	1994
130	Lambert	Cecille	BSC5	Mars	1992
...					

AcademicProgram:

Pgm#	PgmName
BSC1	Bachelor of Science in MIS
BSC2	Bachelor of Science in Computer Science
BSC3	Bachelor of Science in Electronic Engineering
BSC4	Bachelor of Science in Mathematics
BSC5	Bachelor of Science in Computer Science & Mathematics
BSC6	Bachelor of Science in Computer Science & Electronics
BSC7	Bachelor of Science in Chemistry
BSC8	Bachelor of Science in Physics
...	

Department:

Dept#	DeptName	DeptHead#	DeptDiv#
MTH	Department of Mathematics	S10	D01
CSC	Department of Computer Science	S15	D01
PHY	Department of Physics	S05	D01
MGT	Department of Management Studies	S20	D01
MSC	Department of Music	S30	D02
...			

Course:

Crs#	CrsName
CS100	Introduction to Computer Science
CS210	Data Structures
CS220	Visual Programming
CS330	Software Engineering
CS360	Database Systems
...	
M100	Calculus I
M110	Mechanics
M200	Calculus II
M210	Linear Algebra
...	

Figure 7.3 Sample data for a cross section of a college/university database. (Continued)

Hall:

Hall#	HallName
Chan	Chancellor Hall
Len	Lenheim Hall
Mars	Mary Seacole Hall
Urv	Urvine Hall
...	

Employee:

Emp#	EmpName	EmpDept#
S05	Prof. Christine Farr	MSC
S10	Dr. Paul Phillips	MSC
S15	Dr. Scott Foster	CSC
S20	Prof. Hans Gaur	CSC
S25	Dr. Bruce Lambert	MTH
S30	Dr. Carolyn Henry	MTH
S35	Dr. Enid Armstrong	PHY
S40	Dr. Calvin Golding	PHY
...		

Pgm_Struct:

PSPgm#	PSCrs#	PSCrsSeqn
BSC1	M100	01
BSC1	M200	02
BSC1	CS100	03
BSC1	CS210	04
BSC1	CS220	05
...		
BSC2	CS100	01
BSC2	CS210	02
BSC2	CS220	03
...		
BSC2	M100	16
...		

Division:

Div#	DivName	DivHead#
D01	Division of Pure & Applied Sciences	S25
D02	Division of Arts & Humanities	S30
D03	Division of Education & Psychology	S35
...		

Figure 7.3 (Continued) Sample data for a cross section of a college/university database.

2. Both the entity integrity rule and the referential integrity rule of Chapter 4 (Section 4.1) are exemplified in the data. For instance, notice that all the foreign keys in referencing relations (for example, **Department.DeptHead#**) point to tuples in the related referenced relations (for example **Employee.Emp#**).
3. The **Pgm_Struct** relation represents the intersection relation that implements the M:M relationship between relations **AcademicProgram** and **Course** (review Section 3.7). As such and as the data embodies, relation **Pgm_Struct** forms an M:1 relationship with the relation **AcademicProgram** on one side and the relation **Course** on the other.

7.3.2 Syntax Convention

In the interest of clarity, two commonly used notations have been adopted, both with slight modifications. They are the Ullman notation (Garcia-Molina 2009) and the Date notation (Date 2004). Bear in mind, however, that the syntactical implementation of relational algebra will vary from one product to another. What is important therefore is that you understand the concepts discussed. The BNF (Backus–Naur Form) conventions shown in Table 7.1 will be assumed. Additionally, several of the examples will contain clarifying comments that conform to how comments are made in the C-based programming languages (e.g., // this is a comment, /* and so is this */). Use of the BNF notation and C-based comments will continue through Chapter 8 as well.

Table 7.1 Summary of the BNF Notations

Symbol	*Meaning*
::=	This symbol means "is defined as"
[...]	Denotes optional content; optional items are enclosed in square brackets
<Element>	Denotes that the content is supplied by the user and/or is non-terminal
|	Indicates choice (either or). Example: ItemA | ItemB | ItemC
{<Element>}	Denotes zero or more repetitions. Other notations have been proposed but for this course, we will stick to this traditional notation.
<Element>*	Alternate notation to denote zero or more repetitions
<l>*<m><Element>	Denotes l to m repetitions of the specified element

7.3.3 Select Statement

Based on the Ullman notation, the **SELECT** statement is of the following form:

SelectStatement ::=
SELECT <Condition> (<RelationalExpression>)

Condition ::=
[NOT] <Attribute1> **Theta** <Attribute2 | Literal> {AND | OR <Condition>}

RelationalExpression ::=
<RelationName> | <SelectStatement> | <ProjectStatement> | <ProductStatement> | <ThetaJoinStatement> | <UnionStatement> | <IntersectStatement> | <MinusStatement> | <DivisionStatement>

Based on this notation, a select statement is comprised of the verb SELECT, a *condition*, and a *relational expression*. Please note the following clarifications:

1. *Theta* represents a valid Boolean operator, i.e., any of the following: =, >, <, <>, <=, >=
2. A *relational expression* is either a relation or an expression that results in the creation of a virtual relation.
3. A *condition* is an expression that evaluates to true or false. The condition specified is typically used as a filter; tuples of interest must meet the stipulated condition. Complex conditions may be built by using brackets and connectors AND, OR, NOT as in normal Boolean logic.

Let us consider an example: Suppose that we wanted to select all students enrolled in program B.Sc. in MIS. Based on the sample data shown in Figure 7.3, the relational algebra statement (based on the Ullman notation) that would be required is shown in Example 7.1a:

EXAMPLE 7.1A: SELECTING STUDENTS ENROLLED IN B.SC. IN MIS

SELECT StudPgm# = 'BSC1' (Student)

An alternate syntax of the **SELECT** statement as described in Date (2004) specifies the relational expression ahead of the condition as follows:

Select ::=
<RelationalExpression> [WHERE <Condition>]

Condition ::=
[NOT] <Attribute1> **Theta** <Attribute2 | Literal> {AND | OR <Condition>}

RelationalExpression ::=
<RelationName> | <SelectStatement> | <ProjectStatement> | <ProductStatement> | <ThetaJoinStatement> | <UnionStatement> | <IntersectStatement> | <MinusStatement> | <DivisionStatement>

Based on this format, a select statement begins with a relational expression and is optionally followed by the keyword WHERE and a condition. As you will soon see, this syntax is much closer to the actual implementation in SQL and is easy to understand. To revisit the previous example, Example 7.1b shows the relational algebra statement (based on the Date notation) to select all students enrolled in program B.Sc. in MIS:

EXAMPLE 7.1B: SELECTING STUDENTS ENROLLED IN B.SC. IN MIS

Student WHERE StudPgm# = 'BSC1'

7.3.4 Projection Statement

Based on the Ullman notation, the **PROJECT** statement starts with the verb PROJ, followed by a list of attributes of interest, and the relational expression from which the attributes are drawn:

Project ::=

PROJ <Attribute> {,<Attribute>} (<RelationalExpression>)

Attribute ::=

<AttributeName> | <Relation.AttributeName>

Based on this notation, Example 7.2a shows two projections on the **Student** relation. Each statement is preceded by a clarifying comment.

EXAMPLE 7.2A: TWO PROJECTIONS ON THE STUDENT RELATION

// To extract attributes **Stud#** and **StudPgm#** only from relation **Student:**

PROJ Stud#, StudPgm# (Student)

// To extract attributes **Stud#** and **StudPgm#** only from the selection of students enrolled in B.Sc. in MIS:

PROJ Stud#, StudPgm# (SELECT StudPgm# = 'BSC1' (Student))

// Note: Do not mix the notations like this (Ullman-Date):

PROJ Stud#, StudPgm# (Student WHERE StudPgm# = 'BSC1')

Based on the Date notation, an alternate syntax for the **PROJECT** statement that is much closer to the SQL implementation is as follows:

Project ::=

<Attribute> {,<Attribute>} FROM <RelationalExpression>

Attribute ::=

<AttributeName> | <Relation.AttributeName>

Based on this definition, Example 7.2b shows alternate statements for the projections of Example 7.2a. Again note the similarity between the two notations but also remember that SQL more closely resembles the Date notation.

EXAMPLE 7.2B: TWO ALTERNATE PROJECTIONS ON THE STUDENT RELATION

// To extract attributes Stud#, StudPgm# only from relation Student:
Stud#, StudPgm# FROM Student

// To extract attributes Stud#, StudPgm# only from the selection of students enrolled in B.Sc. in MIS:
Stud#, StudPgm# FROM Student WHERE StudPgm# = 'BSC1'

// Note: Do not mix the notations like this (Date-Ullman):
Stud#, StudPgm# FROM (SELECT StudPgm# = 'BSC1' (Student))

7.3.5 Natural Join Statement

The Ullman notation and Date notation essentially agree on the format of the natural join. The natural **join** statement specifies a relational expression, the verb JOIN, and another relational expression:

JoinStatement ::=
<RelationalExpression> JOIN <RelationalExpression> {JOIN <RelationalExpression>}

Recall from Chapter 4 that joins are very important in a relational database. In fact, normalization theory is built on the presumption that different relations will be joined to derive meaningful information whenever it is convenient to do so. Here are a few clarifying points about joins:

1. The join operation presumes that there is a relation (say R1) containing a foreign key that references a second relation (say R2). The natural join presumes further that the attribute playing the role of foreign key (in say R1) has the identical name as another attribute that fulfills the role of primary key in the referenced relation (say R2). Of course, both foreign key and primary key must be defined on the same domain.
2. Since each relational expression participating in a join may itself be a join-expression, what we really have is a recursive definition that may result in multiple joins.
3. In more sophisticated environments (for example, DB2 and Oracle), the attribute playing the role of foreign key could have a different name from the referenced attribute, so long as its characteristic features are the same (as mentioned in previous chapters, it is good design practice to use distinct attribute names across different relations in the database). In such cases, an *equijoin* (to be clarified shortly) would normally take the place of a natural join.

4. The DBMs traverses both relations and joins on matching keys, i.e., the foreign key in the first relation is used to reference the primary key in the second relation.
5. The natural join is commutative (Rl JOIN R2 produces the same result as R2 JOIN Rl). Moreover, in specifying it, it is not necessary to identify the participating attributes, since they would have been identified when the foreign key constraint is defined.
6. The mathematical symbol used for the join is $\bowtie$. However, this course will stick to the convention of using the more descriptive word, JOIN.

Revisit Figures 7.1, 7.2, and 7.3 and observe that the foreign key **Student.StudHall#** corresponds to the primary key **Hall.Halll#**. Now suppose that we had a snapshot of the **Student** relation as follows:

StudentCopy {Stud#, StudFName, StudLName, Hall#} PK [Stud#]

With this snapshot in place, we may create a natural join between **StudentCopy** and **Hall** as illustrated in Example 7.3.

EXAMPLE 7.3: EXAMINING NATURAL JOIN

StudentCopy JOIN Hall

/* This statement produces a relation with each tuple of the relation **StudentCopy** concatenated with a given tuple of the relation **Hall**. The concatenated tuples from each of the two input relations (**StudentCopy** and **Hall**) agree on the attribute value of **Hall#**. The final result therefore eliminates one of the two columns **StudentCopy.Hall#** and **Hall.Hall#**, since their values are identical. */

Following on, let us examine a more specific example: Suppose that it is desirable to obtain a list showing the names of students (**StudLName**) and their corresponding hall names (**HallName**). Here, we need to take the natural join of **StudentCopy** with **Hall**, and then make a projection on the attributes of interest. The solution is shown in Example 7.4. Notice that the projection is written first. In other words, the projection is on the natural join.

EXAMPLE 7.4: PROJECTION ON A NATURAL JOIN

/* We first must link **StudLName** from relation **StudentCopy** to **HallName** from relation **Hall**, via a natural join. Then project on the required attributes: */

// Solution based on Ullman notation
PROJ StudLName, HallName (StudentCopy JOIN Hall)
// Solution based on Date notation
StudLName, HallName FROM StudentCopy JOIN Hall

7.3.6 Cartesian Product

Both the Ullman notation and the Date notation essentially agree on the format of the Cartesian (or cross) product. The **PRODUCT** statement is of the following form:

ProductStatement ::=
<RelationalExpression> TIMES <RelationalExpression>

The resulting virtual relation has all tuples of the first relation concatenated with all tuples of the second relation. The Cartesian product by itself is of limited usefulness, since it produces an avalanche of spurious (i.e., phony) tuples. However, when combined with a selection and/or projection, the operation is extremely useful.

Consider the scenario where it is desirous to obtain combinations of department name and name of the related department head. Looking closely at the sample database specification, we need to pull data from two different relations, namely **Department** and **Employee**. How do we pull that off? Well, notice further that the foreign key **Department.DeptHead#** is the essential link since it references **Employee.Emp#**. One possible solution, therefore, is to take the cross product of **Department** and **Employee**, select out the tuples that obey the condition **Department.DeptHead# = Employee.Emp#**, and then project on the attributes of interest. This solution is shown in Example 7.5a. A more elegant solution involves taking an equijoin, followed by a projection. This second solution (Example 7.5b) is clarified in the next sub-section.

EXAMPLE 7.5A: OBTAINING DEPARTMENT NAMES AND THEIR RESPECTIVE HEADS

Solution 1: Cross product followed by a selection and a projection:

// Solution based on Ullman notation
PROJ DeptName, EmpName (SELECT DeptHead# = Emp# (Department TIMES Employee))
// Based on Date notation
DeptName, EmpName FROM (Department TIMES Employee) WHERE DeptHead# = Emp#

Solution 2: Specifies an equijoin, followed by a projection. See Example 7.5b ahead.

7.3.7 Theta-Join

The *Theta-join* is presented in some texts as the general case of the **JOIN** operation. The term *theta* is used here as a euphemism to mean *general*, i.e., the join condition can involve the use of any relational operator. The BNF form for the theta-join based on the Ullman notation is as follows:

ThetaJoinStatement ::=
SELECT <Condition> (<RelationalExpression> TIMES <RelationalExpression>)

Condition ::=
[NOT] <Attribute1> **Theta** <Attribute2 | Literal> {AND | OR <Condition>}

Notice that this specification clearly conveys the principle that a join is obtained by imposing a selection on a cross product. Theta here represents any valid relational operator. The BNF form for the theta-join based on the Date notation also conveys this principle:

ThetaJoinStatement ::=
<RelationalExpression> TIMES <RelationalExpression> WHERE <Condition>

Condition ::=
[NOT] <Attribute1> **Theta** <Attribute2 | Literal> {AND | OR <Condition>}

An alternate notation for representing this has been described in Russell (2006) and is particularly clear; it is paraphrased below:

ThetaJoinStatement ::=
<RelationalExpression> JOIN (<Condition>) <RelationalExpression>

Condition ::=
[NOT] <Attribute1> **Theta** <Attribute2 | Literal> {AND | OR <Condition>}

Incidentally, this latter notation is really a simplification of the Ullman notation, and is fully reconcilable with either the Date or Ulman notation, bringing additional clarity to the specification of the theta-join. This is illustrated in Example 7.5b. Moreover, as you will see in Chapter 12, the ANSI version of the SQL join is comparable to this notation; for convenience (and perhaps want of a better characterization), we will therefore refer to it as the ANSI join notation.

Whatever the notation, the result of the theta-join is a set of tuples from the cross product of the first relation with the second relation that satisfy the stated join condition. Moreover, with this definition, we can bring further clarity to the equijoin and the natural join as follows:

- The equijoin is simply the specific case where **theta** (the operator) is the equal operation. It is applicable in situations where matching attributes do not have identical names; typically, one attribute (or combination of attributes) constitutes a foreign key in one relation, while the other attribute (or combination of attributes) constitutes a candidate key in the other relation.

- The natural join is a special case of the equijoin, where the attributes compared from both relations are not just defined on the same domain but have the same name. Further, the natural join will eliminate one of the two (sets of) identical attributes from the final result-set.

Let us revisit the problem in Example 7.5a — constructing a list showing combinations of department name and name of the related department head for each department. This second solution involves an equijoin and is shown in Example 7.5b.

EXAMPLE 7.5B: OBTAINING DEPARTMENT NAMES AND THEIR RESPECTIVE HEADS

Solution 1: Cross product followed by a selection and a projection:

// Based on Ullman notation:
PROJ DeptName, EmpName (SELECT DeptHead# = Emp# (Department TIMES Employee))

// Based on Date notation:
DeptName, EmpName FROM (Department TIMES Employee) WHERE DeptHead# = Emp#

Solution 2: Specifies an equijoin, followed by a projection:

// Based on the Ullman notation combined with the ANSI notation:
PROJ DeptName, EmpName (Department JOIN (DeptHead# = Emp#) Employee))

// Based on the Date notation combined with the ANSI notation:
DeptName, EmpName FROM Department JOIN (DeptHead# = Emp#) Employee

It is imperative that you understand the difference between a cross product and a join, and you know when to use each operation. Observe the following:

1. Examples 7.5a and 7.5b illustrate that a natural join or equijoin is equivalent to a Cartesian product followed by a selection. Proof of this principle is not necessary for this course. However, the principle can be easily illustrated as in Figure 7.4.
2. If there is at least one pair of matching keys in both relations, a natural join (if matching attribute names are identical) or an equijoin (if matching attribute names have different names) is preferred to a cross product. The match must be between a foreign key (in the referencing relation) and a primary key (in the referenced relation).
3. The Cartesian product, as defined in the relational model, is associative. However, its mathematical counterpart (from which it is drawn) is not. The reason for this is partly due to the fact that in the definition of a relation, no emphasis is placed on the order of the attributes (review Section 3.3.1)

Department		
Dept#	**DeptName**	**DeptHead#**
D1	Mathematics	S1
D2	Computer Science	S2
D3	Music	S3

Employee	
Emp#	**EmpName**
S1	Bruce
S2	Henry
S3	Jacobson

If Department has d tuples and Employee has s tuples, the cross product of the two relations produces d * s tuples, many of which will be spurious.

Department TIMES Employee				
Dept#	**DeptName**	**DeptHead#**	**Emp#**	**EmpName**
D1	Mathematics	S1	S1	Bruce
D1	Mathematics	S1	S2	Henry
D1	Mathematics	S1	S3	Jacobson
D2	Computer Science	S2	S1	Bruce
D2	Computer Science	S2	S2	Henry
D2	Computer Science	S2	S3	Jacobson
D3	Music	S3	S1	Bruce
D3	Music	S3	S2	Henry
D3	Music	S3	S3	Jacobson

Department TIMES Employee WHERE DeptHead# = Emp# is equivalent to Department JOIN (DeptHead# = Emp#) Employee				
Dept#	**DeptName**	**DeptHead#**	**Emp#**	**EmpName**
D1	Mathematics	S1	S1	Bruce
D2	Computer Science	S2	S2	Henry
D3	Music	S3	S3	Jacobson

Figure 7.4 Illustrating that a Join is obtained by a cross product followed by a selection.

7.3.8 Inner and Outer Joins

The foregoing discussion and examples on natural join and theta join are based on the concept of an *inner join*. Essentially, an inner join is a join that is solely based on the specified join condition; by default, most joins are actually inner joins unless otherwise specified. However, what if there are tuples in either relation that do not satisfy the join condition, but are required in the result? The *outer join* is suited for these situations; it provides a way of including tuples that strictly do not satisfy the join condition to be included in the result set. There are three possibilities to be considered:

- The left outer join includes all the tuples from the relation on the left. Tuples that do not have a match based on the join condition are listed with corresponding attribute values on the right being null.
- The right outer join includes all the tuples from the relation on the right. Tuples that do not have a match based on the join condition are listed with corresponding attribute values on the left being null.
- The full outer join takes all tuples from each relation with nulls taking the place of unmatched attributes from either relation. In essence, the result of the full outer join is the union of the left outer join and the right outer join.

To illustrate outer joins, revisit Figure 7.3 of sample data for the cross section of a college database. Now suppose that it is desirable to obtain the names of all students and their related programs of study. This requires a join of relations **Student** and **AcademicProgram**, with attribute **Student.StudPgm#** acting as the foreign key that references **AcademicProgram.Pgm#**. Suppose further that there are students who have not yet declared their academic majors; for these students, **Student.StudPgm#** would be null. Figure 7.5 illustrates these considerations. Notice from the figure that the **Student** relation contains two student tuples (surnames Murray and Scott) that have not been assigned to academic programs. The figure shows the results from the following

Student:					**AcademicProgram:**	
Stud#	**StudLName**	**StudFName**	**StudPgm#**	**StudHall#**	**Pgm#**	**PgmName**
100	Foster	Bruce	BSC1	Chan	BSC1	Bachelor of Science in MIS
105	Jones	Bruce	BSC2	Chan	BSC2	Bachelor of Science in Computer Science
110	James	Enos	BSC1	Urv	BSC3	Bachelor of Science in Electronics
115	James	Yvonne	BSC1	Mars	BSC4	Bachelor of Science in Mathematics
120	Douglas	Henry	BSC2	Urv	BSC5	Bachelor of Science in CS & Mathematics
125	Henry	Suzanne	BSC2	Urv	BSC6	Bachelor of Science in CS & Electronics
130	Lambert	Cecille	BSC5	Mars	BSC7	Bachelor of Science in Chemistry
140	Murray	Douglas		Chan	BSC8	Bachelor of Science in Physics
150	Scott	Cecille		Mars		
...						

Student JOIN (StudPgm# = Pgm#) AcademicProgram: // This is the inner join

Stud#	StudLName	StudFName	StudPgm#	Pgm#	PgmName	StudHall#
100	Foster	Bruce	BSC1	BSC1	Bachelor of Science in MIS	Chan
105	Jones	Bruce	BSC2	BSC2	Bachelor of Science in Computer Science	Chan
110	James	Enos	BSC1	BSC1	Bachelor of Science in MIS	Urv
115	James	Yvonne	BSC1	BSC1	Bachelor of Science in MIS	Mars
120	Douglas	Henry	BSC2	BSC2	Bachelor of Science in Computer Science	Urv
125	Henry	Suzanne	BSC2	BSC2	Bachelor of Science in Computer Science	Urv
130	Lambert	Cecille	BSC5	BSC5	Bachelor of Science in CS & Mathematics	Mars
...						

Student LEFT JOIN (StudPgm# = Pgm#) AcademicProgram: // This is the left outer join

Stud#	StudLName	StudFName	StudPgm#	Pgm#	PgmName	StudHall#
100	Foster	Bruce	BSC1	BSC1	Bachelor of Science in MIS	Chan
105	Jones	Bruce	BSC2	BSC2	Bachelor of Science in Computer Science	Chan
110	James	Enos	BSC1	BSC1	Bachelor of Science in MIS	Urv
115	James	Yvonne	BSC1	BSC1	Bachelor of Science in MIS	Mars
120	Douglas	Henry	BSC2	BSC2	Bachelor of Science in Computer Science	Urv
125	Henry	Suzanne	BSC2	BSC2	Bachelor of Science in Computer Science	Urv
130	Lambert	Cecille	BSC5	BSC5	Bachelor of Science in CS & Mathematics	Mars
140	Murray	Douglas				Chan
150	Scott	Cecille				Mars
...						

Student RIGHT JOIN (StudPgm# = Pgm#) AcademicProgram: // This is the right outer join

Stud#	StudLName	StudFName	StudPgm#	Pgm#	PgmName	StudHall#
100	Foster	Bruce	BSC1	BSC1	Bachelor of Science in MIS	Chan
105	Jones	Bruce	BSC2	BSC2	Bachelor of Science in Computer Science	Chan
110	James	Enos	BSC1	BSC1	Bachelor of Science in MIS	Urv
115	James	Yvonne	BSC1	BSC1	Bachelor of Science in MIS	Mars
120	Douglas	Henry	BSC2	BSC2	Bachelor of Science in Computer Science	Urv
125	Henry	Suzanne	BSC2	BSC2	Bachelor of Science in Computer Science	Urv
130	Lambert	Cecille	BSC5	BSC5	Bachelor of Science in CS & Mathematics	Mars
				BSC3	Bachelor of Science in Electronics	
				BSC4	Bachelor of Science in Mathematics	
				BSC6	Bachelor of Science in CS & Electronics	
...						

Figure 7.5 Illustrating inner and outer Joins.

operations: inner equijoin between **Student** and **AcademicProgram**; left outer join between **Student** and **AcademicProgram**; and right outer join between **Student** and **AcademicProgram**. The full outer join is not shown in the figure; however, it is simply the union of the left outer join and right outer join.

7.3.9 UNION, INTERESECT, and MINUS Statements

The **UNION, INTERSECT, and MINUS** statements are similarly represented in the Ullman notation as well as the Date notation. The respective syntax forms are as follows:

<RelationalExpression> UNION <RelationalExpression>
<RelationalExpression> INTERSECT <RelationalExpression>
<RelationalExpression> MINUS <RelationalExpression>

To briefly clarify how these operations work, consider three relations R1, R2, and R3. Following are a few important observations about these operations:

1. These are binary operations and, in each case, both relations must have corresponding attributes that are defined on the same domain. For instance, in order to take the union, intersection, or difference of R1 with respect to R2, both relations must have attributes defined on the same domain. Now think about this for a moment, why would you want to define two different relations with attributes defined on the same domain? Wouldn't such an action be somewhat superfluous? Certainly! In practice, we would not design such base relations in a database. The implication here is that these operations are typically applied to logical relations, not base relations. In other words, they are used in constructing queries; this point will become clear in Chapter 12. Figure 7.6 provides a simple illustration on these operations.

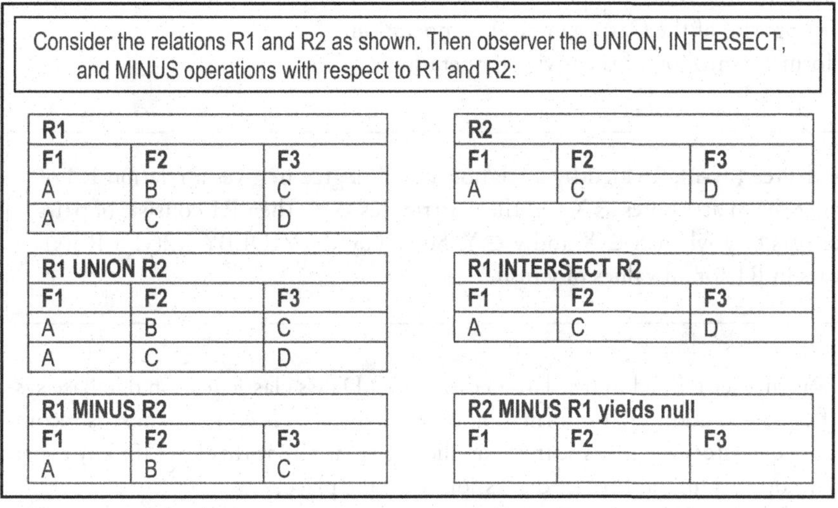

Consider the relations R1 and R2 as shown. Then observer the UNION, INTERSECT, and MINUS operations with respect to R1 and R2:

R1		
F1	F2	F3
A	B	C
A	C	D

R2		
F1	F2	F3
A	C	D

R1 UNION R2		
F1	F2	F3
A	B	C
A	C	D

R1 INTERSECT R2		
F1	F2	F3
A	C	D

R1 MINUS R2		
F1	F2	F3
A	B	C

R2 MINUS R1 yields null		
F1	F2	F3

Figure 7.6 Illustrating union, intersection, and difference.

2. **Union** and **intersect** are associative as well as commutative; **minus** is neither associative nor commutative. To illustrate, the following constructions about R1, R2, and R3 should resonate well with you:
   - R1 UNION R2 UNION R3 ≡ R1 UNION R3 UNION R2 ≡ R2 UNION R1 UNION R3, and so on; the order in which the union is taken is unimportant.
   - R1 INTERSECT R2 INTERSECT R3 ≡ R1 INTERSECT R3 INTERSECT R2 ≡ R2 INTERSECT R1 INTERSECT R3, and so on; the order in which the intersection is taken is unimportant.
   - R1 MINUS R2 ≠ R2 MINUS R1; the order in which the difference is taken is significant.
3. The MINUS and INTERSECT operations are not supported in MySQL. In DB2 and SQL Server, the MINUS operation is implemented as the EXCEPT operation. In Oracle and Informix, the MINUS verb is used.

7.3.10 Division Statement

The **DIVIDE BY** statement is similarly represented in the Ullman notation as well as the Date notation; the format is as follows:

```
R1 DIVIDEBY R2
```

In this format, relation R1 contains [m + n] attributes and relation R2 has n of the [m + n] attributes. The operation of division is as follows:

- Every attribute of R2 must be an attribute of R1.
- The resulting relation will have remaining attributes of R1.
- A tuple occurs in R1 DIVIDEBY R2 if it occurs in R1, concatenated with every tuple in R2.

Figure 7.7 provides two illustrations that should help you gain insight into the division operation. The first demonstrates how the division operation works; the second demonstrates that the division operation is the opposite of the Cartesian product operation.

Here is a more formal definition of the division operation:

> A relation R1 of degree (m+n), divided by a relation R2 of degree n, gives a relation R3 of degree m. Consider the m attributes as X and the n attributes as Y. Then R1 consists of pairs (x,y), and R2 consists of y where x ∈ X and y ∈ Y. So R1(x,y) DIVIDEBY R2(y) = R3(x), where (x,y) appears in R1 for all values of y in R2.

This definition was further refined in (the latest editions of) Date's classic text on database system (Date 2004). If you need additional information on the subject matter, Date's text is recommended. However, since mastery of this is scarcely applicable (if at all) at this level (in any event, division is a secondary operation), no further discussion will take place.

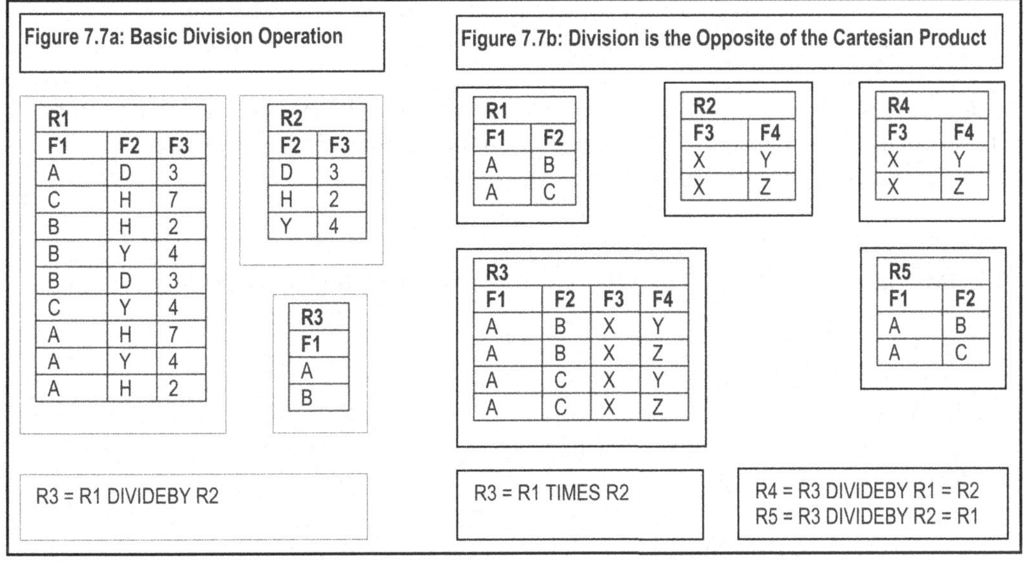

Figure 7.7 Illustrating the division operation.

7.4 Aliases, Renaming, and the Relational Assignment

In addition to the basic operations, three additional (advanced) operations defined on relations are:

- Alias Operation
- Rename Operation
- Assignment Operat

7.4.1 The Alias Operation

Aliasing is giving an alternate name to a relation or relational expression. The format we will use for specifying an alias is as follows:

<Relation> ALIASES <RelationalExpression>

The statement in Example 7.6 creates an alias for the **Department** relation. With the alias in place, we may refer to the relation by its original name or its alias.

EXAMPLE 7.6: ALIASING THE DEPARTMENT RELATION
Dept ALIASES Department

Aliasing is useful in simplifying a data retrieval problem. Three common scenarios are when referring to different tuples in the same relation, when treating one relation as two separate relations, or when it is desirable to use a more convenient relation name in an application. Later in the course, as you learn SQL (in an Oracle environment), you will see that aliases are implemented as *synonyms* in Oracle.

Suppose that we need to find all students who share dates of birth with at least one other student. The solution is obtained by taking the cross product of **Student** with itself and then searching for matches on **StudDoB**. This is shown in Example 7.7.

EXAMPLE 7.7: FINDING STUDENTS WHO SHARE BIRTH DATES

```
// Solution based on Ullman Notation:

Stud2 ALIASES Student
SELECT (Student.Stud# <> Stud2.Stud#) AND (Student.StudDoB = Stud2.StudDoB)
   (Student TIMES Stud2)

// Solution based on Date Notation:

Stud2 ALIASES Student
Student TIMES Stud2 WHERE Student.Stud# <> Stud2.Stud# AND Student.StudDoB
   = Stud2.StudDoB
```

7.4.2 The Assignment Operation

The assignment operator is a colon followed by an equal sign (:=) and is used to have the system remember the results of other operation(s), which may be required for later use. The format of the assignment is:

```
<Relation>: = <RelationalExpression>
```

The example in the upcoming section illustrates how relational assignments are done. Later in the course, you will learn that through various SQL statements (INSERT, UPDATE, and SELECT), you can create a relation from other existing relation(s).

7.4.3 The Rename Operation

The **RENAME** operation renames specified attributes of a specified relation R1 into a new relation R2. We have established that it is desirable to have unique attribute names in each relation. Further, in the case of nested operations, it is a good habit to have an unambiguous way of referring to attributes of the results of an inner expression from an outer expression. The **RENAME** operation is ideal for these situations. The format of the **RENAME** operation (based on Date's notation) is outlined below:

```
<RelationalExpression> RENAME <Attribute> AS <Attribute>
{, <Attribute> AS <Attribute>}
```

As with selection, brackets may be used to avoid ambiguity. Ullman's notation has been omitted here; though slightly different in syntax, its result is similar.

Let us revisit the previous example for a moment. Notice that in taking the cross product of a relation with itself, we are obliged to qualify the attributes in order to avoid ambiguity. One alternative to that strategy is to rename attributes of interest. In Example 7.8, the solution of the previous example is further simplified by application of attribute renaming and the relational assignment operations.

EXAMPLES 7.8: REVISIT FINDING STUDENTS WHO SHARE BIRTH DATES

/* Referring to the previous example, the solution could be revised by renaming selected attributes and using the assignment operator: */

// Solution based on Ullman Notation:

```
Stud2 ALIASES Student
Stud3: = Stud2 RENAME Stud# AS Stud#2, StudDoB AS StudDoB2
SELECT (Stud# <> Stud#2) AND (StudDoB = StudDoB2) (Student TIMES Stud3)
```

// Solution based on Date Notation:

```
Stud2 ALIASES Student
Stud3: = Stud2 RENAME Stud# AS Stud#2, StudDoB AS StudDoB2
Student TIMES Stud3 WHERE Stud# <> Stud#2 AND StudDoB = StudDoB2
```

7.5 Other Operators

Other operations to facilitate computation usually exist. A brief summary of some of the common additional operators is provided here (based on Date's notation). These will be further clarified, once you get exposure to a DSL such as SQL (which we will discuss later in the course).

Extend: The **EXTEND** operator takes a relation as input and returns a replica of it with one additional column (attribute), as defined by the user. The BNF notation for the syntax is as follows:

```
EXTEND <RelationalExpression> ADD <ScalarExpression> AS <NewAttribute>
[WHERE <Condition>]
```

Referring to the sample college database, if it is desirous to list courses in program "BSC1", with an additional descriptive column, then the statement of Example 7.9 would be helpful.

EXAMPLE 7.9: ASCRIBING A DESCRIPTIVE COLUMN VALUE FOR A SPECIFIC SET OF ROWS

```
EXTEND Pgm_Struct ADD 'Bachelor of Science in MIS' AS MyPgm
WHERE PSPgm# = 'BSC1'
```

Aggregate: The *aggregate operators* are used to summarize (aggregate) the values of a column of a relation. The standard operators are COUNT, COUNTD, SUM, SUMD, AVGD, MAX, MIN. COUNT, SUM, and AVG return numeric values; MAX and MIN return alphanumeric or numeric values, depending on the column in question. COUNTD, SUMD, and AVGD avoid duplicate values for that column. Each operator is specified with a column (attribute) as follows:

```
<AggregateOperator> (<Attribute>)
```

The aggregate operators are typically used in a scenario where data is grouped. This is illustrated in the upcoming example.

Grouping: By grouping data from a relation, we can derive summarized information for the purpose of analysis. Grouping and aggregation usually go together. For the purpose of discussion, let us assume that the relational algebra syntax for defining a group is as shown below:

```
<RelationalExpression> GROUPBY <AttributeList>
```

Here, <Attribute List> is simply a list of attributes, with the comma as the separator. Grouping is normally embedded as part of some larger relational algebra expression. With the above notation, we can extract summary information from relations as in the following example.

To illustrate how grouping and aggregation works, take another look at the **Pgm_Struct** relation of the sample college database, and consider the situation where it is desirable to list each academic program and the corresponding number of courses comprising the program. In this example, you want to group the data by academic program and aggregate the number of courses for each group. Example 7.10 shows a relational algebra solution for the problem.

EXAMPLE 7.10: LISTING ACADEMIC PROGRAMS AND THEIR CORRESPONDING NUMBER OF COURSES

// Solution based on Ullman Notation:

```
PROJ PSPgm#, COUNT (Nbr) ((EXTEND Pgm_Struct ADD (1) AS Nbr)
GROUPBY PSPgm#)
```

// Solution based on Date Notation:

```
PSPgm#, COUNT (Nbr) FROM ((EXTEND Pgm_Struct ADD (1) AS Nbr)
GROUPBY PSPgm#)
```

7.6 Summary and Concluding Remarks

This chapter covers the fundamentals of relational algebra as a precursor to understanding SQL, the universal database language. Let us summarize what we have covered in the chapter:

- The relational algebra is said to be prescriptive (also said to be operational) — you specify precisely how an activity is to be carried out.
- The eight basic operations of relational algebra are Union, Difference, Selection, Product, Projection, Join, Intersection, and Division. Each operation follows a specific syntax.
- In addition to the basic operations, other relational operations include Aliasing, Assignment, and Renaming.
- The chapter also makes brief mention of Extend and Aggregation operators.

The next chapter discusses relational calculus as an equivalent alternative to relational algebra. You will soon realize that your knowledge and expertise in one area are complemented by your knowledge and expertise in the other.

7.7 Review Questions

1. Why is relational algebra important?
2. Briefly describe the basic operations of relational algebra.
3. Using the college database described in this chapter, practice writing relational algebra statements that will yield certain desirable results.
4. The following (Figure 7.8) summarizes an abridged specification of tables comprising a music database:
   4.1 This database specification may be refined by introducing a sixth relational table, and adjusting three other tables to each have a foreign key that references this additional table. Identify the additional table that is required, and clearly describe the adjustments that need to be made to three tables in order to have a normalized database.
   4.2 Propose an ESG or RAL that provides specifications for the six relational tables of the music database.
   4.3 Provide some sample data for the music database. Your data should demonstrate that you understand the important role of foreign keys.

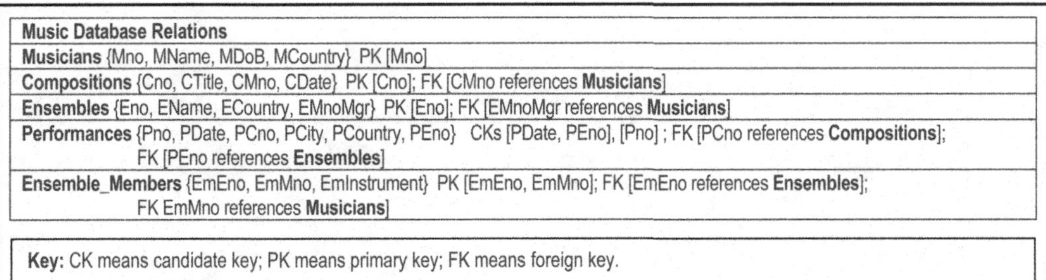

Music Database Relations
Musicians {Mno, MName, MDoB, MCountry} PK [Mno]
Compositions {Cno, CTitle, CMno, CDate} PK [Cno]; FK [CMno references **Musicians**]
Ensembles {Eno, EName, ECountry, EMnoMgr} PK [Eno]; FK [EMnoMgr references **Musicians**]
Performances {Pno, PDate, PCno, PCity, PCountry, PEno} CKs [PDate, PEno], [Pno] ; FK [PCno references **Compositions**]; FK [PEno references **Ensembles**]
Ensemble_Members {EmEno, EmMno, EmInstrument} PK [EmEno, EmMno]; FK [EmEno references **Ensembles**]; FK EmMno references **Musicians**]

Key: CK means candidate key; PK means primary key; FK means foreign key.

Figure 7.8 Preliminary specification for a music database.

5. Based on your response to Question 4, write relational algebra statements to achieve the following objectives:
   5.1 List registered musicians from CUB or CAN (where "CUB" and "CAN" are abbreviated codes for Cuba and Canada, respectively).
   5.2 List each ensemble (code and name) that includes a VIOLIN or GUITAR player.
   5.3 Give the code and name of every ensemble that includes a VIOLIN player but not a GUITAR player.
   5.4 List all compositions (code and title) by musician DAVID FOSTER.
   5.5 List all performances (performance number, composition number, composer number, composer name, and country name) of compositions that have been performed in the country of the composer.

References and Recommended Readings

Connolly, Thomas, & Carolyn Begg. 2015. *Database Systems: A Practical Approach to Design, Implementation and Management* 6th ed. Boston, MA: Pearson. See chapter 5.

Date, Christopher J. 2004. *Introduction to Database Systems* 8th ed. Menlo Park, CA: Addison-Wesley. See chapter 7.

Elmasri, Ramez, & Shamkant B. Navathe. 2011. *Fundamentals of Database Systems* 6th ed. Boston, MA: Pearson. See chapter 6.

Garcia-Molina, Hector, Jeffrey Ullman, & Jennifer Widom. 2009. *Database Systems: The Complete Book* 2nd ed. Boston, MA: Pearson. See chapter 5.

Kifer, Michael, Arthur Bernstein, & Philip M. Lewis. 2005. *Database Systems: An Application-Oriented Approach* 2nd ed. New York: Addison-Wesley. See chapter 5.

Russell, Gordon. 2006. *Database eLearning*. Accessed July 2006. http://www.grussell.org/

Ullman, Jeffrey D., & Jennifer Widom. 2008. *A First Course in Database Systems* 3rd ed. Boston, MA: Pearson. See chapter 4.

Chapter 8

Relational Calculus

This chapter discusses relational calculus as an alternative to relational algebra (discussed in Chapter 7. The partial college database, introduced in the previous chapter (Figures 7.1, 7.2, and 7.3), will be used as a frame of reference. It must be constantly borne in mind that relational algebra and relational calculus are mutual equivalents; most database management system (DBMS) suites will implement one or the other, or aspects of both, depending on what is convenient to the developers. The chapter proceeds under the following subtopics:

- Overview
- Calculus Notations and Illustrations
- Quantifiers: Free and Bound Variables
- Substitution Rule and Standardization Rules
- Introductory Query Optimization
- Domain-Related Calculus
- Summary and Concluding Remarks

8.1 Overview

In relational calculus, we simply specify what data is required, not how to obtain the required relation(s). Relational algebra, on the other hand, provides a collection of explicit operations in SELECT, JOIN, PROJECT, etc., which can be used to tell the system how to derive desired relation(s).

Suppose that we are interested in obtaining a list of program names and associated course names for the Bachelor of Science in management information systems (MIS). Example 8.1 provides some narrative that should help you appreciate the difference between relational algebra and relational calculus.

DOI: 10.1201/9781003275725-10

EXAMPLE 8.1: RETRIEVAL CONCEPTS FOR PROGRAM NAMES AND RELATED COURSE NAMES: ALGEBRA VERSUS CALCULUS

The stepwise relational algebra solution can be described as follows: Take the equijoin of relation **Pgm_Struct, AcademicProgram**, and **Course**; select tuples with **Pgm#** = 'BSC1'; then take a projection on attributes **PgmName** and **CrsName.**

The relational calculus formulation can be described as follows: Get **PgmName** and **CrsName** from **Pgm_Struct, AcademicProgram, Course** such that there exists a program structure tuple (PSP, PCS) in **Pgm_Struct**, a program P in **AcademicProgram**, and course(s) C in **Course**, where PSP = P, PSC = C, and PSP = 'BSC1'.

In relational calculus, the user describes what is required; the system is left to decide how to service the user's request. Relational calculus is therefore said to be descriptive (also declarative), while relational algebra is said to be prescriptive (also operational). As you shall see, this distinction is only superficial. More fundamentally, relational calculus and relational algebra are mutual equivalents: any algebra formation has a calculus equivalent and vice versa. Relational calculus is derived from a branch of mathematics called predicate calculus. It uses the idea of *tuple variable* (range variable). A tuple variable refers to a row of a relation at any given time. The relation is called the range. This relational calculus has its historical roots in older languages such as QUEL (used in the Ingress environment) and ALPHA, which was introduced by Codd.

Since a tuple consists of attribute values, it is also important to be able to access specific attributes. Referencing an attribute is done by the following notation:

<TupleVariable.AttributeName> | <AttributeName>

To illustrate, let PS be a tuple of relation **Pgm_Struct**. Then **PS.PSPgm#** refers to the value of the attribute **PSPgm#** in relation **Pgm_Struct** for some tuple PS.

Tuple variables may be implicit or explicit. An implicit tuple variable assumes the name of the relation on which it is defined; an explicit tuple variable has an alternate name that is different from the relation on which it is defined.

Example 8.2a provides two SQL statements depicting an implicit tuple variable followed by an explicit tuple variable. Immediately following this, Example 8.2b provides two QUEL statements, the first with an implicit tuple variable and the second with an explicit tuple variable.

EXAMPLE 8.2A: ILLUSTRATING IMPLICIT AND EXPLICIT TUPLE VARIABLES IN SQL

```
SELECT Pgm_Struct.PSPgm# FROM Pgm_Struct WHERE Pgm_Struct.PSPgm# =
`BSC1`;
// The above statement is equivalent to the following:
SELECT PS.PSPgm# FROM Pgm_Struct PS WHERE PS.PSPgm# = `BSC1`;
```

EXAMPLE 8.2B: ILLUSTRATING IMPLICIT AND EXPLICIT TUPLE VARIABLES IN QUEL

```
RETRIEVE (Pgm_Struct.PSPgm#) WHERE Pgm_Struct.PSPgm# = 'BSC1'
// The above statement is equivalent to the following:
RANGE OF PS is Pgm_Struct
RETRIEVE (PS.PSPgm#) WHERE PS.PSPgm# = 'BSC1'
```

Relational calculus of this sort is sometimes referred to as tuple calculus. An alternate domain calculus, based on domains, has been developed by Lacroix and Pirotte (1977). Query by example (QBE) is a language developed on this (domain calculus). This course focuses on the former, and not the latter.

8.2 Calculus Notations and Illustrations

As in the previous chapter, the Backus–Naur Form (BNF) notation for expressing syntactical components of a language will be employed. In the examples that will follow to the end of the chapter, the semicolon is used to punctuate the calculus statements in the interest of clarity. However, bear in mind that the original QUEL language did not require such punctuations.

8.2.1 Essential Notations

In order for you to understand and write relational calculus statements, there are seven essential components that you need to be comfortable with. The salient syntactical components of relational calculus are as follows:

1. **Relational Operators** include =, <, >, <=, >=, =>, <> while *connectives* include AND, NOT, OR

```
RelationalOperator ::= = | < | <= | > | >= | => | <>
Connective ::= AND | OR | NOT
```

2. **Comparisons** are done variable-with-variable, attribute-with-literal, or attribute-with-attribute separated by operators.

```
Comparison::= <CompFormat1> | <CompFormat2> | <CompFormat3>
CompFormat1 ::= <Variable> <Operator> <Variable>
CompFormat2 ::= <Attribute> <Operator> <Attribute>
CompFormat3 ::= <Attribute> <Operator> <Literal>| <ScalarExpression>
Attribute ::= <TupleVariable.AttributeName> | <AttributeName>
```

Observe: A comparison evaluates to true or false (similar to a relational algebra condition). The implication operator (=>) works as follows: P => Q means if condition P holds, then so does condition Q. Two alternate interpretations are applicable here: Firstly, if P is true, so is Q; otherwise, P is false. Secondly, either Q is true or P is false, i.e. P => Q means Q or P′. Also note that the compliment (negation) of a comparison or expression X may be represented as X′.

3. **Boolean Expressions** can be formulated by a combination of comparisons using connectors AND, OR, NOT:

> **BooleanExpression::=** [NOT] <Comparison> {<Connector> <BooleanExpression>}
> **Connector::=** AND | OR

Note: A Boolean expression is a logical construction that evaluates to true or false. Boolean expressions may be bracketed in the interest of clarity. This is necessary in situations where there are nested expressions.

4. **De Morgan's Law** for logical expressions is applicable; for Boolean expressions P, Q, R, . . . the following constructions apply:

> (P AND Q AND R …)' = P' OR Q' OR R' …
> (P OR Q OR R …)' = P' AND Q' AND R' …

5. **Range Definition** facilitates tuple variables; the syntax is as follows:

> **RangeDefinition ::= RANGE OF <TupleVariable> IS <Relation> | (<Relational Expression>)**

6. **Relational Expression:** A tuple calculus relational expression is of the following form:

> **RelationalExpression ::=** <TargetList> [WHERE <WFF>];
> **TargetList ::=** [<Attribute> =] <Attribute> [AS <Attribute>] | <TupleVariable> {,<TargetList>}

Two clarifying points are worth stating here:

- Firstly, whenever the tuple variable alone is specified, all attributes associated with the variable are implied (remember that a tuple variable is defined on a relation that has attributes).

- Secondly, the relational expression defines a projection on attributes in the target list from the Cartesian product of all referenced relations (from the target list). The WFF (*well-formed formula*) dictates a selection from the Cartesian product.

7. **Well-Formed Formula** (WFF): A well-formed formula (WFF) is the term used to describe a logical expression, which could be a simple comparison, or a more complex Boolean expression; it may take any of the following formats:

WFF ::= <Comparison> | NOT <WFF> | <Comparison> AND <WFF> |
<Comparison> OR <WFF> | If <Comparison> Then <WFF> |
EXISTS <Variable> (<WFF>) | FORALL <Variable> (<WFF>)

Be sure to invest some time in understanding these very important definitions. To assist you in this effort, let us look at a few clarifying examples.

8.2.2 Some Examples

The following examples are designed to help reinforce your understanding of the relational calculus. To get started, please review the partial college database of the previous chapter (Figures 7.1, 7.2, and 7.3).

EXAMPLE 8.3: LIST STUDENTS (NAMES AND IDS) WHO SHARE SURNAME WITH OTHER STUDENTS

Solution using QUEL:

```
RANGE OF Y IS Student;
RANGE OF X IS Student;
RETRIEVE (X.StudLName, X.Stud#) WHERE (X.Stud# <> Y.Stud#) AND
   (X.StudLName = Y.StudLName);
```

Note:

The QUEL verb RETRIEVE describes retrieval. The pure calculus follows the RETRIEVE verb. From this point on, we will drop the RETRIEVE verb. Also note that the original QUEL did not use a semicolon to punctuate statements. It is done here simply to improve readability.

EXAMPLE 8.4A: LIST PROGRAMS (SHOWING PGM#) THAT INCLUDE THE COURSE M100

```
RANGE of PS is Pgm_Struct;
(PS. PSPgm#) WHERE PS.PSCrs# = 'M100';
```

EXAMPLE 8.4B: LIST PROGRAMS (PGM# AND PGMNAME) THAT INCLUDE THE COURSE M100

```
RANGE OF PS is Pgm_Struct;
RANGE OF P IS AcademicProgram;
(PS. PSPgm#, P.PgmName) WHERE (PS.PSPgm# = P.Pgm# AND PS.PSCrs# = 'M100');
```

EXAMPLE 8.5: LIST ALL PROGRAM CODES AND RELATED COURSE CODES

```
RANGE OF PS IS Pgm_Struct;
(PS.PSPgm#, PS.PSCrs#);
```

EXAMPLE 8.6: LIST ALL PROGRAM NAMES AND RELATED COURSE NAMES

```
RANGE OF PS IS Pgm_Struct;
RANGE OF P IS AcademicProgram;
RANGE OF C IS Course;
(P.PgmName, C.CrsName) WHERE (EXISTS PS (PS.PSPgm# = P.Pgm# AND PS.PSCrs#
   = C.Crs#));
```

EXAMPLE 8.7: ALTERNATE SOLUTION FOR EXAMPLE 8.4A

```
// List programs (showing Pgm#) that include the course M100
RANGE OF PS IS Pgm_Struct;
M100PGM = PS.PSPgm# WHERE PS.PSCrs# = 'M100';
```

8.3 Quantifiers: Free and Bound Variables

Carrying on, we need to introduce two quantifiers and clarify what is meant by *free* and *bound* variables. The mathematical notations will be introduced, followed by the relational calculus notations. The mathematical notations are indicated in Figure 8.1:

Existential Quantifier:		Universal Quantifier:	
$\exists x$:	There exists x	$\forall x$:	For all x
	For some x		For any x
	For at least 1 x		For each x

Figure 8.1 Quantifiers.

Here is another way to interpret the quantifiers: If P(x) is a condition in tuple variable x, then the following two statements apply:

- (∃x) (P(x)) means ∃x satisfying P(x)
- (∀x) (P(x)) means all variables in range of x satisfies the condition P(x)

As expressed in the previous section, the relational calculus notations for the universal and existential quantifiers are as follows:

```
FORALL <Tuple-Variable> <WFF>
EXISTS <Tuple-Variable> <WFF>
```

8.3.1 Well-Formed Formula

The concept of a WFF was introduced in the previous section (see Clause 7 of Section 8.2). Based on the definition given there, a WFF may be verbally clarified as follows:

a. A simple comparison (condition) is a WFF.
b. If F is a WFF, so are NOT (F) and (F).
c. If F1, F2 are WFFs then so are (F1 AND F2), (F1 OR F2) and (If F1 THEN F2).
d. If the tuple variable x occurs *freely* in a WFF F then ∃x (F) and ∀x (F) are WFFs.
e. Nothing else is a WFF.

In layman's terms, a WFF is a simple or complex comparison involving attributes and scalar values. As can be confirmed from the forgoing sections and examples, WFFs are necessary for constructing appropriate data retrieval statements from the database.

8.3.2 Free and Bound Variables

A tuple variable is bounded if it occurs with either an existential or a universal quantifier. Consider the scenario below:

∀p (p.grade = 'Good')is equivalent to∀q(q.grade = 'Good')

This is an example of a bound occurrence of tuple variable *p*. The condition on *p* is either true or false, even if a particular value of *p* is not substituted. Note also that *p* can be replaced by *q*. Now consider the following:

p.grade = 'Good'is not equivalent toq.grade = 'Good'

This is an example of a free occurrence of variable *p*. The variable occurs in a simple condition (comparison) — when a particular value of *p* is substituted we can have a difference. Note that *p* cannot be replaced by *q*.

From the foregoing illustration, observe that when a tuple variable is bounded, it has a different meaning and must be treated differently from when it is free. The following are some rules for free and bound variables:

a. All occurrences of variables in a simple condition (comparison) are free.
b. Any free/bound occurrence in WFF F is also free/bound in NOT F, (F).
c. Any free/bound occurrence in WFF F1, F2 is free/bound in (F1) AND (F2) as well as (F1) OR (F2).
d. Every free occurrence of x in F is bound in ∃x (F) or ∀x (F); the occurrences of other variables are not affected.
e. A tuple variable cannot be both free and bound in the same statement.

So, to paraphrase and summarize, a tuple variable is bound in an expression if it is associated with a quantifier; once bounded, it remains in this state for the expression. To illustrate, in the expression [*EXISTS P (PS.PSPgm# = P.Pgm#)* . . .], P is bounded and PS is free. The following six examples provide additional clarification.

EXAMPLE 8.8: LIST ACADEMIC PROGRAMS (CODE AND NAME) THAT INCLUDES COURSE M100

// The following solution is unacceptable because PS is used to mean two different things (both bound and free)
RANGE OF P is AcademicProgram; RANGE OF PS IS Pgm_Struct;
PS.PSPgm#, P.PgmName WHERE (PS.PSPgm# = P.Pgm#) AND EXISTS PS (PS.PSCrs# = 'M100');

// The correct solution follows
RANGE OF P is AcademicProgram; RANGE OF PS IS Pgm_Struct; RANGE OF PS2 IS Pgm_Struct;
PS.PSPgm#, P.PgmName WHERE (PS.PSPgm# = P.Pgm#) AND EXISTS PS2 (PS2.PSCrs# = 'M100') AND (PS2.PSPgm# = PS.PSPgm#);

EXAMPLE 8.9: LIST SURNAMES OF STUDENTS BORN AFTER 1990

// Assume the date format of YYYYMMDD
RANGE OF S IS Student;
S.StudLName WHERE S.StudDoB > 19901231;

EXAMPLE 8.10: FIND PROGRAMS (NAMES) THAT DO NOT INCLUDE COURSE M100

RANGE OF PS IS Pgm_Struct;
RANGE OF P IS AcademicProgram;
P.PgmName WHERE EXISTS PS (PS.PSCrs# <> 'M100' AND PS.PSPgm# = P.Pgm#);

EXAMPLE 8.11: LIST COURSES (CODES AND NAMES) THAT OCCUR IN ALL DISCIPLINES

RANGE OF C IS Course;
RANGE OF P IS AcademicProgram;
RANGE OF PS IS Pgm_Struct; RANGE OF PS2 IS Pgm_Struct;
C.Crs#, C.CrsName WHERE FORALL P EXISTS PS
((PS.PSPgm# = P.Pgm#) AND EXISTS PS2 ((PS2.PSCrs# = PS.PSCrs#) AND (PS2.PSPgm# <> PS.PSPgm#)) AND (PS.PSCrs# = C.Crs#));

EXAMPLE 8.12: GIVE THE NAMES OF PROGRAMS WITH COURSE M100

RANGE OF P IS AcademicProgram;
RANGE OF PS IS Pgm_Struct;
P.PgmName WHERE EXISTS PS (PS.PSPgm# = P.Pgm# AND PS.PSCrs# = 'M100');

EXAMPLE 8.13: GIVE THE NAMES OF DEPARTMENT HEADS THAT ARE NOT DIVISION HEADS

RANGE OF S IS Staff;
RANGE OF D IS Department;
RANGE OF DV IS Division;
S.EmpName WHERE EXISTS D EXISTS DV (D.DeptHead# = S.Emp# AND D.DeptDiv# = DV.Div# AND D.DeptHead# <> DV.DivHead#);

8.4 Substitution Rule and Standardization Rules

It is sometimes necessary to introduce "dummy" (apparently redundant) tuple variables, in conformance with an old rule that forbids retrieval of quantified (bounded) variable(s); Examples 8.8 and 8.11 are cases in point. Further, for many data sub-languages (including SQL), if x and y are tuple variables and W is a WFF involving x and y, then

x.attribute WHERE (∃y) (W(x,y)) is equivalent to x.attribute WHERE W(x,y)

This rule often simplifies the database query statement by avoiding the use of the existential quantifier and will become clear when we discuss SQL later in the course. For instance, revisit Examples 8.10 and 8.12. In each case, we are retrieving tuple variable P based on a condition involving tuple variables PS and P. When you write the SQL solution for this, you do not have to include the existential quantifier.

Additionally, due to the limitation of some DBMS suites in their support of relational calculus notations, it is often useful to apply certain substitution and standardization rules. These are mentioned in Figure 8.2. In particular, take note of the fact that the universal quantifier can be replaced by skillfully using the existential quantifier. It is perhaps for this reason the universal quantifier is seldom supported (if at all) in the commonly used DBMS suites.

Substitution Rule for Quantifier:

If x is free in WFF F Then
$\forall x\ (F(x)) \equiv \forall y\ (F(y))$ and
$\exists x\ (F(x)) \equiv \exists y\ (F(y))$

Standardization Rules WFFs:

Let x, y be tuple variables and let A,B be WFFs

1. (A)'' ≡ A
2. (x=y)' <=> x <> y

3a. (A or B)' <=> A' AND B'
3b. (A AND B) ' <=> A' OR B'
4a. (∀x (A))' <=> ∃x (A)'
4b. (∃x (A))' <=> ∀x (A)'
4c. ∀x (A) <=> (∃x)' (A)'

5. If A then B <=> (A)' or B
6. If x does not occur in A, Then A AND ∀x (B) <=> ∀x (A AND B)
7. If x does not occur in B. Then (∃x (A) or B) <=> ∃x (A OR B)

Figure 8.2 Substitution and standardization rules.

8.5 Introductory Query Optimization

We have stated that every relational calculus expression has a relational algebra equivalent and vice versa. At an introductory level, query optimization often involves the transformation from calculus to algebra to optimized algebra. The rules for conversion are outlined in Table 8.1. Rule 5 of Table 8.1is a bit tricky and requires a level of expertise seldom attained from an introductory course such as this. Moreover, as indicated in the previous section, you can always avoid using the universal quantifier by skillfully applying the existential quantifier, and in turn, you can often set up your SQL queries to avoid using the existential quantifier (more on this in Chapter 12). For these reasons, we will focus our attention on the application of Rules 1–4 from the table.

Example 8.14 applies Rules 1–4 to Example 8.12 in order to obtain the corresponding relational algebra solution for the stated problem. The first half of the solution shows the application of Rules 1–3 to obtain the non-optimized algebra solution; the second half depicts application of Rule 4 to obtain the optimized algebra solution.

Table 8.1 Introductory Query Optimization Rules

Step	*Activity*
1	Take the Cartesian product of all relations used
2	Select the required tuples
3	Project on desired attributes
4	Optimize by • Taking selection (restriction) first (before product); • Replacing cross product(s) with natural join(s) or equijoin(s) where possible; • Taking the projection on the result.
5	Apply quantifiers from right to left as follows: • For the quantifier "EXIST RX" (where RX is a tuple variable that ranges some relation R), project the current intermediate result to eliminate all attributes of R. • For the quantifier "FORALL RX" (where RX is a tuple variable that ranges some relation R), divide the current intermediate result by the (possibly restricted) relation associated with RX.

EXAMPLE 8.14: REPLACE THE CALCULUS SPECIFICATION IN EXAMPLE 8.12 WITH AN ALGEBRA SPECIFICATION

```
// The non-optimized algebra solution follows:

// Rule 1:
R1:= Pgm_Struct TIMES AcademicProgram;

// Rule 2 (via Date Notation):
R2 := R1 WHERE PSCrs# = 'M100' AND PSPgm# = Pgm#;

// Rule 3 (via Date Notation):
R3:= PgmName FROM R2;

// The optimized algebra solution follows:

// Rule 4 based on Date notation and ANSI notation combined:
PgmName FROM ((Pgm_Struct WHERE PSCrs# = 'M100') JOIN (PSPgm# = Pgm#)
   AcademicProgram);

// Rule 4 based on Date notation :
PgmName FROM (Pgm_Struct WHERE PSCrs# = 'M100') TIMES AcademicProgram)
   WHERE PSPgm# = Pgm#;
```

The two solutions are illustrated in Figure 8.3. Note that consistent with the discussion of the previous section, existential quantifiers are ignored in the translation process as long as the quantified variables are not implicated as part of retrieved attribute list.

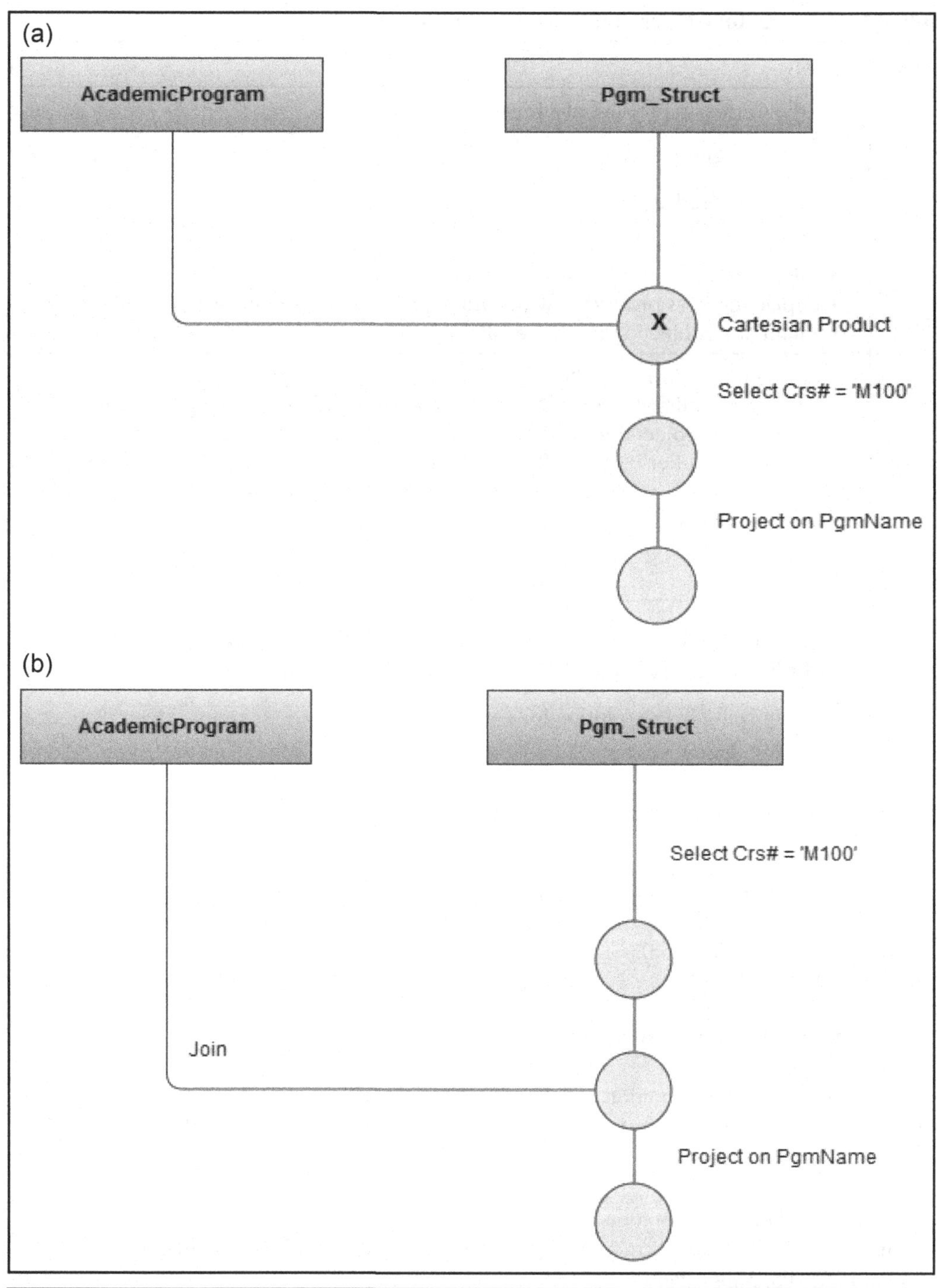

Figure 8.3 (a) Graphical representation of non-optimized solution to Examples 8.12 and 8.14. (b) Graphical representation of optimized solution to Examples 8.12 and 8.14.

There is much more to query optimization than presented here; in fact, this topic remains an intriguing area of focus and research for software engineering companies that construct and market DBMS suites, as well as academics with an interest in database system (DBS). For that reason, the topic will be revisited later in the course (Chapter 17).

8.6 Domain-Oriented Relational Calculus

Domain-related relational calculus involves the manipulation of *domain variables* instead of tuple variables. Queries based on domain relational calculus (DRC) are of the following format:

$$\{<X_1, X_2, \ldots X_n> \mid P(<X_1, X_2, \ldots X_n>)\}$$

In the first part of the construct, $<X_1, X_2, \ldots X_n>$ represents a set of domain variable(s) or constant(s). The separating slash (|) is read "such that." The second part, $P(<X_1, X_2, \ldots X_n>)$, represents a set of one or more DRC formula(s) that evaluate(s) to true or false. Multiple DRC formulas may be connected via the connectives AND, OR, or NOT. Additionally, the symbols from traditional set theory such as ∈ (meaning "is a member of"), ⊂ (meaning "subset of"), ∀ (the universal quantifier), and ∃ (the existential quantifier) are applicable.

EXAMPLE 8.15: DOMAIN CALCULUS ILLUSTRATION

// The following construct represents students with the surname Burton
{<Stud#, StudFName, StudLName> | <Stud#, StudFName, StudLName> ∈ Student
AND StudLName = 'Burton'}

EXAMPLE 8.16: ANOTHER DOMAIN CALCULUS ILLUSTRATION

// The following constructs represents and the names of their assigned halls
{<Stud#, StudFName, StudLName, StudHall#, HallName> |
<Stud#, StudFName, StudLName, StudHall#> ∈ Student
AND ∃ Hall# (<Hall#, HallName> ∈ Hall)
AND StudHall# = Hall#}

There is a lot more to DRC that could occupy the curious reader. Indeed, query by example (QBE) is an attractive implementation of DRC. The user simply completes a table on the screen in order to make a query request. Since this course focuses on the fundamentals, further discussion is omitted. Suffice it to say that domain calculus has its advantages in complex comparisons and is particularly useful when one is willing to concentrate on domain variables rather than tuple variables. Products such as Microsoft Access and Microsoft SQL Server Enterprise Manager employ the aspects of QBE.

8.7 Summary and Concluding Remarks

It's now time to summarize what we have covered in this chapter:

- Relational calculus is said to be descriptive, meaning, you describe precisely the activity required. The chapter concentrated on tuple calculus, i.e., calculus related to the manipulation of tuple variables.
- There are a number of standard relational calculus notations that allow you to succinctly describe data to be retrieved from relational tables. Retrieval statements essentially involve manipulating tuple variables using well-formed formulas (WFFs). A WFF is essentially a simple or complex condition.
- WFFs often involve the use of free and bound variables. The rules for these must be strictly followed.
- Basic query optimization is the process of obtaining the most efficient relational algebra equivalent for a relational calculus statement. In so doing, the basic rules for query optimization must be observed.
- Domain-oriented calculus is the calculus related to the manipulation of domain variables. It is an alternative to tuple calculus.

The language QUEL very closely resembled the relational calculus. The more contemporary SQL exhibits both relational calculus and relational algebra features. SQL will be the focus in the next division of the text. However, before delving into the language, the next chapter provides you an opportunity to review the salient features of the relational database model, the (historical) musings that led to its introduction, and some ramifications of the model.

8.8 Review Questions

1. Why is relational calculus important? Explain why relational algebra is said to be prescriptive, and relational calculus is said to be descriptive. Provide an example to illustrate.
2. Describe the salient components of relational calculus.
3. Using the college database described in this chapter, practice writing relational calculus statements that will yield certain desirable results.
4. Explain the concept and process of basic query optimization. Use an appropriate example to illustrate.

References and/or Recommended Readings

Codd, Edgar F. 1971 (November). "A Data Base Sublanguage Founded on the Relational Calculus." In Proceedings of the 1971 ACM SIGFIDET Workshop on Data Description, Access and Control. San Diego, CA.

Codd, Edgar F. 1972. "Relational Completeness of Database Sub-languages." In *Data Base Systems: Courant Computer Science Symposia Series 6*. Eaglewood Cliffs, NJ: Prentice Hall.

Connolly, Thomas, & Carolyn Begg. 2015. *Database Systems: A Practical Approach to Design, Implementation and Management* 6th ed. Boston, MA: Pearson. See chapter 5.

Date, Christopher J. 2004. *Introduction to Database Systems* 8th ed. Menlo Park, CA: Addison-Wesley. See chapter 8.

Elmasri, Ramez, & Shamkant B. Navathe. 2011. *Fundamentals of Database Systems* 6th ed. Boston, MA: Pearson. See chapter 6.

Kuhns, J. L. "Answering Questions by Computer: A Logical Study." In *Report RM-5428-PR.* Santa Monica, CA: Rand Corporation, 1967.

Lacroix, Michel, & Alain Pirotte. 1977 (October). "Domain-Oriented Relational Languages." In Proceedings of the 3rd International Conference on very Large Data Bases.

Ullman, Jeffrey D., & Jennifer Widom. 2008. *A First Course in Database Systems* 3rd ed. Boston, MA: Pearson. See chapter 4.

Chapter 9

Reflective Look at the Relational Database Model

We have covered much ground in our study of database systems. We have also established the importance of the relational model and its significant contribution to the field of database systems. We now pause to conduct a more enlightened discussion of this contribution and its effect on the field. This rather short chapter proceeds under the following subtopics:

- The Relational Model Summarized
- Ramifications of the Relational Model
- Summary and Concluding Remarks

9.1 The Relational Model Summarized

Table 9.1 summarizes the salient features of the relational model that have been established so far. These features define the minimum requirements of a relational system. Other desirable features of a database system discussed in Chapters 1 and 2 are still important. As it has turned out, meeting those standards was (and is) not easy. For years, they have eluded and continue to elude software engineering firms aspiring to construct and market database management system (DBMS) suites. As you will soon see, the standards bar has been raised even higher for contemporary products.

DOI: 10.1201/9781003275725-11

Table 9.1 Salient Features of Relational Model

Features Category	*Clarification*
Data structures	Data structure supporting the following: • Domains • Normalized relations • Attributes (including candidate/primary keys) that comprise the structure of the normalized relations • Rows (tuples) of time-dependent data for each relation
Data integrity	Data integrity rules that include: • Entity integrity rule • Referential integrity rule
Data manipulation	Data manipulation features that include support of the following: • Relational algebra and/or relational calculus • Relational assignment • Entry, maintenance, and deletion of data
Other features	Discussed in Sections 1.2, 1.5, and 2.5

9.2 Ramifications of the Relational Model

The relational model has very far-reaching implications, and makes very stringent demands on the software engineering industry, to deliver DBMS products that meet a minimum set of standards. Let us briefly look at some of these ramifications.

9.2.1 Codd's Early Benchmark

In 1982, Edgar F. Codd (referred to by many as the father of database systems), in a paper entitled "Relational Database: A Practical Foundation for Productivity" (see Codd 1982), proposed that a system could be regarded as relational if it supported at least the following:

- Relational database model
- The operations RESTRICT, PROJECT, and (natural) JOIN, without requiring any prior definition of access paths to support these operations

Codd further asserted the following:

- The operations may be supported explicitly or implicitly.
- The system must internally optimize user requests for desirable performance.

Codd's arguments centered on the data independence, structural simplicity, and relational processing defined in the relational model and implemented in relational database management systems. Based on those minimum requirements, DBMS suites were classified into one of the four categories: tabular, minimally relational, relationally complete, and fully relational. These classifications are clarified in Table 9.2.

Table 9.2 Categorization of DBMS Suites

Category	*Requirement*
Tabular	The system supports tabular data structure only, but not set-level operations.
Minimally relational	The system is tabular and supports the operations RESTRICT, PROJECT, and JOIN.
Relationally complete	The system is tabular and supports all of the operations of relational algebra.
Fully relational	The system supports all aspects of the model including domains and integrity rules.

Not many products were able to survive the rigors of that benchmark. Interestingly, two of the products that survived the test are still doing well in the industry today; they are Oracle and DB2. A third product — Ingres — has gradually faded but not before spawning and/or impacting development of other products such as structured query language (SQL) Server (from Microsoft), Sybase (now integrated into SAP Adaptive Server Enterprise), and PostgreSQL.

9.2.2 Revised Definition of a Relational System

In 1985, Codd, in a paper captioned "Is Your DBMS Really Relational?" revised the definition of a relational system. Christopher Date later made some important recommendations about the revised model (see Date 1990). The revised model with its recommendations is presented in Table 9.3.

Additionally, Codd proposed 12 rules for determining how relational a DBMS product is. With these 12 rules added to the redefined model, a DBMS is considered to be fully relational if it satisfies all structural, integrity, and manipulative features and fulfills the 12 rules. This was followed by a so-called *zero-rule* that essentially summarized the 12 rules. The rules are presented here.

9.2.2.1 Codd's Zero-Rule

According to Date (1990), *a system qualifies as a **relational**, **database**, and **management** system (highlights are deliberate) if and only if it uses its **relational** facilities exclusively to **manage** its **database**.*

This is a loaded statement that will become clearer as you progress through the course. For instance, after studying Chapters 12–14, you will have a better appreciation of what it means to use the relational facilities of a database to manage the said database. Moreover, it is not sufficient to take a system that is fundamentally not relational, add an interface that facilitates PROJECT, JOIN, SELECT (only), and then claim that the system is relational. Instead, everything must be relational — the view mechanism, the catalog, the structure, the integrity features, and the operations supported.

Table 9.3 Codd's Revised Definition of a Relational System

Features Category	*Clarification*
Data structures	Structural features include the following: • Relations including base relations, views, queries, and snapshots • Attributes that comprise the structure of the normalized relations • Domains • Primary keys • Foreign keys • Queries • User-defined data types • Rows (tuples) of time-dependent data for each relation
Data integrity	Integrity features including: • Entity integrity rule • Referential integrity rule • Primary key inheritance rules • Type inheritance and conversion rules (user defined) • User-defined integrity rules • Ensure closure and uniqueness of attributes
Data manipulation	Manipulation features including: • Entry, update, and deletion of data • Theta-select • Projection on certain attributes, while omitting others • Theta-join and natural join • Outer join • Divide • Union • Intersection • Difference • Relational assignment • Rename • Extend • Summarize
Other features	Discussed in Sections 1.2, 1.5, and 2.5

9.2.2.2 Codd's 12 Rules

1. **The Information Rule:** All information in the database must be represented as relational tables, subject to established integrity constraints, structural features, and manipulation features (mentioned earlier).
2. **The Guaranteed Access Rule:** All data stored in the database must be logically addressable by specifying the relation, the related attribute(s), and the primary–key–value.
3. **Systematic Treatment of Null Values:** The DBMS is required to have a consistent way of representing and treating so-called "missing information" that is different from regular data values and independent of the data types supported.
4. **Active Online Catalog Based on Relational Model:** The system should host a comprehensive relational catalog that is accessible to authorized users via the regular query language. Chapter 14 will provide more clarification on the importance of the system catalog.

5. **Comprehensive Data Sublanguage Rule:** The system must support at least one relational language that meets the following criteria:
   a. Possessing a linear syntax
   b. Exhibiting the capacity to be used both interactively and within application programs
   c. Providing adequate support of data definition language (DDL), data manipulation language (DML), and data control language (DCL) operations

   This requirement is adequately fulfilled in SQL, the universal standard database language that will be covered in Chapters 10–14.
6. **The View Updating Rule:** All views that are theoretically updateable must be updateable by the system (this will be clarified in Chapter 13).
7. **High-level Insert, Update, and Delete:** The system must support set-wise DML operations such as INSERT, UPDATE, and DELETE operations. Chapter 11 will demonstrate that SQL fulfills this requirement.
8. **Physical Data Independence:** The system should isolate all application programs (and end-user accesses) from the physical structure of the database. Changes in one should not affect the other.
9. **Logical Data Independence:** The system should isolate all application programs (and end-user accesses) from the logical structure of the database. Non-disruptive changes in one should not affect the other. This is not an absolute rule; rather, it should be taken in the context that presumes responsible database administration and usage. To clarify, changes such as renaming or removing an attribute or relational table may prove to be disruptive for specific application programs attempting to access the database; such changes would force adjustments in the affected application programs. On the application side, it is expected that application programs will manipulate legitimate database objects.
10. **Integrity Independence:** Where applicable, integrity constraints should be specified separately from application programs and stored in the system catalog. It must be possible to change such constraints as required, without effect on the applications that access the database. This is not to say that integrity checks should be ignored at the application programming level; quite the opposite. The database-level integrity check should be a second level after the application programming level.
11. **Distribution Independence:** Existing applications should continue to operate successfully when distributed versions of the DBMS are first introduced or upgraded.
12. **The Non-subversion Rule:** If the system provides a low-level (record-at-a-time) interface, then it should not be possible to use this interface to undermine or bypass relational security or integrity constraints of the system.

9.2.3 Far-Reaching Consequences

These constraints set a very high standard for relational DBMS (RDBMS) suites to attain. In fact, for a considerable period of time, the industry did not see a product that irrefutably met all of these requirements. On the other hand, many proposed products have fallen by the wayside due to failure to come close enough to the established standards. With incremental improvements to products such as (but not only) DB2, Oracle, Informix, and Sybase over several years, the industry can now boast of products meeting these standards (but not without room for improvement). Additionally, relatively newer products such as SQL Server and MySQL have made successful entries to the marketplace. Chapters 22–25 take some time to familiarize yourself with some of these products.

Is the benchmark too high for RDBMS products? Not at all. It defines an ideal that software engineering firms can strive to attain. It also establishes a firm mathematical basis for the relational model. In this regard, the work of Codd, Date, Fagin, and others cannot be over applauded. To a certain extent, the standards have protected the consuming public from rogue companies that might have tried to exploit us by marketing inferior database products under the false claim of them being relational. We have seen many such attempts, but for the most part, they have not gone very far.

As you continue your pursuit of this course, you will discover a rather interesting phenomenon: Many of the standards described in the revised benchmark for an RDBMS have been implemented in SQL, the universal database language, and leading DBMS products such as DB2, Oracle, Sybase (now part of SAP), Informix, MySQL, and MS SQL Server. This is comforting information.

9.3 Summary and Concluding Remarks

Here is a summary of what has been covered in this chapter:

- The benchmarks for a relational database system may be defined in terms of the data structure requirements, the data integrity requirements, and the data manipulation requirements.
- Over the years, the database systems industry has embraced Codd's benchmarks (first introduced in 1982 and subsequently revised in 1985) as the standards for relational DBMS suites. The revised standards also include Codd's 12 rules and Date's zero-rule.
- Even though these standards are more than three decades old, they still remain binding on the industry. In fact, many of the smaller DBMS products still struggle to meet them.

Although Codd died in 2003, his legacy will no doubt continue to live and guide the field of database systems well into the foreseeable future. Date, a colleague of Codd, continues to be a renowned author and consultant in the field. So thankfully, we are in good hands.

9.4 Review Questions

1. Describe in your own words, what is meant by a relational DBMS.
2. Conduct a critical evaluation and comparative analysis of three leading DBMS products that you are familiar with. Use Codd's revised definition of a relational database system as your benchmark.
3. Conduct a critical evaluation of E.F. Codd's contribution to the field of database systems.

References

Codd, Edgar F. 1982 (February). "Relational Database: A Practical Foundation for Productivity." *Communications of the ACM* vol. 25, no. 2. pp. 109–117.

Codd, Edgar F. 1985a (October). "Is Your DBMS Really Relational?" *Computer World*, October 14, 1985.

Codd, Edgar F. 1985b (October). "Does Your DBMS Run By the Rules?" *Computer World*, October 21, 1985.

Date, Christopher J. 1990. *Introduction to Database Systems* Vol. I 5th ed. Menlo Park, CA: Addison-Wesley. See chapter 15.

Date, Christopher J. 2004. *Introduction to Database Systems* 8th ed. Menlo Park, CA: Addison-Wesley.

THE STRUCTURED QUERY LANGUAGE

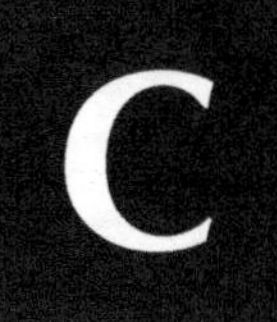

The next six chapters will focus on the Structured Query Language (SQL). This language has become the universal standard database language. It is therefore imperative that as a student of computer science, or a practicing IT professional, you are not just familiar, but have a working knowledge of the language. The objectives of this division are:

- To provide a solid overview of SQL as a database language
- To help you gain a good working knowledge of the main SQL data definition statements
- To help you gain a good working knowledge of the main SQL data manipulation statements
- To help you gain insights and a good working knowledge of logical views and system database security
- To discuss and illustrate the importance and usefulness of the system catalog
- To discuss some limitations of SQL

As you will soon see, SQL is a very powerful data sublanguage (DSL), and it is easy to learn. Chapters to be covered include:

- Chapter 10 — Overview of SQL
- Chapter 11 — SQL Data Definition Statements
- Chapter 12 — SQL Data Manipulation Statements
- Chapter 13 — Logical Views and System Security
- Chapter 14 — The System Catalog
- Chapter 15 — Some Limitations of SQL

With the foundation laid so far, you will hopefully find that SQL is a very easy language to learn. Moreover, it is a very powerful language! Knowing how to use it to solve database-related or software engineering problems will give you a sense of confidence and accomplishment not hitherto experienced. Have fun!

DOI: 10.1201/9781003275725-12

Chapter 10

Overview of SQL

The Structured Query Language (SQL) has become the universal language of choice for database management system (DBMS) products. A study of this language is therefore imperative for the student of computer science or computer information systems. This and the next few chapters will help you acquire a working knowledge of the language, as implemented in the Oracle environment. One fact you need to be immediately cognizant of is that there are different implementations of SQL. However, the implementations usually have more in common than differences; therefore, once you have mastered the language in one DBMS environment, adjusting to another environment is a trivial matter. This brief chapter provides you with an overview of the language, advancing through the following captions:

- Important Facts
- Advantages of SQL
- Summary and Concluding Remarks

10.1 Important Facts

Structured Query Language (SQL) is an example of a DSL — consisting of data definition language (DDL), data control language (DCL), and data manipulation language (DML) as defined in Chapter 2. First developed by IBM in the 1970s, SQL is the universal language of databases, adopted by the American National Standards Institute (ANSI) since 1986.

SQL may be described as an interactive query language as well as a database programming language. Commands can be entered directly at the command prompt, or embedded in application programs, written in some other high-level language (HLL); this is standard practice.

Through one of its standard committees, ANSI sets and revises standards for SQL on a consistent basis. The latest set of standards for SQL that is available for public access is SQL-2016. SQL-2016 is an enhancement of SQL-2011, which is an enhancement of SQL-2007.

In this course, the focus is essentially concentrated on core SQL statements that for the most part cut across barriers of SQL standards. However, it is perhaps worthwhile to note that the SQL

DOI: 10.1201/9781003275725-13

standards are updated roughly every 4 years and are available from the International Standards Organization (ISO). For more information, on the SQL standards see the references listed in Section 10.5.

SQL is a non-procedural (declarative) language that closely mirrors the relational calculus as discussed in Chapter 8; there are also features that mirror the relational algebra of Chapter 7. Although the language was originally introduced by IBM, no organization has a monopoly on it. Different implementations of SQL have their own idiosyncrasies and flavors. Some of the major DBMS suites are Oracle, MySQL, PostgreSQL, Microsoft SQL Server, DB2, Informix, SAP Adaptive Server Enterprise (SAP-ASE), and MongoDB.

10.1.1 Commonly Used DDL Statements

Table 10.1 provides a list of commonly used DDL statements of SQL. You will observe that the statements are self-explanatory. This makes learning of the language very easy, particularly when compared to more cryptic traditional programming languages. Another feature that you will notice about SQL is that it is symmetrical and follows very consistent standards (programming language experts favor the term *orthogonality* to describe this feature). Thus, for most database objects, you can create the object, alter (i.e., modify) it, or drop (i.e., remove) it from the system. The main database objects/resources that you will be working with are tables, indexes, constraints, sequences, views, and synonyms; in the case of Oracle, there are also databases, tablespaces. Of course, there are other objects/resources that will be covered later in the course

10.1.2 Commonly Used DML and DCL Statements

The commonly used DML and DCL statements of SQL are indicated in Table 10.2. The DML statements are often referred to as SUDI (select, update, delete, and insert) statements. The DCL statements fall into three categories: those that affect how DML operations take place (mainly COMMIT and ROLLBACK); those that relate to system privileges; and those that affect the environmental settings of the end user

10.1.3 Syntax Convention

In the next four chapters (Chapters 11–14), the core DDL, DML, and DCL statements of SQL will be discussed. The convention that will be employed for each statement is to present the syntax in Backus–Naur Form (BNF) for the most common (often abridged) format(s) of the statement. However, for more complex statements, in the interest of comprehensive coverage, you will be directed to the more detailed syntax (in Appendix 4). All the examples provided will be based on the commonly used forms of the statements discussed.

Table 10.1 Commonly Used DDL Statements of SQL

Generic DDL Statements	
Statement	*Explanation*
Create-Database	Creates a database. An Oracle database is a complex object that stores various other database objects and resources.
Alter-Database	Modifies the database.
Drop-Database	Removes the database from the system catalog.
Create-Tablespace:	Creates a tablespace. A tablespace is a logical container within a database. It can contain several datafiles that contain the actual database objects.
Alter-Tablespace	Modifies a tablespace
Drop-Tablespace	Removes a tablespace from the system catalog.
Create-Table	Creates a database table. The table may be a relational table, an object table, or an XML table; the default is relational.
Alter-Table	Alters the physical and/or logical structure of a table.
Drop-Table	Removes a table from the system catalog. The table and all its data are deleted. Due to referential integrity, the typical DBMS forbids you to delete a table that contains referenced tuples.
Create-Index	Creates an index on a table. There are different types of indexes, but the default is B-tree.
Alter-Index	Modifies the structure of an index.
Drop-Index	Removes an index from the system catalog.
Create-Constraint	Creates a constraint on a database table. As you will later see, there are different types of constraints.
Alter-Constraint:	Modifies the terms of a constraint.
Drop-Constraint	Removes a constraint from the system catalog.
Create-View	Creates a logical view of data contained in physical database tables. Remember, a view is a virtual relation.
Alter-View	Allows modifications to a logical view.
Drop-View	Removes a logical view from the system catalog.
Create-Sequence	Creates a sequence. A sequence is a special database object which is used to generate unique numbers. Sequences are useful in coding specific attributes (for instance numeric values for primary key attributes or surrogates).
Alter-Sequence	Modifies a sequence.
Drop-Sequence	Removes a sequence from the system catalog.
Create-Synonym	Creates an alias of a database object.
Drop-Synonym	Removes a synonym from the system catalog. Note that synonyms cannot be altered, since they are logical objects.

Table 10.2 Commonly Used DML and DCL Statements of SQL

DML Statements	
Statement	*Explanation*
Select	Retrieves data from database tables in a manner that is consistent with user specifications.
Update	Updates data contained in physical database tables.
Delete	Removes data from physical database tables.
Insert	Inserts data into physical database tables.
DCL Statements	
Statement	*Explanation*
Commit	Forces permanent storage of all changes appearing in related *transaction*(s), i.e., sets of related activities, erasing all temporary save-points and transaction locks.
Rollback	Returns the database to some previous safe-state that protects the integrity of the system; incomplete transactions are not allowed.
System privilege statements	A set of statements that govern database security.
Environment setting statements	A set of statements that govern environmental settings of database users.

10.2 Advantages of SQL

SQL brings a number of significant advantages to the software engineering industry. Some of these advantages are as follows:

Rapid Software Development: SQL enhances rapid development of business application systems by its powerful and easy-to-learn statements. In fact, most RAD tools, DBMS suites, and computer-aided software engineering (CASE) tools support the language.

Higher Software Quality: SQL brings a higher software quality to the software engineering arena. By using more powerful SQL statements than would be possible in traditional high-level languages, the software engineer is likely to produce a shorter code with fewer errors. The declarative nature of the language is useful in this regard.

Higher Productivity: SQL brings higher productivity to businesses by providing superior database management features than possible in systems developed with traditional high-level languages alone. These superior database management features include the following:

- Faster access to data
- Larger files
- Set-at-a-time access instead of record-by-record access
- Powerful data aggregation facilities to provide meaningful end-user information
- Facilities for the enforcement of data integrity constraints

- Facilities (such as logical views and various internal functions) for logically reorganizing data to provide useful information to end users
- Facilities for the enforcement of database security constraints

Data Independence: SQL helps DBMS suites to meet the objective of data independence, with its associated benefits (review Chapter 1).

Standardization: SQL facilitates standardization among competing software development tools, since they all are forced to support the language.

10.3 Summary and Concluding Remarks

Let us summarize what we have covered in this chapter:

- SQL is the universal relational database language. It consists of various statements for creating and administering a database. These statements can be classified as DDL statements, DML statements, and DCL statements.
- SQL can be used at the command prompt or embedded in other high-level language programs.
- SQL brings a number of significant advantages to the software engineering industry.

SQL is not without a fair number of limitations. In order to fully understand the limitations, you need to have a working knowledge of the language first. With this in mind, a discussion of these limitations is left for Chapter 15.

The next chapter discusses the main DDL statements of SQL. As you will see, they are easy to learn and use. Another observation you should soon make is that an understanding of fundamental database theory (as covered in the foregoing chapters of this text) goes a long way in enhancing your grasp of SQL, as well as your alacrity in knowing how to apply the language.

10.4 Review Questions

1. What are the most common DDL statements of SQL? Briefly explain the purpose of each.
2. What are the main DML statements of SQL? Briefly explain the purpose of each.
3. Explain the three categories of DCL statements in SQL.
4. Identify some of the advantages of SQL.

Recommended Readings

Eisenberg, Andrew, & Jim Melton. 2000 (March). "SQL Standardization: The Next Step." *SIGMOD Record* vol. 29, no. 1, pp. 63–67.

International Standards Organization. 2011. "Information Technology: Database Languages SQL Technical Reports." Accessed June 2013. http://www.iso.org/iso/home.html

Kulkarni, Krishna. 2003. *Overview of SQL:2003*. San Jose, CA: Silicon Valley Laboratory, IBM Corp. Accessed August 2008. http://www.wiscorp.com/SQL2003Features.pdf

Oracle Corporation. 2015. *Database SQL Language Reference*. Accessed June 2015. http://docs.oracle.com/database/121/SQLRF/toc.htm. See chapters 1 & 2.

Whitemarsh Information Systems Corp. 2015. "SQL Standards." Accessed June 2015. http://www.wiscorp.com/SQLStandards.html

Zenke, Fred. 2012 (March). "What's new in SQL 2011." *SIGMOD Record* vol. 41, no. 1, pp. 67–73.

Chapter 11

SQL Data Definition Statements

The main SQL definition statements in an Oracle environment are shown in Table 11.1. These statements relate to the six basic types of database objects in Oracle — tables, indexes, views, constraints, synonyms, and sequences. Other more advanced types of database objects include databases, tablespaces, datafiles, users, user profiles, functions, and procedures.

Table 11.1 Commonly Used SQL Definition Statements

Create-Table	Create-Index	Create-Constraint	Create-View
Alter-Table	Alter-Index	Alter-Constraint	Alter-View
Drop-Table	Drop-Index	Drop-Constraint	Drop-View
Create-Database	Create-Tablespace	Create-Sequence	Create-Synonym
Alter-Database	Alter-Tablespace	Alter-Sequence	Drop-Synonym
Drop-Database	Drop-Tablespace	Drop-Sequence	

This chapter will focus on tables, synonyms, sequences, and indexes; a discussion of the more advanced objects will follow in subsequent chapters. The chapter will proceed under the following subtopics:

- Overview of Oracle's SQL Environment
- Basic Concepts in a Typical Oracle Database Environment
- Database Creation
- Database Management
- Tablespace Creation
- Tablespace Management
- Oracle Cloud Features
- Table Creation Statement
- Dropping or Modifying a Table

DOI: 10.1201/9781003275725-14

- Working with Indexes
- Working with Sequences
- Working with Synonyms
- Summary and Concluding Remarks

11.1 Overview of Oracle's SQL Environment

The Oracle 19C suite represents a culmination of efforts in earlier releases such as 10G, 11G, 12C, and 18C. The end result is a powerful database management system (DBMS) suite that supports various forms of operational databases (representing transactional processing) and analytic databases (such as data warehouses, big data lakes, and graphs). The product is packed with an impressive plethora of features, including but not limited to the following:

- Multitenant architecture that facilitates a wide range of database configurations including pluggable databases (PDBs) and container databases (CDBs).
- Sophisticated security mechanism that includes traditional access control, data encryption, data masking, database vault, and active directory integration.
- Automatic indexing to improve the performance of database queries.
- Database in-memory strategy that allows the aspects of the database to be kept in memory in order to improve performance.
- Comprehensive application development support for the JavaScript Object Notation (JSON) format, as well as various programming languages including PL/SQL, SQL, Python, PHP, Java, C, C#, and JavaScript.

As you will see in Chapter 22, the Oracle DBMS suite provides a rich collection of components to manage a database environment of any kind — operational or analytic, large or small, general purpose or special, standalone or distributed, cloud based, or locally controlled. Following is a brief description of three commonly used components:

- SQL*Plus is a command-based interface for supporting SQL, along with a few additional commands for batch-scripting and other maneuverings in the Oracle environment. While SQL*Plus is still supported in Oracle 19C, this component is not as widely used as in versions prior to Oracle 12C.
- SQL Developer (SQLD) is a client application — a front-end tool — that allows the user to access, manipulate, and manage one or more databases. The application allows you to create one or more connections for each database (local or remote) of interest. Once the connection is established, the database can be accessed through the graphical user interface (GUI) provided.
- Application Express (APEX) is an alternate application that facilitates access of a cloud-based Oracle database. While this application does not provide as much functionality and flexibility as the SQLD, it provides the flexibility of running on hand-held devices such as mobile telephones.
- Oracle Enterprise Manager (OEM) provides all the services that SQLD provides, but also provides additional database administration services (see Chapter 16) such as tuning, backup and recovery, and configuration management.

Oracle implements its own host language, called PL/SQL (acronym for Procedural Language Extensions to SQL). It is a simple language with a predominantly Pascal-like syntax (Pascal is an older programming language that was widely used in the 1990s). Although coverage of the syntax of this language is beyond the scope of this course, a few examples will contain the PL/SQL code (particularly in Chapter 12). Because of the simplicity of the syntax, you should be able to read the code and understand it, so there is no need to panic.

Recall that the BNF notation was first introduced in Chapter 7. Throughout the remainder of the course, a slight modification of this notation will be used for specifying the syntax of SQL statements. The symbols used are shown in Table 11.2. Additionally, from time to time, you will observe the inclusion of clarifying comments that conform to how comments are made in the C-based programming languages (e.g., // this is a comment, /* and so is this */). These comments serve to clarify the syntactic representations or examples that they appear in.

Table 11.2 BNF Notation Symbols

Symbol	*Meaning*
::=	This symbol means "is defined as"
[. . .]	Denotes optional content; optional items are enclosed in square brackets
<Element>	Angular braces denote that the content is supplied by the user and/or is non-terminal
|	Indicates choice (denotes "or"). Example: **<ItemA>** | **<ItemB>** | **<ItemC>**
/* **Or** */	Indicates choice between complex alternatives. This option is used when syntactic elements are complex to the point where using a single slash (| as in the preceding case) would not provide the clarification needed. Example: **<ItemA>** **<ItemB>** /* Or */ **<ItemC>**
{<Element>}	Denotes zero or more repetitions. Other notations have been proposed but for this course, we will stick to this traditional notation.
<Element>*	Alternate notation to denote zero or more repetitions
<l>*<m><Element>	Denotes **l** to **m** repetitions of the specified element

11.2 Basic Concepts in a Typical Oracle Database Environment

In the Oracle environment, database creation is quite involved, requiring you to be aware of several intricate details about the Oracle DBMS itself. Figure 11.1 shows an entity–relationship diagram (ERD; using the UML notation) that summarizes the basic Oracle database configuration that you should be aware of. Here is the essence of what the figure depicts:

- A database comprises logical components called *tablespaces*.
- The *tablespace* (sometimes abbreviated TBS) is the logical holding area for database objects; you will learn to create and manipulate several of these objects, some of which are represented in Table 11.1.
- Each tablespace also hosts one or more *datafiles*.
- A *datafile* is the physical storage area for the database objects. However, these objects are logically accessible through the tablespace.
- Each datafile typically consists of multiple data segments; each segment consists of multiple blocks, which are in turn made up of bytes of data.

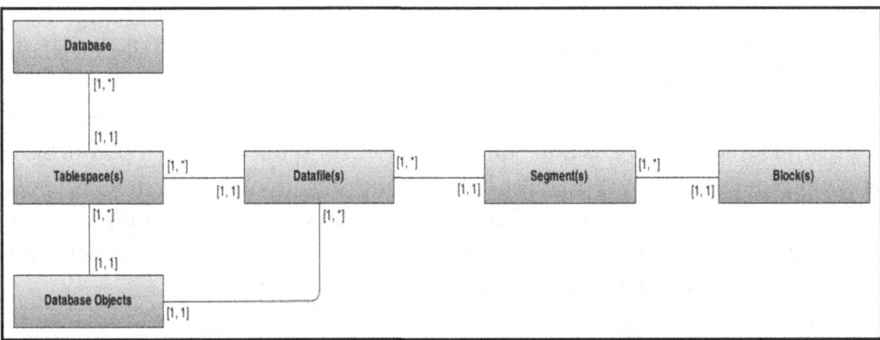

Figure 11.1 Overview ERD of the Oracle database configuration.

Before proceeding further, there are a few concepts that you need to be familiar with. We will revisit them later in the course (for instance, Chapters 13 and 22), but here are some basic clarifications that will suffice for now:

Database Server and Clients: The Oracle database server software (which is the DBMS suite) is typically installed and configured on an appropriately designated machine in your affiliated network environment. Other machines in the networking environment will have some form of Oracle client software running on them. One popular example of such software is Oracle SQL Developer. You will be shown how to configure your client to access the database server.

Database Abstractions: To improve on scalability, starting with Oracle 12C, there is an additional layer of abstraction that was not present in earlier versions. The Oracle database environment consists of a Container Database (CDB) and the Pluggable Database (PDB). The CDB is for hosting global and/or inheritable database resources. The PDB contains objects and resources specific to that database instance; moreover, there may be multiple PDBs for any given database environment.

Important Configuration Files: You have the option of installing the Oracle DBMS suite on your machine instead of a mere client software. Remember, this is optional; if you do, be sure to observe the installation guidelines available on the Oracle website. And having installed the DBMS suite, it is a good idea to become familiar with the Oracle directory structure (this is quite a complex hierarchy of folders/directories and files; see Chapter 16). It would be a good idea to locate the home directory for any local database you have created or will create: the *initialization parameter file* (a file storing initial parameters for the database), the *network configuration file* (a file storing network settings for the database environment), and the *control file* (a binary file storing important information about the database).

User Account: You cannot access an Oracle database environment without a *user account*. In a classroom setting, your instructor typically will create an account for you and inform you of your credentials for accessing the database. In a workplace environment, your database administrator (DBA; or someone with DBA privilege) will create the account for you and inform you of the pertinent credentials. The third alternative is that you install the Oracle DBMS yourself and assume the role of DBA.

Schema and Privilege: The DBMS creates a *schema* with the same name as the user account. A *privilege* is a right to access certain resources in the database environment. Initially, your DBA and/or instructor will determine a minimal set of privileges for you to access the database system. Subsequently, as you create database objects, your schema will own these objects, so you will have existence privileges to them (meaning, you can do whatever you wish with them).

Session: When you log onto an Oracle database, a database *session* is created, and this allows you to communicate with the database server.

The Oracle-Certified Professional Database Administrator (OCP-DBA) is a comprehensive certification program supported by Oracle. It is a highly coveted certification that many strive to achieve. Covering the OCP-DBA is not the purpose of this course. However, upon completion, you will be in good stead to explore this option should this be your desire, or go in a different direction with your database systems education.

11.3 Database Creation

A database may be created in one of the four ways: it may be created manually through issuance of the **Create-database** statement; it may be created via the Database Configuration Assistant (DBCA); it may be created through using the Oracle Enterprise Manager (OEM); finally, it may be created during installation of the Oracle Database server software. In each case, the **Create-Database** statement is employed (manually or automatically). However, there are certain procedures and intricate details that have to be carefully observed. In the interest of getting you started, these intricacies will be de-emphasized for now. Figure 11.2 shows the syntax of the **Create-Database** statement; Example 11.1 provides an illustration of its usage. In order to use this statement, you must have the DBA privilege; this you may acquire by some other authorized user granting it to you, or by being the one responsible for installing and configuring the DBMS. Chapter 13 provides more discussions on privileges.

EXAMPLE 11.1: ILLUSTRATING THE USE OF THE CREATE-DATABASE STATEMENT

```
CREATE DATABASE SampleDB
CONTROLFILE REUSE
LOGFILE
GROUP 1 ('C:\Oracle\Oradata\SampleDB\Log0101.log', 'D:\Oracle\Oradata\SampleDB\
   Log0101.log') SIZE 50K,
GROUP 2 ('C:\Oracle\Oradata\SampleDB\Log0201.log', 'D:\Oracle\Oradata\SampleDB\
   Log0202.log') SIZE 50K
MAXLOGFILES 5
MAXLOGHISTORY 100
MAXDATAFILES 10
MAXINSTANCES 2
ARCHIVELOG
CHARACTER SET AL32UTF8
NATIONAL CHARACTER SET AL16UTF16
DATAFILE
'C:\Oracle\Oradata\SampleDB\System0101.dbf' AUTOEXTEND ON,
'D:\Oracle\Oradata\SampleDB\System0201.dbf ' AUTOEXTEND ON
NEXT 10M MAXSIZE UNLIMITED
DEFAULT TEMPORARY TABLESPACE Temp_TBS
UNDO TABLESPACE Undo_TBS
SET TIME_ZONE = '+02:00';
```

```
Create-Database ::=
CREATE DATABASE <DatabaseName>
[USER SYS IDENTIFIED BY <Password> ]
[USER SYSTEM IDENTIFIED BY <Password>]
[CONTROLFILE REUSE]
[LOGFILE [GROUP <n> (<FileSpec> {,<FileSpec>})
            { GROUP <n> (<FileSpec> {,<FileSpec>})}
[MAXLOGFILES <n>] [MAXLOGMEMBERS <n>] [MAXLOGHISTORY <n>]
[MAXDATAFILES <n>] [MAXINSTANCES <n>]
[ARCHIVELOG|NOARCHIVELOG] [FORCE LOGGING]
[CHARACTER SET <Charset>] [NATIONAL CHARACTER SET <Charset>]
[DATAFILE <FileSpec> [<Auto-extend-Clause>] {, <FileSpec> [<Auto-extend-Clause>]}
[EXTENT MANAGEMENT LOCAL]
[<Default-temporary-tablespace-Clause>]
[<Undo-tablespace-Clause>]
[Set-Time-Zone-Clause]

Auto_Extend-Clause ::=
AUTOEXTEND ON|OFF NEXT <n> K|M MAXSIZE <n>|UNLIMITED K|M

Default-Temporary-Tablespace-Clause ::=
DEFAULT TEMPORARY TABLESPACE <TablespaceName>
[TEMPFILE <FileSpec>
[EXTENT MANAGEMENT LOCAL]
[UNIFORM SIZE <n> K|M]

Undo-Tablespace-Clause ::=
UNDO TABLESPACE <TablespaceName> DATAFILE File-spec

Set-Time-Zone-Clause ::=
SET TIME_ZONE = '<Time-Zone-Spec>'

FileSpec ::= '<PathtoFile>' SIZE <Integer> K|M [REUSE]
```

Note:

1. File-Spec is a path to a physical operating system file that you specify, based on your environment and database standards.
2. If you specify locally managed TBS (EXTENT MANAGEMENT LOCAL), you must specify the default temporary TBS.
3. If you specify locally managed TBS and the **Datafile-Clause**, you must specify the default temporary TBS and a datafile for that TBS.
4. If you specify locally managed TBS but do not specify the **Datafile-Clause**, you can omit the default temporary TBS clause. Oracle will create a temporary TBS called TEMP with a 10M datafile.
5. The default temporary TBS size is (uniform) 1M.
6. The default log-file size (if you do not specify the **Log-File-Clause**, Oracle creates two log-files) is 100M.
7. Time zone is specified as a number of hours and minutes ahead of (+) the standard GMT, or by using a predetermined time-zone name (as obtained from the view **v$TIMEZONE_NAMES**).

Figure 11.2 Syntax of the Create-Database statement.

As you can see, the **Create-Database** statement has several parameters. Following is a brief clarification on each of the main parameters of the command:

USER SYS: Each Oracle database has a super user account called **Sys**. The IDENTIFIED BY clause allows you to specify a password for that user account.

USER SYSTEM: Each Oracle database has a user account called **System**, not as powerful as **Sys**, but quite close, with all DBA privileges. You may also specify the password for this account.

CONTROLFILE REUSE: Every Oracle database contains a control file (a binary file) storing the database name, names and locations of datafiles and redo log files, timestamp of the database creation, and other control information. This clause allows control files associated with the database to be reused.

LOGFILE: This clause allows for specifying the location(s) where redo log file(s) for the database will be stored. Redo logs are essential in keeping track of changes made to the database. In the event of any activity that threatens the existence of the database, they facilitate its timely recovery.

MAXLOGFILES: The database creator is allowed to stipulate the maximum number of log file groups that will exist for the database.

MAXLOGMEMBERS: The database creator may state the maximum number of log files that will exist within a log group.

MAXHISTORY: If the database is in ARCHIVELOG mode with Real Application Clusters (RAC), this clause allows specification of the maximum number of archived redo logs that may be used in an automatic recovery of Oracle RAC.

MAXINSTANCES: This clause allows stipulation of the maximum number of instances of the database that can be simultaneously mounted. The allowed range of possible values is [1 . . . 1055]; the default is dependent on the underlying operating system.

MAXDATAFILES: Allows stipulation of an anticipated number of datafiles for the database; however, under desired circumstances, the database size can exceed this number.

ARCHIVELOG | NOARCHIVELOG: ARCHIVELOG mode means that the database will generate archived logs from the redo logs. This means that the DBMS makes copies of all historical transactions that have occurred. NOARCHIVELOG mode means that the database will not generate archive logs; no historical transaction is stored. Consequently, only current redo log data can be used in the event of an instance failure.

FORCELOGGING: The DBMS will log all changes in the database except for changes in temporary tablespaces. This state is persistent across database instances and changes only when the control file is recreated.

CHARACTERSET: Allows stipulation of the character set that will be used for the database. Oracle supports a wide range of character sets, depending on the intended location of the database and the underlying operating system. Unless you are constructing the database for some unusual purpose, it is advisable to take the default for this parameter. If in doubt, consult the Oracle product documentation.

DATAFILE: This clause facilitates specification of the location(s) and constraints for datafile(s) associated with the database. This/these datafile(s) will typically be stored in a default tablespace called System. Alternately, you may forego using this clause at database creation, and subsequently use the **Create-Tablespace** statement after the database has been created. This alternative is discussed in Section 11.5.

The **Auto-Extend-Clause** is used to facilitate automatic growth of the database whenever a tablespace approaches its defined limit; this will be further clarified in Section 11.5.

The **Default-TemporaryTablespace-Clause** creates a default tablespace for the database. Any user account not assigned a tablespace will be assigned this tablespace as a default. Alternately, if the database is not assigned a temporary tablespace, then the **System** tablespace is used as the default. However, the more standard practice is for user accounts to be assigned to specific tablespaces.

Initialization Parameter File: Every Oracle database installation has an initialization parameter file that stores important parameters for the database. The file is typically named according to the construct **init<DatabaseName>.ora** (for example, **initTestDB.ora** where the database name is **TestDB**) and is stored in the home directory of the database. It is a good idea to locate this file and familiarize yourself with its content. However, you typically do not have to alter its content unless you are very experienced with Oracle databases. The parameter file may be altered directly, or via the **Alter-System** statement after logging onto the database. To use the **Alter-System** statement for managing initialization parameters, check to ensure that there is a server parameter file (SPFILE) version; if one does not exist, it can be created using the **Create-SPFile** statement. For more on this, see the Oracle documentation.

The **Undo-Tablespace-Clause** allows for stipulating a tablespace to be used for undoing data (similar to rolling back). When using this option, ensure that the UNDO_MANAGEMENT initialization parameter is set to AUTO (which is the default setting). Alternately, you may omit this clause on the database creation provided that you have the UNDO_TABLESPACE initialization parameter set to the desired undo tablespace.

Finally, the **Set-Time-Zone-Clause** is used for setting the time zone for the database. This can be done either by specifying the time as a displacement in hours and minutes from the Universal Time Coordinated (UTC, formerly called the Greenwich Mean Time) or by specifying a predefined time-zone-name stored in the Oracle catalog and accessible via the logical view called **V$TIMEZONE_NAMES**. Again, you may consult the Oracle product documentation for more insights; however, this will be much clearer after reading Chapter 14.

Observation:

In some systems, database creation is as simple as creating a directory (library or folder depending on the operating system used). Unfortunately, this is not the case in Oracle. For more details on this matter, see the Oracle Database Documentation Library (Oracle 2019).

11.4 Database Management

Once the database has been created, it must be populated with database resources (objects and services) to store and/or facilitate reference of data from the system. These resources include tablespaces, tables, indexes, views, synonyms, procedures, triggers, packages, sequences, users, and roles.

As a good DBA and/or software engineer, you will also need to carry out performance tuning on your database. This may involve reorganizing database tables and indexes, deleting unnecessary indexes or moving other objects. You will also be required to periodically perform backup and recovery procedures, or make alterations to the database itself. This and the next four chapters will provide important information relevant to the management of a database; additionally, Chapter 16 provides an overview of database administration. To enhance your overall appreciation, the syntax for the **Alter-Database** statement is provided in Figure 11.3. Like the **Create-Database** statement, you must have the DBA privilege in order to use the **Alter-Database** statement. A full discussion of the statement is not considered necessary for this introductory course, but is readily available in the Oracle product documentation. Suffice it to say, this statement is used extensively to effect structural as well as status changes to the database. Notice that each of the several clauses in the statement is optional.

EXAMPLE 11.2: SIMPLE DATABASE MODIFICATIONS

```
ALTER DATABASE SampleDB MOUNT;
ALTER DATABASE SampleDB OPEN READ ONLY;
ALTER DATABASE SampleDB Backup Controlfile To TRACE AS 'C:\Oracle\EFDB06\
   SampleTrace';
```

There is also a **Drop-Database** statement that allows you to delete the database (syntax omitted). As an alternative, you can use the DBCA to delete a database. Of course, in order to create, modify, or delete a database, you must have the DBA privilege.

```
Alter-Database ::=
Alter Database [<DatabaseName>]
Startup-Clauses | Recovery-Clauses | Datafile-Clauses | Logfile-Clauses | Controlfile-Clauses |
Standby-Database-Clauses | Default-Settings-Clauses | Conversion-Clauses |
Redo-Thread-Clauses | Security-Clause;

Startup-Clauses::= Startup1 | Startup2 | Startup3
Startup1 ::= MOUNT [STANDBY | CLONE DATABASE]
Startup2 ::= OPEN READ ONLY
Startup3 ::= OPEN READ WRITE [RESETLOGS | NORESETLOGS] [MIGRATE]

Recovery-Clauses ::= General-Recovery-Clause | Managed-Standby-Recovery | END BACKUP

General-Recovery-Clause ::=
RECOVER [AUTOMATIC [FROM '<PathtoFile>']]
Full-Database-Recovery-Clause | Partial-Database-Recovery-Clause | LOGFILE '<PathtoFile >'
TestCorruption | RecoveryOption1 | RecoveryOption2

TestCorruption ::= [TEST] | [ALLOW <n> CORRUPTION] | [NOPARALLEL] | [PARALLEL <n>]
RecoveryOption1 ::= RECOVER [AUTOMATIC [FROM '<PathtoFile >']] CANCEL
RecoveryOption2 ::= RECOVER [AUTOMATIC [FROM '<PathtoFile >']] CONTINUE [DEFAULT]

Full-Database-Recovery-Clause ::= Standby1 | Standby2 | Standby3 | Standby4
Standby1 ::= [STANDBY] DATABASE UNTIL CANCEL
Standby2 ::= [STANDBY] DATABASE UNTIL TIME <Timestamp>
Standby3 ::= [STANDBY] DATABASE UNTIL CHANGE <n>]
Standby4 ::= [STANDBY] DATABASE USING BACKUP CONTROLFILE

Partial-Database-Recovery-Clause ::= PartialR1 | PartialR2 | PartialR3 | PartialR4
PartialR1 ::= TABLESPACE <Tablespace> {, <Tablespace>}
PartialR2 ::= DATAFILE '<PathtoFile>' | <File-Number> {, '<PathtoFile >' | <File-Number>}
PartialR3 ::= STANDBY TABLESPACE <Tablespace> {, <Tablespace>}
              UNTIL [CONSISTENT WITH] CONTROLFILE
PartialR4 ::= STANDBY DATAFILE '<PathtoFile >' | <File-Number> {, '<PathtoFile >' | <File-Number>}
              UNTIL [CONSISTENT WITH] CONTROLFILE

Managed-Standby-Recovery-Clause ::=
RECOVER MANAGED STANDBY DATABASE Recover-Clause | Cancel-Clause | Finish-Clause

Cancel-Clause ::=
CANCEL [IMMEDIATE] [WAIT | NOWAIT]

Finish-Clause ::=
DISCONNECT [FROM SESSION] [NOPARALLEL] | [PARALLEL <n>]
FINISH [SKIP [STANDBY LOGFILE]][WAIT | NOWAIT]
```

Figure 11.3 Syntax for Alter-Database statement. (Continued)

```
Recover-Clause ::=  RClause1 | RClause2 | RClause3 | RClause4 | RClause5 | RClause6 | RClause7 | RClause8 |
                    RClause9 | RClause10 | RClause11 | RClause12 | RClause13 | RClause14
RClause1 ::= DISCONNECT [FROM SESSION]
RClause2 ::= TIMEOUT <n>
RClause3 ::= NOTIMEOUT
RClause4 ::= NODELAY
RClause5 ::= DEFAULT DELAY
RClause6::= DELAY <n>
RClause7::= NEXT <n>
RClause8::= EXPIRE <n>
RClause9 ::= NO EXPIRE
RClause10 ::= NOPARALLEL
RClause11::= PARALLEL <n>
RClause12 ::= THROUGH [THREAD <n>] SEQUENCE <n>
RClause13 ::= THROUGH ALL ARCHIVELOG
RClause14::= THROUGH ALL | LAST | NEXT SWITCHOVER

Datafile-Clause ::= DFSpec1 | DFSpec2 | DFSpec3 | DFSpec4 | DFSpec5 | DFSpec6 | DFSpec7 | DFSpec8 | DFSpec9 |
                    DFSpec10
DFSpec1 :: = CREATE DATAFILE <File-Number> | '<File-Name>' {,<File-Number> | '<File-Name>'}
             [AS NEW] | [AS <File-spec> {, <File-spec>}]
DFSpec2 :: = DATAFILE <File-Number> | '<File-Name>' {, <File-Number> | '<File Name>'} ONLINE
DFSpec3 :: = DATAFILE <File-Number> | '<File-Name>' {, <File-Number> | '<File Name>'} OFFILINE [DROP]
DFSpec4 :: = DATAFILE <File-Number> | '<File-Name>' {, <File-Number> | '<File Name>'} RESIZE <n> [K|M]
DFSpec5 :: = DATAFILE <File-Number> | '<File-Name>' {, <File-Number> | '<File Name>'} Autoextend-Clause
DFSpec6 :: = DATAFILE <File-Number> | '<File-Name>' {, <File-Number> | '<File Name>'} END BACKUP
DFSpec7 :: = TEMPFILE <File-Number> | '<File-Name>' {, <File-Number> | '<File Name>'}
             ONLINE | OFFLINE | Autoextend_Clause
DFSpec8 :: =TEMPFILE <File-Number> | '<File-Name>' {, <File-Number> | '<File Name>'} DROP [INCLUDING DATAFILES]
DFSpec9 :: = TEMPFILE <File-Number> | '<File-Name>' {, <File-Number> | '<File Name>'} RESIZE <nr> [K|M]
DFSpec10 :: = RENAME FILE '<PathtoFile>' {, '<PathtoFile>' } TO '<PathtoFile>' {, '<PathtoFile>' }

Autoextend-Clause ::= AutoOnOption | AutoOffOption
AutoOnOption ::= AUTOEXTEND OFF
AutoOffOption ::= AUTOEXTEND ON [NEXT <nr> [K|M] [Maxsize-Clause]]

Maxsize-Clause ::= FixedSize | VariableSize
VariableSize ::= MAXSIZE UNLIMITED
FixedSize ::= MAXSIZE <nr> [K|M]

Logfile-Clauses ::= LogClause1 | LogClause2 | LogClause3 | LogClause4 | LogClause5 | LogClause6 | LogClause7 |
                    LogClause8 | LogClausen9 | LogClause10
LogClause1 ::=  ARCHIVELOG | NOARCHIVELOG
LogClause2::=  [NO] FORCE LOGGING
LogClause3::=  ADD [STANDBY] LOGFILE [THREAD <n>] [GROUP <n>] Logfile-Spec {, [GROUP <n>] Logfile-Spec}
LogClause4::=  DROP [STANDBY] LOGFILE Logfile-Spec
LogClause5::=  ADD [STANDBY] LOGFILE MEMBER '<PathtoFile>' [REUSE]
               {, '<PathtoFile>' [REUSE]} TO Logfile-Spec {, Logfile-Spec}
LogClause6::=   DROP [STANDBY] LOGFILE MEMBER '<PathtoFile>' {, '<PathtoFile>'}
LogClause7::=   ADD SUPPLEMENTAL LOG DATA [(PRIMARY KEY | UNIQUE INDEX
                {, PRIMARY KEY | UNIQUE INDEX}) COLUMNS]
LogClause8::=   DROP SUPPLEMENTAL LOG DATA
LogOClause9::=   RENAME FILE '<PathtoFile>' {, '<PathtoFile>'} TO '<PathtoFile>' {, '<PathtoFile>'}
LogClause10::=   CLEAR [UNARCHIVED] LOGFILE Logfile-Spec {, Logfile-Spec} [UNRECOVERABLE DATAFILE]
```

Figure 11.3 (Continued) Syntax for Alter-Database statement. (Continued)

```
Logfile-Spec ::= LogOption1 | LogOption2 | LogOption3
LogOption1 ::= GROUP <n>
LogOption2 ::= ( '<PathtoFile>' {, '<PathtoFile>'})
LogOption3 ::=  '<PathtoFile>'

Controlfile-Clauses ::=  CFOption1 | CFOption1 | CFOption3
CFOption1 ::=  CREATE STANDBY CONTROLFILE AS '<PathtoFile>' [REUSE]
CFOption2 ::=  BACKUP CONTROLFILE TO Tracefile-Clause
CFOption3 ::=  BACKUP CONTROLFILE TO  '<PathtoFile>' [REUSE]

Tracefile-Clause ::=
TRACE [AS '<PathtoFile> [REUSE] [RESETLOGS | NORESETLOGS]]

Standby-Database-Clauses ::=  SClause1 | SClause2 | SClause3 | SClause4 | SClause5 | SClause6 | SClause7 |
                              SClause8 | SClause9
SClause1 ::=  ACTIVATE [PHYSICAL | LOGICAL] STANDBY DATABASE [SKIP [STANDBY LOGFILE]]
SClause2::=  SET STANDBY DATABASE TO MAXIMIZE PROTECTION | AVAILABILITY | PERFORMANCE
SClause3 ::=  REGISTER [OR REPLACE] [PHYSICAL | LOGICAL] LOGFILE Logfile-Spec {, Logfile-Spec}
SClause4 ::=  START LOGICAL STANDBY APPLY [[NEW PRIMARY <DBLINK>] | [INITIAL <Scan-value>]]
SClause5 ::=  STOP | ABORT LOGICAL STANDBY APPLY
SClause6 ::=  NOPARALLEL
SClause7 ::=  PARALLEL <n>
SClause8 ::=  COMMIT TO SWITCHOVER TO PHYSICAL | LOGICAL PRIMARY | STANDBY
SClause9::=  [WITH | WITHOUT SESSION SHUTDOWN] [WAIT | NOWAIT]

Default-Settings-Clauses ::=  DSClause1 | DSClause2 | DSClause3 | DSClause4
DSClause1 ::= [NATIONAL] CHARACTER SET <Charset>
DSClause2 ::= Set-Time-Zone-Clause
DSClause3 ::= DEFAULT TEMPORARY TABLESPACE <TablespaceName>
DSClause4 ::= RENAME GLOBAL_NAME TO <Database-Name>.<Domain>

Set-Time-Zone-Clause ::=
SET TIME_ZONE = '± <HH:MM>' | <Time-Zone-Region>

Conversion-Clauses ::=
CONVERT | ResetOption
ResetOption ::= RESET COMPATIBILITY

Redo-Thread-Clauses ::=  EnableThread  |  DisableThread
EnableThread ::= ENABLE [PUBLIC] THREAD <n>
DisableThread ::= DISABLE THREAD <n>

Security-Clause ::=
GUARD ALL | STANDBY | NONE
```

Figure 11.3 (Continued) Syntax for Alter-Database statement.

11.5 Tablespace Creation

As mentioned in Section 11.2, the tablespace is the holding for various database objects (actually, the objects are stored in datafile(s) contained within the tablespace). A database must have at least one tablespace, but typically has several. The following are tablespaces that are typically found in a database:

- System tablespace (prefixed SYSTEM): This is a default tablespace used for system resources.
- Temporary tablespace (prefixed TEMP): Used for intermediate results such as internal sorts for user queries.
- Tools tablespace (prefixed TOOLS): Used for Oracle administrative tools.

- Index tablespace (prefixed INDX): For storing and maintaining indexes.
- Undo tablespace (prefixed UNDOTBS): For automatic logs due to data changes (this is particularly useful for data rollbacks and/or data recoveries after a catastrophe).
- Users tablespace (prefixed USER): For storing user created tables.

A full discussion of Oracle tablespaces is beyond the scope of this course. However, in the interest of a credible introduction, Figure 11.4 provides the syntax of the **Create-Tablespace** statement.

As can be seen from the syntax, there are three categories of tablespaces: a permanent tablespace stores persistent data in datafile(s); a temporary tablespace is used for storing transitory data in temporary files during an Oracle session (which typically starts at login time); and an undo tablespace is useful in supporting the database when it is in automatic undo management mode. Following is an overview of some of the main parameters in the statement:

BIGFILE | SMALLFILE: BIGFILE indicates that the tablespace will contain one datafile with up to 4 billion data blocks. SMALLFILE indicates that this will be a traditional Oracle tablespace with the possibility for multiple datafiles each of which can contain up to 4 million blocks.

DATAFILE: This clause facilitates specification of the location(s) and constraints for datafile(s) associated with the database. This/these datafile(s) will be stored in this tablespace.

MINIMUM EXTENT: This clause specifies the minimum size of extents for the tablespace. The tablespace will grow in increments of the extent size. The size may be specified in kilobytes (K), megabytes (M), gigabytes (G), terabytes (T), petabytes (P), or exabytes (E).

BLOCKSIZE: The size of each data block. If not specified, Oracle will set a standard block size based on the underlying operating system.

LOGGING | NOLOGGING: LOGGING indicates that the creation and subsequent insertions to the tablespace will be logged in the redo logs; NOLOGGING is the opposite.

Force LOGGING ensures that activities in the tablespace are logged. This logging is inherited by other objects stored in the tablespace.

The **Extent-Management-Clause** allows you to specify how the table extents (i.e., growth and usage of disk space) will be managed. You have the option of dictionary-managed tablespace (managed from the Oracle Data Dictionary) or a locally managed tablespace. In the interest of efficiency, the latter approach is recommended.

The **Segment-Management-Clause** allows you to specify whether storage segments will be managed manually or automatically; the latter is recommended.

The **Storage-Clause** facilitates specification of storage parameters for the tablespace. It includes issues such as initial extent size, subsequent extent increments, the number of extents, free-pace list, and other related matters.

EXAMPLE 11.3: ILLUSTRATING TABLESPACE CREATION

```
CREATE TABLESPACE SampleTBS
DATAFILE 'C:\Oracle\Oradata\SampleDB\SampleTBS.dbf' SIZE 500M
AUTOEXTEND ON NEXT 1M
MAXSIZE UNLIMITED
EXTENT MANAGEMENT LOCAL UNIFORM SIZE 100K
SEGMENT SPACE MANAGEMENT AUTO;
```

```
Create-Tablespace::=
CREATE [BIGFILE | SMALLFILE] <Permanent-Tablespace-Clause> | <Temporary-Tablespace-Clause> | <Undo-Tablespace-Clause>;

Permanent-Tablespace-Clause ::=
TABLESPACE <TablespaceName>
[DATAFILE <FileSpec> {, <FileSpec>}]
[MINIMUM EXTENT <n> [K|M|G|T|P|E]] /*NA if locally managed; see Extent-Management-Clause */
[BLOCKSIZE <n> [K]]
[LOGGING | NOLOGGING] [FORCE LOGGING]
[ONLINE | OFFLINE]
 [DEFAULT] [COMPRESS | NOCOMPRESS] <Storage-Clause>]
[<Extent-Management-Clause>]
[<Segment-Management-Clause>]
[FLASHBACK ON|OFF]

Temporary-Tablespace-Clause ::=
TEMPORARY TABLESPACE <TablespaceName>
[TEMPFILE <FileSpec> {,<FileSpec>}]
[TABLESPACE GROUP <TablespaceGroupName> | ' ']
[<Extent-Management-Clause>]

Undo-Tablespace-Clause ::=
UNDO TABLESPACE <TablespaceName>
[DATAFILE <FileSpec> {,<FileSpec>}]
[<Extent-Management-Clause>]
[RETENTION GUARANTEE | NONGUARANTEE]

Extent-Management-Clause ::= DictionaryOption | LocalOption1 | LocalOption2
DictionaryOption ::= EXTENT MANAGEMENT DICTIONARY
LocalOption1 ::= EXTENT MANAGEMENT LOCAL AUTOALLOCATE
LocalOption2 ::= EXTENT MANAGAEMENT LOCAL [UNIFORM [SIZE <n> [K|M|G|T|P|E]]]

FileSpec ::=
'<PathtoFile>' SIZE <n> K|M|G|T|P|E [REUSE] [<Autoextend-Clause>]

Storage-Clause ::=
STORAGE (
[INITIAL <n> [K|M|G|T|P|E]]
[NEXT <n> [K|M|G|T|P|E]]
[MINEXTENTS <n>]
[MAXEXTENTS <In> | UNLIMITED]
[PCTINCREASE <n>] [FREELISTS <n>] [FREELIST GROUPS <n>]
[OPTIMAL <n> [K|M|G|T|P|E]]
[OPTIMAL NULL]
[BUFFER_POOL KEEP | RECYCLE | DEFAULT]
)

Autoextend-Clause ::=
AUTOEXTEND OFF /* Or */
AUTOEXTEND  ON [NEXT <n> K|M|G|T|P|E] [<Maxsize-Clause>]

Maxsize-Clause ::=
MAXSIZE <n> K|M|G|T|P|E  /* Or */
MAXSIZE UNLIMITED

Segment-Management-Clause ::=
SEGMENT SPACE MANAGEMENT MANUAL | AUTO   /* AUTO recommended */
```

Figure 11.4 Syntax for Create-Tablespace statement.

11.6 Tablespace Management

Once the tablespace has been created and populated with database objects, it will need to be managed. One way to do this is via the **Alter-Tablespace** statement. Through this command, you can determine structural as well as status changes to the tablespace. As with the **Create-Tablespace** command, a full discussion of the **Alter-Tablespace** statement is beyond the scope of this course. However, in the interest of credibility, its syntax is provided in Figure 11.5. From the figure, observe that parameters that were specified during the **Create-Tablespace** statement can be modified via the **Alter-tablespace** statement. You will observe this (approximate) symmetry (i.e.,

```
Alter-Tablespace::=
ALTER TABLESPACE <TablespaceName>
[DEFAULT [COMPRESS | NOCOMPRESS] <Storage-Clause>]
[MINIMUM EXTENT <n> K|M|G|T|P|E]
[RESIZE <n> K|M|G|T|P|E]
[COALESCE]
[SHRINK SPACE [KEEP <n> K|M|G|T|P|E]]
[RENAME TO <NewTablespaceName>]
[BEGIN | END BACKUP]
[<Datafile-Tempfile-Clauses>]
[<Tablespace-Logging-Clause>]
[TABLESPACE GROUP <TablespaceGroupName> | ' ']
[<Tablespace-State-Clause>]
[<Autoextend-Clause>]
[FLASHBACK ON | OFF]
[<Tablespace-Retention-Clause>];

Datafile-Tempfile-Clauses ::=
[ADD DATAFILE | TEMPFILE <FileSpec> {,<FileSpec>}]
[DROP DATAFILE | TEMPFILE <FileNumber> | <'FileName'>]
[SHRINK TEMPFILE <FileNumber> | <'FileName'> [KEEP <n> K|M|G|T|P|E]]
[RENAME DATAFILE <'FileName'> {, <'FileName'>} TO <'NewFileName'> {, <'NewFileName'>}]
[DATAFILE | TEMPFILE ONLINE | OFFLINE]

Tablespace-State-Clause ::=
[ONLINE | OfflineSpec]
[READ ONLY | WRITE]
[PERMANENT | TEMPORARY]

Offline-Spec ::= OFFLINE [NORMAL | TEMPORARY | IMMEDIATE]

Tablespace-Logging-Clause ::=
LOGGING | NOLOGGING /* Or*/
[NO] FORCE LOGGING

Tablespace-Retention-Clause ::=
RETENTION GUARANTEE | NOGUARANTEE

FileSpec ::=              // As defined in Create-Tablespace
Storage-Clause ::=        // As defined in Create-Tablespace
Autoextend-Clause ::=     // As defined in Create-Tablespace
Maxsize-Clause ::=        // As defined in Create-Tablespace
```

Figure 11.5 Syntax for Alter-Tablespace statement.

orthogonality) of SQL (with respect to objects and applicable operations) for most of the database resources that we will discuss. Finally, observe that each main clause of the statement is optional.

EXAMPLE 11.4: ALTERING A TABLESPACE TO BE OFFLINE

```
ALTER TABLESPACE SampleTBS
OFFLINE NOLOGGING;
```

Of course, you can also delete a tablespace via the **Drop-Tablespace** statement. Figure 11.6 provides the syntax along with an example. Finally, through the Oracle Enterprise Manager (OEM) component, you can manage tablespaces (and most other database objects) in a GUI environment.

```
Drop-Tablespace ::=
DROP TABLESPACE <Tablespace> [INCLUDING CONTENTS [AND DATAFLES]] [CASCADE CONSTRAINTS];

Example:
DROP TABLESPACE SampleTBS INCLUDING CONTENTS AND DATAFILES;
```

Figure 11.6 Syntax and example for Drop-Tablespace statement.

11.7 Oracle Cloud Features

In recent years, there has been increased emphasis on Oracle cloud services, facilitated through the Oracle Academic Initiative (OAI); this program provides support for higher education in related disciplines such as Software Engineering, Database Systems, and Data Science in over 120 countries. Supported DBMS platforms include Oracle, MySQL, and NoSQL. All services and resources are free to OAI members. Through the Oracle Academy Cloud (OAC) program and the Oracle Application Express (APEX) program, cloud-based database services are provided with no observed restriction on storage space. Moreover, for better or worse (and there are opinions on either side), the cloud-based services provide the database expert with a simplified perspective of the database environment by obscuring the gory details of creating and managing the database and constituent tablespaces. The cloud infrastructure therefore invites the database expert to focus more on conceptual schema objects such as tables, user accounts, logical views, and stored procedures. Other configuration objects such as the database, tablespaces, datafiles, and data blocks are effectively handled by the cloud infrastructure.

11.7.1 Creating and Using a Database Instance via OAC

Oracle allows each OAI account holder to log on and request an OAC account. This will require following several related steps. Once completed, the OAC account holder is able to then log onto the Oracle Cloud Infrastructure and create a cloud-based database. Once created, the database must be secured and additional accounts can be created by the account administrator. The database can be accessed via SQLD or APEX — the two aforementioned applications (mentioned in Section 11.1).

11.7.2 Accessing the OAC Database via SQL Developer

The SQLD component can be configured to access the cloud-based Oracle database as follows: First, the database administrator must log onto the OAC account, then retrieve and securely download a database wallet of the database. The wallet stores the essential URL and other related data for the actual database. Next, the wallet must be placed in an area that is available to client installations of SQLD.

Each installation of SQLD must be configured to have at least one connection that accesses the database via the database wallet. Each connection must have a name, password, and connection type of cloud wallet.

11.7.3 Accessing the OAC Database via Application Express

As an alternative to SQLD, the database administrator may log onto the OAC account and access APEX. During the first access of APEX, the administrator will be prompted to create at least one workspace. Subsequently, the administrator may create other workspaces and user accounts for the various users of the database.

Each user will be able to access the database by accessing the assigned URL and logging on. The administrator may access any of the workspaces from the central OAC account. Once logged in, the user is able to use the GUI features of APEX or issue direct SQL commands.

11.8 Table Creation

Once the database and tablespaces have been configured, the next logical step is to create database objects which will be stored in the tablespace(s). Relational and/or object tables are among the first type of database objects to be created in a database; this is done via the **Create-Table** statement. The **Create-Table** statement is one of the most multi-faceted statements in SQL (next to the **Select** statement). The complete syntax is provided in Appendix 4, while an abridged version of the statement is provided in Figure 11.7.

If you take a quick peek at Appendix 4, you will notice that the **Create-Table** statement has several parameters and clauses. A full discussion of each of these parameters/clauses is beyond the scope of this course, hence the abridged form of the statement as shown in Figure 11.7. Following is a brief clarification on each component shown in the figure.

When you create a table, you are required to specify its name; if the table is to be owned by a schema that is different from the current user, it must be qualified by the schema name. The name given to the table must be unique for the database being constructed, and should observe a predefined object naming convention such as described in Foster (2021) as well as Section 3.9. This is followed by the definition of columns and constraints for the table.

The **Column-Definition-Clause** facilitates column definition. A table must have at least one column, but typically has several columns. Each column definition involves stating the column name, its data-type specification (via the **Data-Type-Spec**), an optional specification of a default initial value for the column in the absence of an explicit stipulation when records are being inserted, and the optional use of keywords NOT NULL (to indicate that the column is not allowed to have a null value) and/or UNIQUE (to indicate that duplicate values for this column is not allowed). The **Data-Type-Spec** allows you to indicate a valid data-type for the column. Figure 11.8 provides a list of the commonly used data-type options (this list is not exhaustive).

The **Constraint-Definition-Clause** gives the opportunity to define integrity constraints for the table. The table may have one or multiple constraints. As shown in Figure 11.7, a constraint

```
Create-Table-Statement::=
CREATE TABLE [<Schema>.] <TableName>
      (<Column-Definition-Clause> {,<Column-Definition-Clause>}
      [,<Constraint-Definition-Clause> {, <Constraint-Definition-Clause>}]);

Column-Definition-Clause ::=
<ColumnName> <Data-Type-Spec> [DEFAULT <ScalarExpression>][NOT NULL][UNIQUE]

Constraint-Definition-Clause ::=
[CONSTRAINT <ConstraintName>] <Primary-Key-Spec> | <Foreign-Key-Spec> | NOT NULL | <Check-Spec> | <Unique-Spec>

Primary-Key-Spec ::=
PRIMARY KEY (<ColumnName> [ASC/DESC] {,<ColumnName> [ASC/DESC] })

Foreign-Key-Spec ::=
FOREIGN KEY (<ColumnName>) REFERENCES [<Schema>.] <TableName> [(<ColumnName>)]
[ON DELETE CASCADE]

Check-Spec ::= <CFormat1> | <CFormat2> | <CFormat3>
CFormat1 ::= CHECK (<Column> BETWEEN <Value1> AND <Value2>)
CFormat2 :::= CHECK (<Column> operator <Value>)
CFormat3 :::= CHECK (<Column> IN <Value> {,<Value>})

Unique-Spec ::= UNIQUE <Column>
```

Figure 11.7 Abridged version of the Create-Table statement.

could be a primary-key specification, foreign-key specification, check specification, not-null specification, or a unique specification. Each table should have a primary-key specification; the other constraints are optional depending on the structure of the table.

The **Primary-Key-Spec** begins with the keywords PRIMARY KEY, followed by a parenthesized list of columns participating in the primary key (with the comma being used as the separator); obviously, the minimum number of columns is 1. Immediately following each column, you have the option of specifying ASC (for ascending, which is the default) or DESC (for descending).

The **Foreign-Key-Spec** starts off with the keywords FOREIGN KEY, followed by the parenthesized column acting as the foreign key. This is followed by the keyword REFERENCES, the name of the table being referenced (optionally qualified by its schema if not owned by the current user), and in parentheses, the name of the column being referenced. Note that if the referenced column has the same name as the referencing column, specifying the former is optional. However, as pointed out earlier in the course, it is highly recommended that you keep column names (i.e., attributes) unique across the database; if you observe this recommendation, then specifying the referenced column becomes mandatory. Whatever the circumstance, in the interest of clarity, it is a good idea to treat the referenced column as mandatory. Finally, notice the optional ON DELETE CASCADE. If specified, this clause has the potential of causing a cascaded deletion of the referencing record (provided that this option is also used for all other foreign keys in the referencing table) when a referenced record is being deleted from the system. This is a rather risky option that could ultimately compromise the integrity of the database; for this reason, it is not recommended.

The **Check-Spec** commences with the keyword CHECK and is followed by a parenthesized constraint that must be satisfied by any value stored in this column. The syntax shows three formats: the column value must fall in a range of values BETWEEN a starting value and a final value; the column value must meet the stated comparison to some other stated value; or the column value must belong to a set of values specified after the IN keyword.

Data Type	Meaning
Scalar Data Types:	
CHAR (<Length> [BYTE|CHAR])	Character or string of specified length; default is BYTE; maximum length is 2000 bytes.
VARCHAR2 (<Length> [BYTE|CHAR])	Variable string of specified maximum possible length; minimum length is 1; maximum length is 4,000 bytes if the system parameter MAX_STRING_SIZE is STANDARD, or 32,767 bytes if system parameter MAX_STRING_SIZE is EXTENDED.
NCHAR (<Length>)	Similar to CHAR, but used for Unicode character set; default length is 1 character.
NVARCHAR2 (<Length>)	Similar to VARCHAR2, but used for Unicode character set.
LONG	Variable length character up to 2GB. CLOB or NCLOB preferred. Included for backward compatibility.
BINARY_FLOAT	32-bit floating point number.
BINARY_DOUBLE	64-bit floating point number.
NUMBER(<Length> [,<Decimal>])	Fixed and floating point (numeric) data; default is 38 digits; valid range is $-1 * 10^{-130}$ to $9.999 * 99^{125}$
DECIMAL(<Length>, <Decimal>)	Same as NUMBER((<Length>, <Decimal).
INTEGER	Defaults to NUMBER(38).
SMALLINT	Same as INTEGER.
FLOAT [(<Length>)]	Similar NUMBER; default length is 38; maximum length or 126.
REAL	Defaults to NUMBER(63).
DOUBLE PRECISION	Same as NUMBER(38)
DATE	Date in the form century, year, month, day, hour, minute, second. Can be displayed in various formats. Range is January 1, 4712 BC to December 31, 9999 AD. Default time is 12:00 AM.
TIMESTAMP [(<Precision>)]	Similar to DATE, but with possibility of fractional seconds precision.
TIMESTAMP [(<Precision>)] [WITH TIME ZONE]	Similar to TIMESTAMP, but stores the time zone displacement. Displacement is the difference (in hours and minutes) between the local time and the Universal Time Coordinate (UTC), also known as the Greenwich Mean Time
RAW (<Length>)	Variable length unstructured data; maximum 2,000 bytes if the system parameter MAX_STRING_SIZE is STANDARD, or 32,767 bytes if system parameter MAX_STRING_SIZE is EXTENDED. BLOB or BFILE preferred. Included for backward compatibility.
LONG RAW (<Length>)	Variable length unstructured data; maximum 2GB. BLOB or BFILE preferred. Included for backward compatibility.
BLOB	Binary Large Object; maximum size of [(4GB – 1) * block-size].
CLOB	Character BLOB; maximum size of [(4GB – 1) * block-size].
NCLOB	Unicode Character BLOB; maximum size of [(4GB – 1) * block-size].
BFILE	Large external file, stored by the operating system; maximum file size of 4GB supported. Oracle stores a pointer to the file.
ROWID	Binary data representing a physical row address of a table's row.
Collection Data Types:	
VARRAY	Variable array; elements are ordered and have a maximum limit.
TABLE	Nested table; elements are not ordered, and there is no limit on the possible number of elements.
Reference Data Type:	
REF	Reference (via a pointer) to data stored in another object table

Figure 11.8 Valid data types.

Having a working knowledge of the **Create-Table** statement is imperative. As you study this statement, please note the following:

1. The semicolon (;) signals the end of the SQL statement and each clause is separated by a comma (,).
2. Pay attention to how null values are treated in the DBMS being used; the treatment tends to vary across different systems. For instance, in DB2, primary keys must be declared NOT NULL; additionally, you are not allowed to define a unique constraint that includes a

column that can have a null value. Note also that null is not the same as zero or blank. Generally speaking, null represents the absence of a value for that particular column. In some instances, the DBMS may insert a place holder for the null column (if this null column is followed by a non-null column).

3. Most DBMS suites provide a utility that allows the user to key data into the table, once it has been created. Earlier versions of Oracle do not. However, through Oracle Enterprise Manager (OEM) and SQL Developer, later versions (starting with Oracle 9i) provide this feature.
4. Data can be entered in the table via the SQL insert statement (which you will learn in Chapter 12) or an application program, written for that purpose.
5. In many modern DBMS suites, a GUI sits on top of SQL definition statements, thus facilitating easier object creation and management. For example, MySQL's GUI interface is called **phpMyAdmin**. In the Oracle environment, resources such as SQL Developer (SQLD) and Oracle Enterprise Manager (OEM) fulfill this role.
6. Some DBMS (e.g., Oracle) suites provide the user with the flexibility of specifying which attribute of a referenced table is to be used in the foreign key characteristic.
7. The operator specified in the **Check-Clause** may be any valid Boolean operator as defined in Chapters 7 and 8. When the third form of the **Check-Clause** is used, the column specified must have one of the values specified.

EXAMPLE 11.5: CREATE RELATIONS ACADEMICPROGRAM, COURSE, HALL, AND PGM_STRUCT OF CHAPTER 7

```
CREATE TABLE AcademicProgram
(Pgm# CHAR (5) NOT NULL,
PgmName VARCHAR2 (30) NOT NULL,
CONSTRAINT ProgPK PRIMARY KEY (Pgm#));

CREATE TABLE Course
(Crs# CHAR (8) NOT NULL,
CrsName VARCHAR2 (30) NOT NULL,
CONSTRAINT CoursePK PRIMARY KEY (Crs#));

CREATE TABLE Pgm_Struct
(PSPgm# CHAR (5) NOT NULL,
PSCrs# CHAR (8) NOT NULL,
PSCrsSeqn NUMBER (3),
CONSTRAINT ProgStructPK PRIMARY KEY (PSPgm# ,PSCrs#),
CONSTRAINT ProgStructFK1 FOREIGN KEY (PSPgm#) REFERENCES
  AcademicProgram (Pgm#),
CONSTRAINT ProgStructFK2 FOREIGN KEY (PSCrs#) REFERENCES Course
  (Crs#));

CREATE TABLE Hall
(Dept# CHAR(4) NOT NULL,
HallName VARCHAR2(30),
CONSTRAINT HallPK PRIMARY KEY (Hall#));
```

EXAMPLE 11.6: CREATE RELATIONS STUDENT, JOBDESCRIPTION, DEPARTMENT, AND EMPLOYEE

```
CREATE TABLE Student
(Stud# NUMBER (7) NOT NULL,
StudLName VARCHAR2 (15) NOT NULL,
StudFName VARCHAR2 (15) NOT NULL,
StudMName VARCHAR2 (15),
StudSex CHAR (1) NOT NULL,
StudDoB NUMBER (8),
StudPgm# CHAR (5),
StudHall# CHAR(4),
StudGPA NUMBER (4,2),
CONSTRAINT StudentPK PRIMARY KEY (Stud#),
CONSTRAINT StudentFK1 FOREIGN KEY (StudPgm#) REFERENCES
AcademicProgram (Pgm#),
CONSTRAINT StudentFK2 FOREIGN KEY (StudHall#) REFERENCES Hall (Hall#),
CONSTRAINT StudentCheck1 CHECK (StudSex IN ('M', 'F')),
CONSTRAINT StudentCheck2 CHECK (StudDoB BETWEEN 19000101 AND 21991231),
CONSTRAINT StudentCheck3 CHECK (StudGPA >= 0));

CREATE TABLE JobDescription
(JobCode CHAR(4) NOT NULL,
JobTitle VARCHAR2(30),
JobNarrative VARCHAR2(100),
JobAbbrev VARCHAR2(6),
CONSTRAINT JobDescriptionPK PRIMARY KEY (jobCode));

CREATE TABLE Department
(Dept# NUMBER(4) NOT NULL,
DeptName VARCHAR2(30),
CONSTRAINT DepartmentPK PRIMARY KEY (Dept#));

CREATE TABLE Employee
(Emp# NUMBER (7) NOT NULL,
EmpLName VARCHAR2 (15) NOT NULL, EmpFName VARCHAR2 (15) NOT NULL,
EmpMName VARCHAR2 (15), EmpSex CHAR (1) NOT NULL,
EmpDoB NUMBER (8),
EmpDept# NUMBER (4),
EmpSalary NUMBER (11,2), EmpJobCode CHAR(4),
CONSTRAINT EmployeePK PRIMARY KEY (Emp#),
CONSTRAINT EmployeeFK1 FOREIGN KEY (EmpDept#) REFERENCES Department
(Dept#),
CONSTRAINT EmployeeFK2 FOREIGN KEY (EmpJobCode) REFERENCES
JobDescription (JobCode),
CONSTRAINT EmployeeCheck1 CHECK (EmpSex IN ('M', 'F')),
CONSTRAINT EmployeeCheck2 CHECK (EmpDoB BETWEEN 19000101 AND
21991231));
```

Once your table has been created, you may view the structure of the table via the **Describe** statement, which has the following format:

```
DESCRIBE <TableName>;
```

EXAMPLE 11.7: LIST THE STRUCTURE OF THE TABLES CREATED

```
DESCRIBE AcademicProgram;
DESCRIBE Course;
DESCRIBE Pgm_Struct;
DESCRIBE Student;
```

You can also create a table from an existing table by feeding the output of a **Select** statement as the input to a **Create-Table** statement. The format of the **Create-Table** statement for this is as follows:

```
CREATE TABLE <TableName> [(<Column> {,<Column>}] AS <Sub-query>;
```

The sub-query component of the statement is specified by a **Select** statement. An example of this is provided below; this will be much clearer after you have been introduced to the **Select** statement (next chapter).

EXAMPLE 11.8: TWO ILLUSTRATIONS OF CREATING A TABLE FROM AN EXISTING TABLE

```
/* The following statements create two snapshot relations: the first one stores
Computer Science courses only; the second stores Computer Science and Information
Systems majors only. Sample data comes from Figure 7.3 of Chapter 7. */

CREATE TABLE CSCourses AS SELECT * FROM Course WHERE Crs# LIKE 'CS%';

CREATE TABLE CSMajors (Stud#, StudLName, StudFName, StudSex) AS SELECT *
   FROM Student WHERE (StudPgm# = 'BSC1' OR StudPgm# = 'BSC2');
```

11.9 Dropping or Modifying a Table

As you would expect, a table may be dropped (deleted from the system) via the **Drop-Table** statement. The syntax follows:

```
DROP TABLE [<Schema>.] <TableName> [CASCADE CONSTRAINTS];
```

To drop all referential integrity constraints that refer to primary and unique keys in the dropped table, specify the **Cascade-Constraints-Clause**. If you omit this clause and such referential integrity constraints exist, Oracle will return an error and will not drop the table. Also note that when a table is removed, all references to it are removed from the system catalog. Any attempt to access a nonexistent table will result in an execution error.

EXAMPLE 11.9: REMOVE THE CSCOURSES TABLE FROM THE SYSTEM

DROP TABLE CSCourses CASCADE CONSTRAINTS;

The structure of a table can be modified via the **Alter-Table** statement. Figure 11.9 provides an abridged syntax for the **Alter-Table** statement; a more detailed syntax is available in Appendix 4.

```
Alter-Table ::=
ALTER TABLE <TableName>
[ADD <Column-Definition>]
[MODIFY <Column-Definition>]
[ADD <Constraint-Definition>]
[DROP <Column> {,<Column>}]
[DROP PRIMARY KEY]
[DROP FOREIGN KEY (<Column> {,<Column>})]
[DROP/DISABLE/ENABLE CONSTRAINT <ConstraintName>];

// Column-Definition and Constraint-Definition are as defined in Create-Table
```

Figure 11.9 Abridged version of the Alter-Table statement.

EXAMPLE 11.10: INCLUDE AN ADDITIONAL ATTRIBUTE FOR COURSE-ABBREVIATION IN THE COURSE TABLE

ALTER TABLE Course ADD CrsAbbr CHAR (4);

This example introduces the idea of including additional attributes in a table after it has been created and put into production. Recall that **Course** was originally created in Example 11.5 with two columns. Note the following:

1. The **Course** table now has three columns — **Crs#, CrsName,** and **CrsAbabbr.**
2. Existing records in the table are all amended to have the additional attribute.
3. The records are not ALTERED at the time of the statement, but noted in the system catalog. At the next read of the table, the DBMS appends NULLed attribute values to the records. The next write to disk writes the expanded records if the additional null values have been updated to the non-null.
4. It is a good habit to assign default values to attributes appended subsequent to table creation and usage, provided that this is supported by the DBMS.

EXAMPLE 11.11: ADJUST THE STUDENT TABLE TO INCLUDE THE ASSIGNED HALL AND DEPARTMENT AS A FOREIGN KEY

```
ALTER TABLE Student ADD StudHall# CHAR (4)
ADD FOREIGN KEY (StudHall#) REFERENCES Hall (Hall#)
ALTER TABLE Student ADD StudDept# NUMBER (4)
ADD FOREIGN KEY (StudDept#) REFERENCES Department (Dept#);
```

Oracle also allows you to rename a table (or any valid database object) via the **Rename** command. You can also remove all rows from a table via the **Truncate-Table** command. The abridged syntax for each statement is provided in Figure 11.10:

Rename-Statement ::= RENAME <ObjectName> To <ObjectName>;

Truncate-table-Statement ::= TRUNCATE TABLE <TableName>;

Figure 11.10 Syntax for the Rename and Truncate statements.

EXAMPLE 11.12: RENAME THE HALL TABLE TO DORMITORY AND EMPTY THE STUDENT TABLE

```
RENAME Hall TO Dormitory;
TRUNCATE TABLE Student;
```

11.10 Working with Indexes

An index is a database object that is used to speed up the retrieval of tuples from base relations. The index is independent of the table it indexes. It stores address pointers to tuples of the base relation. Once created, indexes are automatically maintained by the DBMS and used to service certain user access requests as required. However, they must be initially created by an authorized member of the database team.

Despite the benefits, indexes should not be created indiscriminately. Following are some general circumstances, any combination of which may warrant the introduction of index(es):

- A column or column combination is frequently used in query conditions — within a table or in join queries across multiple tables.
- The table is large, and there are frequent queries retrieving a fraction — roughly less than 15% — of the tuples. However, this benchmark may be lower for higher table scan rates or higher for more clustered data.
- A column or column combination is frequently used in the **Order-By** clause of queries.
- If a column or column combination contains a wide range of values, this may be a candidate for a regular index; if the range of values is narrow, this may be suitable for a bitmap index.

- If a column or column combination serves as a candidate key or will contain values that are relatively unique, this may warrant the introduction of an index.
- A column contains a large number of null values.

There are also occasions when index creation is not recommended. Generally speaking, an index is not necessary when any of the following circumstances holds:

- The table is small.
- Most queries are expected to retrieve a high percentage of the tuples.
- The table is frequently updated.
- Columns are not often used in database queries.

The foregoing guidelines should impact your decision on the creation of indexes for most of the scenarios you will likely encounter. Additionally, here are two caveats:

1. Notwithstanding the aforementioned guidelines, there will likely be scenarios that challenge the application of these guidelines. In such cases, good judgment should follow careful analysis and discussion.
2. Revisit Section 11.7 and Examples 11.5 and 11.6. Observe that primary key constraints were given specific names. Typically, the DBMS creates an index based on the primary key constraint. It is recommended that you assign a name for the constraint; otherwise, a cryptic name will be assigned.

Consider the **Student** table from the college database of the earlier discussion. Let us suppose that in addition to the attributes of Figure 7.1b, there is an attribute **StudSSN** representing the student's social security number. In a typical college database environment, one would expect that the **Student** table is frequently updated, thus suggesting that it should not be indexed. However, it is also reasonable to anticipate that this table will be the target of numerous queries that retrieve less than 15% of the records, thus suggesting that it should be indexed. Moreover, apart from keying the table on **Stud#** (the student's ID number), it would be prudent to also facilitate alternate access paths to the student data such as an index on **StudSSN**. Given these conflicting demands, Example 11.13 provides a set of possible indexes for the table.

EXAMPLE 11.13: RECOMMENDED INDEXES FOR THE STUDENT TABLE

Student {Stud#, StudLName, StudFName, StudSex, StudAddr, StudPgm#, StudHall#, StudDoB, StudSSN…} PK [Stud#]

Recommended Indexes:

- StudentPK on [Stud#]: This would be the primary key
- StudentNX2 on [StudSSN]: This would be a unique index
- StudentNX3 on [StudLName, StudFName]: This would be a non-unique index
- StudentNX4 on [StudPgm#, Stud#]: This would be a non-unique index

An experienced software engineer or DBA can reasonably determine what indexes are to be created in a database, as this could seriously affect the performance of the database. Once created, the DBMS will automatically maintain these indexes. However, at a subsequent time, these indexes can be revisited and a decision made as to whether they should be kept or removed. When queries are executed, if there are indexes available to support the queries, the DBMS will automatically make use of them. This results in significantly improved performance on the queries. In the absence of these query-supporting indexes, the DBMS is made to work harder on servicing these queries.

Figure 11.11 shows an abridged syntax for the **Create-Index** statement; a more detailed syntax is provided in Figure 11.12; this is followed by the syntax for the **Alter-Index** statement in Figure 11.13 and an example.

```
Create-Index ::=
CREATE [UNIQUE] [BITMAP] INDEX <IndexName> ON <TableName> (<Column> [ASC/DESC]
{,<Column> [ASC/DESC]}) [CLUSTER] [COMPUTE STATISTICS];
```

Figure 11.11 Abridged version of the Create-Index statement.

After specifying the index name and participating columns, the other parameters (on the **Create-Index** as well as the **Alter-Index** statement) are optional. A full discussion of each clause/parameter is beyond the scope of this course. However, the Oracle product documentation provides copious clarifications and illustrations (see recommended readings). Here are a few noteworthy points about these two statements:

1. Three broad categories of indexes are cluster indexes, regular table indexes, and bitmap indexes.
2. The default order is ASC(ending).
3. The (left–right) ordering of the columns in the index is significant in the usual major–minor convention. This means that the values on the outermost (i.e., rightmost) column of the index change most frequently; the values on the innermost (i.e., leftmost) column change the least frequently.
4. The CLUSTER option specifies that the index being created is a clustered index. Index values are clustered within a local area on the storage medium; this means that the records are stored sorted, based on their key values (the B-tree leaf nodes contain data pages as opposed to index-rows).
5. The UNIQUE option specifies that no two rows in the indexed table will be allowed to have the same value for the index column(s).
6. The COMPUTE STATISTICS option instructs the DBMS to maintain important statistics about the index that can be subsequently used in performance tuning (discussed further in Chapter 16) to refine the performance of the database.
7. The default algorithm for Oracle indexes is B-tree; however, the user has the option of creating a bit-map index. The bit-map is like a matrix representing row-IDs and columns. It is useful in situations where many column values are identical (for instance where the primary key is composite).
8. Subsequent to its creation, the features of the index may be modified via the **Alter-Index** statement.

```
Create-Index ::=
CREATE [UNIQUE | BITMAP] INDEX [<Schema>.] <IndexName>
ON Cluster-Index-Clause | Table-Index-Clause | Bitmap-Join-Index-Clause;

Cluster-Index-Clause ::=
CLUSTER [<Schema>.] <ClusterName> Index-Attributes

Index-Attributes ::=
[Physical-Attributes-Clause ] [LOGGING | NOLOGGING] [ONLINE] [REVERSE]
[SORT | NOSORT] [Parallel-Clause] [Key-Compression] [Compute-Stat] [Tablespace-Spec]
{ Index-Attributes }

Key-Compression ::= NOCOMPRESS | Compress-Option
Compress-Option ::= COMPRESS <Integer>

Compute-Stat ::=
COMPUTE STATISTICS

Tablespace-Spec ::=
TABLESPACE <Tablespace> | DEFAULT

/* Physical-Attributes-Clause and Parallel-Clause are defined in Create-Table */

Table-Index-Clause ::=
[<Schema>.] <Table> [<Table-Alias>]
( Index-Expr [ASC/DESC] {,Index-Expr [ASC/DESC] } )
[Global-Partitioned-Index | Local-Partitioned-Index] [Index-Attributes] [Domain-Index-Clause]

Index-Expr ::=
<Column> | <Column-Expression>

// Column expressions occur in function-based indexes

Global-Partitioned-Index ::=
GLOBAL PARTITION BY RANGE (<Column>{,<Column>})
(Index-Partitioning-Clause>

Index-Partitioning-Clause ::=
PARTITION [<Partition>] VALUES LESS THAN (<Value> {,<Value>})
[Segment-Attributes-Clause]

// Segment-Attributes-Clause defined in Create-Table

Local-Partitioned-Index ::=
LOCAL [On-Range-Partitioned-Table | On-List-Partitioned-Table | On-Hash-Partitioned-Table |
On-Comp-Partitioned-Table]

On-Range-Partitioned-Table ::=
(PARTITION [<Partition> [Segment-Attributes-Clause]]
{,PARTITION [<Partition> [Segment-Attributes-Clause]]} )

// On-List-Partition-Table is as for On-Range-Partition-Table
On-Hash-Partitioned-Table ::= OHPOption1 | OHPOption2
OHPOption1 ::= STORE IN (<TableSpace> {,<TableSpace>})
OHPOption2 ::= On-Range-Partitioned-Table
```

Figure 11.12 The Create-Index statement. (Continued)

```
On-Comp-Partitioned-Table ::=
STORE IN (<TableSpace> {,<TableSpace>})
(PARTITION [<Partition> {Segment-Attribute-Clause} [Index-Subpartition-Clause]]
{,PARTITION [<Partition> {Segment-Attributes-Clause}[Index-Subpartition-Cause]]})

Index-Subpartition-Clause ::= ISCOption1 | ISCOption2
ISCOption1 ::= STORE IN (<TableSpace>{,<TableSpace>})
ISCOption2 ::= (SUBPARTITION[<Subpartition>[TABLESPACE<Tablespace>]]
               {, SUBPARTITION[<Subpartition>[TABLESPACE<TableSpace>]]})

// Segment-Attributes-Clause is defined in Create-Table

Domain-Index-Clause ::=
INDEXTYPE IS <IndexType> [Parallel-Clause]
[PARAMETERS ('<ODCI-Parameters>')]

Bitmap-Join-Index-Clause ::=
[<Schema>.] <Table> ([[<Schema>.]<Table>.] | [<Table-Alias>.] <Column>[ASC/DESC]
{,[[<Schema>.]<Table>.] | [<Table-Alias>.]<Column>[ASC/DESC]]})
FROM [<Schema>.]<Table> [<Table-Alias>]
{,[<Schema>.]<Table> [<Table-Alias>]}
WHERE <Condition>[Local-Partitioned-Index] Index-Attributes
```

Figure 11.12 (Continued) The Create-Index statement.

EXAMPLE 11.14: CREATE INDEXES ON ACADEMICPROGRAM AND COURSE TABLES RESPECTIVELY

```
// It will be possible to gather statistics on this index:
CREATE INDEX PgmX ON AcademicProgram (PgmName) COMPUTE STATISTICS;

// There will be no statistics on this index:
CREATE INDEX CrsX ON Course (CrsName);
```

Finally (and this should not surprise you), you can drop an index via the **Drop-Index** statement. The syntax for this statement is shown in Figure 11.14. The optional FORCE option applies only to domain indexes, which we have not discussed (and will not).

Alter-Index ::=
ALTER INDEX [<Schema>.]<IndexName>
{Deallocate-Unused-Clause | Allocate-Extent-Clause | Parallel-Clause |
Physical-Attributes-Clause | LOGGIN | NOLOGGING} | Rebuild-Clause | ENABLE | DISABLE | UNUSABLE |
COALESCE | Rename-Index | Index-Parms1 | Monitor-Usage | Updated-Block-Ref |
Alter-Index-Partitioning};

Index-Parms1 ::=
PARAMETERS ('<ODCI-Parameters>')

Rename-Index ::=
RENAME TO <NewIndexName>

Monitor-Usage ::=
MONITORING | NONMONITORING USAGE

Update-Bloc-Ref ::=
UPDATE BLOCK REFERENCES

Alter-Index-Partitioning ::=
Modify-Index-Default-Attributes | Modify-Index-Partition | Rename-Index-Partition |
Drop-Index-Partition | Split-Index-Partition | Modify-Index-Subpartition

Rebuild-Clause ::=
REBUILD
[PARTITION <PartitionName>] |
[SUBPARTITION <SubpartitionName>] |
[REVERSE | NOREVERSE]
[<Parallel-Clause>]
[TABLESPACE <TablespaceName>]
[PARAMETERS ('<ODCI-Parameters>')]
[ONLINE]
[COMPUTE STATISTICS]
[<Physical-Attributes-Clause>]
[NOCOMPRESS | COMPRESS [<n>]]
[LOGGING | NOLOGGING]

Modify-Index-Default-Attributes ::=
MODIFY DEFAULT ATTRIBUTES [FOR PARTITION <Partition>]
{Physical-Attributes-Clause | LOGGING | NOLOGGING | TableSpace-Spec}

// **Physical-Attributes-Clause** and **Parallel-Clause** are defined in Create-Table

Modify-Index-Partition ::=
MODIFY PARTITION <Partition>
{Physical-Attributes-Clause | LOGGING | NOLOGGING | Allocate-Extent-Clause |
Deallocate-Unused-Clause}
Index-Parms2 | COALESCE | UNUSEABLE | Update-Block-Ref

Index-Parms2 ::=
PARAMETERS ('<Alter-Partition-Parms>')

// **Physical-Attributes-Clause** is defined in Create-Table
// **Allocate-Extent-Clause** and **Deallocate-Unused-Clause** defined in Alter-Table

Figure 11.13 The Alter-Index statement. (Continued)

Rename-Index-Partition ::= RIPOption1 | RIPOption2
RIPOption1 ::= RENAME PARTITION <partition> TO <NewName>
RIPOption2 ::= RENAME SUBPARTITION <Subpartition> TO <NewName>

Drop-Index-Partition ::=
DROP PARTITION <Partition>

Split-Index-Partition ::=
SPLIT PARTITION <Partition> AT (<Value> {,<Value>})
[INTO (Index-Partition-Description, Index-Partition-Description)] [Parallel-Clause]

Index-Partition-Description ::=
PARTITION [<Partition> {[Segment-Attributes-Clause | Key-Compression]}]

// **Segment-Attributes-Clause** is defined in Create-Table

Modify-Index-Subpartition ::=
MODIFY SUBPARTITION <Subpartition> UNUSABLE | Allocate-Extent-Clause |
Deallocate-Unused-Clause

// **Allocate-Extent-Clause** and **Deallocate-Unused-Clause** are defined in Alter-Table

Figure 11.13 (Continued) The Alter-Index statement.

Drop-Index ::=
DROP INDEX [<Schema>.] <IndexName> [FORCE];

Figure 11.14 The Drop-Index statement.

Like the **Create-Index** and **Alter-Index** statements, there are consequences associated with the use of the Drop-Index statement; these are summarized below:

1. Dropping an index removes it from the system catalog and makes it unavailable to any pre-existing queries that were using it.
2. The FORCE option applies only to domain indexes, which are not focused on in this course.
3. You should drop indexes only after studying their related system statistics and/or determining that they are not useful to the database (Chapter 16 will revisit this matter).
4. When a table is dropped, all its related indexes are automatically dropped also. However, the converse does not hold: dropping an index has no effect on its related underlying table.

EXAMPLE 11.15: DROP THE CRSX INDEX THAT WAS CREATED IN EXAMPLE 11.14

DROP INDEX CrsX;

11.11 Working with Sequences

A sequence is a database object that automatically generates unique numbers. It is typically used to create primary key values, particularly if the primary key is a single attribute. For composite primary keys, or other alphanumeric codes, the sequence could still be useful in generating a unique number, which is to be concatenated with some other data to comprise a code. The sequence provides two significant advantages:

- Its use could lead to shorter application code.
- When sequence values are cached, processing efficiency is enhanced.

11.11.1 Creating and Using Sequences

The syntax for the **Create-Sequence** statement is shown in Figure 11.15. As you can see, the statement is quite straightforward and the clauses are self-explanatory.

```
Create-Sequence ::=
CREATE SEQUENCE <SequenceName> [INCREMENT BY <Number>]
[START WITH <Number2>] [MAXVALUE <Number3> | NOMAXVALUE]
[MINVALUE <Number4> | NOMINVALUE] [CYCLE | NOCYCLE]
[CACHE <Number5> | NOCACHE];
```

Figure 11.15 The Create-Sequence statement.

The sequence is accessed via two pseudo-columns CURRVAL and NEXTVAL. Typically, it is accessed via a **Select** statement to retrieve its current value, or via an **Insert** statement to generate the next unique value to be used for insertion of data into a table (both statements will be discussed in the next chapter). To view the current value, do a selection on DUAL (DUAL is the general-purpose pseudo-table used when a scalar value that is not stored in a specific database table is to be displayed).

EXAMPLE 11.16: ILLUSTRATING SEQUENCE CREATION AND USAGE

```
CREATE SEQUENCE DeptSeqn INCREMENT BY 05
START WITH 0 MAXVALUE 9095 NOCYCLE;
// The following statement uses the sequence to assist in inserting a row into the Department
   table
INSERT INTO Department (Dept#, DeptName) VALUES (DeptSeqn.NEXTVAL,
   &DeptName);
// The following statement displays the value of the sequence
SELECT DeptSeqn.CURRVAL FROM DUAL;
```

11.11.2 Altering and Dropping Sequences

The attributes of a sequence may be modified via the **Alter-Sequence** statement. Its syntax is similar to that of the **Create-Sequence** statement and is shown in Figure 11.16, followed by the syntax for the **Drop-Sequence** statement. The following guidelines apply to sequence modification:

- Only the owner of a sequence (or a user with alter privilege to the sequence) can modify it.
- Only future sequence numbers are affected by the modification.
- To restart a sequence, you would delete it and then recreate it. Alternately, you may create or alter the sequence to be in CYCLE mode. However, care must be taken to ensure recycling will not disrupt data previously stored in the table.

```
Alter-Sequence ::=
ALTER SEQUENCE <SequenceName> [INCREMENT BY <Number>]
[START WITH <Number2>] [MAXVALUE <Number3> | NOMAXVALUE]
[MINVALUE <Number4> | NOMINVALUE] [CYCLE | NOCYCLE]
[CACHE <Number5> | NOCACHE];

Drop-Sequence ::=
DROP SEQUENCE <SequenceName>;
```

Figure 11.16 The Alter-Sequence and Drop-Sequence statements.

EXAMPLE 11.17: ILLUSTRATING MODIFICATION/ DELETION OF A SEQUENCE

```
// Cause the sequence to cycle after reaching the maximum
ALTER SEQUENCE DeptSeqn CYCLE;
// . . .
// Remove the sequence
DROP SEQUENCE DeptSeqn;
```

11.12 Working with Synonyms

A synonym is a virtual object (i.e., the alias of an object), used to fulfill any or both of the following purposes:

- Shortening the length of an object's name
- Referring to an object owned by another schema (user) without having to qualify the object name with the name of the other schema

The synonym is managed by two statements: The Create **Create-Synonym** and the **Drop-Synonym** statements; the syntax for each is shown in Figure 11.17; the parameters are self-explanatory. Note that there is no **Alter-Synonym** statement. This is so because the synonym is merely a reference to a preexisting object; any required modification must necessarily be on the original object and not its reference.

```
Create-Synonym ::=
CREATE [OR REPLACE] [PUBLIC] SYNONYM <SynonymName> FOR <[Schema.]ObjectName]>;

Drop-Synonym ::=
DROP SYNONYM <SynonymName>;
```

Figure 11.17 The Create-Synonym and Drop-Synonym statements.

The term *synonym*, as used here, is comparable to the term *alias*, as used in Chapter 7. However, the Oracle synonym is wider and more implementation specific in two respects: Firstly, a synonym is applicable to database objects, not just relational tables, aliasing as described in Chapter 7, related to relations. Secondly, a synonym as described here is unconstrained by schema boundaries, aliasing as described in Chapter 7, did not consider schema boundaries. As you view the syntax for the synonym, please note points of clarification:

1. The OR REPLACE option ensures that any preexisting synonym with the same name will be replaced when the statement executes.
2. The PUBLIC option indicates that the synonym is publicly accessible to all users; otherwise, it is private to the user who created it.

As you will see in Chapter 12, synonyms are particularly useful when referencing objects not owned by your schema. Moreover, in Chapter 13, you will learn that synonyms can be used to help bolster the security mechanism for a database environment. For instance, when you create a synonym for an object owned by another schema, access to the referenced object is only assured if your schema has been granted specific privileges to the intended object.

The following example creates a synonym called **ValidDate** for an object of the same name in another schema. The presumption here is that the object **ValidDate** (which could be any valid Oracle database object, including a function) resides in a schema called **Jones**, and synonym is being created in the current schema.

EXAMPLE 11.18: ILLUSTRATING SYNONYM CREATION AND DELETION

```
// The following statement creates a synonym called ValidDate
CREATE PUBLIC SYNONYM ValidDate FOR Jones.ValidDate;
// The following statement deletes the synonym
DROP SYNONYM ValidDate;
```

11.13 Summary and Concluding Remarks

It's time to summarize what was covered in this chapter:

- Oracle DBMS suite provides a rich collection of components to manage a database environment of any kind — operational or analytic, large or small, general-purpose or special, stand-alone or distributed, cloud-based or locally controlled. Three widely used components are SQL Developer (SQLD), Application Express (APEX), and Oracle Enterprise Manager (OEM).

- A typical Oracle database consists of various database objects contained in various tablespaces. Physically, a tablespace contains datafiles which are in turn comprised of segments, which consists of multiple data blocks.
- It is helpful to have a basic understanding of database configuration as it relates to database server and client components, configuration files, user accounts, schemas, privileges, and database sessions.
- The **Create-Database** statement allows you to create an Oracle database. Alternately, you may use the Oracle DBCA.
- The **Alter-Database** statement allows you to change features of a database, and the **Drop-Database** statement allows you to delete the current database. Alternately, you may use the Oracle DBCA.
- The **Create-Tablespace** statement allows you to create a tablespace. The **Alter-Tablespace** statement allows you to change features of a tablespace, and the **Drop-Tablespace** statement allows you to delete a tablespace.
- If the database is hosted and administered via the Oracle Cloud Infrastructure (OCI), this significantly simplifies the design and administration requirements for database experts, allowing them to focus more on conceptual schema resources.
- The **Create-Table** statement allows you to create a table. The **Alter-Table** statement allows you to change features of a table, and the **Drop-Table** statement allows you to delete a table.
- The **Create-Index** statement allows you to create an index. The **Alter-Index** statement allows you to change features of an index, and the **Drop-Index** statement allows you to delete an index.
- The **Create-Sequence** statement allows you to create a sequence. The **Alter-Sequence** statement allows you to change features of a sequence, and the **Drop-sequence** statement allows you to delete a sequence.
- The **Create-Synonym** statement allows you to create a synonym. The **Drop-Synonym** statement allows you to delete a synonym.

Each of the database objects discussed in this chapter can be created and fully managed using the OEM or OSQLD. Whatever your choice, it is imperative that you make an effort to learn the basic syntax of SQL. The upcoming chapter continues this journey by discussing the common DML and DCL statements of the language.

11.14 Review Questions

1. Identify and briefly clarify the basic components and configuration issues in a typical Oracle database environment.
2. What are the commonly used SQL definition statements?
3. Use the Oracle DBCA component to create a database. Practice writing SQL statements to create simple databases.
4. Practice writing SQL statements to modify different aspects of a database.
5. Write SQL statements to add a tablespace to your database. Create a second tablespace to hold the tables described in your sample college database.
6. Practice writing SQL statements to modify different aspects of a tablespace.
7. Write SQL statements to create the tables described in the sample college database described in Chapter 7 (Figure 7.1).

8. Write SQL statements to define indexes on your database tables.
9. Write appropriate SQL statements to do the following:
   - Add a constraint to a table.
   - Modify a constraint.
   - Enable or disable a constraint.
10. Practice writing SQL statements to modify or drop indexes.
11. By considering the **Student** table in your sample college database, write SQL statements to create a sequence for this table. Describe and demonstrate how this sequence could be used.
12. What is a synonym and how is it used in a database? Practice writing SQL statements to create and drop a synonym.

References and Recommended Readings

Connolly, Thomas, & Carolyn Begg. 2015. *Database Systems: A Practical Approach to Design, Implementation and Management* 6th ed. Boston, MA: Pearson. See chapter 7.

Coronel, Carlos, & Steven Morris. 2015. *Database Systems: Design, Implementation & Management* 11th ed. Boston, MA: Cengage Learning. See chapter 7.

Elmasri, Ramez, & Shamkant B. Navathe. 2011. *Fundamentals of Database Systems* 6th ed. Boston, MA: Pearson. See chapter 4.

Foster, Elvis C. 2021. *Software Engineering: A Methodical Approach* 2nd ed. New York: CRC Press. See chapter 13.

Oracle Corporation. 2019. *Database SQL Language Reference*. Accessed June 2019. https://docs.oracle.com/en/database/oracle/oracle-database/19/sqlrf/index.html. See chapters 8, 12, and 13.

Shah, Nilesh. 2004. *Database Systems Using Oracle* 2nd ed. Boston, MA: Pearson. See chapters 2–4.

Thomas, Biju, Gavin Powell, Robert Freeman, & Charles Pack. 2014. *Oracle Certified Professional on 12C Certification Kit*. Indianapolis, IN: Wiley.

Chapter 12

SQL Data Manipulation Statements

There are four core data manipulation language (DML) statements in structured query language (SQL), namely INSERT, UPDATE, DELETE, and SELECT. These statements apply to base tables and views (views will be discussed in the next chapter). The chapter proceeds under the following subheadings:

- Insertion of Data
- Update Operations
- Deletion of Data
- Commit and Rollback Operations
- Basic Syntax for Queries
- Simple Queries
- Queries Involving Multiple Tables
- Queries Involving the Use of Functions
- Queries Using the LIKE and BETWEEN Operators
- Nested Queries
- Queries Involving Set Operators
- Queries with Runtime Variables
- Queries Involving SQL Plus Format Commands
- Embedded SQL
- Dynamic Queries
- Summary and Concluding Remarks

The chapter will continue to use examples based on the college database described in Chapter 7, so feel free to take some time to review this.

12.1 Insertion of Data

There are three general formats of the **Insert** statement; the abridged form of the syntax for each format is provided in Figure 12.1. By now you should be comfortable reading the Backus–Naur Form (BNF) representation without any prodding; as usual, the keywords are in upper case.

DOI: 10.1201/9781003275725-15

```
Insert ::= Insert1 | Insert2 | Insert3

Insert1 ::=
INSERT INTO <TableName> [(<Column> {, <Column>})]
VALUES (<Literal> {, <Literal>});

Insert2 ::=
INSERT INTO <TableName> [(<Column> {, <Column>})]
VALUES ([&|:] <PgmVariable> {, [&|:] <PgmVariable>});

Insert3 ::=
INSERT INTO <TableName> [(<Column> {, <Column>})]
<Subquery>;

// Subquery represents a Select statement — to be defined later
```

Figure 12.1 Abridged form of the Insert statement.

As indicated in the figure, there are three commonly used formats of the statement. Here are some applicable guidelines:

1. The first format is used when the attribute values to be inserted are explicitly provided with the statement (typically by an SQL user). A literal is specified for each attribute. A literal is simply a scalar value; if this value is alphanumeric, then it must be specified within single quotes; numeric values are specified without quotation marks.
2. The second format is suited for situations where the SQL **Insert** statement is incorporated into a high-level language (HLL) program, so that the attribute values to be inserted are implicitly provided from program variables; the term used to describe this scenario is *embedded SQL*. Oracle ships with a default host language called PL/SQL, which is part of the SQL*Plus environment. In PL/SQL, an optional ampersand (&) that precedes the program variable denotes an execution-time variable; the user will be prompted to specify a value when the SQL statement is executed. If no ampersand precedes the variable, then the SQL statement must appear as an embedded SQL statement within a PL/SQL program block; the values for the insertion are obtained from the specified program variables. In some high-level languages (for example, C++), the program variable must be preceded by a colon (:). This matter will be revisited toward the end of the chapter (when embedded SQL is discussed).
3. The third format is used when tuples to be inserted are to come from the result of some query; queries will be discussed later in the chapter.
4. If all the column names are omitted, this is equivalent to specifying them all in the same order as they were specified in the **Create-Table** statement. This is rather risky, particularly if you are uncertain about that order (of column names); for this reason, it is recommended that you explicitly specify the column names.

EXAMPLE 12.1: ADDING A NEW DEPARTMENT

```
// Add a new department to the database with the following details:
// Code = D500; Name = Engineering; Head = S20:
INSERT INTO Department (Dept#, DName, DHead#)
Values ('D500', 'Engineering', 'S20');
```

EXAMPLE 12.2: SAVE ALL STUDENTS ENROLLED IN THE MIS PROGRAM TO A TEMPORARY TABLE

```
// Solution involving two steps

CREATE TABLE MajorsMIS (Stud# NUMBER(7) NOT NULL, SName VARCHAR2(15),
   FName VARCHAR2(15), PRIMARY KEY (Stud#));
INSERT INTO MajorsMIS (Stud#, SName, FName)
SELECT Stud#, SName, FName FROM Student WHERE StudPgm# = 'BSC1';

// Alternate solution involving a single step

CREATE TABLE MajorsMIS (Stud#, SName, FName, Sex) AS SELECT * FROM Student
WHERE (StudPgm# = 'BSC1');
```

From the above example, the following should be noted:

1. The first statement of the first solution creates an empty table called **MajorsMIS** and then inserts data into it. The last line of the second statement in the solution shows a subquery (the **Select** statement) that pulls from the **Student** table, students that are enrolled in the academic program that has the code 'BSC1' (which from Figure 7.3 of Chapter 7, represents BSc in MIS). The result of the **Select** statement is placed into the temporary table **MajorsMIS**. When used in this way, the **Insert** statement can insert a set of several rows at once; it illustrates what was referred to as set-at-a-time insertion in Chapter 9.
2. The alternate solution creates the temporary table **MajorsMIS** with data by including a subquery in the **Create-Statement**.
3. Table **MajorsMIS** is an example of a temporary relation or snapshot as discussed in Chapter 3.

The following example illustrates a PL/SQL procedure to insert data into the **Course** table, where the values come from program variables. Stored procedures are not directly covered in the course; for that, you need to learn Oracle's PL/SQL. However, you will see several examples of stored procedures later in the chapter; this should give you an appreciation for them and a motivation to want to learn more about the topic.

EXAMPLE 12.3: STORED PROCEDURE TO HANDLE DATA INSERTION OF NEW COURSES

```
Create Procedure InsertCourse (ThisCourse IN Course.Crs#%Type, ThisCrsName IN
Course.CrsName%Type)

IS
BEGIN
INSERT INTO Course (Crs#, CrsName)
VALUES (ThisCourse, ThisCrsName);
EXCEPTION
WHEN OTHERS THEN /* catch all errors */
DBMS_Output.Put_Line ('There is an execution error');
END;
```

The above procedure can be called from the SQL prompt, or from within a PL/SQL application program by issuing the call statement as in the following example:

Call InsertCourse ('CS101', 'Computer Applications');

An alternate way of specifying the above procedure is to remove the parameters and write the **Insert** statement with *execution-time variables*. An execution-time variable is a variable with an ampersand (&) preceding it. This indicates to the PL/SQL parser that the user will be prompted to supply a value for the variable. The SQL statement is then constructed on the user's behalf and sent to the SQL parser. You will see examples of the use of such variables in upcoming examples.

12.2 Update Operations

The **Update** statement is used for updating rows of a table. The update may be on a single row or a set of rows, depending on the condition specified in the **Where-Clause**. The abridged general formats of the **Update** statement are given as follows (Figure 12.2).

```
Update1 ::= Update1 | Update2 | Update3

Update1 ::=
UPDATE <TableName>
SET <Column> = <Scalar-Expression> | <Literal> {, <Column> = <Scalar-Expression> | <Literal>}
[WHERE <Condition>];

Update2 ::=
UPDATE <TableName>
SET <Column> = [&|:]<PgmVariable> {, <Column> = [&|:]<PgmVariable>}
[WHERE <Condition>];

Update3 ::=
UPDATE <TableName>
SET (<Column> {,<Column>}) = (<Subquery>)
[WHERE <Condition>];
```

Figure 12.2 Abridged form of the Update statement.

The figure conveys that like the **Insert** statement, the **Update** statement has three widely used formats. Please note the following clarifications:

1. A scalar expression is any combination of column(s), literal(s), arithmetic operator(s), and SQL function(s) that evaluates to a scalar value (i.e., a number or an alphanumeric value). More will be said about SQL functions later. The simplest form of an expression is a single column or literal.
2. The first form of the update statement is applicable when the update values are to come from literals that the user (programmer) will supply, or other columns of the table, or a combination of columns and literals.
3. The second form of the update statement is used in an embedded SQL scenario, or where execution-time prompt is desired. The use of the ampersand (&) or the colon (:) is as explained in the previous section.

4. Whenever a subquery is specified within an update statement (the third form), it must provide values for the attributes to be updated.

Following is an example of a simple update (assuming that "CSC" is the unique code for the department). The row that satisfies the condition specified in the **Where-Clause** is fetched and updated.

EXAMPLE 12.4: MODIFYING A DEPARTMENT RECORD

```
/* Change the Computer Science Department name to 'Department of Computer Science and Mathematics'; change the department head to Professor Hans Gaur (Employee code of "S20"): */
UPDATE Department SET DeptName = 'Department of Computer Science & Mathematics', DeptHead# = 'S20' WHERE Dept# = 'CSC';
```

Following is an example of a multiple update. All rows satisfying the condition specified are fetched and updated. This also illustrates what was referred to as set-at-a-time update in Chapter 9.

EXAMPLE 12.5: REASSIGNING STUDENTS TO A NEW HALL

```
// Assign all male students in the BSC in MIS to Urvine Hall
UPDATE Student SET StudHall # = 'Urv' WHERE StudPgm# = 'BSC1' AND StudSex = 'M';
```

EXAMPLE 12.6: CHANGING A PRIMARY KEY VALUE

```
// Change the course code CS100 to CS105
UPDATE Course SET Crs# = 'CS105' WHERE Crs# = 'CS100';
```

The matter of changing the primary key value (as illustrated in Example 12.6) is very important and therefore deserves some attention. Please note:

1. Generally speaking, it is not a good practice to change codes that are primary keys since this action increases the possibility of violation of the referential integrity rule (review Section 4.1).
2. The Oracle database management system (DBMS) will allow such updates providing that no violation of the referential integrity rule would result; if a violation would result, the instruction would be disallowed. An alternative to this approach is to restrict the update of primary keys (the system would allow insertion and deletion only), except in the case where the DBMS introduces and maintains surrogates, transparent to the end user. Still another alternative is to allow cascaded updates, but this would be very expensive.

EXAMPLE 12.7: USING EXECUTION-TIME VARIABLE IN UPDATE STATEMENT

// The following statement uses execution time variable in an **Update** statement
UPDATE Course SET CrsName = &ThisCourse WHERE Crs# = 'CS220' ;

EXAMPLE 12.8: USING A SUBQUERY TO PROVIDE DATA FOR THE UPDATE

// The following statement uses a query to update a row. Try explaining what the statement accomplishes.
UPDATE Course SET (CrsName, CrsAbbr) =
(SELECT CrsName, CrsAbbr FROM Course WHERE Crs# = 'CS220')
WHERE Crs# = 'CS490';

12.3 Deletion of Data

To remove rows from a table, the **Delete** statement is used. Alternately, you may use the **Truncate** statement to delete all rows from a table. Figure 12.3 shows the syntax of both statements; this is followed by three examples.

Delete ::= DELETE FROM <TableName> [WHERE <Condition>];

Truncate ::= TRUNCATE TABLE <TableName>;

Figure 12.3 The Delete statement and the Truncate statement.

EXAMPLE 12.9: DELETING A COURSE

// Delete course 'C9999' from the database
DELETE FROM CourseWHERE Crs# = 'C9999';

Note:

The DBMS will not allow referenced tuples to be deleted unless DELETE CASCADES was specified when the related foreign key(s) was (were) specified. The default referential integrity for most DBMS suites is DELETE RESTRICTED.

EXAMPLE 12.10: DELETING MULTIPLE RECORDS

// Delete all students that belong to Mary Seacole Hall
DELET FROM Student WHERE StudHall# = 'Mars';

EXAMPLE 12.11: EMPTYING A TABLE

// Delete all Employee members
DELETE FROM Employee;

// Alternately, you may truncate:
TRUNCATE TABLE Employee;

Observe the power and flexibility of the **Delete** statement. As you consider the above examples, please note the following:

1. Examples 12.10 and 12.11 are illustrations of set-at-a-time deletions as mentioned in Chapter 9. In the case of Example 12.10, all records satisfying the condition are deleted; in the case of Example 12.11, all records are deleted from the table.
2. While the DELETE and TRUNCATE may appear to achieve the same ends, there are two important differences that you should be aware of: Firstly, DELETE is subject to a COMMIT or ROLLBACK operation, while TRUNCATE is irreversible (though one could recover from a restoration of a previous system backup). Secondly, the **Delete** statement has a **Where-Clause**; the **Truncate** statement does not.
3. The **Truncate** statement is fast and risky; you should only use it on test data that you are absolutely certain you will never need again. As a matter of caution, it does not hurt to simply stick with the **Delete** statement.

12.4 Commit and Rollback Operations

The **Commit** statement and the **Rollback** statement are both DCL statements that have relevance to data manipulation. They are useful in transaction management. The **Commit** statement forces permanent storage of all changes appearing in related transaction(s), erasing all temporary save-points and transaction locks. The **Rollback** statement returns the database to some previous safe-state that protects the integrity of the system; incomplete transactions are not allowed. This is particularly important when multiple related tuples are written to different tables (for instance, an invoice may consist of a summary tuple, written to one table, and several detail tuples, written to one or more additional tables).

Two related statements are the **Set-Transaction** statement and the **Save-Point** statement. The former is used to define the beginning of a transaction block; the latter is used to create a recovery point in case of catastrophes such as power loss or system failure. Figure 12.4 shows the syntax for these four related statements.

```
Commit ::= COMMIT;

Rollback ::= ROLLBACK [TO SAVEPOINT <SavePointName>];

Set-Transaction ::=
SET TRANSACTION <Mode>;
Mode ::= READ ONLY | READ WRITE

Save-Point ::=
SAVEPOINT <Save-point-name>;
```

Figure 12.4 Syntax for Commit, Rollback, Set-Transaction, and Save-Point statements.

These statements are sometimes described as transaction management statements. Here are a few points to note about them:

1. These statements are typically used in a programming block rather than at the SQL prompt.
2. The default transaction mode for Oracle is READ WRITE. However, for application programs that are required to read data only, this should be changed to READ ONLY to prevent record locking. This is particularly important in a multi-user environment.
3. These transaction management statements work the same way for insertions, updates, or deletions.

The idea of a transaction was first introduced in Chapter 2, and again mentioned in Chapter 10. Let us take a closer look at the term: A transaction is a set of related operations that is treated as a coherent whole. To illustrate, let us revisit the idea of processing invoices for an organization, as introduced in Chapter 4 and elucidated in Chapter 5 (review Section 5.4.2). Drawing from those earlier discussions, processing of purchase orders and invoices would typically involve manipulation of the following relational tables: **Supplier**, **InventoryItem**, **PurchaseOrdSummary**, **PurchaseOrdDetail**, **PurchaseInvSummary**, and **PurchaseInvDetail**. Assuming that the first two tables are preloaded with relevant data on suppliers and inventory items of interest to the organization, the following are two transactions that can be anticipated:

- Generating a purchase order would involve writing a summary record to the **PurchaseOrdSummary** table and one or multiple related detail records to the **PurchaseOrdDetail** table. The summary record would reference a **Supplier** record, and each detail record would reference its related **PurchaseOrdSummary** record as well as an **InventoryItem** record.
- A purchase invoice is typically received from a supplier along with, or shortly after, receiving goods and/or services that were requested on a purchase order. Recording the invoice would involve writing a summary record to the **PurchaseInvSummary** table, and one or more related detail records to the **PurchaseInvDetail** table. The summary record would reference its related **PurchaseOrdSummary** record; each detail record would reference its related **PurchaseInvSummary** record as well as an **InventoryItem** record.

EXAMPLE 12.12: PL/SQL PROCEDURE FOR INSERTING AN INVOICE

```
CREATE PROCEDURE InsertInvoice
AS
Success BOOLEAN; /* A boolean variable */
BEGIN
// . . . possible additional code
SAVEPOINT InvoicePoint;
SET TRANSACTION READ WRITE;
/* Call procedure to insert row in table PurchaseInvSummary */
InsertSummary;
/* Call procedure to insert related detail rows in table PurchaseInvDetail. Commit if no
   errors; otherwise,
rollback */
InsertDetail (Success);
IF Success THEN
COMMIT;
ELSE
ROLLBACK TO SAVEPOINT InvoicePoint;
END IF;
// . . .
END
```

12.5 Basic Syntax for Queries

From an end-user perspective, one of the most important services of the database system is the facility to retrieve information and put it to use. This is made possible via the SQL **Select** statement. The **Select** statement is by far one of the most powerful and complex statements in SQL. It is used to retrieve data from database tables, but as you will soon see, it has many optional clauses. Figure 12.5 provides an abridged general form of the **Select** statement, showing the more commonly used clauses. A brief summary of each clause is as follows.

The **Function-Spec-Clause** may involve any combination of a vast list of SQL functions, some of which will be mentioned in subsequent sections. The following are some categories of functions that are applicable:

- Row functions
- Date functions
- Data conversion functions
- Aggregation and analytic functions
- Programmer-defined functions

The **Item-List** represents the attributes (columns) to be listed in the query result. Each item may be in the form of a column-name, literal, scalar expression involving the combination of column(s) and/or literal(s) or a substitution name (i.e., alias) for an attribute. A scalar expression is an appropriate combination of column(s) and/or literal(s), punctuated by arithmetic operators, which results in a scalar value (typically numeric or alphanumeric). The valid scalar operators are multiplication (*), division (/), addition (+), subtraction (-), and concatenation (||).

```
Select-Statement ::=
SELECT [<Function-Spec-Clause>] [DISTINCT] [,] <Item-List>
[INTO <Retrieval-List]
FROM <Relation-List>
[WHERE <Condition>]
[GROUP BY <Group-List>]
[HAVING <Condition>]
[ORDER BY <Order-List>]
[Start-With-Clause]
[Connect-By-Clause];

Item-List::=
<Attribute> | <Scalar-Expression>  [[AS]  <Alias>] {,<Attribute> | <Scalar-Expression> [[AS] <Alias>] }

Retrieval-List::=
:<ProgramVariable> {,:<ProgramVariable>}
// The number of variables must match the number items in the items list.

Relation-List ::=
<Table> | <Select-Statement> [<TupleVariable>] {, <Table> | <Select-Statement> [<TupleVariable>]}

Condition ::=
Comparison | NotCondition | AndCondition | OrCondition | ExistsItem

NotCondition ::= NOT <Condition>
AndCondition ::= <Condition> AND < Condition >
OrCondition ::= <Condition> OR <Condition>
ExistsItem ::= EXISTS (<Subquery>)

Comparison::=  Comparison1 | Comparison2 | Comparison3 | Comparison4 | Comparison5

Comparison1 ::= <Variable> <Operator> <Variable>
Comparison2 ::= <Attribute> <Operator> <Attribute>
Comparison2 ::= <Attribute> <Operator> <Literal> | <Scalar-Expression>
Comparison4 ::= <Attribute> IN (Values-List | Subquery)
Comparisin5 ::= <Attribute> BETWEEN (<Scalar-Expression1> AND <Scalar-Expression2>)
Values-List ::= <Literal> | <Scalar-Expression> {,<Literal> | <Scalar-Expression>}

Attribute ::=
<TupleVariable.Column> | <Column>

Operator ::=  = | < | <= | > | >= | <> | LIKE | IN | BETWEEN
Order-List::=       <Column> [ASC/DESC] {,<Column> [ASC/DESC]}
Group-List::=       <Column> {,<Column>}

Start-With-Clause::=
START WITH <Condition> CONNECT BY [NOCYCLE} <Condition>

Connect-By-Clause::=
CONNECT BY [NOCYCLE] <Condition> [START WITH <Condition>]
```

Figure 12.5 Abridged syntax for the Select statement.

The DISTINCT keyword ensures that in the event where duplicate records may occur in the result-set, only unique records will be provided in the final output. The *result-set* is simply a term used to represent the records that meet the selection criterion/criteria.

The **Into-Clause** is applicable in an embedded SQL scenario where the SQL **Select** statement is incorporated into a high-level language (HLL) program. When used, there should be a named program variable corresponding to each attribute in the item-list. We will examine this alternative later in the chapter.

The **From-Clause** allows specification of the relation(s) and/or relational expression(s) from which the data are retrieved. Recall from Chapters 7 and 8 that a relational expression is simply an expression that results in a relation. In SQL, the relational expression is implemented as a subquery (i.e., a nested **Select** statement). The simplest possible scenario for the **From-Clause** is where the data comes from a single table. However, as you will soon see, more complex scenarios typically involve relational expression(s) defined on a single table or multiple tables. Additionally, complex queries often involve the use of tuple variables. The tuple variable is defined by simply stating its name immediately following the relational expression that it is defined on. This clause is mandatory, and will therefore be used in various upcoming examples.

The **Where-Clause** is used to impose a condition that must be met by all records in the result-set. The condition specified after the WHERE keyword is a well-formed formula (WFF) as discussed in Chapter 8 except for the following:

a. In SQL, the construct IF <Comparison> THEN WFF is not supported.
b. The syntax for using the existential quantifier s slightly different — the tuple variable is implicit rather than explicit.
c. The universal quantifier is not widely supported in SQL implementations. However, in light of available substitution rules (review Section 8.4) we will not pursue its use any further.

The **Group-By-Clause** facilitates the grouping of data. This clause is often used in concert with the **Having-Clause**. Both clauses will be revisited later in the chapter.

The **Order-By-Clause** allows for the ordering of data in the result-set. The default ordering is ASC (for ascending); the alternate ordering is DESC (for descending). You may specify multiple columns, in which case, the comma is used to separate the columns.

The **Start-With-Clause** and **Connect-By-clause** are called **Hierarchical-Query-Clause.** They subscribe to the notion of parent–child relationships and are not considered essential. Further exploration is optional.

Query Condition: Except for a few omissions, the query condition closely mirrors the definition of a WFF that has been discussed in Chapter 8. This is not an accident; it is deliberate.

The **Select** statement is used to write queries and subqueries. As we proceed, you will see that there is a wide range of scenarios for queries. We will start by looking at simple queries and then pull in additional features as we proceed.

12.6 Simple Queries

A simple query may be considered as a query that retrieves data from a single table. We will use the following format for such queries:

SimpleQuery ::= SELECT <Item-List> FROM <TableName> [WHERE <Condition>];

EXAMPLE 12.13: RETRIEVING MULTIPLE RECORDS MEETING A SPECIFIED CRITERION

// List name and ID of students enrolled in the MIS program
SELECT StudLName, StudFName, Stud# FROM Student WHERE StudPgm#= 'BSC1';

EXAMPLE 12.14: RETRIEVING ALL RECORDS FROM A TABLE

// List all courses offered
SELECT Crs#, CrsName, CrsDept# from Course;

// To include all attributes from the table, we could simply write the following

SELECT * FROM Course;
// The asterisk in the statement is used as a *wildcard*

Refer to the supplier–items data set of earlier discussions (Chapters 3 and 4). Suppose that we have the following relations:

InventoryItem {Item#, ItemName, ItemPrice, Weight, …} PK [Item#]
Supplier {Supp#, SuppName, LocationCode, …} PK [Supp#]
Schedule {Supp#, Item#, Qty ...} PK [Supp#, Item#]

Suppose further that we wish to show shipment amount on items received. Let us assume that each shipment is a box of 12 items (an arbitrary assumption just for the sake of illustration). We could have the query illustrated in the upcoming example.

EXAMPLE 12.15: INTRODUCING A CALCULATED COLUMN IN A QUERY

// Show the equivalent amount for each scheduled shipment of items
SELECT Sup#, Item#, Qty, 'Amount=', Qty * 12 FROM Schedule;

// The result would look like this:

Sup#	*Item#*	*Qty*	
S001	I001	025	Amount = 300
S001	I002	012	Amount = 144
…	…	…	…

Getting back to the sample college database, suppose that we desire to produce a list of students (showing the student's full name and Student ID) from Chancellor Hall, sorted on first name within surname. The solution is shown in Example 12.16.

EXAMPLE 12.16: PREPARING AND PRESENTING A SORTED LIST

```
// Show students from Chancellor Hall, sorted by name
SELECT StudLName || ' ' || StudFName AS FullName, Stud# FROM Student
WHERE StudHall#= 'Chan' ORDER BY StudLName, StudFName;
```

Note:

The use of the concatenation operator in the scalar expression (SName || ' ' || FName) and the introduction of the alias **FullName**. Please be careful not to confuse "alias" as used here, with the term described in Chapter 7. What we are talking about here is a column alias, whereas in Chapter 7, our focus was on aliasing an entire relation (implemented in Oracle via synonyms).

12.7 Queries Involving Multiple Tables

Quite often, it will be required to pull data from multiple tables in order to service a particular query. The situation may warrant a natural join, a theta-join, an outer join, or a Cartesian product. Oracle supports two different approaches to dealing with queries from multiple tables — the traditional approach and the newly introduced American National Standards Institute (ANSI) approach.

12.7.1 The Traditional Method

The traditional method for treating queries involving multiple tables is to specify the join condition in the **Where-Clause.** Let us take a few examples:

Referring to supplier–items database that was clarified in Section 12.6, suppose that we need to multiply the schedule quantity by the item price in order to obtain the value of each shipped box, and that it is desirable to display supplier-name and item-name in addition to mere codes. Notice that the **Schedule** relation does not contain the supplier-name nor the item-name. It will therefore be necessary to pull these attributes from the **Supplier** table and the **InventoryItem** table, respectively. The solution requires us to join **Schedule** with **Supplier** and **InventoryItem,** respectively. And based on previously discussed database theory, relational tables are joined by linking the foreign key in the referencing table to the corresponding primary key in the referenced table. Example 12.17 shows a solution to this problem.

EXAMPLE 12.17: CONSTRUCTING A JOIN QUERY WITH CALCULATED COLUMN

```
// Prepare information for Shipment Report

SELECT SH.Supp#, S.SuppName, SH.Item#, I.ItemName, SH.Qty, SH.Qty * I.ItemPrice
   AS Value
FROM Schedule SH, Supplier S, InventoryItem I WHERE SH.Supp# = S.Supp# AND
   SH.Item# = I.Item#;

// The result of the above query would look like this:
```

Supp#	*SuppName*	*Item#*	*ItemName*	*Qty*	*Value*
S001	ABC Co.	I001	Crank Shaft	10	250.00
...	...	...	...	...	...

Note:

The above example showcases a natural join. It also includes a scalar expression (**SH.Qty * I.ItemPrice**) and a column alias (**Value**). Note the introduction of explicit tuple variables **SH**, **S**, and **I**, and the new column, **Value**. It is a good habit to introduce tuple variables when the information queried is to come from multiple tables, and/or comparisons are to be made in order to service the query.

The next few examples are based on the partial college database of earlier discussions. Whenever necessary, be sure to refresh your memory by reviewing Figures 7.1, 7.2, and 7.3 of Chapter 7.

EXAMPLE 12.18: JOIN QUERY FOR PROGRAM-NAME AND COURSE-NAME COMBINATIONS

```
// Show program-name and course-name combinations
SELECT P.PgmName, C.CrsName FROM Pgm_Struct PS, AcademicProgram P, Course C
WHERE PS.PSPgm# = P.Pgm# AND PS.PSCrs# = C.Crs#;
```

EXAMPLE 12.19: JOIN QUERY FOR DEPARTMENT HEADS WHO ARE NOT DIVISION HEADS

```
// Show the names of department heads who are not division heads
SELECT E.Emp#, E.EmpName FROM Employee E, Department D, Division DV
WHERE D.DHead# = E.Emp# AND D.D_Div# = DV.Div# AND DV.DvHead# =
   E.Emp# AND D.DHead# <> DV.DvHead#);
```

Note:

This is an example of a theta-join. The solution makes use of the following facts: **Division table** stores a foreign key [**DvHead#]** that links it to the **Employee** table; the **Department** table stores a foreign key [**DHead#**] that also links it to the **Employee** table; and every department belongs to a division.

EXAMPLE 12.20: QUERY FOR STUDENTS WHO SHARE THE SAME DATE OF BIRTH

```
SELECT S1.Stud#, S1.StudDoB FROM Student S1, Student S2
WHERE S1.Stud# <> S2.Stud# AND S1.StudDoB = S2.StudDoB;
```

Note:

This is an example of a Cartesian product of a relation with itself. Compare this with Example 7.8 in Section 7.4 and Example 8.3 in Section 8.2.

Caution:

If you wish to join relation R1 with relation R2, be sure to consider whether or not the FK can have null values, as this will affect the result-set. If in doubt, use an *outer join,* as clarified in the next example.

Now suppose that we desire a list of all students and their assigned hall-name. Based on the theory discussed in Chapter 5 (review Section 5.5), you should appreciate that the **Student** table is a designative entity with respect to the **Hall** table. What this means is that **Student** has a foreign key that links to **Hall**'s primary key. With some thought on multiplicity (review Section 3.6.4), appreciate that not all students will be assigned to a specific hall; some of them may live off campus. Therefore, there may very well be several students with no hall designation (i.e., null value for assigned hall). Example 12.21 shows an incorrect solution followed by a correct solution to the problem. The incorrect solution involves an inner join; the correct solution employs an outer join.

EXAMPLE 12.21: JOIN QUERY FOR STUDENT-NAME AND HALL-NAME COMBINATIONS

```
// The inner join would produce a spurious result-set
SELECT StudLName, HallName FROM Student S, Hall H WHERE S. StudHall# =
   H.Hall#;

// The outer join would produce the desired result-set
SELECT StudLName, HallName FROM Student S, Hall H WHERE S. StudHall# =
   H.Hall#(+);
```

This matter of inner join versus outer join deserves a bit more clarification. Carefully observe Example 12.21 and note the following:

1. The result-set from inner join will exclude all students who have not been assigned to a hall. To avoid this situation, the outer join is specified by simply including a parenthesized plus sign (+) on the attribute of the table where there may not be a corresponding row (in this case H.Hall#), when writing the join condition.
2. The plus sign indicates that rows from the implied relation may have null values (or non-matching values) for the attribute of interest; these rows should be included. If the plus sign is on the right, we have a *left outer join*; if the plus sign is on the left, we have a *right outer join*.
3. Using this approach, you can have the parenthesized plus sign (+) on either side, but not on both sides of the join condition. This means that you can have a left outer join, a right outer join, but not a full outer join. To obtain a full outer join, you have to take the union of a left outer join with a right outer join.
4. Left outer join means all rows from the table on the left are to be included; right outer join means all rows from the table on the right are to be included.

12.7.2 The ANSI Method

Oracle supports the more newly introduced ANSI syntax for handling queries from multiple tables. With this new syntax, the **From-Clause** is modified so that you are forced to explicitly specify the join. The syntax for the modified **From-Clause** is shown in Figure 12.6.

```
From-Clause ::= From1 | From2 | From3 | From4 | From5

From1 ::= FROM <Table> NATURAL [INNER] JOIN <Table>
      {NATURAL [INNER] JOIN <Table>}

From2 ::= FROM <Table> [INNER] JOIN <Table> USING <Column-list>
         {[INNER] JOIN <Table> USING <Column-list>}

From3 ::= FROM <Table> [Alias] [INNER] JOIN <Table> [Alias] ON <Condition>
         {[INNER] JOIN <Table> [Alias] ON <Condition>}

From4 ::= FROM <Table> CROSS JOIN <Table>

From5 ::= FROM <Table> [Alias] [NATURAL] LEFT|RIGHT|FULL [OUTER] JOIN <Table> [Alias] [ON <Condition>]
         {[NATURAL] LEFT|RIGHT|FULL [OUTER] JOIN <Table> [Alias] [ON <Condition>]}
```

Figure 12.6 Modified From-Clause for ANSI Join.

From the syntax, note the use of keywords LEFT, RIGHT, FULL, INNER, and OUTER; you will also notice five formats in the syntax. The clarifications for these formats and keywords are as follows:

1. The first format describes the natural join. For this, the join column(s) in each table must be identical. Moreover, you are not allowed to qualify column names.
2. The second format is useful where there is at least one column in each participating table that has the same name; you then need to specify which columns should be used for the joining. If you have a properly designed database with unique attribute-names in each table, there will not be many occasions to use this format.
3. The third format is the most widely used, because it is the most flexible. You specify the join condition after the keyword **ON**.
4. The fourth format is for taking Cartesian (cross) products; no join condition is required.
5. The fifth format is for outer joins; there are three types:
   - The left outer join is a join between two tables that returns rows based on the join condition, and unmatched rows from the table on the left.
   - The right outer join is a join between two tables that returns rows based on the join condition, and unmatched rows from the table on the right.
   - The full outer join is a join between two tables, returning rows based on the join condition, and unmatched rows from the table on the left, as well as the table on the right.
6. When the keyword LEFT | RIGHT | FULL is used, the keyword OUTER is implied and is therefore optional. Conversely, when the keyword LEFT | RIGHT | FULL is omitted, the keyword INNER is implied.
7. The left outer join means that all rows from the table on the left are kept; the right outer join means that all rows from the table on the right are kept; the full outer join means that all rows are kept from both tables (the one on the left and the one on the right).
8. One significant advantage of the ANSI syntax over the traditional is that you can separate join conditions from other conditions that can still be specified in the **Where-Clause**. Another advantage is a much easier achievement of a full outer join (in the traditional approach, you have to take a union of a left outer with a right outer).

In Example 12.22, the query of Example 12.18 is repeated using the ANSI syntax (show all program-name and course-name combinations. And in Example 12.23, the traditional join of Example 12.21 is replaced with an ANSI equivalent.

EXAMPLE 12.22: ALTERNATE JOIN QUERY FOR PROGRAM-NAME AND COURSE-NAME COMBINATIONS

```
// Show program-name and course-name combinations
SELECT P.PgmName, C.CrsName FROM PgmStruct PS JOIN AcademicProgram P ON
   PS.PSPgm# = P.Pgm# JOIN Course C ON PS.PSCrs# = C.Crs#;
```

The problem of Example 12.21 is revisited in Example 12.23 using the ANSI syntax (student-name and hall-name combinations, including students not assigned to halls).

EXAMPLE 12.23: ALTERNATE JOIN QUERY FOR STUDENT-NAME AND HALL-NAME COMBINATIONS

```
// This requires a left outer join
SELECT S.StudLName, H.HallName FROM Student S LEFT JOIN Hall H ON
   S.StudHall# = H.Hall#;
```

12.8 Queries Involving the Use of Functions

SQL allows the use of several functions in order to provide the user (typically a programmer) with flexibility in specifying queries. We will briefly discuss the following categories of SQL functions:

- Row Functions
- Date Functions
- Data Conversion Functions
- Programmer-defined Functions
- Aggregation Functions
- Analytic Functions

12.8.1 Row Functions

Row functions are functions that act on rows (tuples) of a query result, typically affecting the value of specific columns of a given row. Table 12.1 provides a list of commonly used row functions that can be used within a query specification. The list is by no means exhaustive. You are encouraged to familiarize yourself with these functions through experimentation.

Example 12.24 illustrates how some of the row functions may be used (note the use of the pseudo-table DUAL when displaying data not stored in a traditional relational table).

EXAMPLE 12.24: ILLUSTRATING SOME ROW FUNCTIONS

```
SELECT ROUND (136.876, 2) ROUND(136.876, 0), ROUND(136.876, -1) FROM
DUAL;
// Produces the result 136.88, 137, 140
SELECT TRUNC (136.876, 2) TRUNC(136.876, 0), TRUNC(136.876, -1) FROM
   DUAL;
// Produces the result 136.87, 136, 130
SELECT Stud#, SName, RPAD(DECODE(Sex, 'M', 'Male', 'F', 'Female'), 6) AS Sexx
   FROM Student;
// Displays "Male" instead of "M" and "Female" instead of "F"
SELECT Stud#, SUBSTR(SName,1,1) || SUBSTR(FName,1,1) AS Initl FROM Student;
// Displays the initials of students
SELECT GREATEST(123, 457, 899, 898998, 23000) FROM DUAL;
// Displays 898998 which is the highest value in the specified list
SELECT SName, NVL(StudPgm#, 'Trial') FROM Student;
// Displays surname and program code, with "Trial" for unassigned students
SELECT LPAD ('This is a lovely day! ', 46, 'Yes! ') Today FROM DUAL;
// Displays Yes! Yes! Yes! Yes! Yes! This is a lovely day!
```

Table 12.1 Commonly Used SQL Row Functions

Function	*Explanation*
NVL(<ScalarExpr1>, <ScalarExpr2>)	Replaces null values in the specified Expression 1 with the value of Expression 2.
NVL2(<ScalarExpr1>, <ScalarExpr2>, <ScalarExpr3>)	If Expression 1 is non-null, Expression 2 is returned; otherwise, Expression 3 is returned.
COALESCE(<ScalarExpr> {,<ScalarExpr>})	Returns the first non-null expression from the list; returns null if each expression in the list is null.
CONCAT(<ScalarExpr1>, <ScalarExpr2>)	Concatenates the two expressions specified. The expressions must be alphanumeric — of data-type CHAR, NCHAR, VARCHAR2, NVARCHAR2, CLOB, or NCLOB.
LOWER(<ScalarExpr>)	Converts the alphanumeric argument to lower case.
UPPER(<ScalarExpr>)	Converts the alphanumeric argument to upper case.
INITCAP(<ScalarExpr>)	Converts the first character of each word in the alphabetic (CHAR, NCHAR, VARCHAR2, or NVARCHAR2) argument to upper case.
LPAD(<ScalarExpr1>, <n> [, <ScalarExpr2>])	Returns Expression 1 left-padded up to length n with characters from Expression 2 (repeated as often as required); if Expression 2 is not specified, the default character is a single space. The expressions must be alphanumeric.
RPAD(<ScalarExpr1>, <n> [, <ScalarExpr2>])	Returns Expression 1 right-padded up to length n with characters from Expression 2 (repeated as often as required); if Expression 2 is not specified, the default character is a single space. The expressions must be alphanumeric.
LTRIM(<ScalarExpr1> [, <ScalarExpr2>])	Removes from the left of Expression 1 all characters of Expression 2; if Expression 2 is not specified, the default character is a single space. Both expressions must be alphanumeric.
RTRIM(<ScalarExpr1> [, <ScalarExpr2>])	Similar to LTRIM except that trimming occurs on the right.
TRIM([[LEADING|TRAILING|BOTH] [<ScalarExpr2>] FROM] <ScalarExpr1>)	Removes from the left, right, or both ends of Expression 1 all characters of Expression 2; if Expression 2 is not specified, the default character is a single space. Both expressions must be alphanumeric. The default trim position is BOTH.

(Continued)

Table 12.1 (Continued) Commonly Used SQL Row Functions

Function	*Explanation*
SUBSTR(<ScalarExpr>, <Start>,[<Length>])	Returns a substring of the specified length, starting at the start position. If the length is omitted, all character from **Start** to the end of the string will be returned. **ScalarExpr** must be alphanumeric.
LENGTH(<ScalarExpr>)	Returns the length of the alphanumeric string specified.
ABS(<ScalarExpr>)	Returns the absolute value of the numeric argument or non-numeric argument that can be implicitly converted to numeric.
MOD(<ScalarExpr1 >, <ScalarExpr2>)	Returns the remainder of Expression 1 divided by Expression 2; if Expression 2 is 0, then Expression 1 is returned.
CEIL(<ScalarExpr>)	Rounds the numeric argument up to the nearest integer.
FLOOR(<ScalarExpr>)	Rounds the numeric argument down to the nearest integer.
ROUND(<ScalarExpr>, [<Precision>])	Rounds the numeric argument to the specified precision. If the precision is negative, round to the left of the decimal point. The default precision is 0 decimal places.
TRUNC(<ScalarExpr>, [<Precision>])	Truncates to the specified precision. If the precision is negative, truncate to the left of the decimal point. The default precision is 0 decimal places.
GREATEST(<ScalarExpr> {,<ScalarExpr>})	Returns the highest value from a list of arguments; arguments are assumed to be of the same data-type; the first item is used to determine the data-type of the others and the return type.
LEAST(<ScalarExpr> {,<ScalarExpr>})	Similar to GREATEST but returns the lowest value of the list.
SQRT(<ScalarExpr>)	Returns the square root of the numeric argument (or argument that can be converted to numeric form).
VSIZE(<ScalarExpr>)	Returns the size in bytes of the argument.
DECODE(<ScalarExpr>, <Search>, <Result> {,<Search> , <Result>})	Replaces the search argument with the result for the specified column or expression.
POWER(<ScalarExpr1>, <ScalarExpr2>)	Returns **ScalarExpr1** raised to the power of **ScalarExpr2**. The arguments must be numeric; moreover, if **ScalarExpr1** is negative, then **ScalarExpr2** must be an integer.

You may be surprised by a negative precision used with the rounding function. It simply means that rounding takes place to the left (instead of right) of the decimal point. The following examples should provide clarification:

- ROUND (134.4, -1) = 130
- ROUND (134.4, -2) = 100
- ROUND (134.4, -3) = 0
- ROUND (158.6, -1) = 160
- ROUND (158.6, -2) = 200
- ROUND (158.6, -3) = 0

12.8.2 Date Functions

Date functions constitute a special group of row functions in Oracle. Oracle has an internal representation of date, which stores the day, month, year (including century), hour, minute, and second. The default date format is DD-MON-YY. Additionally, Oracle provides a number of date manipulation functions, some of which are summarized in Table 12.2; some illustrations are provided in Example 12.25a.

EXAMPLE 12.25A: SELECTED ILLUSTRATIONS ON USING DATE FUNCTIONS

SELECT SYSDATE, CURRENT_DATE FROM DUAL; // Displays the current date twice
SELECT SName, TRUNC ((SYSDATE – StudDoB)/365.25) AS Age FROM Student;
// Displays students' name and age, assuming StudDoB is stored internally as type date
SELECT ADD_MONTHS (SYSDATE, 20) FROM DUAL;
// Displays date 20 months beyond current date
SELECT SName, MONTHS_BETWEEN (StudDoB, SYSDATE) AS AgeInMonths FROM Student;
// Displays students' name and age in months, assuming StudDoB is stored internally as Date type
SELECT LAST_DAY('17-AUG-01') FROM DUAL; // Displays "31-AUG-01"
SELECT ROUND ('22-JUL-98', 'MONTH'), ROUND ('22-JUL-98', 'YEAR') FROM DUAL;
// Displays "01-AUG-98" and "01-JAN-99"
SELECT TRUNC ('22-JUL-98', 'MONTH'), TRUNC ('22-JUL-98', 'YEAR') FROM DUAL; // Displays "01-JUL-98" and "01-JAN-98"
SELECT EXTRACT (YEAR FROM SysDate) AS Current_Year FROM DUAL;
// Displays the current year

Table 12.2 Some SQL Date Manipulation Functions

<table>
<tr><th>Function</th><th>Explanation</th></tr>
<tr><td>MONTHS_ BETWEEN(<Date1>, <Date2>)</td><td>Returns the number of months between two dates.</td></tr>
<tr><td>ADD_MONTHS(<Date>, <Months>)</td><td>Returns a new date after adding a specified number of months.</td></tr>
<tr><td>NEXT_DAY(<Date>)</td><td>Returns the new date after a specified date.</td></tr>
<tr><td>LAST_DAY(<Date>)</td><td>Returns the last day of the month specified.</td></tr>
<tr><td>ROUND(<Date> [,<Format>])</td><td>Truncates the date given to the nearest day, month, or year,depending on the format (which is 'DAY' or 'MONTH' or 'YEAR').</td></tr>
<tr><td>TRUNC(<Date> [,<Format>])</td><td>Truncates the date given to the nearest day, month, or year, depending on the format (which is 'DAY' or 'MONTH' or 'YEAR').</td></tr>
<tr><td>EXTRACT(<DateField)> FROM <DateTimeExpr> | <InternalExpr>)</td><td>Extracts and returns the value for the specified date-field from the stated date-time expression or internal expression — expression based on the DATE or TIMESTAMP data-type (review Figure 11.8 of the previous chapter). The syntax for DateField is stated below. #</td></tr>
<tr><td>NEW_TIME(<Date>, <Zone1>, <Zone2>)</td><td>Converts a date and time in time zone Zone1 to date and time in time zone Zone2. Time formats are in the form XST or XDT (S for Standard, D for Daylight saving) with two exceptions: GMT and there is no NDT for Newfoundland.</td></tr>
<tr><td>CURRENT_DATE</td><td>Returns the current system date; typically treated as a pseudo-column.</td></tr>
<tr><td>SYSDATE</td><td>An alternative to CURRENT_DATE; similar behavior.</td></tr>
<tr><td colspan="2"># Date-Field ::=[YEAR | MONTH | DAY | HOUR | MINUTE | SECOND][TIMEZONE_HOUR | TIMEZONE_MINUTE][TIMEZONE_REGION | TIMEZONE_ABBR]</td></tr>
</table>

To change the default format for system, you can use the **Alter-Session** statement and change the pseudo-column **nls_date_format** as in Example 12.25b.

EXAMPLE 12.25B: CHANGING THE SYSTEM FORMAT FOR DATE

```
ALTER SESSION SET nls_date_format = 'YYYY-MM-DD HH24:MI:SS';
// If you now display the system date, you will observe the new format of the system date
```

12.8.3 Data Conversion Functions

Data conversion functions are used to convert data in the following ways:

- Character to number
- Number to character
- Character to date
- Date to character

Figure 12.7 indicates four common data conversion functions (other less common functions are not discussed in this course). Table 12.3 shows date formats and numeric formats, which are often used with these functions. Example 12.26 provide illustrations of how these functions are used.

```
TO_CHAR(<ScalarExpr>[,<Format>])
   // Converts a number or date to a VARCHAR2 value, based on the format (specified in single quotes)

TO_NUMBER(<ScalarExpr> [,<Format>])
   // Converts a string with valid digits to number, based on the format specified

TO_DATE(<ScalarExpr>[,<Format>])
   // Converts a string to date, based on the format specified (default DD-MON-YY)

CAST (<ScalarExpr> | <MultisetSpec>) AS <DataType>

MultisetSpec ::=
MULTISET (<Subquery>)
   // Cast facilitates data belonging to one data-type to be construed as data of another data-type.
   // The effect is typically on the result-set, and not the source.
```

Figure 12.7 SQL commonly used data conversion functions.

EXAMPLE 12.26: ILLUSTRATING DATA CONVERSION FUNCTIONS

```
SELECT EmpID, TO_CHAR(Salary, '$99,999.99') AS Salary FROM Emp;
// Displays Salary in the format shown
SELECT Stud#, TO_DATE(StudDoB, 'YYYY-MM-DD') StudDoB FROM Student;
// Converts StudDoB from YYYYMMDD to standard date format
SELECT Stud#, TO_CHAR(TO_DATE(StudDoB, 'YYYY-MM-DD'), 'YYYY-MM-DD')
   StudDoB FROM Student;
// Displays StudDoB in YYYY-MM-DD format
SELECT TO_CHAR(SYSDATE, 'YYYY-MM-DD') CurrentDate FROM DUAL;
// Displays the current date in YYYYMMDD format
SELECT SUBSTR(TO_CHAR(SYSDATE, 'YYYY-MM-DD'), 1, 4) CurrentYear FROM
   DUAL;
// Displays the current year as YYYY
SELECT CAST (CURRENT_DATE AS VARCHAR2(20)) Current_Date FROM DUAL;
// Converts the current date to VARCHAR2 and displays it
```

Table 12.3 Some Valid Date and Numeric Formats

Valid Date Formats	
Date format	**Clarification**
YYYY	Four digit year
Y, YY, or YYY	Last 1, 2, or 3 digits of year
YEAR	Year spelled out
Q	Quarter of the year
MM	Two digits for month
MON	First three characters of month
MONTH	Month spelled out
WW or W	Week number of year or month
DDD, DD, or D	Day of year, month, or week
DAY	Day spelled out
DY	Three-letter abbreviation of day
DDTH	Ordinal number of day, e.g., 7th
HH, HH12, HH24	Hour of day, hour (0–12), or hour (0–23)
MI	Minute (0–59)
SS	Second (0–59)
Valid Numeric Formats	
Numeric format	**Clarification**
9	Number of 9s determine width
0	Displays leading zeros
$	Displays floating dollar sign
L	Displays floating local currency
.	Displays decimal point
,	Displays thousand indicator as specified
PR	Displays negative numbers in parentheses

12.8.4 Programmer-Defined Functions

Oracle supports programmer-defined functions; these are used in the same manner as regular SQL system functions. Programmer-defined functions are written in Oracle's PL/SQL and stored on the server. For more information on how to define them, refer to your Oracle documentation. By way of illustration, Example 12.27 provides the PL/SQL code for a function that calculates the value of the expression y^x, where y and x are numeric data.

EXAMPLE 12.27: PL/SQL FUNCTION TO DETERMINE *Y* TO THE POWER *X*

```
/* This is a sample function. */
Create Function YPowerX (y Number, x Number) Return Number
AS
k Number(3);
theResult Number;
Begin
If (x = 0) then
theResult := 1;
Else
theResult := 1;
For k In 1..x Loop
theResult := theResult * y;
End Loop;
End If;
Return theResult;
End;
```

12.8.5 Aggregation Functions

SQL provides several aggregate functions to provide a summarized view of data. Aggregate functions (also called group functions) act on rows of data to return a single summarizing row. Some commonly used aggregate functions are mentioned in Table 12.4; this is followed by a few examples.

Of course, there are other aggregation functions not included here; the table shows the commonly used ones. The following guidelines apply:

1. The DISTINCT keyword indicates to SQL to return only distinct rows for the specified column expression of query. Example 12.29 clarifies this.
2. The ALL keyword means that all rows for the specified column expression in the query will be returned.
3. SUM, AVG, STDDEV, and VARIANCE work with numeric values.
4. For MAX and MIN, the DISTINCT option is irrelevant.
5. COUNT works with the DISTINCT option always, except for the case where COUNT (*) or COUNT (1) is used.

EXAMPLE 12.28: QUERYING THE NUMBER OF COURSES OFFERED

```
// How many courses are offered by the institution?
SELECT COUNT (*) FROM Course;
// The following alternate solution is more efficient
SELECT COUNT (1) FROM Course;
```

Table 12.4 Commonly Used SQL Aggregate Functions

Function	*Explanation*
AVG([DISTINCT | ALL] <Scalar-Expr>) [OVER (<Analytic-Clause)]	Returns average of tuple values for the expression specified. The column-related expression must be numeric or convertible to numeric form.
COUNT([DISTINCT | ALL |*] <Scalar-Expr>)[OVER (<Analytic-Clause)]	Returns number of tuples spanned by the query.
SUM([DISTINCT | ALL] <Scalar-Expr>) [OVER (<Analytic-Clause)]	Returns total of tuple values for the expression specified. The column-related expression must be numeric or convertible to numeric form.
MAX([DISTINCT | ALL] <Scalar-Expr>) [OVER (<Analytic-Clause)]	Returns largest tuple value for the expression specified. The expression is typically column-related.
MIN([DISTINCT | ALL] <Scalar-Expr>) [OVER (<Analytic-Clause)]	Returns smallest tuple value for the expression specified. The expression is typically column-related.
STDDEV([DISTINCT | ALL] <Scalar-Expr>)[OVER (<Analytic-Clause)]	Returns standard deviation of tuple values for the expression specified. The column-related expression must be numeric.
VARIANCE([DISTINCT | ALL] <Scalar-Expr>)[OVER (<Analytic-Clause)]	Returns variance (which is the square of the standard deviation) of tuple values for the expression specified. The column-related expression must be numeric.
Note: Each of the functions in this list may be used as an aggregation function or an analytics function, hence the **Analytic-Clause,** which will be discussed in the upcoming section.	

EXAMPLE 12.29: QUERYING THE NUMBER OF PROGRAMS OFFERING COURSE M100

```
// How many programs offer the course M100?
SELECT COUNT (DISTINCT PSPgm#) FROM Pgm_Struct WHERE PSCrs# = 'M100';
```

Examples 12.30 and 12.31 are based on the supplier–items data set of earlier discussions (Chapters 3–5) and also based on the simplified structure presented in Section 12.6.

EXAMPLE 12.30: ILLUSTRATING AGGREGATION BASED ON SPECIFIC SEARCH CRITERIA

```
// How many boxes of item 'I100' is supplied?

// What is the maximum quantity and the minimum quantity for this item?
SELECT SUM (Qty) AS SumQty, MAX(Qty) AS MaxQty, MIN(Qty) AS MinQty FROM
   Schedule WHERE Item# = 'I100';
```

Aggregate functions are also often used in situations where data is to be grouped in order to set up control breaks. This is illustrated in the following examples.

EXAMPLE 12.31: ILLUSTRATING AGGREGATION BASED ON SPECIFIC TABLE COLUMN(S)

// Produce a list showing item number and quantity of items shipped
SELECT Item#, SUM (Qty) AS SumQty FROM Schedule GROUP BY Item#;

// Sample result would appear as follows:

Item#	*SumQty*
I100	600
I101	700
...	...

The aggregation functions are often used when grouping data (hence the alternate name, group functions) in order to set up control (summary) lines in reports. For several of the upcoming examples, consider an extended data set for the college database of earlier discussions, particularly with respect to the following relational table structures:

Department {Dep#, DeptName} with PK [Dept#]
Employee {Emp#, EmpLName, EmpFName, EmpMName, EmpSex, EmpDoB, EmpDept#, EmpJobCode, EmpSalary} with PK [Emp#]
JobDescription {JobCode, JobTitle, JobAbbrev} PK [JobCode]
Assume that **Employee.EmpDept#** is a foreign key that references **Deptartment.Dept#**
Assume that **Employee.EmpJobCode** is a foreign key that references **JobDescription. JobCode**

EXAMPLE 12.32: ILLUSTRATING AGGREGATION OF DEPARTMENTAL DATA

Objective: Develop a list from the employee table, showing for each department, the total salary, average salary, minimum salary, maximum salary and standard deviation.

// This solution produces unrounded aggregation numbers
SELECT EmpDept#, SUM (EmpSalary) AS TotalSal, AVG (EmpSalary) AS Average, MAX (EmpSalary) AS
MaxSal, MIN (EmpSalary) AS MinSal, STDDEV (EmpSalary) AS StdDev
FROM Employee GROUP BY EmpDept# ORDER BY EmpDept#;
// This improved solution rounds the aggregation results to 2 decimal places
SELECT EmpDept#, ROUND (SUM(EmpSalary), 2) AS TotalSal, ROUND (AVG(EmpSalary), 2) AS AvgSal,
ROUND (MAX(EmpSalary), 2) AS MaxSal, ROUND (MIN(EmpSalary), 2) AS MinSal,
ROUND (STDDEV(EmpSalary), 2) AS StdDev
FROM Employee GROUP BY EmpDept# ORDER BY EmpDept#;

Observe:

It is sometimes not permissible to use the **Group-by-Clause** and the **Where-Clause** in the same query (or subquery), particularly if the **Where-Clause** is intended to specify a join. For instance, you may think, it would be nice to rewrite the solution above so that the actual name of each department is displayed. Attempting to join tables **Employee** and **Department** like this in the same aggregation query would not work. If you thought that approach would and wrote a solution like the one shown below, it would be rejected by the SQL compiler.

```
// This query will not work:
SELECT EmpDept#, SUM (EmpSalary) AS TotalSal, AVG (EmpSalary) AS Average, MAX
   (EmpSalary) AS
MaxSal, MIN (EmpSalary) AS MinSal, STDDEV (EmpSalary) AS StdDev, D.DeptName
FROM Employee E, Department D WHERE E.EmpDept# = D.Dept# GROUP BY
   EmpDept#;
```

There are ways to get around this hurdle: One recommended approach is to create a logical view or snapshot and then write a query on the view or snapshot; this strategy will be further discussed in Chapter 15. Another recommended approach is to use analytic functions; this will be clarified in the upcoming section.

The **Having-Clause** may be used to restrict groups. **HAVING…** works after the groups have been selected; **WHERE…** works on rows before group selection.

EXAMPLE 12.33A ILLUSTRATING AGGREGATION OF DEPARTMENTAL DATA AND USING THE HAVING-CLAUSE

Objective: Develop a list from the employee table, showing for each department, the total salary, average salary, minimum salary, maximum salary, and standard deviation; show only departments with a total salary of at least $60,000.

```
// This solution will not run; TotSal is a calculated attribute and cannot be reused as shown
SELECT EmpDept#, SUM(EmpSalary) AS TotalSal, AVG(EmpSalary) AS AvgSal,
   MAX(EmpSalary) AS
MaxSal, MIN(EmpSalary) AS MinSal, STDDEV(EmpSalary) AS StdDevSal
FROM Employee GROUP BY EmpDept# HAVING TotalSal >= 120000;

// The correct solution
SELECT EmpDept#, SUM(EmpSalary) AS TotalSal, AVG(EmpSalary) AS AvgSal,
   MAX(EmpSalary) AS
MaxSal, MIN(EmpSalary) AS MinSal, STDDEV(EmpSalary) AS StdDevSal
FROM Employee GROUP BY EmpDept# HAVING SUM(EmpSalary) >= 120000;
```

```
// Refined solution
SELECT EmpDept#, SUM(EmpSalary) AS TotalSal, AVG(EmpSalary) AS AvgSal,
   MAX(EmpSalary) AS
MaxSal, MIN(EmpSalary) AS MinSal, STDDEV(EmpSalary) AS StdDevSal
FROM Employee GROUP BY EmpDept# HAVING SUM(EmpSalary) >= 120000;
```

Note:

The question of when a column should be included in a query that involves grouping of data has often troubled inexperienced users of SQL. Here are two guidelines:

1. Traditionally speaking, if you could not group on the column, then you would not include that column in the **Select** clause of the aggregation query.
2. Recent developments provide more flexibility in mixing regular columns and aggregate (calculated) columns in the same query. A rather interesting strategy is to employ an analytic function, discussed in the upcoming section.

12.8.6 Analytic Functions

An analytic function calculates aggregation value(s) based on a group of rows and returns multiple rows for the group. The group is defined by the **Analytic-Clause** and is called a window. Several of the aggregation functions also qualify as analytic functions, provided that the **Analytic-Clause** is applied with the function; in fact, all the functions listed in Table 12.4 are also analytic functions. Figure 12.8 provides the syntax for the **Analytic-Clause**.

While the similarity between analytic functions and aggregation functions is noteworthy, you should also know that the two categories are not identical. There are two important differences between analytic functions and aggregation functions that you should note:

```
AnalyticFunctionSpec ::=
<AnalyticFunction> ([<Argument(s)>]) OVER (<Analytic-Clause>)

AnalyticClause ::=
[<Query-Partitioning-Clause>]
[<Order-By-Clause> [<Windowing-Clause>]]

Query-PartitioningClause ::=
PARTITION BY <Scalar-Expr> {,<Scalar-Expr>} /* Or */
PARTITION BY (<Scalar-Expr> {,<Scalar-Expr>})

// The OrderByClause is as previously defined in figure 12-5
// The scalar expression specified is typically a column-based expression
// The WindowingClause has been omitted; see the Oracle Product Documentation
```

Figure 12.8 Syntax for using analytic function.

1. Aggregation functions return a single value. In contrast, analytic functions do not reduce the number of rows returned by the query; they return the rows covered by the related query (or subquery) along with the analytic value being sought.
2. Each analytic function call is characterized by the use of the OVER keyword followed by the Analytic-Clause.

EXAMPLE 12.33B: ILLUSTRATING AGGREGATION OF DEPARTMENTAL DATA AND USING ANALYTIC FUNCTION

Objective: Produce a list from the employee table, showing employees grouped by department. Include the employee number, name, salary, the total salary of the department, average departmental salary, minimum departmental salary, and maximum departmental salary:

```
// Basic solution
SELECT Emp#, EmpLName || ' -- ' || EmpFName || ' ' || SUBSTR(EmpMName, 1, 1)
   AS EmpFullName, EmpSalary, EmpDept#, SUM(EmpSalary) OVER (PARTITION
   BY EmpDept#) AS TotalSal, AVG(EmpSalary) OVER (PARTITION BY EmpDept#)
   AS AvgSal, MAX(EmpSalary) OVER (PARTITION BY EmpDept#) AS MaxSal,
   MIN(EmpSalary) OVER (PARTITION BY EmpDept#) AS MinSal
FROM Employee;

// Improved solution
SELECT E.Emp#, E.EmpLName || ' -- ' || E.EmpFName || ' ' || SUBSTR(E.EmpMName,
   1, 1) AS EmpFullName, E.EmpSalary, E.EmpDept#, D.DeptName, SUM(E.
   EmpSalary) OVER (PARTITION BY E.EmpDept#) AS TotalSal, AVG(E.EmpSalary)
   OVER (PARTITION BY E.EmpDept#) AS AvgSal, MAX(E.EmpSalary) OVER
   (PARTITION BY E.EmpDept#) AS MaxSal, MIN(E.EmpSalary) OVER (PARTITION
   BY E.EmpDept#) AS MinSal
FROM Employee E, Department D WHERE E.EmpDept# = D.Dept#;
```

12.9 Queries Using the LIKE and BETWEEN Operators

The LIKE operator is used to test for the existence of string patterns in a column. The BETWEEN … AND operator is used to test for column values within a range of values. The following examples based on the sample database of earlier discussions illustrate how these operators are used.

EXAMPLE 12.34: LOOKING FOR A STRING PATTERN IN AN ALPHANUMERIC COLUMN

Objective: Assuming that Computer Science course codes are prefixed with the acronym "CS," list all such courses in the database.

```
// List the names of CS courses
SELECT CrsName FROM Course WHERE Crs# LIKE 'CS%';
```

Note:

The wildcard % is used to denote that the string being searched for must begin with the character(s) specified; the characters that follow do not matter.

EXAMPLE 12.35: LOOKING FOR A STRING PATTERN IN AN ALPHANUMERIC COLUMN

Objective: Assuming that a course code is always preceded by its acronym, the course level (1, 2, 3, or 4 for the year the course is taught) and a two-digit sequence number get names of all second year.

```
// Display the names of second year courses
SELECT CrsName FROM Course WHERE Crs# LIKE '%2__';
```

EXAMPLE 12.36: RETRIEVING NUMERIC DATA IN A SPECIFIED RANGE

Objective: Referring to the Oracle employee–department data set (of Section 12.8.5), list employees within a salary range of 50,000 and 120,000.

```
// List employees within a salary range of 50,000 and 120,000
SELECT Emp#, EmpLName, EmpFName, EmpMName, EmpSalary FROM Employee
WHERE EmpSalary BETWEEN 50000 AND 120000;
```

12.10 Nested Queries

A subquery is a nested query, i.e., a query within another query. A subquery is particularly useful when a query on a given table depends on data in the table itself. Additionally, nested queries are alternatives to queries involving theta joins (not recommended for complex join conditions). Here are a few principles to remember:

1. A subquery may return a single row or multiple rows of data.
2. Several layers of nesting may be constructed. However, this is not recommended unless it is warranted. For instance, do not introduce nested queries (as in Example 12.38) where a simple equijoin will suffice.
3. Within nested queries, a scalar variable is confined (known only) to its Select block at each level. However, variables (scalar) within a higher block, A, are known to the inner blocks of A. For instance, if a query solution consists of blocks A, B, and C, where A consists of B and C, then:
   - Variables of block C are not known to block B and vice versa
   - Variables of blocks B and C are known to block A
   - Variables of block A are not known to blocks B and C

There are two common formats for nested queries, both of which will be discussed. The typical format is shown below:

SELECT <Item-List> FROM <Relation-List> WHERE <Column> | <Scalar-Expression> <Boolean-Operator>
(<Subquery>);

In this format, the inner query (i.e., subquery) comes after the Boolean operator that connects it to the outer query, which is specified first. The following rules apply:

1. The subquery is another **Select** statement, which cannot include an **Order-By-Clause**.
2. The inner query is executed first, and its result is passed to the outer query.
3. Within the **Where-Clause**, a scalar expression involving at least one column may be specified instead of a column.

An alternate format for a subquery allows for at least one relational expression from the specified relations-list to be itself a subquery; the format is as follows:

SELECT **<Item-List>** FROM **<Relations-List>** [WHERE **<Condition>**];

This second format appears to be more flexible than the first; it is therefore often used for more complex scenarios. In this second format, the following rules apply:

1. The relations-list includes at least one subquery.
2. It is possible to have this format combined with the first.
3. The inner query/queries is/are executed first, and the result(s) is/are passed to the outer query.

The operators IN, ANY, and ALL are often used in nested queries to test for column values within a set of values; the set of values may be expressed explicitly or implied from the result of a subquery. Additionally, some nested queries may involve the use of the EXISTS keyword (representing the existential quantifier).

12.10.1 Nested Queries Involving Use of the IN Operator

The IN operator is useful when checking for the presence of a data pattern in another data set. It can be used as an efficient replacement for multiple OR conditions or as an alternative to the equijoin. The following three examples illustrate the usage of the IN operator.

EXAMPLE 12.37: ILLUSTRATING USE OF IN INSTEAD OF MULTIPLE ORS

Objective: Assuming the sample data of Figure 7.2, list all students enrolled in MIS, Computer Science, or Mathematics.

```
// Solution via the IN operator
SELECT * FROM Student WHERE StudPgm# IN ('BSC1', 'BSC2', 'BSC4');

// Alternate solution via multiple ORs
SELECT * FROM Student WHERE StudPgm# = 'BSC1' OR StudPgm# = 'BSC2' OR
   StudPgm# = 'BSC4';
```

EXAMPLE 12.38: ILLUSTRATING USE OF IN INSTEAD OF JOIN QUERY

Objective: Assuming that a student's program code is never null, produce a list showing student name and associated major (program) name.

```
// Solution that applies the IN operator
SELECT S.SName || S.FName AS FullName, P.PgmName FROM Student S,
   AcademicProgram P
WHERE S. StudPgm# IN (SELECT Pgm# FROM AcademicProgram) AND S.StudPgm#
   = P.Pgm#;

// Following is an equivalent and improved solution that applies the equijoin:
SELECT S.SName || S.FName AS FullName, P.PgmName FROM Student S,
   AcademicProgram P
WHERE S.StudPgm# = P.Pgm#;
```

EXAMPLE 12.39: ILLUSTRATING BOTH TYPICAL AND ALTERNATE FORMS OF NESTED QUERIES

Objective: Referring to the expanded data set described in Section 12.8.5, produce a list of employees who earn the maximum salary in their respective departments.

```
// Here is the typical solution
SELECT EmpLName || ' ' || EmpFName || ' ' || EmpMName AS EmpFullName,
   EmpJobCode, EmpSalary, EmpDept# FROM Employee
WHERE EmpSalary IN (SELECT MAX (EmpSalary) FROM Employee GROUP BY
   EmpDept#);

// Here is an alternate solution that produces the same result
SELECT DISTINCT E1.EmpLName || ' ' || E1.EmpFName || ' ' || E1.EmpMName AS
   EmpFullName, E1.EmpJobCode, E1.EmpSalary, E1.EmpDept#
FROM Employee E1, (SELECT MAX (EmpSalary) AS MaxSal, EmpDept# FROM
   Employee GROUP BY EmpDept#) E2 WHERE E1.EmpSalary = E2.MaxSal;
```

Note:

In queries of this form, do not use the **equal operator** (=) as the connecting operator unless you are certain that the inner subquery produces only one row. In any event, the **IN operator** is safer.

12.10.2 Nested Queries Involving Use of ANY or ALL Operator

The **ANY** (or **SOME) operator** and **ALL operator** both work nicely with nested queries. The expression on the left is compared to *any* row or *all* rows from the subquery on the right. In using these operators, the following convention must be observed:

- < ANY(...) means less than the maximum value
- = ANY(...) is equivalent to IN (...)
- > ANY(...) means greater than the minimum value
- > ALL(...) means greater than the maximum value
- < ALL(...) means less than the minimum value

The next three examples illustrate use of the ANY or ALL operator. In each case, an alternate query that does not employ the operator is also provided.

EXAMPLE 12.40: ILLUSTRATING QUERY THAT USES THE ANY OPERATOR

Objective: Staying with the expanded dataset described in Section 12.8.5, produce a list of employees who earn less than the maximum salary of a secretary.

```
// Assume that the job code for a secretary is 'JD50'
SELECT EmpLName, EmpFName, EmpSalary FROM Employee
WHERE EmpSalary < ANY(SELECT EmpSalary FROM Employee WHERE
   EmpJobCode = 'JD50');

// Alternate Query
SELECT E1.EmpLName, E1.EmpFName, E1.EmpSalary FROM Employee E1, (SELECT
   EmpJobCode,
MAX(EmpSalary) as MaxSal FROM Employee GROUP BY EmpJobCode) E2
WHERE E2.EmpJobCode = 'JD50' AND E1.EmpSalary < E2.MaxSal;

// Alternate Query
SELECT E1.EmpLName, E1.EmpFName, E1.EmpSalary FROM Employee E1, (SELECT
   EmpJobCode,
MAX(EmpSalary) as MaxSal FROM Employee GROUP BY EmpJobCode) E2
WHERE E2.EmpJobCode = 'JD50' AND E1.EmpSalary < E2.MaxSal ORDER BY
   EmpLName, EmpFName;
```

EXAMPLE 12.41: ILLUSTRATING QUERY THAT USES THE ALL OPERATOR

Objective: Still referring to the expanded data set described in Section 12.8.5, produce a list of employees who earn less than the minimum salary of a secretary.

```
// Assume that the job code for a secretary is 'JD50'
SELECT EmpLName, EmpFName, EmpSalary FROM Employee
WHERE EmpSalary < ALL(SELECT EmpSalary FROM Employee WHERE EmpJobCode
   = 'JD50');

// Alternate Query
SELECT E1.EmpLName, E1.EmpFName, E1.EmpSalary FROM Employee E1, (SELECT
   EmpJobCode,
MIN (EmpSalary) as MinSal FROM Employee GROUP BY EmpJobCode) E2
WHERE E2.EmpJobCode = 'JD50' AND E1.EmpSalary < E2.MinSal;

// Alternate Query
SELECT E1.EmpLName, E1.EmpFName, E1.EmpSalary FROM Employee E1, (SELECT
   EmpJobCode,
MIN(EmpSalary) as MinSal FROM Employee GROUP BY EmpJobCode) E2
WHERE E2.EmpJobCode = 'JD50' AND E1.EmpSalary < E2.MinSal ORDER BY
   EmpLName, EmpFName;
```

EXAMPLE 12.42: ILLUSTRATING QUERY THAT COMPARES DATA SUBSETS FROM A LARGER DATA SET

Objective: Still referencing to the expanded data set described in Section 12.8.5, produce a list of employees who earn more than the average earning for his/her department (show both the employee's salary and his/her department's average salary).

```
// Basic solution
SELECT E1.Emp#, SUBSTR(E1.EmpFName,1,1) || '. ' || SUBSTR(E1.EmpMName, 1, 1)
   || '. ' || E1.EmpLName AS EmpName, E1.EmpSalary, E1.EmpDept#, E2.AvgSal
FROM Employee E1, (SELECT EmpDept#, AVG(EmpSalary) AS AvgSal FROM Employee
   GROUP BY EmpDept#) E2
WHERE (E1.EmpDept# = E2.EmpDept# AND E1.EmpSalary > E2.AvgSal) ORDER BY
   E1.EmpDept#;

// Alternate Query with Rounding
SELECT E1.Emp#, SUBSTR(E1.EmpFName,1,1) || '. ' || SUBSTR(E1.EmpMName, 1, 1)
   || '. ' || E1.EmpLName AS EmpName, E1.EmpSalary, E1.EmpDept#, E2.AvgSal
FROM Employee E1, (SELECT EmpDept#, ROUND(AVG(EmpSalary),2) AS AvgSal
   FROM Employee GROUP BY EmpDept#) E2
WHERE (E1.EmpDept# = E2.EmpDept# AND E1.EmpSalary > E2.AvgSal) ORDER BY
   E1.EmpDept#;
```

```
// Alternate Query with analytic function
SELECT DISTINCT E1.Emp#, SUBSTR(E1.EmpFName,1,1) || '. ' || SUBSTR(E1.
   EmpMName, 1, 1) || '. ' || E1.EmpLName AS EmpName, E1.EmpSalary, E1.EmpDept#,
   E2.AvgSal
FROM Employee E1, (SELECT EmpDept#, AVG(EmpSalary) OVER (PARTITION BY
   EmpDept#) AS AvgSal FROM Employee) E2
WHERE E1.EmpDept# = E2.EmpDept# AND E1.EmpSalary > E2.AvgSal ORDER BY
   E1.EmpDept#;
```

12.10.3 Nested Queries Using the Existential Quantifier

SQL also allows the specification of nested queries that involve the use of the existential quantifiers (EXISTS and NOT EXISTS). The universal quantifier is not really necessary since it can be substituted (review the substitutions rules of Section 8.4). Moreover, in most cases you are likely to encounter, the convention of not retrieving a bounded tuple variable, along with its consequences, ensures that SQL statements can be specified without the use of quantifiers. That said, SQL nonetheless supports the use of the existential quantifiers, albeit in a manner that is slightly different from what was discussed in Chapter 8 — the tuple variable is usually implicit.

EXAMPLE 12.43: ILLUSTRATING USE OF THE EXISTENTIAL QUANTIFIER

Objective: Let us revisit the problem of determining program names of programs that include the course M100: Write the SQL solution using the EXISTS keyword.

```
// The relational calculus solution would be:
RANGE OF P IS AcademicProgram;
RANGE OF PS IS Pgm_Struct;
P.PgmName WHERE EXISTS PS (PS.PSPgm# = P.Pgm# AND PS.Crs# = 'M100');

// Basic SQL solution using nested query
SELECT Pgm#, PgmName FROM AcademicProgram
WHERE Pgm# IN (SELECT PSPgm# FROM Pgm_Struct WHERE PSCrs#= 'M100');

// Alternate SQL solution using nested query with existential quantifier
SELECT Pgm#, PgmName FROM AcademicProgram
WHERE EXISTS (SELECT * FROM Pgm_Struct PS WHERE PS.PSPgm# =
   AcademicProgram.Pgm#
AND PS.PSCrs# = 'M100');

// Alternate SQL solution using equijoin
SELECT PS.PSPgm#, P.PgmName FROM AcademicProgram P, Pgm_Struct PS
WHERE PS.PSPgm# = P.Pgm# AND PS.PSCrs# = 'M100';
```

12.11 Queries Involving Set Operation

SQL supports the binary set operations UNION, UNION ALL, INTERSECT, and MINUS. Recall from Chapter 7 that for these binary relational algebra operations, a basic requirement is that (in each case) the two participating relations must have corresponding attributes defined on the same domain. Since from a design perspective, it is highly unlikely that a database will have many base relations that meet this criterion, these set operators are usually used in queries where two subqueries produce results that meet the compatibility criterion. The general syntax, therefore, is:

```
<Query1> <SetOperator> <Query2>;
```

Also recall from Chapter 7 that the set operations are relational algebra operations. By way of review (except for UNION ALL), the set operators are explained below:

- UNION: Returns all rows from both queries, but duplicate rows are not displayed.
- UNION ALL: Returns all rows from both queries, including duplicate rows.
- INTERSECT: Returns rows that appear in both query results.
- MINUS: Returns rows that appear in the first query result, but not in the second.

EXAMPLE 12.44: ILLUSTRATING QUERY USING UNION OF DATA SETS

Objective: Assuming the sample data of Figure 7.3, list all students enrolled in MIS, Computer Science, or Mathematics, but this time using the UNION of related datasets.

```
// Solution 1: Using UNION of sets
SELECT * FROM Student WHERE StudPgm# = 'BSC1'
UNION SELECT * FROM Student WHERE StudPgm# = 'BSC2'
UNION SELECT * FROM Student WHERE StudPgm# = 'BSC4';

// Solution 2: Using the OR operator
SELECT * FROM Student WHERE StudPgm# = 'BSC1' OR StudPgm# = 'BSC2'
OR StudPgm# = 'BSC4';

// Solution 3: Using the IN operator
SELECT * FROM Student WHERE StudPgm# IN ('BSC1', 'BSC2', 'BSC4');
```

EXAMPLE 12.45: ILLUSTRATING QUERY USING INTERSECTION OF DATA SETS

Objective: Find all students from Lenheim Hall who are Mathematics majors.

```
SELECT * FROM Student WHERE StudHall# = 'Len '
INTERSECT SELECT * FROM Student WHERE StudPgm# = 'BSC4';

// Alternate solution follows:
SELECT * FROM Student WHERE StudHall# = 'Len ' AND StudPgm# = 'BSC4';
```

EXAMPLE 12.46: ILLUSTRATING QUERY USING DIFFERENCE OF DATA SETS

Objective: Find all male students from Lenheim Hall who are not mathematics majors.

```
SELECT * FROM Student WHERE StudHall# = 'Len ' AND StudSex = 'M'
MINUS SELECT * FROM Student WHERE StudHall# = 'Len ' AND StudSex = 'M'
AND StudPgm# = 'BSC4';

// Alternate solution follows:
SELECT * FROM Student WHERE StudHall# = 'Len ' AND StudSex = 'M' AND
   StudPgm# <> 'BSC4';
```

Although optional, the set operators can be quite useful. In deciding how to structure your query, the following constructs are worth remembering:

- A condition involving the AND connector is equivalent to the intersection of the two related sets.
- A condition involving the OR connector is equivalent to the union of the two related sets.
- A condition involving the construct *<Proposition P1> but not <Proposition P2>* is equivalent to the difference between the set implied by P1 and the set implied by P2.

12.12 Queries with Runtime Variables

Earlier in the chapter, it was established that through the Pl/SQL language, Oracle supports runtime substitution variables that may be specified in an SQL statement. This applies to data insertion, update, and deletion statements; it also applies to data retrieval statements (i.e., queries). The rules are unchanged. The intent is not to teach you PL/SQL here; you would need to pursue an OCP-DBA course for that. However, the following examples should enhance your appreciation of the language.

EXAMPLE 12.47: ILLUSTRATING QUERY WITH RUNTIME VARIABLE

```
// Produce a list of courses, starting from a particular course which the user will specify:
SELECT * FROM Course WHERE Crs# >= &InputCode;
```

When using execution-time variables, there are a few guidelines worth remembering. Following are the salient issues:

1. The ampersand (&) preceding the variable (**InputCode** in Example 12.47) indicates that user will be prompted to specify a value for the variable. This value will then be used to complete the query.
2. If the expected input is alphanumeric, then you have the option of specifying the ampersand and input variable in single quotes (as in '&**InputCode**'); if you do this, then you will be able to specify the actual input without quotation marks. Alternately, you may specify the ampersand and input variable without quotes (as in &**InputCode**), in which case, you must enclose your actual input in single quotes when prompted to do so.
3. If the expected input is numeric, no ampersand is required.
4. A double ampersand indicates (to Oracle) that the input is to be obtained from the previous input value.

EXAMPLE 12.48: ILLUSTRATING ANOTHER QUERY WITH RUNTIME VARIABLE

```
// This statement lists courses starting at the name the user specifies to the end of the table
SELECT * FROM Course WHERE CrsName >= &StartPoint ORDER BY CrsName;

/* This second statement uses the input from the first statement to display the specific
course(s) meeting the specified LIKE criterion */
SELECT * FROM Course WHERE CrsName LIKE '%' || &&StartPoint || '%';
```

12.13 Queries Involving SQL*Plus Format Commands

The Oracle SQL environment provides certain format commands that can be used to affect the appearance of outputs from queries. Table 12.5a shows some of the commonly used format commands, while Table 12.5b provides a list of valid format codes that may be used in a format specification.

Table 12.5a Commonly Used Format Commands

Function	*Explanation*
TTITLE <Text> [ON|OFF]	Sets up title for a query output
BTITLE <Text> [ON|OFF]	Sets up footnote for a query
BREAK ON <Column>	Specifies a control break
COLUMN <Column> <Option>	noindentSpecifies how a column is to be displayed. The **Column Option** consists of column commands and optional formats. Column commands include: • CLEAR • FORMAT <Format-Specification> • HEADING <Text> • JUSTIFY LEFT/CENTER/RIGHT

Table 12.5b Valid Format Codes

Format Code	*Explanation*
An	Alphanumeric, e.g., A20
9	Numeric, e.g., 999.99
0	Forces leading zero, e.g., 009.99
$	Floating dollar sign, e.g., $999.99
.	Decimal point
,	Thousand separator, e.g., 999,999.99

EXAMPLE 12.49: ILLUSTRATING USE OF FORMATTING COMMANDS

```
TTITLE 'Student Listing' ON;
COLUMN SName HEADING 'Last_Name' FORMAT A15 JUSTIFY LEFT;
BREAK ON StudPgm#;
...
/* The following statement produces a list of students, sorted on surname within major, with
   a control break on major. The list will also have a heading as specified above. */
SELECT StudPgm#, SName, FName, Stud# FROM Student ORDER BY SPgm, SName;
```

To clear the format command settings, you use the format clearing statements. The syntax of two such commands are as follows:

```
CLEAR BREAKS | COLUMNS | BUFFER | SCREEN;
TTITLE | BTITLE | REHEADER | REFOOTER OFF;
```

12.14 Embedded SQL

When developing applications for industrial or commercial purposes, SQL is typically featured as a database language that is used by the software engineer when it is convenient to do so. What this often means is that an application program may be written in a high-level language (such as C++, Java), but containing *embedded SQL statements* to take care of database access issues. The general convention for embedded SQL is as follows:

- Wherever a column, literal, or scalar expression is applicable in an SQL statement, a program variable, or scalar expression involving programming variables, may be applied.
- In many languages, the convention for specifying a program variable is to precede the variable with a colon (:). One exception to this rule is the Oracle SQL*Plus environment where the ampersand is used or the Oracle PL/SQL scenario where nothing precedes the program variable.
- In the case where data is to be retrieved into program variables, the **Select** statement is specified with an **Into-Clause**. The required syntax is shown below.
- In the case where a query will produce multiple rows, a *cursor* is set up. Although cursors are implemented differently from one host language to the other, the general principle is somewhat similar: A full discussion of cursors is beyond the scope of this course; suffice it to say that a cursor is a holding area for rows retrieved by a query. Cursors are normally used in stored procedures (in a host language), where multiple rows are retrieved for subsequent usage.

It is not practical (or even possible) to have a discussion on how each HLL handles embedded SQL statements here, so be sure to check the specific language you are using to ascertain how this is done in that particular language. However, you are likely to find that your language is using some variation of what has been described above.

Embedded SQL was mentioned earlier in this chapter (see Sections 12.1 and 12.2); in fact, Examples 12.3 and 12.12 illustrate embedded SQL relating to data insertion and transaction control, respectively. The host language used in those examples is Oracle's PL/SQL. The general syntax for such queries is shown in Figure 12.9, followed by illustrations in Examples 12.50–12.52.

```
SELECT <Item-List> INTO [:]<ProgramVariable> {, [:]<ProgramVariable>} FROM <Relation-List>
[WHERE <Condition>]
[GROUP BY <Group-List>]
[HAVING <Condition>]
[ORDER BY <Order-List>];
```

Figure 12.9 Basic syntax for embedded query.

Embedded SQL is widely used in industrial software development. As you view the syntax and examples, please note the following:

1. As mentioned, some languages require a colon to precede each program variable of the **Into-Clause** while others do not.
2. The syntax shown is directly applicable for queries that return a single row. For queries returning multiple rows, the syntax is tweaked to implement the cursor concept described earlier.

EXAMPLE 12.50: RETRIEVING SINGLE ROW INTO PROGRAM VARIABLES

Objective: Referring to the Oracle data set of earlier discussion (Section 12.8.5), Figure 12.10 illustrates two PL/SQL program blocks that each retrieves a single row into program variables. The first block uses column-type variables corresponding to columns of the table being accessed; the second block more elegantly uses a row-type program variable that corresponds to an entire row from each table being accessed.

```
// Block A: Using Column-type Program Variables
Declare
 ThisDept Emp.DeptNo%Type;
 ThisEmp  Emp.EmpNo%Type;
 ThisEmpName Emp.EName%Type;
BEGIN
-- Accept ThisEmp prompt 'Enter Employee Number:';
 SELECT EmpNo, EName, DeptNo INTO ThisEmp, ThisEmpName, ThisDept FROM Emp
 WHERE EmpNo = &ThisEmp;
 DBMS_output.put_line (ThisEmp);
 DBMS_output.put_line (ThisEmpName);
 DBMS_output.put_line (ThisDept);
END;
```

```
// Block B: Using Row-type Program Variables
Declare
 ThisDeptRec Dept%Rowtype;
 ThisEmpRec  Emp%Rowtype;
 ThisEmp Emp.EmpNo%Type;
 ThisDept Dept.DeptNo%Type;

BEGIN
-- Accept ThisEmp prompt 'Enter Employee Number:';
 SELECT * INTO ThisEmpRec FROM Emp WHERE EmpNo = &ThisEmp;
 SELECT * INTO ThisDeptRec FROM Dept WHERE DeptNo = &ThisDept;
 DBMS_output.put_line (ThisDeptRec.Deptno || ' - ' || ThisDeptRec.DName);
 DBMS_output.put_line (ThisEmptRec.EmpNo || ' - ' || ThisEmpRec.EName);
END;
```

Figure 12.10 PL/SQL blocks to retrieve a single row into program variables.

EXAMPLE 12.51: RETRIEVING MULTIPLE ROWS

Objective: Figure 12.11 illustrates a PL/SQL program block that retrieves multiple rows via a cursor. The block throws an exception whenever there is a failure in accessing the data.

```
Declare
 ThisEmp Emp.EmpNo%Type;
 ThisEmpRec Emp%RowType;
 RecordFound boolean;
 SalaryNull Exception;

Cursor EmpCursor is
 SELECT * FROM Emp WHERE EmpNo >= &ThisEmp;

BEGIN
 RecordFound := True;
 Open EmpCursor;
 While RecordFound Loop
  Fetch EmpCursor Into ThisEmpRec;
  If EmpCursor%NOTFOUND Then
    RecordFound := False;
  End If;
  If ThisEmpRec.Sal IS NULL Then
    Raise SalaryNull;
  End If;
  DBMS_Output.Put_Line(ThisEmpRec.Empno || ' ' || ThisEmpRec.Ename);
 End Loop;
 Close EmpCursor;

 Exception
 When No_Data_Found then  /* Pre-defined (implicit) exception */
  DBMS_Output.Put_Line ('No data found');
 When SalaryNull then    /* User-defined (explicit) exception */
  DBMS_Output.Put_Line ('Salary is null');
 When Others then /* catch all errors */
  DBMS_Output.Put_Line ('There is an execution error');
END;
```

Figure 12.11 PL/SQL block to retrieve multiple rows into a cursor.

EXAMPLE 12.52: ANOTHER ILLUSTRATION ON RETRIEVING MULTIPLE ROWS

Objective: Figure 12.12 performs the same activity as the previous, but here a PL/SQL procedure is used, and the looping strategy is different. As in the previous figure, the block throws an exception whenever there is a failure in accessing the data.

```
CREATE OR REPLACE PROCEDURE QueryEmp (ThisEmp In Emp.EmpNo%Type)
AS
 ThisEmpRec Emp%RowType;
 SalaryNull Exception;

 Cursor EmpCursor is
 SELECT * FROM Emp WHERE EmpNo >= &ThisEmp;

BEGIN
 Open EmpCursor;
 <<Hunt>>
 Loop
  Fetch EmpCursor Into ThisEmpRec;
  Exit Hunt When EmpCursor%NOTFOUND;
  If ThisEmpRec.Sal IS NULL Then
     Raise SalaryNull;
  End If;
  DBMS_Output.Put_Line (ThisEmpRec.Empno || ' ' || ThisEmpRec.Ename);
 End Loop;
 Close EmpCursor;

 Exception
  When No_Data_Found then  /* Implicit exception */
   DBMS_Output.Put_Line ('No data found');
  When SalaryNull then    /* Explicit exception */
   DBMS_Output.Put_Line ('Salary is null');
  When Others then /* catch all errors */
   DBMS_Output.Put_Line ('There is an execution error');
END;
```

Figure 12.12 PL/SQL procedure to retrieve multiple rows into a cursor.

EXAMPLE 12.53: EMBEDDED SQL IN C++

Objective: Now consider how SQL may be embedded in a C++ program. The syntax in Figure 12.9 applies; moreover, C++ syntax requires that each program variable in the **Into-Clause** is preceded by a colon.

Action: Figure 12.13 provides an illustration. The code shows how you may control insertion of data or search for specific data in a table. For this illustration, please assume the table **Resource_User_Access** with the following properties:

- Resource Code (RU_ResCode which references another table not shown)
- User Code (RU_UsrCode which references another table not shown)
- Access Sequence Number (RU_SeqN)
- Rights List (RU_RightsList)
- Candidate keys: [RU_ResCode, RU_SeqN]; [RU_ResCode, RU_UsrCode]

```
// This section of code checks for a user-resource entry in the Resource_User_Access table
string queryR = NULL; // the query result
string queryS = NULL; // the query string that will be constructed
string sRes, sUser // the resource code and user code that will be known to the program.
string statusMessage

// Construct the SQL query statement and then run the query
queryS = "SELECT RU_ResCode, RU_UsrCode INTO :queryR FROM Resource_User_Access " +
"WHERE RU_ResCode = " + "\'" + sRes + "\'" + " AND RU_UsrCode =" + "\'" + sUser + "\'" + ";";
EXEC SQL queryS;

// Now check the success. If queryR is still null, this means that the is no match; otherwise, a match was found.
If (queryR == NULL) statusMessage = "Sorry, you are not authorized to access this resource.";
else statusMessage = "Yes, you may proceed with using the resource.";

//*******************************************************************************************************************
// This section of code writes data to the Resource_User Access table
string insertS; // the query string that will be constructed
string resCode; // the Resource Code known to the program
string usrCode; // the user code known to the program
int seqN; // the sequebnce number known to the program

// Construct the SQL insertion statement
insertS = "INSERT INTO Resource_User_Access (RU_ResCode, RU_UsrCode, RU_SeqN, ...) " +
"VALUES (:resCode, :usrCode, :seqN, ...);
EXEC SQL SET TRANSACTION READ WRITE;
EXEC SQL insertS;
EXEC SQL COMMIT;
//*******************************************************************************************************************
```

Figure 12.13 Illustrating embedded SQL in C++.

Note:

Obviously, there is more to imbedded SQL in C++, but this example should give you confidence that the required rigor is by no means beyond your ability to master.

12.15 Dynamic Queries

Dynamic queries are used a lot in complex programming environments, where it is either difficult or cumbersome to write a single SQL statement to correspond to each request that the end user is likely to make. Instead of attempting this feat, the application programmer or software engineer writes a sophisticated program (typically in a high-level language), which constructs or generates SQL statements that are pertinent to the user request, and passes them on to an SQL parser. Three scenarios for this kind of programming come readily to mind:

- A sophisticated user needs to conduct ad hoc queries on an underlying database. The user is shielded from implementation details of the database, but is aware of underlying base tables or logical views, and the corresponding data fields (physical and/or logical) that they contain. The user is allowed to determine what details are to be included in his/her query.

- A front-end system (such as Delphi, C++ Builder, Visual Basic) is being used to construct a user interface for a software system that accesses one or more underlying databases. The software developer incorporates embedded SQL statements in various application queries based on user inputs. This request is then sent to the relevant database server to be serviced. The database server processes the request and sends the response back to the client.
- A database administrator needs to periodically back up several components of an Oracle database. Rather than repeatedly issuing backup statements for the different items (for instance, tablespaces) he/she may write an SQL script to dynamically generate the required SQL statements needed to perform backup of each component.

Figure 12.14 illustrates an inquiry screen for an application developed in Delphi. The end user is accessing a database for a list of publishers that the institution conducts business with. On the screen, you will notice a push button labeled **Search**, a radio group (with entries **ByCode** and **ByName**), and an input field. This application program works as follows:

- The end user can specify any search argument in the input field to indicate the starting publisher code or the name of interest (blank means start at the beginning of the list of publishers).
- The user will use the radio group to select whether information will be displayed sorted by publisher code or publisher name.
- When the user clicks the **Search** push button, the program will examine the entries made, build an appropriate SQL statement to fetch the information requested, invoke the SQL parser to request this information from the underlying database (which could be in a different back-end system such as Oracle, DB2). The code for building the SQL statement is shown in the figure.
- The database server that receives this request will process it and return the information to the program. The program will then load the grid shown with the information, display it on screen, and then await the next request from the end user.

Caution:

In recent years, Web-based hackers have mastered a technique called *SQL injection*, whereby embedded SQL statements of the form illustrated in Figure 12.14 can be intercepted and distorted or used for malicious purposes. Time does not allow for a full discussion of this topic here. However, four recommended strategies for safeguarding against SQL injection are summarized as follows:

- **Data Sanitization:** Filtering irrelevant data from the input
- Use of adequately fortified **application firewalls**
- Limiting **database privileges** on end-user accounts to the minimum requirements (this matter will be elaborated in Chapter 13)
- **Replacing queries with direct user inputs** with predefined parameters

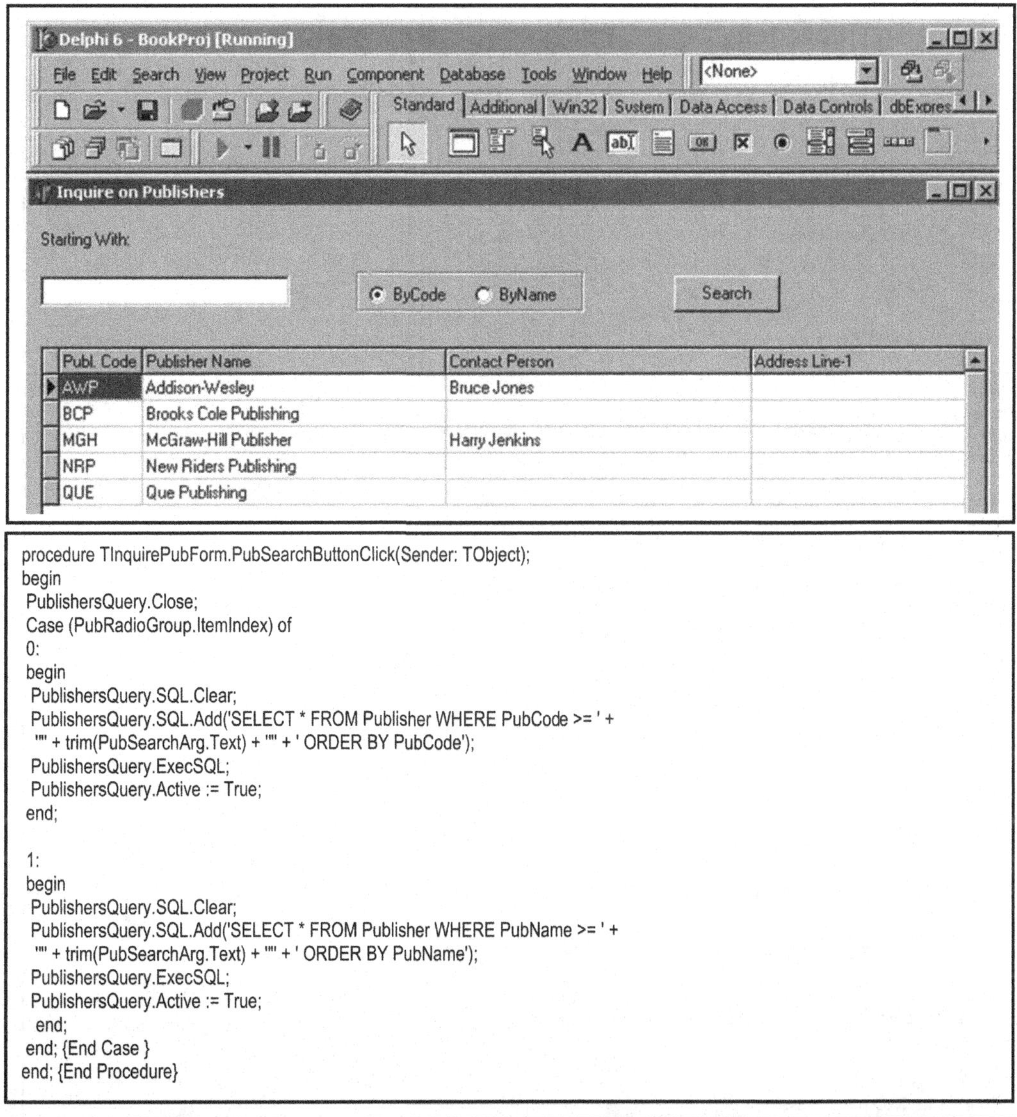


```
procedure TInquirePubForm.PubSearchButtonClick(Sender: TObject);
begin
 PublishersQuery.Close;
 Case (PubRadioGroup.ItemIndex) of
 0:
 begin
  PublishersQuery.SQL.Clear;
  PublishersQuery.SQL.Add('SELECT * FROM Publisher WHERE PubCode >= ' +
   '''' + trim(PubSearchArg.Text) + '''' + ' ORDER BY PubCode');
  PublishersQuery.ExecSQL;
  PublishersQuery.Active := True;
 end;

 1:
 begin
  PublishersQuery.SQL.Clear;
  PublishersQuery.SQL.Add('SELECT * FROM Publisher WHERE PubName >= ' +
   '''' + trim(PubSearchArg.Text) + '''' + ' ORDER BY PubName');
  PublishersQuery.ExecSQL;
  PublishersQuery.Active := True;
  end;
 end; {End Case }
end; {End Procedure}
```

Figure 12.14 Dynamic SQL for end-user database access from a Delphi front-end system.

Figure 12.15 shows how a similar end-user facility could be developed using PHP. You will notice that the PHP code is considerably longer than the Delphi code. This is due in part to the fact that the Delphi RAD tool takes responsibility for generating a high-level hidden code. The important point to note from both illustrations is methodology involved in constructing dynamic SQL: You have to use program variables of the host language to construct the SQL statement required for the scenario of relevance.

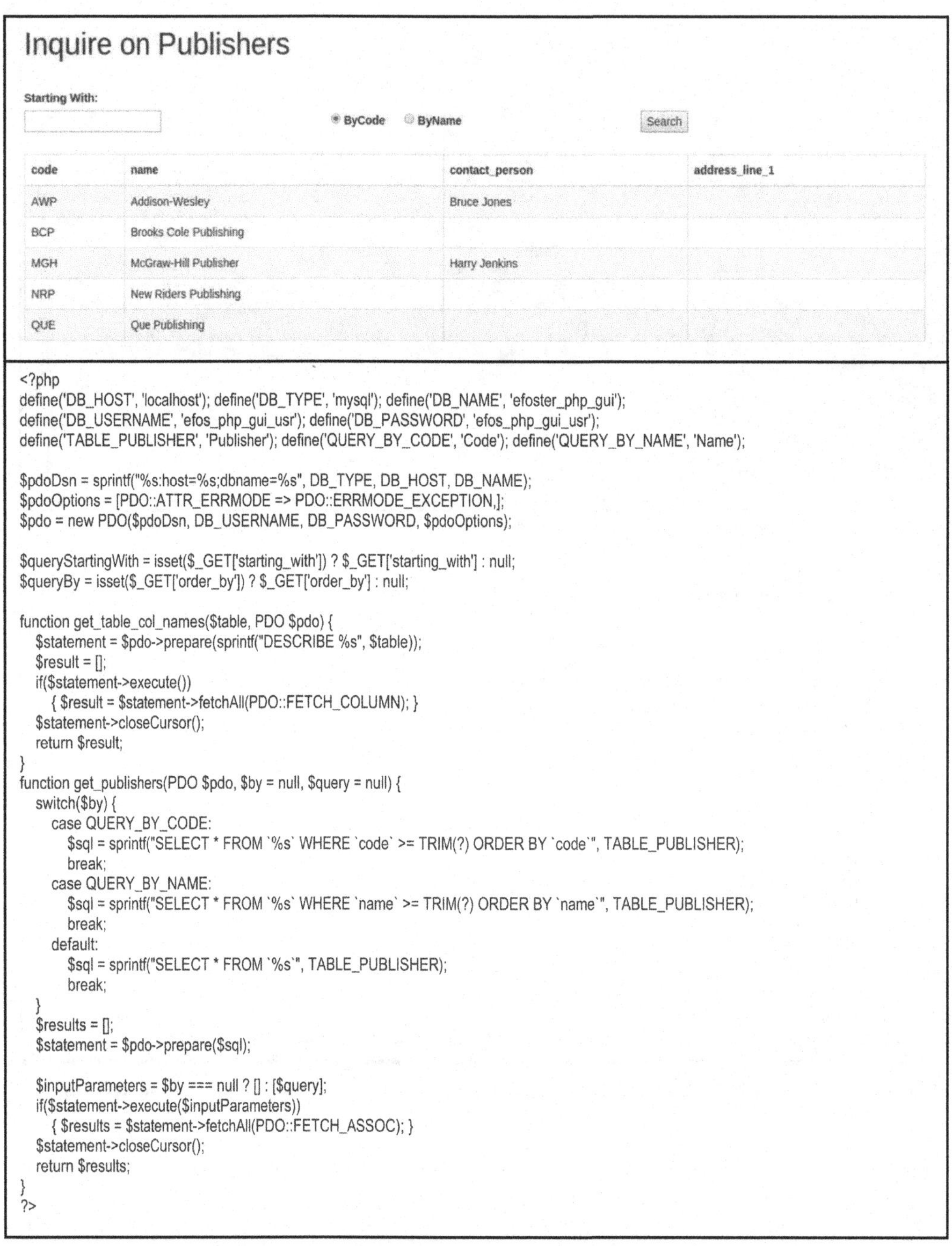


```
<?php
define('DB_HOST', 'localhost'); define('DB_TYPE', 'mysql'); define('DB_NAME', 'efoster_php_gui');
define('DB_USERNAME', 'efos_php_gui_usr'); define('DB_PASSWORD', 'efos_php_gui_usr');
define('TABLE_PUBLISHER', 'Publisher'); define('QUERY_BY_CODE', 'Code'); define('QUERY_BY_NAME', 'Name');

$pdoDsn = sprintf("%s:host=%s;dbname=%s", DB_TYPE, DB_HOST, DB_NAME);
$pdoOptions = [PDO::ATTR_ERRMODE => PDO::ERRMODE_EXCEPTION,];
$pdo = new PDO($pdoDsn, DB_USERNAME, DB_PASSWORD, $pdoOptions);

$queryStartingWith = isset($_GET['starting_with']) ? $_GET['starting_with'] : null;
$queryBy = isset($_GET['order_by']) ? $_GET['order_by'] : null;

function get_table_col_names($table, PDO $pdo) {
   $statement = $pdo->prepare(sprintf("DESCRIBE %s", $table));
   $result = [];
   if($statement->execute())
      { $result = $statement->fetchAll(PDO::FETCH_COLUMN); }
   $statement->closeCursor();
   return $result;
}
function get_publishers(PDO $pdo, $by = null, $query = null) {
   switch($by) {
      case QUERY_BY_CODE:
         $sql = sprintf("SELECT * FROM `%s` WHERE `code` >= TRIM(?) ORDER BY `code`", TABLE_PUBLISHER);
         break;
      case QUERY_BY_NAME:
         $sql = sprintf("SELECT * FROM `%s` WHERE `name` >= TRIM(?) ORDER BY `name`", TABLE_PUBLISHER);
         break;
      default:
         $sql = sprintf("SELECT * FROM `%s`", TABLE_PUBLISHER);
         break;
   }
   $results = [];
   $statement = $pdo->prepare($sql);

   $inputParameters = $by === null ? [] : [$query];
   if($statement->execute($inputParameters))
      { $results = $statement->fetchAll(PDO::FETCH_ASSOC); }
   $statement->closeCursor();
   return $results;
}
?>
```

Figure 12.15 Dynamic SQL for end-user database access from a PHP front-end system. (Continued)

```
<!DOCTYPE html> <html> <head>
    <title>Inquire on Publishers</title>
    <link rel="stylesheet" href="https://maxcdn.bootstrapcdn.com/bootstrap/3.3.5/css/bootstrap.min.css" />
    <style type="text/css"> row.top-buffer { margin-top:25px; }
      label[for=order_by-Code] { margin-right:12px;}
      label[for=order_by-Name], label[for=order_by-Code] { margin-left:4px;}
    </style>
  </head>
  <body>
    <div class="container">
      <div class="row top-buffer">
        <div class="col-md-12">
          <h1>Inquire on Publishers</h1>
        </div>
      </div>
      <div class="row top-buffer">
        <div class="col-md-12">
          <label for="starting_with">Starting With:</label>
        </div>
      </div>
      <div class="row">
        <form action="" enctype="application/x-www-form-urlencoded" method="get">
          <div class="col-md-4">
            <input type="text" name="starting_with" id="starting_with" value="<?= htmlentities($queryStartingWith); ?>" />
          </div>
          <div class="col-md-4">
            <input type="radio" name="order_by" id="order_by-Code" value="<?= QUERY_BY_CODE; ?>" <?= $queryBy ==
                        QUERY_BY_CODE ?
                        'checked=checked' : ''; ?> /><label for="order_by-Code">ByCode</label>
            <input type="radio" name="order_by" id="order_by-Name" value="<?= QUERY_BY_NAME; ?>" <?= $queryBy ==
                        QUERY_BY_NAME ? 'checked=checked' : ''; ?> /><label for="order_by-Name">ByName</label>
          </div>
          <div class="col-md-4">
            <input type="submit" name="submit" id="submit" value="Search" />
          </div>
        </form>
      </div>
      <div class="row top-buffer">
       <div class="col-md-12">
          <table class="table table-bordered table-hover">
            <thead>
              <tr>
                <?php foreach(get_table_col_names(TABLE_PUBLISHER, $pdo) as $tableColumnName): ?>
                  <th><?= $tableColumnName; ?></th>
                <?php endforeach; ?>
              </tr>
            </thead>
            <tbody>
              <?php foreach(get_publishers($pdo, $queryBy, $queryStartingWith) as $publisherRow): ?>
                <tr>
                  <td><?= $publisherRow['code']; ?></td>
                  <td><?= $publisherRow['name']; ?></td>
                  <td><?= $publisherRow['contact_person']; ?></td>
                  <td><?= $publisherRow['address_line_1']; ?></td>
                </tr>
              <?php endforeach; ?>
            </tbody>
          </table>
        </div>
      </div>
    </div>
  </body>
</html>
```

Figure 12.15 (Continued) Dynamic SQL for end-user database access from a PHP front-end system.

Figure 12.16 provides another illustration of dynamic SQL. A detailed explanation is beyond the scope of this course; however, a cursory clarification is in order, and will suffice:

- The SPOOL statement (sixth line down) causes the output of the subsequent query to be redirected to the file specified.
- The SELECT statement causes a series of rows to be written to the spool file. These rows actually contain SQL statements to backup tablespaces found in the current database.
- The spool file can therefore be subsequently run as an SQL script to backup these tablespaces.

```
SET VERIFY OFF;
SET FEEDBACK OFF;
SET TERMOUT OFF;
SET ECHO OFF;
SET PAGESIZE 0;
SPOOL C:\Oracle\Admin\Backup\BackupTS.sql;
SELECT 'ALTER TABLESPACE ' || Tablespace_Name || ' BEGIN BACKUP; ',
 HOST copy ' || File_Name || ' D:\OracleBackup\EFDB ', ' ALTER TABLESPACE ' || Tablespace_Name ||
 ' END BACKUP' FROM DBA_DATA_FILES;

SPOOL OFF; SET VERIFY ON; SET FEEDBACK ON;
SET TERMOUT ON; SET ECHO ON;
SET PAGESIZE 20;

/* The instructions will be stored in C:\Oracle\Admin\Backup\BackupTS.sql. We can now run this script */
@ C:\Oracle\Admin\Backup\BackupTS.sql;
```

Figure 12.16 Automatic backup of Tablespaces in a database via dynamic SQL statements.

12.16 Summary and Concluding Remarks

Relatively speaking, this has been a rather lengthy chapter. In recognition, the following paragraphs provide a summary of the various related topics covered, followed by some concluding remarks.

The **Insert** statement facilitates insertion of data into a specific table. The statement allows data insertion in one of three ways:

- Insertion by specifying literal column values for a row of a table
- Insertion via execution-time variables for a row of a table
- Insertion by redirecting the result of a query into a table (multiple record insertion is supported by this strategy)

The **Update** statement facilitates update of rows of a table. Depending on the condition specified (in the **Where-Clause**), the update might affect a single row, multiple rows, or all the rows (if no condition is specified, or all the rows meet the condition specified). The data to be used for the update can be specified in one of the three ways:

- By specifying literal column values for the row(s) of the table that meet(s) the condition specified
- By specifying execution-time variables corresponding to stated columns, for the row(s) of the specified table that meet(s) the condition specified

- By specifying a subquery that includes columns corresponding to the stated columns, for the row(s) of the specified table that meet(s) the condition specified

The **Delete** statement facilitates deletion of rows from a specified table. Depending on the condition specified (in the **Where-Clause**), the deletion might be for a single row, multiple rows, or all the rows (if no condition is specified, or all the rows meet the condition specified).

The **Commit** and **Rollback** statements facilitate transaction management, and are useful in situations where it is desirable to have a transaction wholly committed, or not committed, in order to preserve the integrity of the database.

The **Select** statement is one of the most powerful and widely used statements in SQL. It facilitates retrieval of data from one or more tables and presenting the information to the end user. This process is called querying. It is very flexible and can therefore be used in several different ways:

- Simple queries involve data retrieval (via the **Select** statement) from one table only.
- Queries involving multiple tables can be constructed in one of the two ways:
  - The traditional method requires you to specify the join condition in the **Where-Clause** of the **Select** statement.
  - The American National Standard Institute (ANSI) method requires you to specify the join using join-related keywords. The ANSI method is more verbose but is also more flexible than the traditional method.
- Queries involving the use of functions provide additional functionality and flexibility in data retrieval. The functions can be classified into five categories: row functions, date functions, data conversion functions, aggregation functions, analytic functions, and programmer-defined functions.
- Queries can be constructed using the special operators LIKE, BETWEEN, and IN.
- You can have queries containing other queries — nested queries. These queries may involve the use of keywords such as ANY, ALL, EXISTS, and NOT EXISTS; some of them may also make use of the IN operator.
- You can define queries involving the use of set operations — UNION, UNION ALL, INTERSECT, and MINUS.
- You can define queries that make use of execution-time variables.
- You can define queries that make use of various SQL*Plus format commands.
- Typically, software applications do not use SQL only, but embed SQL statements in programs written in another HLL.
- A dynamic query is a query that uses data fed to it to construct a **Select** statement that is then executed. It avoids hard-coding values, thereby providing more flexibility to the end users who use it.

Complete discussion of all aspects of the **Select** statement could easily take up a considerable portion of a book. This chapter captures and discusses the salient features, which if mastered, will place you on a solid footing in any environment that requires expertise in SQL.

One thing that should be clear is the remarkable power of the language. This can be appreciated if you attempt to write a Java or C++ program to replace any of the examples provided in the chapter. No wonder it is the universal database language. But go on to the next chapter, as there is much more to learn about the language.

12.17 Review Questions

1. Describe three ways in which data may be inserted into base table.
2. Write PL/SQL blocks to allow you to quickly populate the sample college database (created in Chapter 11) with data. Test these out in Oracle.
3. Write PL/SQL blocks to allow you to quickly modify data in your database. Test these out in Oracle.
4. Write PL/SQL blocks to allow you to easily delete data from your database. Test these out in Oracle.
5. Explain the purpose of the **Commit** and **Rollback** statements. Demonstrate how they may be used.
6. Practice writing the following:
   - Simple queries
   - Queries Involving Multiple Tables
   - Queries Involving the Use of Functions
   - Queries Using LIKE, BETWEEN, and IN Operators
   - Nested Queries
   - Queries Involving Set Operators
   - Queries with Runtime Variables
7. Differentiate between embedded SQL code and dynamic SQL code. Describe a scenario that would warrant the use of each.

References and/or Recommended Readings

Connolly, Thomas, & Carolyn Begg. 2014. *Database Systems: A Practical Approach to Design, Implementation and Management* 6th ed. Boston, MA: Pearson. See chapter 8.

Coronel, Carlos, & Steven Morris. 2015. *Database Systems: Design, Implementation & Management* 11th ed. Boston, MA: Cengage Learning. See chapter 8.

Elmasri, Ramez, & Shamkant B. Navathe. 2016. *Fundamentals of Database Systems* 7th ed. Boston, MA: Pearson. See chapters 6–9.

Oracle Corporation. 2019. *Database SQL Language Reference.* Accessed June 2019. https://docs.oracle.com/en/database/oracle/oracle-database/19/sqlrf/index.html. See chapter 9.

Thomas, Biju, Gavin Powell, Robert Freeman, & Charles Pack. 2014. *Oracle Certified Professional on 12C Certification Kit.* Indianapolis, IN: Wiley.

Chapter 13

Logical Views and System Security

Two very powerful and important features of Structured Query Language (SQL) are the facility to create and manage logical views, and the capability to manage security issues of a database. This chapter discusses these two related issues. The chapter proceeds under the following subtopics:

- Traditional Logical Views
- System Security
- Materialized Views
- Summary and Concluding Remarks

13.1 Traditional Logical Views

As pointed out in Chapter 3, a logical view is a virtual relation that allows end-users to access information in a manner that is consistent with their requirements. Unlike base relations, a logical view stores no data, only the access path to data. Any creditable database management system (DBMS) will allow the creation and manipulation of logical views. Following are a few important points about views:

- SQL views allow for logical interpretation of information in the database.
- Views include virtual, named (but not base), and derived relations; these significantly help to comprise the external schema of the database (review Chapter 2).
- Like named relations, views are created and dropped using the **Create-View** and **Drop-View** statements. There is an **Alter-View** statement, but its purpose is primarily for addressing view constraints, not for modifying the structure of a view. The **Create-View** statement has an OR REPLACE option that effectively serves to structurally modify a preexistent logical view. Logical views are virtual relations that are created and dropped as the situation dictates.
- Views are stored in the system catalog.

DOI: 10.1201/9781003275725-16

- Data modification (insert, update, and delete operations) can be applied through views in the normal (SQL) manner, providing that the view is updateable (more on this later). The view statements are compiled or translated (depending on the DBMS), and the implied base relations are updated at execution time.
- Views can also be queried in a manner that is identical to how base relations are queried.

13.1.1 View Creation

An abridged version of the **Create-View** statement appears in Figure 13.1. For most practical circumstances, this will suffice. However, the full syntax is shown in Figure 13.2.

```
View-Definition::=
CREATE [OR REPLACE] [FORCE/NOFORCE] VIEW [<Schema>.] <ViewName> [(Column {,Column})] AS <Sub-query>
[WITH CHECK OPTION [CONSTRAINT <Constraint-name>]] | [WITH READ ONLY];
```

Figure 13.1 Abridged form of Create-View statement.

As one would expect, the **Create-View** statement is quite flexible in its provision of various options to meet different needs. The following are some basic points of clarification:

1. Any derivable table (which can be obtained via **Select** statement) can be defined as a view. In earlier versions of SQL the subquery of a logical view was not allowed to include an **Order-By-Clause**, aggregation function, or analytic function. This stipulation was not a serious omission, as there were alternatives around these constraints. Nonetheless, you will be pleased to know that these restrictions have been removed.
2. The view is in a way similar to an insertion with a **Select-Clause**. The difference is that the insertion inserts actual data into a base relation; the view on the other hand stores virtual data, i.e., it stores the access path to actual data that resides in the underlying base relation(s).
3. The OR REPLACE option replaces an existing view with the same name, thereby indirectly fulfilling the role of view alteration.
4. The FORCE option creates the view even if the underlying relation(s) does/do not exist, assuming that the schema owner has appropriate privileges; the default is NOFORCE.
5. The WITH CHECK OPTION applies to the condition specified in the subquery (in the **Where-Clause**): It allows insertion and update of rows based on the related condition. This CHECK OPTION may be given an optional constraint name.
6. The READ ONLY option ensures that the view cannot be used for update of the underlying base relation.

A logical view is a virtual relation that stores an access path to data. As such, the view does not store actual data, though the illusion is given that it does; the data actually resides in the underlying base relations(s) that the view accesses.

```
Create-View ::=
CREATE [OR REPLACE] [FORCE/NOFORCE] VIEW [<Schema>.] <ViewName>
[Column-Spec | Object-View-Clause | XML-Type-View-Clause | View-Constraint-Spec]
AS <Sub-query>
[WITH CHECK OPTION [CONSTRAINT <Constraint-name>]] | [WITH READ ONLY];

View-Constraint-Spec ::= VCOption1 | VCOption2
VCOption1 ::= (Out-of-line-Constraint)
VCOption2 ::= (<Alias> Inline-Constraint {, Inline-Constraint} {, <Alias> Inline-Constraint {, Inline-Constraint})

/* Inline-Constraint and Out-of-line-Constraint are defined in Create-Table (review chapter 11) */

Object-View-Clause ::= OVOption1 | OVOption2
OVOption1 ::= OF [<Schema>.] <TypeName> UNDER [<Schema>.] <Super-View>
              (Out-of-line-Constraint | [<Attribute> Inline-Constraint]
              {, Out-of-line-Constraint | [<Attribute> Inline-Constraint]})
OVOption2 ::= OF [<Schema>.] <TypeName> WITH OBJECT IDENTIFIER DEFAULT |
              [<Attribute> {, <Attribute>}]
              (Out-of-line-Constraint | [<Attribute> Inline-Constraint]
              {, Out-of-line-Constraint | [<Attribute> Inline-Constraint]})

XML-Type-View-Clause ::=
OF XMLTYPE XML-Schema-Spec
WITH OBJECT IDENTIFIER DEFAULT |[(<Expression> {, <Expression>})]

XML-Schema-Spec ::=
[XMLSCHEMA <XML-Schema-URL>] ELEMENT [<XML-Schema-URL> #] <Element>

Column-Spec ::=
(<Column> {, <Column>}
```

Figure 13.2 The detailed Create-View statement.

EXAMPLE 13.1: ILLUSTRATING LOGICAL VIEW FOR COMPUTER SCIENCE COURSES ONLY

Objective: Create a logical view **CSCourses** that stores Computer Science Courses only (assume that the first two characters of the Course Code for Computer Science is "CS").

```
CREATE VIEW CSCourses (Crs#, CrsName)
AS SELECT Crs#, CrsName FROM Course WHERE Crs# LIKE 'CS%';
```

EXAMPLE 13.2: ILLUSTRATING LOGICAL VIEW FOR PROGRAM COMPOSITIONS

Objective: Create a logical view that stores for each academic program, a full breakout of all courses included in that program, as they would appear in a college bulletin.

```
CREATE OR REPLACE VIEW CollegeBulletin (PsPgm#,
PgmName, PsCrsSeqn, PsCrs#, CrsName)
AS SELECT PsPgm#, PgmName, PsCrsSeqn, PsCrs#, CrsName
FROM Pgm_Struct PS, AcademicProgram P, Course C
WHERE PS.PsPgm# = P.Pgm# AND PS.PsCrs# = C.Crs# ORDER BY PsPgm#, PsCrsSeqn;
```

13.1.2 View Modification and Removal

To modify a logical view, use the **Alter-View** statement. Notice that this statement is very limited; it is confined to constraints on the view. To change the actual structure of the view, you must use the OR REPLACE option of the Create-View statement. To remove the view from the system catalog, use the **Drop-View** statement. The syntax for each statement is shown in Figure 13.3.

```
Alter-View ::=
ALTER VIEW [<Schema>.] <ViewName>
[ADD Constraint-Clause] |
[MODIFY CONSTRAINT <Constraint> RELY | NORELY] |
[DROP CONSTRAINT <Constraint>] |
[DROP PRIMARY KEY] |
[DROP UNIQUE (<Column> {, <Column}] | COMPILE;

// Constraints-Clause is as defined in Create-Table (review chapter 11)

Drop-View ::=
DROP VIEW [<Schema>.] <ViewName> [CASCADE CONSTRAINTS];
```

Figure 13.3 The Alter-View and Drop-View statements.

As you view the syntax, here are a few points to note:

1. Due to the aforementioned limitations of the **Alter-View** statement, in practice, it is seldom used.
2. When a view is dropped, it is removed from the system catalog.
3. The **Cascade-Constraints-Clause** is used to drop all referential integrity constraints that refer to primary and/or unique keys in the view to be dropped. If omitted and such constraints exist, the **Drop-View** statement will fail.
4. Dropping a base relation automatically drops all associated views on that base relation.

EXAMPLE 13.3: ILLUSTRATING VIEW REMOVAL

```
// Remove the view CSCourses from the system
DROP VIEW CSCourses;
```

13.1.3 Usefulness and Manipulation of Logical Views

Logical views are very useful, particularly during development of software system(s) that access related database(s). The following are some advantages of views:

- Simplification of the perception of end users — the users concentrate only on data that is of concern to them.
- Aiding system security — users have access only to data that concerns them and cannot access or manipulate in any way, data to which they are not authorized.

- Provision of some amount of logical data independence in the face of restructuring of the database.
- Facilitation of assorted external views of the data stored.

Data manipulation operations on *updateable views* are converted to equivalent operations on the underlying base table(s). In the case of data retrieval (via the **Select** statement), the conversion is straightforward and traces directly to operations on the underlying base relation(s). In the case of data changes (insertion, update or deletion), more care is required, as explained below.

A logical view is said to be *updateable* if records of the underlying base relation can be updated through the view. An updateable logical view must meet all of the following criteria:

- The view must be defined to include a candidate key of the underlying base relation.
- The view should not involve a JOIN, UNION, or INTERSECT operation, but with one caveat: You can actually update an underlying base table through a join logical view, by specifying data (through the view) that will affect one (and only one) of its underlying base tables. However, this is not recommended.
- The **Select** statement should not contain the keyword DISTINCT.
- The **Select** statement should not include use of any aggregate function, analytic function, a **Group-By-Clause** or a **Having-Clause**.
- The view should not contain derived columns.
- The view should not include the **READ ONLY** option.

13.2 System Security

Recall from Section 1.2 that security is a primary objective of a database system. At a minimum, there are three levels of system security that you should be aware of:

- Access to the system
- Access to the system resources
- Access to system data

Let us briefly examine each level and see how they are facilitated in SQL. For the most part, this is examined in the context of an Oracle environment. However, the principles covered are also applicable in non-Oracle database environments as well.

13.2.1 Access to the System

Oracle allows access to the system through user profiles and user accounts. A profile is a working environment for a group of user accounts. When you create a database, Oracle creates a default profile called DEFAULT. When you create a user account, if you do not specify a profile, Oracle assigns the DEFAULT profile to the user account. A user account consists of a user name, a password, and other optional parameters. Each user account has an associated profile.

Figure 13.4 shows the syntax for the **Create-Profile** statement, and Example 13.4 illustrates how it is used. As can be seen from the syntax, the command is used to help create a working environment for the user.

```
Create-Profile ::=
CREATE PROFILE <ProfileName> LIMIT Resource-Parms | Password-Parms {Resource-Parms | Password-Parms};

Resource-Parms :=
[SESSIONS_PER_USER <n> | UNLIMITED | DEFAULT] /* No. of concurrent sessions per user */
[CPU_PER_SESSION <n> | UNLIMITED | DEFAULT] /* CPU time limit for a session, expressed in hundredth of seconds*/
[CPU_PER_CALL <n> | UNLIMITED | DEFAULT] /* CPU time limit for a call (parse, execute or fetch), expressed in
    hundredth of seconds*/
[CONNECT_TIME <n> | UNLIMITED | DEFAULT] /* The total elapsed time limit for a session, expressed in minutes */
[IDLE_TIME <n> | UNLIMITED | DEFAULT] /* Inactive time during a session, expressed in minutes */
[LOGICAL_READS_PER_SESSION <n> | UNLIMITED | DEFAULT]
[LOGICAL_READS_PER_CALL <n> | UNLIMITED | DEFAULT]
[COMPOSITE_LIMIT <n> | UNLIMITED | DEFAULT] /* The total resource cost for a session, expressed in service units.
    Oracle calculates the total service units as a weighted sum of CPU_PER_SESSION, CONNECT_TIME,
    LOGICAL_READS_PER_SESSION, and PRIVATE_SGA. */
[PRIVATE_SGA [<n> [K|M]] | UNLIMITED | DEFAULT]

Password-Parms :=
[FAILED_LOGIN_ATTEMPTS <Expression> | UNLIMITED | DEFAULT] /* the number of failed attempts to log in to the user
    account before the account is locked */
[PASSWORD_LIFE_TIME <Expression> | UNLIMITED | DEFAULT] /* The number of days the same password can be
    used for authentication. The password expires if it is not changed within this period */
[PASSWORD_REUSE_TIME <Expression> | UNLIMITED | DEFAULT] /* The number of days before which a password
    cannot be reused */
[PASSWORD_REUSE_MAX <Expression> | UNLIMITED | DEFAULT] /* The number of password changes required before
    the current password can be reused*/
[PASSWORD_LOCK_TIME <Expression> | UNLIMITED | DEFAULT] /* The number of days an account will be locked after
    the specified number of consecutive failed login at */
[PASSWORD_GRACE_TIME <Expression> | UNLIMITED | DEFAULT] /* The number of days after the grace period begins
      during which a warning is issued and login is allowed. The password expires after the grace period. */
[PASSWORD_VERIFY_FUNCTION <Function> | NULL | DEFAULT] /* Function to verify password */
```

Figure 13.4 The Create-Profile statement.

EXAMPLE 13.4: ILLUSTRATING USER PROFILE CREATION

Objective: The following statement creates a user profile called **InventoryProfile**. The parameters not specified will have default values.

```
CREATE PROFILE InvoryProfile LIMIT
SESSIONS_PER_USER 3
IDLE_TIME 15
FAILED_LOGIN_ATTEMPTS 3
PASSWORD_LIFE_TIME 1
PASSWORD_REUSE_TIME 0
PASSWORD_REUSE_MAX 4
PASSWORD_LOCK_TIME UNLIMITED;
```

As expected, Oracle allows you to modify or drop a profile via the **Alter-Profile** statement or the **Drop-Profile** statement, respectively. The syntax for each is shown below (Figure 13.5).

Note:

The CASCADE option on the **Drop-Profile** statement instructs Oracle to reassign all user accounts formerly assigned to the profile, to the DEFAULT profile.

```
Alter-Profile ::=
ALTER PROFILE <ProfileName> LIMIT Resource-Parms | Password-Parms {Resource-Parms | Password-Parms}

Resource-Parms := // As defined in Create-Profile

Password-Parms := // As defined in Create-Profile

Drop-Profile ::=
DROP PROFILE <ProfileName> [CASCADE];
```

Figure 13.5 The Alter-Profile and Drop-Profile statements.

In addition to profiles, Oracle allows the DBA (or some user with system admin privileges) to create and remove user accounts. Each user account is assigned to a profile. If the profile is not specified at account creation, Oracle assigns the account to the DEFAULT profile. The required syntax for the **Create-User** statement appears in Figure 13.6.

```
Create-User ::= CUOption1 | CUOption2 | CUOption3
CUOption1 ::= CREATE USER <UserName> IDENTIFIED BY <Password> User-Spec;
CUOption2 ::= CREATE USER <UserName> IDENTIFIED EXTERNALLY User-Spec;
CUOption3 ::= CREATE USER <UserName> IDENTIFIED GLOBALLY AS '<ExternalName>' User-Spec;

User-Spec ::=
[DEFAULT TABLESPACE <Tablespace>]
[TEMPORARY TABLESPACE <Tablespace>]
[QUOTA <n> [K|M] ON <Tablespace>]
[QUOTA UNLIMITED ON <Tablespace>]
[PROFILE <Profile>]
[PASSWORD EXPIRE]
[ACCOUNT LOCK | UNLOCK]
```

Figure 13.6 The Create-User statement.

In observing the syntax, bear in mind the following points of clarification:

1. Oracle supports a local database user, a global user (authenticated by Enterprise Directory Service), or an external operating system user.
2. To activate the password expiration (defined in the user's assigned profile), specify PASSWORD EXPIRE.
3. Use the ACCOUNT LOCK/UNLOCK option to lock the account (make it inaccessible by the user) or unlock the account.
4. When you create a user, Oracle creates a schema by the same name as the user. All objects created by that user will be owned by the user's schema.

EXAMPLE 13.5: ILLUSTRATING CREATION OF USER ACCOUNT

// The following statement creates a user called **Bremar**

```
CREATE USER Bremar IDENTIFIED BY Brem1199
DEFAULT TABLESPACE SampleTBS
QUOTA UNLIMITED ON SampleTBS
PROFILE InvoryProfile
PASSWORD EXPIRE
ACCOUNT UNLOCK;
```

Of course, user accounts may be modified or dropped from the system. The **Alter-User** statement allows you to change all the parameters on the **Create-User** statement, but gives the added flexibility of changing some additional settings for the account. A condensed version of the syntax is shown in Figure 13.7, which also includes the syntax for the **Drop-User** statement.

Note the almost perfect symmetry of the Alter-User statement with the Create-User statement. At this point, you should not be surprised at this. Here are a few additional points of clarification.

1. You can use this command to assign *roles* to the user account. We will discuss *roles* shortly.
2. You can also use the **Proxy-Clause** to grant other privileges to the user (for more on this, see the Oracle documentation for the **Alter-User** statement).
3. The CASCADE option on the **Drop-User** statement drops all objects owned by the user's schema before dropping the user.

```
Alter-User ::= AUOption1 | AUOption2 | AUOption3
AUOption1 ::=     ALTER USER <UserName> IDENTIFIED BY <Password>
              [REPLACE <OldPassword>] Alter-User-Spec;
AUOption2 ::=     ALTER USER <UserName> IDENTIFIED EXTERNALLY Alter-User-Spec;
AUOption3 ::=     ALTER USER <UserName>
              IDENTIFIED GLOBALLY AS '<ExternalName>' Alter-User-Spec;

Alter-User-Spec ::=
[DEFAULT TABLESPACE <Tablespace>]
[TEMPORARY TABLESPACE <Tablespace>]
[QUOTA <n> [K|M] ON <Tablespace>]
[QUOTA UNLIMITED ON <Tablespace>]
[DEFAULT ROLE [ALL EXCEPT] <Role> {,<Role>}]
[DEFAULT ROLE NONE]
[PROFILE <Profile>]
[PASSWORD EXPIRE]
[ACCOUNT LOCK | UNLOCK]
[Proxy-Clause] /* See Oracle Documentation */

Drop-User ::=
DROP USER <User> [CASCADE]
```

Figure 13.7 The Alter-User and Drop-User statements.

EXAMPLE 13.6: ILLUSTRATING ALTERING AND DROPPING OF USER ACCOUNT

```
// Change the password for Scott
ALTER USER Scott IDENTIFIED BY "Frosty_2015";

// Remove the user called Bremar

DROP USER Bremar CASCADE;
```

13.2.2 Access to System Resources

There are two kinds of privileges to be managed, namely system privileges and object privileges. System privileges include all valid SQL statements; additionally, there are other system-wide privileges and roles that will be revisited shortly. With respect to object privileges, we are concerned with access to system objects such as tables, views, indexes, sequences, synonyms, procedures, and other objects that the DBMS may support. Three possible actions (considered as privileges) may apply to each of these objects: creation, alteration, and dropping.

The **Grant** statement is used for granting access of resources to user(s) and/or role(s), while the **Revoke** statement is used for revoking access of resources from user(s) and/or role(s). The basic syntax of each command is shown in Figure 13.8.

```
Grant ::=
GRANT [ALL PRIVILEGES] | [<Privilege> | <Role> {,<Privilege> | <Role>} ]
[(<Column> {,<Column>}] [ON <Object Name>]
To <User> | <Role> | PUBLIC {,<User> | <Role>}
[WITH ADMIN OPTION] [WITH GRANT OPTION];

Revoke ::=
REVOKE ALL|<Privilege>|<Role> {,<Privilege>|<Role>} [On <Object Name>]
FROM <User>|<Role>|PUBLIC {, <User>|<Role>} [CASCADE CONSTRAINTS];
```

Figure 13.8 The Grant and Revoke statements.

Any privilege issued by the **Grant** statement can be rescinded by the **Revoke** statement. Here are some additional guidelines:

1. The ON Clause is used if the privilege relates to a database object.
2. The privilege is typically a command name or a role; the recipient of the privilege may be a user or a *role* (roles will be clarified shortly).
3. The WITH ADMIN OPTION enables the recipient (user) to be able to grant/revoke this system privilege to/from other users or roles.
4. The WITH GRANT OPTION enables the recipient (user) to be able to grant/revoke this object privilege to/from other users or roles.

EXAMPLE 13.7: ILLUSTRATING GRANTING OF OBJECT PRIVILEGES TO A USER

Objective: Create user account **BruceJones** and grant certain privileges to it; also allow user account **Scott** to be able to run queries on the **AcademicProgram** table.

```
CREATE USER BruceJones IDENTIFIED BY BJ999123;
GRANT CREATE TABLE, CREATE VIEW, CREATE PROCEDURE TO BruceJones;
GRANT SELECT ON AcademicProgram TO BruceJones, Scott WITH GRANT OPTION;
```

EXAMPLE 13.8: ILLUSTRATING REVOKING OF OBJECT PRIVILEGES FROM A USER

Objective: Remove query access of **AcademicProgram** from **Bruce Jones** and all privileges from **Stalker**.

```
REVOKE SELECT ON AcademicProgram FROM BruceJones;
REVOKE ALL FROM Stalker;
```

13.2.3 *Managing Access to System Resources via Development Privileges and Roles*

In a typical database environment (or software engineering environment where database access is paramount), there will be personnel fulfilling the role of application development. An application developer should have the following system privileges:

Create Session	Create Table	Create Sequence	Create View
Create Trigger	Create Synonym	Create Procedure	Create Tablespace
Analyze Any	Insert Any Table	Select Any Dictionary	

The **Create-Session** privilege is required for logging onto Oracle and starting a work session. The **Analyze-Any** privilege is useful for database performance analysis and tuning (discussed in Chapter 16). The **Select-Any-Dictionary** privilege is required in order to query the system catalog (to be discussed in the upcoming chapter). The other privileges are all self-explanatory, and need no further clarification at this point.

In managing multiple users in different user groups, *roles* are particularly useful. You can define a role to consist of several privileges, and then grant the role to users. Additionally, roles can be granted to other roles. A role is created via the **Create-Role** statement, modified by the **Alter-Role** statement, and removed via the **Drop-Role** statement. Figure 13.9 shows abridged formats of these statements.

```
Create-Role ::=
CREATE ROLE [<Schema>.] <RoleName>
[NOT IDENTIFIED] |
[IDENTIFIED BY <Password>] |
[IDENTIFIED USING [<Schema>.] <Package>] |
[IDENTIFIED EXTERNALLY | GLOBALLY];

Alter-Role ::=
ALTR ROLE [<Schema>.] <RoleName>
[NOT IDENTIFIED] |
[IDENTIFIED BY <Password>] |
[IDENTIFIED USING [<Schema>.] <Package>] |
[IDENTIFIED EXTERNALLY | GLOBALLY];

Drop-Role ::=
DROP ROLE [<Schema>.] <RoleName>;
```

Figure 13.9 Create-Role, Alter-Role, and Drop-Role statements.

Note:

To create a role, you need to have the **Create-Role** system privilege. Only role owners and users with **Alter-Any-Role** and **Drop-Any-Role** system privileges can modify or delete a role. If a user has the **Create-Role** system privilege but not the other two, he/she can alter/delete only the role(s) he/she has created.

EXAMPLE 13.9: ILLUSTRATING ROLE CREATION AND CONFIGURATION

Objective: Create a role called **Developer** with appropriate privileges and grant the role to user account BruceJones.

```
CREATE ROLE Developer;
GRANT CREATE SESSION, CREATE TABLE, CREATE SEQUENCE, CREATE
   VIEW, CREATE ROLE,
CREATE PROCEDURE, CREATE TRIGGER, CREATE TABLESPACE, CREATE
   SYNONYM,
INSERT ANY TABLE, ANALYZE ANY, SELECT ANY DICTIONARY TO Developer;
GRANT Developer TO BruceJones WITH ADMIN OPTION;
```

Take a closer look at the **Grant** statement again and notice the power and flexibility that it provides. In its broadest sense, you can grant a set of (system and/or object) privileges to a set of recipients (consisting of roles and/or users). Roles may be treated as the source or target of other privileges. You can therefore grant a list of system and/or object privileges to a user or role; you can also grant a role to a user or another role. Additionally, you can use terms such as ANY, ALL, NONE, and EXCEPT in assigning privileges. You should also note that the acronym SYSDBA represents the database administrator role in Oracle installations; this is a powerful role that is typically assigned only to senior member(s) of the database system team. Another commonly used role is PUBLIC, representing all users. There are several others; in the upcoming chapter, you will learn how to query the system catalog to find out information on roles and other related things.

EXAMPLE 13.10: ILLUSTRATING PRIVILEGE MANIPULATIONS

Objective: The following examples illustrate the use of use of ANY, ALL, NONE, and EXCEPT keywords.

```
// Give user account BruceJones default access to all system roles
ALTER USER BruceJones DEFAULT ROLE ALL;
// Give user Bremar all system privileges except SYSDBA
ALTER USER Bremar DEFAULT ROLE ALL EXCEPT SYSDBA;
//Grant roles Connect and Developer to user KK_Club
ALTER USER KK_Club DEFAULT ROLE Connect, Developer;
// Grant user BruceJones the privilege to create a role, or manipulate any role
GRANT CREATE ROLE, ALTER ANY ROLE, DROP ANY ROLE to BruceJones;
// Deny user Stalker access to any system resource
REVOKE ALL PRIVILEGES FROM Stalker;
ALTER USER Stalker DEFAULT NONE;
```

13.2.4 Access to System Data

Access to system data can be managed in one of the three ways:

- Through object privileges
- Through logical views
- Through intricate database design

13.2.4.1 Security via Object Privileges

Object privileges apply to specific database objects and are sometimes referred to as SUDI (select, update, delete, and insert) privileges. Table 13.1 provides a list of possible object privileges and the relevant types of database objects to which they apply.

Table 13.1 Object Privileges

Privilege	*Objects Applicable*
ALTER	Table and Sequence
DELETE	Table, View, Sequence, Procedure
EXECUTE	Procedure
INDEX	Table
INSERT	Table and View
REFERENCE	Table
SELECT	Table, View, Sequence
UPDATE	Table and View

Revisit the **Grant** statement and observe that it is applicable for the issuance of object privileges. When used for this purpose, the following rules apply:

- The **ON-Clause** is required.
- The privilege specified must be an object privilege, or the keyword ALL.

EXAMPLE 13.11: ILLUSTRATING USE OF OBJECT PRIVILEGES TO CONTROL DATA ACCESS

Objective: Referring to the partial college database of earlier discussions, the following statements manipulate object privileges related to tables **Student**, **Department**, and **Pgm_Struct**.

```
// For these statements, assume the prior creation of the user accounts specified:
GRANT SELECT, UPDATE ON Student TO BruceJones;
GRANT ALL PRIVILEGES ON Student TO Boss;
GRANT SELECT ON Pgm_Struct TO PUBLIC; /* granted to all users */
GRANT UPDATE ON Department TO BruceJones, Developer; /*granted to user and
   role */
```

EXAMPLE 13.12: ILLUSTRATING OBJECT PRIVILEGE REVOCATION

Objective: The following statement revokes UPDATE privilege on the **Student** table from users **BruceJones** and **Boss**.

```
REVOKE UPDATE ON Student FROM BruceJones;
REVOKE UPDATE ON Student FROM Boss;
```

13.2.4.2 Security via Views

As stated in the previous section, views can be used to enhance system security by allowing users to access only what is relevant to them. Conversely, views can be used to prevent users from accessing data for which they have no access privilege. In fact, it is typical in many database scenarios to prevent end users from having direct access to the physical database tables, and instead grant controlled access to the end-user constituency through logical views. The upcoming example provides an illustration of this strategy.

EXAMPLE 13.13: ILLUSTRATING USE OF LOGICAL VIEWS TO CONTROL DATA ACCESS

Objective: Referring to the college database of earlier discussions, develop a set of logical views that allow department heads to have access to student information if and only if the student is enrolled in a major offered by that department.

```
CREATE VIEW CSMajors (Stud#, StudLName, StudFName, StudSex, Major)
AS SELECT S.Stud#, S.StudLName, S.StudFName, S.StudSex, P.PgmName AS Major
FROM Student S, AcademicProgram P
WHERE S.StudPgm# IN ('BSC1', 'BSC2', 'BSC5') AND S.StudPgm# = P.Pgm#;
...
CREATE VIEW MathMajors (Stud#, StudLName, StudFName, StudSex, Major)
AS SELECT S.Stud#, S.StudLName, S.StudFName, S.StudSex, P.PgmName AS Major
FROM Student S, AcademicProgram P
WHERE S.StudPgm# IN ('BSC4', 'BSC5') AND S.StudPgm# = P.Pgm#;
...
// Assume BruceJones is Chair for Computer Science
REVOKE ALL PRIVILEGES ON CSMajors FROM PUBLIC;
GRANT SELECT ON CSMajors TO BruceJones;
...
// Assume TimMaitland Chair for Mathematics
REVOKE ALL PRIVILEGES ON MathMajors FROM PUBLIC;
GRANT SELECT ON MathMajors TO TimMaitland;
```

13.2.4.3 Security via Database Design

In addition to views, a database designer may design database tables with security attributes that will subsequently be used to control user access. Only authorized users will have access to these security attributes, but they can be used to block other users from accessing sensitive data. However, you would be required to create and maintain some additional tables (an example of this approach appears in Foster (2012).

13.3 Materialized Views

Oracle supports database objects called *materialized views*. A materialized view is a database object that stores the results of a query. It differs from the traditional logical view in that whereas the logical view stores the definition of the query, the materialized view stores the result of the query. A materialized view would therefore qualify as a snapshot relation (review Chapter 3). A full discussion of materialized views is beyond the scope of this course; however, a brief introduction is worthwhile.

The **From-Clause** of the subquery that feeds a materialized view can name tables, views, and other materialized views. Collectively, these are called *master tables* (a replication term) or *detail tables* (a data warehouse term). Databases that contain the master tables are called the *master databases*. Materialized views are used in replication environments, as well as in data warehousing (to be discussed in Chapter 19). In replication environments, the materialized views commonly created are *primary key views*, *rowed views*, *object views*, and *sub-query views*. For data warehousing purposes, the materialized views commonly created are materialized aggregate views, single-table materialized aggregate views, and materialized join views.

To create a materialized view in your own schema, you need the Create Materialized View system privilege. To create a materialized view in another user's schema, you need the Create Any Materialized View system privilege. Other system privileges needed are:

- Create Table or Create Any Table
- Select Any Table

In order to create a materialized view that is refreshable on each commit, in addition to the aforementioned privileges, it is necessary to have the ON COMMIT REFRESH system privilege or the ON COMMIT REFRESH object privilege on master tables not owned by the schema.

If you desire to create a materialized view with query rewrite enabled (see syntax below), in addition to the above-mentioned privileges, the following must hold:

- The owner of the master tables must have the Query Rewrite system privilege.
- If the schema owner does not own the master tables, then the schema owner must have the Global Query Rewrite privilege or the Query Rewrite object privilege on each table outside the schema.
- If the materialized view is being defined on a pre-built container (see syntax below), the creator must have the Select privilege WITH GRANT OPTION on the container table.

The user whose schema contains the materialized view must have sufficient quota in the target tablespace to store the materialized view's master table and index, or must have the Unlimited Tablespace system privilege.

13.3.1 Creating a Materialized View

The **Create-Materialized-View** statement is used for creating materialized views. The syntax is shown in Figure 13.10. This course foregoes a discussion of the **Create-Materialized-View** statement; for that, you are encouraged to consult the Oracle documentation (Oracle 2019). Suffice it to say, the **Create-MV-Refresh-Clause** allows for stipulating how and when the materialized view will be refreshed.

```
Create-Materialized-View ::=
CREATE MATERIALIZED VIEW [<Schema>.] <ViewName> [(<ColumnName> {, <ColumnName>})]
[OF [<Schema>.] <ObjectType>] [(Scoped-Table-Ref-Constraint)]
Properties-Spec | Prebuilt-Spec
[Using-Index-Clause] [Create-Mv-Refresh]
[FOR UPDATE] [ENABLE | DISABLE QUERY REWRITE]
AS Subquery;

Using-Index-Clause ::=  No-Index-Option | Index-Option
No-Index-Option ::= USING NO INDEX
Index-Option ::=  USING INDEX {Physical-Attributes-Clause | Tablespace-Spec}

Tablespace-Spec ::=  TABLESPACE <Tablespace>

Scoped-Table-Ref-Constraint ::=
SCOPE FOR (<Ref-Column> | <Ref-Attribute>) IS [<Schema>.] <ScopeTable>
{, SCOPE FOR (<Ref-Column> | <Ref-Attribute>) IS [<Schema>.] <ScopeTable>}

Properties-Spec ::=
Physical-Properties Materialized-View-Props

Prebuilt-Spec ::=
ON PREBUILT TABLE [WITH | WITHOUT REDUCED PRECISION]

Physical-Properties ::= // as defined in Create-Table statement (see chapter 11)

Physical-Attributes-Clause ::= // as defined in Create-Table statement (see chapter 11)

Materialized-View-Props ::=
[Column-Properties] [Table-Partitioning-Clause] [CACHE | NOCACHE [Parallel-Clause]
[BUILD IMMEDIATE | DEFERRED]

Column-Properties ::= // as defined in Create-Table statement (see chapter 11)

Table-Partitioning-Clause ::= // as defined in Create-Table statement (see chapter 11)

Parallel-Clause ::= // as defined in Create-Table statement (see chapter 11)

Create-Mv-Refresh ::=
REFRESH
[FAST | COMPLETE | FORCE]  [ON DEMAND | COMMIT]
[START WITH <Date>] [NEXT <Date>]
[[WITH PRIMARY KEY] | [WITH ROWID]]
[USING DEFAULT [MASTER | LOCAL] ROLLBACK SEGMENT]
[USING [MASTER | LOCAL]  ROLLBACK SEGMENT <Rollback-Segment>]

Subquery ::=  // as defined in Select statement (review chapter 12)
```

Figure 13.10 Create-Materialized-View statement.

EXAMPLE 13.14: ILLUSTRATING CREATION OF MATERIALIZED VIEWS TO CONTROL DATA ACCESS

Objective: Create materialized views corresponding to the logical views of Example 13.13.

```
CREATE MATERIALIZED VIEW CSMajors_MV REFRESH ON COMMIT
AS SELECT S.Stud#, S.StudLName, S.StudFName, S.StudSex, P.PgmName AS Major
   FROM Student S, AcademicProgram P
WHERE S.StudPgm# IN ('BSC1', 'BSC2', 'BSC5') AND S.StudPgm# = P.Pgm#;
...
CREATE MATERIALIZED VIEW MathMajors_MV REFRESH ON COMMIT
AS SELECT S.Stud#, S.StudLName, S.StudFName, S.StudSex, P.PgmName AS Major
   FROM Student S, AcademicProgram P
WHERE S.StudPgm# IN ('BSC4', 'BSC5') AND S.StudPgm# = P.Pgm#;
...
// Assume BruceJones is Chair for Computer Science
REVOKE ALL PRIVILEGES ON CSMajors_MV FROM PUBLIC;
GRANT SELECT ON CSMajors_MV TO BruceJones;
...
// Assume TimMaitland Chair for Mathematics
REVOKE ALL PRIVILEGES ON MathMajors_MV FROM PUBLIC;
GRANT SELECT ON MathMajors_MV TO TimMaitland;
```

13.3.2 Altering or Dropping a Materialized View

As you no doubt expect, there is an **Alter-Materialized-View** statement and a **Drop-Materialized-View** statement. The respective syntactical structures are shown in Figure 13.11, while an illustration of its use is shown in Example 13.15.

EXAMPLE 13.15: ILLUSTRATING ALTERATION OF MATERIALIZED VIEWS

```
// The following statements change methods used for materialized view refresh.

ALTER MATERIALIZED VIEW CSMajors_MV
REFRESH USING DEFAULT MASTER ROLLBACK SEGMENT;
// ...
ALTER MATERIALIZED VIEW MathMajors_MV
REFRESH WITH PRIMARY KEY;
```

```
Alter-Materialized-View ::=
ALTER MATERIALIZED VIEW [ <Schema>.] <Materialized-View>
[Properties-Option] [Alter-IOT-Clause]
 [USING INDEX Physical-Attributes-Clause]
 [[MODIFY Scoped-Table-Ref-Constraint] | Alter-Mv-Refresh]
 [[ENABLE | DISABLE QUERY REWRITE] | COMPILE | [CONSIDER FRESH]];

Properties-Option ::=
Physical-Attributes-Clause | COMPRESS | NOCOMPRESS | CACHE | NOCACHE | Allocate-Extent-Clause |
Alter-Table-Partitioning | Parallel-Clause | Logging-Clause | LOB-Option | Modify-LOB-Option

LOB-Option ::= LOB-Storage-Clause {, LOB-Storage-Clause}
Modify-LOB-Option ::= Modify-LOB-Storage-Clause {, Modify-LOB-Storage-Clause}

Physical-Attributes-Clause ::= // as defines in Create-Table statement (see chapter 11)
LOB-Storage-Clause ::= // as defines in Create-Table statement (see chapter 11)
Modify-LOB-Storage-Clause ::= // as defines in Alter-Table statement (see chapter 11)
Alter-Table-Partitioning ::= // as defines in Alter-Table statement (see chapter 11)
Parallel-Clause ::= // as defines in Create-Table statement (chapter 11)
Allocate-Extent-Clause ::= // as defines in Alter-Table statement (chapter 11)
Alter-IOT-Clause ::= // as defines in Alter-Table statement (chapter 11)
Scoped-Table-Ref-Constraint ::= // as defined in Create-Materialized-View
Alter-Mv-Refresh ::= // as defined for Create-Mv-Refresh in Create-Materialized-View

Logging-Clause ::= LOGGING | NOLOGGING

Drop-Materialized-View ::=
DROP MATERIALIZED VIEW [<Schema>.] <Materialized-View>;
```

Figure 13.11 Alter-Materialized-View and Drop-Materialized-View statements.

13.4 Summary and Concluding Remarks

Let us summarize what we have covered in this chapter:

- A logical view is a virtual relation that allows end users to access information in a manner that is consistent with their requirements. The view is created by the **Create-View** statement, which simply allows a few required keywords to be superimposed on a query (see Figure 13.2).
- The view is treated just like a normal base table. It can be designed to be updateable or read-only.
- The **Alter-View** statement facilitates modification of the view and the **Drop-View** statement facilitates deletion of the view.
- Views are very beneficial in enhancing logical data independence, facilitation of assorted external views of the database, simplification of the perception of end users, and enhancing system security.
- SQL facilitates the enforcement of a stringent security mechanism at three levels: access to the system, access to system resources, and access to data.
- Access to the system is controlled by profiles and user accounts. SQL provides statements for creating, altering, and dropping of profiles as well as user accounts.

- Access to system resources is controlled by privileges and roles. You can grant privileges and/or roles to users via the **Grant** statement and revoke them via the **Revoke** statement. You can lump privileges together by creating a role via the **Create-Role** statement, and then granting the privileges to the role. The role can then be granted to or revoked from other users and/or roles. Of course, a role can be altered (via the **Alter-Role** statement) and dropped (via the **Drop-Role** statement).
- Access to system data can be controlled through object privileges to base tables, logical views and object privileges to them, or intricate database design.
- A materialized view is a database view that stores both the definition and the result of the related subquery. It therefore qualifies as a snapshot relation. You can create, alter, and drop materialized views via the **Create-Materialized-View** statement, the **Alter-Materialized-View** statement, and the **Drop-Materialized-View** statement, respectively.

Logical views constitute a very important part of a database, providing several conveniences that translate to improved efficiency, flexibility, and productivity. As you will see in the upcoming chapter, they also form an integral part of the database system catalog.

13.5 Review Questions

1. What is a logical view? How does a logical view differ from a base relation? Discuss the importance and usefulness of logical views in a database.
2. When are views updateable and when are they not? Discuss.
3. Practice writing SQL statements to define logical views for various scenarios. Use the sample college database as your model.
4. Briefly explain the three levels of security in a typical database system. Explain how Oracle's implementation of SQL facilitates these three levels.
5. Improve on your college database (from Chapter 11) by doing the following:
   a. Create a user profile called **EndUser** for your database. Create a second profile called **Developer**.
   b. Create two roles: one for each profile (you may call them **EndUserR** and **DeveloperR**, respectively). Grant appropriate system privileges to these roles.
   c. Create two users: **BruceEnd** and **BruceDev**. **BruceEnd** must belong to profile **EndUser**, with default role **EndUserR**; **BruceDev** must belong to profile **Developer**, with default role **DeveloperR**. Both users must have your tablespace (created from Chapter 11) as their default tablespace, with QUOTA UNLIMITED.
   d. Grant full access of all your database objects created so far to user **BruceDev**. Grant limited access of your database objects to user **BruceEnd**.
   e. Alternately log on to the system as **BruceDev** and then **BruceEnd**, and check to see whether the privilege restrictions you have set are taking effect.
6. Practice writing SQL statements to manipulate system and object privileges. Test with your account (assumed to be **System** or **SysDBA**), **BruceDev**, and **BruceEnd**.
7. What are materialized views? Discuss their relevance usefulness. When would you use a materialized view versus a traditional logical view, and vice versa?
8. Describe a situation that would warrant the use of a materialized view in your college database. Write an SQL statement to define that materialized view.
9. Practice writing SQL statements to modify materialized views.

References and Recommended Readings

Connolly, Thomas, & Carolyn Begg. 2014. *Database Systems: A Practical Approach to Design, Implementation and Management* 6th ed. Boston, MA: Pearson. See chapter 7.

Date, Christopher J. 2004. *Introduction to Database Systems* 8th ed. Menlo Park, CA: Addison-Wesley. See chapters 9 and 17.

Elmasri, Ramez, & Shamkant B. Navathe. 2011. *Fundamentals of Database Systems* 6th ed. Boston, MA: Pearson. See chapter 5.

Foster, Elvis C. 2012. "Dynamic Menu Interface Designer." In *Athens Institute for Education and Research, Enterprise Management Information Systems.* Selected Papers from the 7th and 8th International Conferences on Information Technology and Computer Science, June 13–16, 2011, May 21–24, 2012. Athens: Athens Institute for Education and Research. See chapter 23.

Oracle Corporation. 2019. *Database SQL Language Reference.* Accessed June 2019. https://docs.oracle.com/en/database/oracle/oracle-database/19/sqlrf/index.html

Chapter 14

The System Catalog

Every reputable database management system (DBMS) contains a system catalog (also called the data dictionary) of some form. This has been mentioned several times earlier in the course. This chapter discusses this very important component of the database system. The chapter proceeds under the following subtopics:

Introduction

- Three Important Catalog Tables
- Other Catalog Tables
- Querying the System Catalog
- Updating the System Catalog
- Summary and Concluding Remarks

14.1 Introduction

The system catalog (data dictionary) is perhaps the most important resource in a database system. This is so because it facilitates and supports all or most of the other database objects. The system catalog typically contains *metadata* about the database. By metadata, we mean data about other data. This catalog itself consists of relational tables, which can be manipulated using Structured Query Language (SQL) statements. The system catalog provides the following benefits:

- The system catalog, by maintaining metadata in the form of other relational tables, facilitates most (if not all) of the other database objects. This fulfills the requirements of the Zero Rule of Chapter 9.
- Through the system catalog, the DBMS is able to deliver on the requirements of physical and logical data independence (Chapters 1 and 9) in a sleek manner.
- Through the system catalog, the DBMS is able to deliver on the requirement of integrity independence (Chapters 4, 5, and 9).
- System and object privileges (discussed in the previous chapter) are stored in special catalog tables, thus facilitating the management of the security mechanisms of the database.

DOI: 10.1201/9781003275725-17

- As you will see later in the course (Chapter 17), the system catalog also facilitates the successful implementation of distributed database systems.

The rest of this chapter will focus on aspects of the system catalog, as implemented in Oracle. Note, however, that all features of the Oracle catalog may not apply to catalogs of other DBMS suites; also, there may be features of others, not included in the Oracle catalog. Note also that a comprehensive discussion of the Oracle system catalog is beyond the scope and intent of this course; the discussion here is necessarily cursory, but detailed enough to give you a good appreciation of the subject matter.

The Oracle system catalog contains system tables for various database objects: Table 14.1 provides some of the most commonly referenced objects that are facilitated by the system catalog. Please note that Oracle catalog tables are traditionally prefixed by the $ symbol and are inaccessible to the user. The terms provided in Table 14.1 are conceptual and euphemistic, provided for the purpose of representing known Oracle database resources; they are not necessarily the names of Oracle catalog tables.

Table 14.1 Commonly Referenced Catalog Objects

Tablespaces	*Datafiles*	*Tables*	*Sequences*
Tab_Columns	Constraints (on tables)	Cons_Columns	Synonyms
Indexes	Users	Roles	Privileges
Tab_Comments	Col_Comments	Views	

Catalog tables are automatically maintained by the DBMS in a manner that is transparent to the user. Oracle allows viewing of data in catalog tables only through views. For each table, three views are often available: the view that has the prefix **DBA** (for all database objects), a view that has the prefix **USER** (for the objects owned by the current user), and a view that has the prefix **ALL** (for all objects that are accessible to the current user). However, there are exceptions to this rule. Additionally, views prefixed by **V$** are dynamic performance views, which can be queried, irrespective of your schema, provided that you have the appropriate privilege. Finally, views prefixed by **GV$** are global dynamic views.

14.2 Three Important Catalog Tables

To illustrate the importance of the system catalog, let us focus our attention on three conceptual catalog tables that we will call **Tables, Tab_Columns**, and **Indexes**. We will focus on three views on these tables: **User_Tables, User_Tab_Columns**, and **User_Indexes.** Remember, these tables are not directly accessible; they can only be indirectly accessed through the various logical views defined on them.

Before proceeding further, take some time to review the discussion about E-relations and P-relations in Section 5.4. In the discussion on the XR model, E-relations are described as relations that store data about entities (implemented as relational tables) comprising a database, and

P-relations are described as relations that store data about the properties (i.e., attributes or columns) of those entities. As stated then and reiterated now, the concept of a system catalog as described here exhibits and expands the foundational concepts of E-relations and P-relations.

14.2.1 The User_Tables View

This catalog view is based on the underlying tables (the DB2 equivalent being **Systables**). It contains a row for every base table in the user's schema. When a user account is created, Oracle creates a schema with the same name as the user name. All objects created by that user are linked to his/her schema, and are eventually stored in a **tablespace,** the default tablespace carrying the name **System** (for a full discussion of tablespaces, see the Oracle production documentation). Each table that you create has a row that is automatically entered in the catalog table called **Tables**, and is accessible via the view **User_Tables**, **DBA_Tables**, or **ALL_Tables**.

The **User_Tables** view includes several columns that provide useful information about the tables created by that user. To observe the structure of **User_Tables**, you may invoke the SQL*Plus **Describe** statement thus:

```
DESCRIBE User_Tables;
```

You will observe that the table contains a number of columns. Among the ones of immediate concern are those indicated in Table 14.2.

Table 14.2 Columns of User_Tables

Column Name	*Description*
Owner	The user who created the object
Table_Name	The user who created the object
Tablespace_Name	Name of tablespace in which it is stored
Num_Rows	Number of rows

14.2.2 The User_Tab_Columns View

This catalog view is based on the underlying conceptual table **Tab_Columns** (DB2 equivalent being **Syscolumns**). It contains a row for every column of every table mentioned in **User_Tables**. Each column that you create for each table has a row that is automatically entered in the catalog table called **Tab_Tables** and is accessible via the view **User_Tab_Columns**, **DBA_Tab_Columns**, or **ALL_Tab_Columns**. You may observe the columns of the **User_Tab_Columns** view by issuing the **Describe** statement with respect to the view. Table 14.3 shows the columns of immediate concern.

Table 14.3 Columns of User_Tab_Columns

Column Name	*Description*
Table_Name	Name of database table
Column_Name	Name of column in the table
Data_Type	Data type of the column
Data_Length	Length of the column (in bytes)
Data_Precision	Number of decimal places of column
Data_Default	Default value of column

14.2.3 The User_Indexes View

This catalog view is based on the underlying conceptual table Indexes (DB2 equivalent being **Sysindexes**). It contains a row for every index in the user's schema. Each index created is automatically cataloged in the Indexes table, which is accessible via views **User_Indexes**, **DBA_Indexes**, or **ALL_Indexes**. As always, you may use the **Describe** statement to observe the columns of the **User_Indexes** view. Some columns of interest are mentioned in Table 14.4.

Table 14.4 Columns of User_Indexes

Column Name	*Description*
Index_Name	Name of index
Table_Owner	User who created the table
Index_Type	Type of index
Table_Name	Name of table being indexed
Num_Rows	Number of rows indexed

Table 14.5 provides a simplified illustration of possible catalog data, assuming that the database consists of the tables mentioned in the sample college database that we have been referencing (since Chapter 7).

Table 14.5 Simplified Illustration of System Catalog for College Database

User_Tables:			
Table_Name	**Tablespace**	**Num_Rows**	...
Student	TBS_FosterE	8	
AcademicProgram	TBS_FosterE	2	
Hall	TBS_FosterE	2	
Department	TBS_FosterE	4	
Division	TBS_FosterE	3	
Course	TBS_FosterE	2	
Employee	TBS_FosterE	2	
Pgm_Struct	TBS_FosterE	2	
User_Indexes:			
Index_Name	**Table_Name**	**Table_Owner**	...
xStud	Student	BruceJones	
xStud2	Student	BruceJones	
xProgram	AcademicProgram	BruceJones	
xProgram2	AcademicProgram	BruceJones	
xHall	Hall	BruceJones	
xDept	Department	BruceJones	
xDept2	Department	BruceJones	
xDiv	Division	BruceJones	
xDiv2	Division	BruceJones	
xCourse	Course	BruceJones	
xCourse2	Course	BruceJones	
xEmployee	Employee	BruceJones	
xEmployee2	Employee	BruceJones	
xPgm_Struct	Pgm_Struct	BruceJones	
xPgm_Struct2	Pgm_Struct	BruceJones	
User_Tab_Columns:			
Column_Name	**Table_Name**	**Data_Type**	...

(*Continued*)

Table 14.5 (Continued) Simplified Illustration of System Catalog for College Database

User_Tables:			
Stud#	Student	Integer	
StudLName	Student	Varchar2	
StudFName	Student	Varchar2	
...			
Pgm#	AcademicProgram	Char	
PgmName	AcademicProgram	Varchar2	
...			
Crs#	Course	Char	
CrsName	Course	Varchar2	
...			
Dept#	Department	Char	
DName	Department	Varchar2	
...			

14.3 Other Important Catalog Tables

There are several other catalog tables all of which are managed in a manner transparent to the database user. Table 14.6 provides a list of commonly used catalog views. This list is by no means exhaustive. However, studying and probing this list will give you a good insight into the usefulness of the catalog views, as you manage the database. Do not feel disconcerted or overwhelmed if you are not familiar with all of the views listed in the figure. These you would normally cover in a course on Oracle database administration. The intent here is to give you a reasonable overview of the role and complexity of the system catalog.

Table 14.6 Commonly Used Catalog Views

View Name	*Contents*
V$SYSTEM_PARAMETER	Database system parameters as defined in the parameter file
V$PARAMETER	Database system parameters as defined in the parameter file
V$SESSION	Information on all current sessions running in the database
V$BACKUP	Backup status for datafiles in the database
V$BACKUP_DATAFILE	Information on files backed up via RMAN
V$BACKUP_REDOLOG	Information on archived log files backed up via RMAN
V$BACKUP_PIECE	Information on backed up pieces, updated via RMAN
V$BACKUP_SET	Information on complete, successful backups up via RMAN
V$DATABASE	Information on databases created on the machine
V$DATAFILE_HEADER	Datafile header information
V$DATAFILE	Information on datafiles associated with the database
V$CONTROLFILE	List of controlfiles for the database and their status
V$CONTROLFILE_ RECORD_SECTION	Controlfile record information (record size, records used, etc.)
V$ARCHIVEED_LOG	Information on archived logs
V$LOGFILE	Online redo log members of groups
V$LOG	Online redo log groups
V$LOG_HISTORY	History of log information
V$ARCHIVE_DEST	Information about the five archive destinations, status, and failures
V$ARCHIVE_PROCESSES	Status on the ten archive processes
V$PWFILE_USERS	List of users entitled to use SYSDBA and SYSOPER privs.
V$THREAD	Information on log files assigned to each instance
V$INSTANCE	List of database instances running
V$OBJECT_USAGE	List of indexes and their usage
V$PROCESS	List of processes running
V$SESSTAT	Indicates statistics for various sessions
V$LATCHNAME	List of DB latches for various sessions
V$TIMEZONE_NAMES	Valid time zones

(Continued)

Table 14.6 (Continued) Commonly Used Catalog Views

View Name	*Contents*
V$ROLLNAME	List of all online undo segments
V$ROLLSTAT	Undo statistics. Can be joined with V$ROLLNAME
V$TABLESPACE	Information (TBS name, number, and backup status) on tablespaces
V$SORT_USAGE	Information (User, Session#, tablespace, segment, extents, etc.) on active sorts in the database. Can be joined with V&Session & V&SQL
V$SQL	SQL statements run by the users
V$TEMP_EXTENT_MAP	Extents of all locally managed temporary tablespaces
V$TEMP_EXTENT_POOL	Temporary space used and cached for the current instance for locally managed temporary tablespaces
V$DATAFILE	Info on data file from the control file
V$UNDOSTAT	10-minute snapshots reflecting the performance of the undo tablespace
V$TEMPFILE	Info on temporary files (similar to V&DATAFILE)
V$DISPATCHER	Info on dispatchers
V$DISPATCHER_RATE	Info on performance statistics for dispatchers
V$QUEUE	Info on the request queue and response queues
V$CIRCUIT	Info on Shared Server virtual circuits.
V$SHARED_SERVER	Info on shared servers in the system.
V$SHARED_SERVER_ MONITOR	Summary info on maximum connections, maximum servers, servers started, servers terminated, high-water level for the shared servers (combined).
V$SESSION	Info on sessions
V$TIME ZONE_NAMES	Time Zones allowed
V$SORT_SEGMENT	Info about every sort segment in a given instance. Uploaded only for temporary tablespaces.
V$BACKUP_SET	Stores information on backup sets.
RC_BACKUP_SET	Stores information on backup sets. Applicable only if Recovery catalog is in place.
DBA_EXTENTS	Extents allocated for all segments in the database
DBA_FREE_SPACE	Free extents in tablespaces

(Continued)

Table 14.6 (Continued) Commonly Used Catalog Views

View Name	*Contents*
DBA_SEGMENTS	Segments created in the database, their size, tablespace, type, storage forms, etc.
DBA_DATA_FILES	Information on datafile(s) for each tablespace. An alternative to joining V$TABLESPACE and V$DATAFILE.
DBA_CACHE_SIZE	Block Size for the db
DBA_PROFILES	List of profiles.
XXX_SOURCE	The source code for programs running
XXX_OBJECTS	Objects belonging to a database
XXX_TABLES	Tables belonging to a database
XXX_TAB_COLUMNS	Columns of database tables
XXX_TAB_PRIVS	Table privileges granted to users and roles
XXX_COL_PRIVS	Column privileges granted to users and roles
XXX_INDEXES	Indexes defined on database tables
XXX_IND_COLUMNS	Columns included in each index
DBA_CONSTRAINTS	List of constraints
DBA_CONS_COLUMNS	Columns for each constraint
XXX_SYS_PRIVS	System privileges granted
DBA_ROLES	List of roles
SESSION_ROLES	Roles for the session
XXX_ROLE_PRIVS	Role privileges granted
XXX_ROLLBACK_SEGS	Information on undo segments
ROLE_SYS_PRIVS	System privileges granted to roles
ROLE_ROLE_PRIVS	Role privileges granted to other roles
ROLE_TAB_PRIVS	Table privileges granted to roles
XXX_TABLESPACES	Information (TBSName, Block-Size, Extent Info) Tablespace in the database
XXX_FREE_SPACE	Free extents available in tablespaces (for each data-file in each tablespace). Locally Managed temporary tablespaces are not included.
XXX_SEGMENTS	Info on segments and their storage parameters

(Continued)

Table 14.6 (Continued) Commonly Used Catalog Views

View Name	*Contents*
XXX_EXTENTS	Info on extents (Size, assorted segments, associated tablespace).
XXX_DATAFILES	Info on data-files belongings do tablespaces
XXX_TEMP-FILES	Info no temporary files belongs to locally managed temporary tablespaces
XXX_USERS	Info (Including default tablespace allocation) on users.
XXX_VIEWS	Info on logical views
XXX_TEMP_FILES	Info on data files belonging to locally managed temporary tablespaces.
XXX_USED_EXTENTS	Info on used extents for tablespaces
XXX_ROLLBACK_SEGS	Info on rollback segments for tablespaces
DBA_UNDO_EXTENTS	Info on undo extents for tablespaces

Note: Prefix XXX means DBA, USER or ALL, e.g., DBA_SYS_PRIVS or USER_SYS_PRIVS.

14.4 Querying the System Catalog

The system catalog can be queried using SQL **Select** statements in a manner similar to any created relational table. This is one of the many remarkable features of the relational DBMS, and a powerful witness to the potency of Codd's Zero Rule (see Chapter 9).

EXAMPLE 14.1: DETERMINING ATTRIBUTES COMPRISING A RELATIONAL TABLE

Objective: What columns does the table **Student** have?

```
SELECT Column_Name FROM User_Tab_Columns WHERE Table_Name =
   'STUDENT';
// Similar information can be obtained via the Describe statement in SQL*Plus
DESCR Student;
```

EXAMPLE 14.2: FINDING TABLES WITH A CERTAIN ATTRIBUTE

Objective: Find the relation(s) contain(s) the attribute Pgm# or any reference to Pgm#.

```
SELECT DISTINCT Table_Name FROM User_Tab_Columns
WHERE Column_Name IN ('Pgm#', 'PGM#') OR Column_Name LIKE '%PGM#%';
```

You are encouraged to try out these and other similar examples based on any sample Oracle database that you may have access to. Here are a few points of clarification:

1. The above query will list all tables and/or logical views that have column-name(s) referencing **Pgm#**.
2. By default, identifiers such as table-names and column-names are stored in the Oracle catalog in uppercase. One way around this is to specify table-names and attribute-names in double quotations during table creation; however, this is seldom done and is not recommended. The use of the **IN** operator in the query above is therefore a precaution that may not be necessary depending on the table creation strategy in practice for the database.
3. Examples 14.2 and 14.3 should help you appreciate the fact that the DBMS keeps track of foreign keys in order to enforce referential integrity. With a comprehensive catalog in place, it is relatively easy to determine if an attribute in a given table is defined on the same domain as another attribute in another table.

EXAMPLE 14.3: FINDING TABLES ATTRIBUTES MEETING A SPECIFIED CRITERION

Objective: What relation(s) contain(s) at least one attribute that is CHAR and a maximum length of 8 bytes?

```
SELECT DISTINCT Table_Name FROM User_Tab_Columns
WHERE Data_Type = 'CHAR' AND Data_Length <= 8;
```

EXAMPLE 14.4: KEEPING TRACK OF TABLES WITH A CERTAIN NAME PATTERN

Objective: Suppose that tables belonging to a certain subsystem or system were all prefixed by the string pattern 'FMS' (which may mean Financial Management System or some other appropriate name). We could easily track such tables:

```
SELECT DISTINCT Table_Name FROM User_Tables WHERE Table_Name LIKE 'FMS%';
```

EXAMPLE 14.5: KEEPING TRACK OF VIEWS AND USERS

```
// List all logical views from the system
SELECT substr (View_Name, 1, 15) View_Name, substr(Owner, 1, 15) Owner FROM
   DBA_Views;
// List all logical views owned by user Foster
SELECT substr(VIEW_NAME, 1, 15) View_Name, substr(Owner, 1, 15) Owner FROM
   DBA_Views
WHERE Owner LIKE '%FOSTER%';
// user accounts, related system Id, and account status
SELECT substr(UserName, 1, 15) User_Name, User_ID, substr(Account_Status, 1, 20)
   User_Status FROM DBA_Users;
```

Observe that in Example 14.5, the SUBSTR (substring) function is used to extract the first 15 or 20 bytes from the columns specified. If you display the structure of the catalog views referenced (**DBS_Views** and **DBA_Users**), you will see that columns used in the queries occupy several bytes of storage. Moreover, in many instances, you will get the information you are looking for in the first few bytes of those respective columns. If you try to run the queries without peeling off bytes of concern, you will likely get a result-set in each case that is very difficult to make sense of. The SUBSTR function helps to make the data in your result-set less cluttered and therefore more understandable in each case.

14.5 Updating the System Catalog

Direct update (via INSERT, UPDATE, DELETE) of system catalog records is not allowed as this would be an avenue for compromising the integrity of the database.

The system catalog is automatically updated by the system DBMS whenever DDL statements such as CREATE TABLE, ALTER TABLE, DROP TABLE, CREATE VIEW, DROP VIEW, CRETE INDEX, DROP INDEX, CREATE SYNONYM, and DROP SYNONYM, are issued. In short, whenever a database object is created or modified, the system catalog is automatically updated by the DBMS.

The closest semblance of direct update of the system catalog from a user perspective is the Comment statement. It allows updates of the **Comments** column in conceptual catalog table **Tab_Comments** or **Col_Comments** — one for storing comments about other database tables, and the other for storing comments about columns for various tables. The Backus-Naur Form (BNF) format for the command follows:

Comment-Statement ::=
COMMENT ON TABLE | COLUMN <TableName> | <ColumnName> IS <String>;

In using this statement, the column-name specified must be qualified by its table-name. Additionally, the string supplied is enclosed in single quotes in the usual manner.

In order to understand how the **Comment** statement works, take some time and familiarize yourself with the structure and purpose of the conceptual catalog tables **Tab_Comments** and **Col_Comments.** Like other conceptual catalog tables, these two tables are accessible only through logical views prefixed by DBA, USER, or ALL. The **Describe** statement will allow you to observe the structure (left as an exercise). You will observe the following columns in respective tables:

User_Tab_Comments {Table_Name, Table_Type, Comments}
User_Col_Comments {Table_Name, Column_Name, Comments}

The **Tab_Comments** table is used to store comments on tables of the database; the **Col_Comments** table is used to store comments on table columns of the database. The **Comment** statement allows modification of these comments.

EXAMPLE 14.6: COMMENTING ON TABLES AND COLUMNS

```
COMMENT ON TABLE Course IS 'The Course Relation';
COMMENT ON COLUMN Course.Crs# IS 'Course Code';
COMMENT ON COLUMN Course.CrsName IS 'Course Name';
...
// Do this for each user-created table in the database
```

As the previous example suggests, it is a good idea to create comments for each user table (and its related columns) comprising the database. To access these comments, you may query any of the following views:

- **DBA_Tab_Comments:** All table comments for the database.
- **USER_Tab_Comments:** All table comments owned by the current user.
- **ALL_Tab_Comments:** All table comments to which the current user has access.
- **DBA_Col_Comments:** All column comments for the database.
- **USER_Col_Comments:** All column comments owned by the current user.
- **ALL_Col_Comments:** All column comments to which the current user has access.

EXAMPLE 14.7: ACCESSING CATALOG COMMENTS FOR TABLES AND COLUMNS

```
// Show comments for tables owned by user Foster
SELECT SUBSTR(Table_Name,1,15) TabName, SUBSTR(comments,1,40) TabComment
FROM DBA_Tab_Comments WHERE Owner LIKE '%FOSTER%';
// Show table comments for tables owned by user Foster
SELECT Substr (Table_Name,1,15) TabName, Substr (Column_Name, 1, 15) ColName,
   Substr (comments,1,40) TabComment
FROM DBA_Col_Comments WHERE Owner LIKE '%FOSTER%';
```

14.6 Summary and Concluding Remarks

It is now time to summarize what has been discussed in this chapter:

- The system catalog is the most important database object in a database system. This is so because it facilitates and supports all or most of the other database objects. The system catalog typically contains *metadata* about the database.
- The Oracle system catalog contains system tables for various database objects including (but not confined to) tablespaces, datafiles, tables, views, table columns, constraints, indexes,

users, roles, and privileges. Oracle does not allow direct access of its catalog tables; rather, it provides views prefixed by **DBA**, **ALL**, **USER**, and **GV$**.

- Three commonly used Oracle catalog views are **User_Tables**, **User_Tab_Columns**, and **User_Indexes**. In actuality, the Oracle catalog contains scores of catalog views (review Section 14.3).
- You can query catalog views just as you would any other table. This is often how DBAs and software engineers get useful information about the database.
- As a rule, Oracle does not allow direct updates of its catalog tables. However, note that the catalog is automatically updated every time the physical or logical structure of the database is modified. The only exception to this rule is the **Comment** statement. This statement allows the specification of comments for database tables and columns.

If you consider what has been said about the system catalog in light of Codd's Zero Rule as well as his 12 rules for relational DBMS suites, you will soon realize the catalog is an absolute necessity if the DBMS is to stand up to the lofty industry expectations. To be more direct, if you are evaluating a DBMS suite and discover that it does not host a comprehensive system catalog, you are pursuing a product that is likely to buckle under significant rigor; you'd be well advised to save your effort for some more meaningful project.

As powerful and useful as SQL is, the language is not without limitations. The next chapter discusses some of these limitations.

14.7 Review Questions

1. Discuss the importance of the system catalog in a database system.
2. State some commonly referenced catalog tables.
3. Discuss three important catalog tables and explain how their use can help in the management of a database.
4. The SQL*Plus **Describe** statement allows you to view the structure of a database table. Write an equivalent **Select** statement on the system catalog that provides the same information. Use one of the tables in the sample college database as your frame of reference.
5. Review as many of the catalog views listed in Table 14.6 as possible and practice writing **Select** statements on them.
6. Explain clearly how the system catalog is maintained. Provide useful examples.

References and/or Recommended Readings

Oracle Corporation. 2019. *Database SQL Language Reference*. Accessed June 2019. https://docs.oracle.com/en/database/oracle/oracle-database/19/sqlrf/index.html

Thomas, Biju, Gavin Powell, Robert Freeman, & Charles Pack. 2014. *Oracle Certified Professional on 12C Certification Kit*. Indianapolis, IN: Wiley.

Chapter 15

Some Limitations of SQL

As can be seen from Chapters 11–14, SQL is a very powerful programming language, ideally suited for the management of databases. However, like all languages, SQL has limitations. This chapter briefly examines some of these limitations. The chapter proceeds as follows:

- Programming Limitations
- Limitations on Views
- Stringent Enforcement of Referential Integrity
- Limitations on Calculated Columns
- If-Then Limitation
- Summary and Concluding Remarks

15.1 Programming Limitations

While SQL is a very sophisticated fourth generation language (4GL) for database management, it does not have many facilities that are normally present in a traditional high-level language (HLL). These include user interface programming and traditional internal processing facilities for basic data structures. The truth is, SQL was never intended for these facilities. SQL is, therefore, most effective in an environment where it is embedded in HLL code. The HLL may be the host language of a DBMS, or some other language that the DBMS supports.

Quite often, in complex software development projects, the developer will encounter situations where a single SQL statement is inadequate to service the needs of the user. What is required is a series of SQL statements (and possibly non-SQL statements). The limitations on logical views (discussed in the following section) provide a case in point. Again, an HLL support is often a perfect antidote for these scenarios.

15.2 Limitations on Views

The limitations on logical views were mentioned in Chapter 13 (Sections 13.1 and 13.2), without much elaboration. Let us revisit this matter here.

DOI: 10.1201/9781003275725-18

15.2.1 Restriction on Use of the ORDER-BY-Clause for Earlier Versions of SQL

Earlier versions of SQL forbade the use of the **Order-By-Clause** in the sub-query for a logical view. This meant that you could not create a view that ordered data. The rationale for this limitation was that a view is a virtual relation, and therefore, ordering of the data would only increase system overheads; also recall from Chapter 3 (Section 3.3.1), that since a view qualifies as a relation (albeit virtual), strictly speaking, ordering of data was antithetical to the definition of a relation. The obvious counterargument to this position was, since a view merely stores the definition of an access path to data stored in base relation(s), why not include data ordering in the definition?

The work-around for this limitation was to create the view (obviously without ordering data), then when the view was being accessed via a query, the **Order-By-Clause** would be included in the query. You will be pleased to know that this criticism did not fall on deaf ears: Contemporary versions of SQL allow for use of the Order-By-Clause in the subquery.

EXAMPLE 15.1: ORDERING DATA IN A LOGICAL VIEW

Objective: Create a virtual College Bulletin from the sample college database of earlier discussions, so that this can be used to support subsequent queries.

```
// The traditional approach follows:

CREATE VIEW Bulletin (PsPgm#, PgmName, PsCrsSeqn, PsCrs#, CrsName)
AS SELECT PsPgm#, PgmName, PsCrsSeqn, PsCrs#, CrsName FROM Pgm_Struct PS,
AcademicProgram P, Course C WHERE PS.PsPgm# = P.Pgm# AND PS.PsCrs# = C.Crs#;
...
SELECT * FROM Bulletin ORDER BY PsPgm#, PsCrsSeqn;

// The contemporary approach follows
CREATE VIEW Bulletin (PsPgm#, PgmName, PsCrsSeqn, PsCrs#, CrsName)
AS SELECT PsPgm#, PgmName, PsCrsSeqn, PsCrs#, CrsName FROM Pgm_Struct PS,
AcademicProgram P, Course C WHERE PS.PsPgm# = P.Pgm# AND PS.PsCrs# = C.Crs#
ORDER BY PsPgm#, PsCrsSeqn;
```

15.2.2 Restriction on Data Manipulation for Views Involving UNION, INTERSECT, or JOIN

As pointed out in Chapter 13, a view is not updateable if it involves a JOIN, a UNION or an INTERSECT operation (Section 13.1.3 provides an exception to this rule). A little thought will reveal that while this is an understandable constraint, it is not always a prudent one, as there are situations that could warrant updateable views involving these operations (for more elaboration, see Date 2004).

The example above is a useful illustration: it should be possible for a user to modify **PgmName** and/or **CrsName** for a logical row in **Bulletin** as follows:

- If **PgmName** is modified, the DBMS should use **Pgm#** to access the correct row in the table **AcademicProgram** and modify its corresponding column for **PgmName**.
- If **CrsName** is modified, the DBMS should use **Crs#** to access the correct row in the table **Course** and modify its corresponding column for **CrsName**.

In either case, the search for a corresponding column must not be made merely on the column name (since queries can rename columns), but the name as well as characteristics of the column (which can be obtained from the system catalog).

Administering changes (insertion, update or deletion of rows) to join logical views of this sort is by no means a trivial matter; hence SQL does not support it. The point to note here is that it is thinkable and indeed doable, though complex.

Similar to join logical views (i.e., views involving the joining of two or more tables), logical views involving UNION and INTERSECT operations are not updateable. To eliminate this limitation, SQL would need to trace records that appear in intersection or union operations back to their normal storage areas in respective tables and update them there. This activity, while doable, would be rather costly, especially as the database grows in size.

15.2.3 Restriction on Use of Aggregation Functions for Earlier Versions of SQL

Like **the Order-By-Clause**, earlier versions of SQL did not support use of aggregation (or analytic) functions in the subquery for a logical view. The work-around was to create the logical view without such functions and subsequently include them in queries on the view in question.

Contemporary versions of SQL do not have this restriction; you are allowed to include aggregation and/or analytic functions in the subquery for your logical view.

EXAMPLE 15.2: INCLUDING AGGREGATION AND/OR ANALYTIC FEATURES IN A LOGICAL VIEW

Objective: The following are two logical views constructed for query Examples 12.32 and 12.33b. In each case, a traditional solution would be more complex.

```
// This view includes aggregation, grouping, and ordering; review example 12.32
CREATE OR REPLACE VIEW Employee_SalaryAggregation
AS SELECT EmpDept#, ROUND(SUM(EmpSalary), 2) AS TotalSal,
ROUND(AVG(EmpSalary), 2) AS AvgSal, ROUND(MAX(EmpSalary), 2) AS MaxSal,
ROUND(MIN(EmpSalary), 2) AS MinSal, ROUND(STDDEV(EmpSalary), 2) AS
StdDev
FROM Employee GROUP BY EmpDept# ORDER BY EmpDept#;

// This view includes aggregation, analytics, joining, and ordering; review example 12.33b
CREATE OR REPLACE VIEW Employee_Salary_Profile02
AS SELECT E.Emp#, E.EmpLName || '--' || E.EmpFName || ' ' || SUBSTR(E.EmpMName, 1,
1) AS EmpFullName, E.EmpSalary, E.EmpDept#, D.DeptName, SUM(E.EmpSalary) OVER
(PARTITION BY E.EmpDept#) AS TotalSal, AVG(E.EmpSalary) OVER (PARTITION BY
E.EmpDept#) AS AvgSal, MAX(E.EmpSalary) OVER (PARTITION BY E.EmpDept#) AS
MaxSal, MIN(E.EmpSalary) OVER (PARTITION BY E.EmpDept#) AS MinSal
FROM Employee E, Department D WHERE E.EmpDept# = D.Dept#;
```

15.3 Stringent Enforcement of Referential Integrity

Most implementations of SQL exhibit a stringent enforcement of the referential integrity rule that forbids you to change the value of a primary key that is being referenced by other records. This is not really a weakness but rather a desirable feature; it is a deliberate strategy to protect the integrity of the database (review Chapters 4 and 5).

EXAMPLE 15.3: UPHOLDING REFERENTIAL INTEGRITY

Objective: Suppose that we have a course — M100 College Algebra — that occurs in all academic programs. We wish to change the course code to M105, thus:

UPDATE Course SET Crs# = 'M105' WHERE Crs# = 'M100';

Note:

Most DBMSs will forbid this update, since M100 would be a referenced tuple. Nor would you be allowed to delete M100 as a course. Allowing a reassignment of the course code in this way, or deletion of the course, would undermine the integrity of the database.

So, what if you were working in a test/development database with phony or test data that you now want to clear out before putting the database into production? One may argue that allowing the removal of referenced tuples in such a scenario would be warranted. Oracle handles this situation quite elegantly, by allowing the designer to temporarily **disable** a constraint, and **enable** it at a subsequent time. If your DBMS does not have this kind of facility, what you may have to do is drop the foreign key constraint(s) from the referencing table(s), make the update/deletion, and then reintroduce the foreign key constraint(s) (of course, in such a circumstance, you could also write a utility program to do that). You may also want to check whether your DBMS has a **Truncate-Table** statement (as described in Section 12.3), and how it works; just be very careful that you do not end up zapping unintended data.

EXAMPLE 15.4: DISABLING AND ENABLING A CONSTRAINT

Objective: You want to delete all sample data from tables in a test/development database before moving it to production, but the referential integrity constraints are inhibiting your progress. What gives?

```
// Step 1: Disable the FK constraints as in the following example
ALTER TABLE Pgm_Struct DISABLE CONSTRAINT PSForeign2;
// Alternate solution
ALTER TABLE Pgm_Struct MODIFY CONSTRAINT PSForeign2 DISABLE;
```

```
//Step 2: Go ahead and empty/purge the tables needing attention

// Step 3: Reinstate the FK constraints as in the following example
ALTER TABLE Pgm_Struct ENABLE CONSTRAINT PSForeign2;
// Alternate solution
ALTER TABLE Pgm_Struct MODIFY CONSTRAINT PSForeign2 ENABLE;

// Step 4: Your database is now ready for production
```

15.4 Limitations on Calculated Columns

Generally speaking, you are not allowed to define a virtual (calculated) column in terms of another calculated column in the same subquery. However, as you may recall from Chapter 12 (for instance, Examples 12.39 through 12.42), you may reference a calculated column from a different subquery so long as the scoping rules (Section 12.10) are followed.

EXAMPLE 15.5: BREAKING UP A COLUMN FOR SUBSEQUENT USAGE

Objective: Consider a table that is keyed on a numeric attribute, **Trans_Date**, that stores the date (YYYYMMDD format). One might want to create a virtual attribute, **Trans_Year**, that stores the year only (first four bytes of **Trans_Date**), and cumulate values for another numeric column of the table, based on **Trans_Year**.

Work-around: One obvious work-around would be as follows:

a. Create a logical view with the virtual column;
b. Write a query on the view, grouping on the virtual column.

Alternate Work-around: An alternate work-around is to simply restate the calculated expression wherever it is needed within the query.

EXAMPLE 15.6: ILLUSTRATING REUSE OF A CALCULATION CONCEPT IN THE SAME QUERY

Objective: Example 12.33a is repeated here to illustrate the point: We desire a list from the employee table (of the Oracle default database), showing for each department, the total salary, average salary, minimum salary, maximum salary, and standard deviation; we want to show only departments with a total salary of at least $120,000.

```
// This solution will not run; TotSal is a calculated attribute and cannot be reused as shown
SELECT EmpDept#, SUM(EmpSalary) AS TotalSal, AVG(EmpSalary) AS AvgSal,
   MAX(EmpSalary) AS
MaxSal, MIN(EmpSalary) AS MinSal, STDDEV(EmpSalary) AS StdDevSal
FROM Employee GROUP BY EmpDept# HAVING TotalSal >= 120000;
```

```
// The correct solution
SELECT EmpDept#, SUM(EmpSalary) AS TotalSal, AVG(EmpSalary) AS AvgSal,
   MAX(EmpSalary) AS
MaxSal, MIN(EmpSalary) AS MinSal, STDDEV(EmpSalary) AS StdDevSal
FROM Employee GROUP BY EmpDept# HAVING SUM(EmpSalary) >= 120000;

// Refined solution
SELECT EmpDept#, SUM(EmpSalary) AS TotalSal, AVG(EmpSalary) AS AvgSal,
   MAX(EmpSalary) AS
MaxSal, MIN(EmpSalary) AS MinSal, STDDEV(EmpSalary) AS StdDevSal
FROM Employee GROUP BY EmpDept# HAVING SUM(EmpSalary) >= 120000;
```

Notice that in the **Having-Clause**, the calculated column **TotSal** cannot be used. You have to specify the expression used for the derived column.

EXAMPLE 15.7: ANOTHER ILLUSTRATION OF REUSING A CALCULATION CONCEPT IN THE SAME QUERY

Objective: Referring to the default Oracle database, suppose that we want to list employees beyond a certain hire year. Following is an incorrect SQL statement followed by a correct SQL statement for the problem.

```
// This attempted query will not run due to the attempted reuse of calculated column
HireYear
SELECT EmpNo, EName, SUBSTR (TO_CHAR(HireDate, 'YYYYMMDD'),1,4)
   HireYear FROM Emp
WHERE HireYear >= 1985;

// This query will run successfully
SELECT EmpNo, EName, SUBSTR(TO_CHAR(Hiredate, 'YYYYMMDD'),1,4)
   HireYear FROM Emp
WHERE SUBSTR(TO_CHAR(Hiredate, 'YYYYMMDD'),1,4) >= 1985;
```

15.5 If-Then Limitation

In Chapter 8 (Section 8.2), it was mentioned that a well-formed formula (WFF) is a comparison (simple or complex) that evaluates to true or false. It was also mentioned that one form of a WFF is as follows.

```
If <Condition> then <WFF>
```

It appears that this format is seldom implemented in typical SQL implementations (though there are deviations of it). The work-around is to apply the appropriate standardization rule (review

Section 8.4 of Chapter 8), which is shown below (assume A, B, and C to represent Boolean expressions or comparisons):

```
If A then B <=> (A)' or B.
```

This is by no means a significant setback. One can get by without ever using an if-then construct, and simply using its equivalent.

15.6 Summary and Concluding Remarks

Here is a summary of what was covered in this chapter:

- SQL is a powerful language, but not without limitations. The first is that SQL is a database language by design and intent. It does not include features for building complex user interfaces, because it was not intended to.
- Earlier versions of SQL forbade use of the Order-By-Clause, aggregation functions, or analytic functions in logical views; however, there were ways one could compensate for these limitations. These limitations have been removed from the current version of SQL.
- Logical views that include the UNION, INTERSECT, or JOIN operations are not updateable; there is one exception with respect to JOIN, but it is not significant.
- Most implementations of SQL exhibit a stringent enforcement of the referential integrity rule. While this may seem too inflexible to the untrained eye, it is absolutely necessary for preserving the integrity of the database.
- You are not allowed to define a calculated column in terms of another calculated column in the same subquery, or to reference a calculated column in the same subquery.
- It appears that the if-then construct is not directly supported in SQL.

Fortunately, there is a work-around for each limitation. This is perhaps why SQL is so popular: its benefits far outweigh its limitations. We can, therefore, that SQL will continue to be the universal database language a long time yet.

This brings us to the end of our study of SQL for this course, and the end of this section of the text. The next six chapters provide an overview of some advanced topics in database systems.

15.7 Review Questions

1. What are the programming limitations of SQL? As a software engineer, how do you make up for these limitations?
2. Discuss the limitations of SQL with respect to logical views. What are the implications of these limitations?
3. What problems relating to referential integrity are often present in implementations of SQL? How are these problems addressed in the Oracle DBMS?
4. Describe the SQL limitation on calculated columns. Carefully explain how it can be circumvented.

Recommended Readings

Date, Christopher J. 2004. *Introduction to Database Systems* 8th ed. Menlo Park, CA: Addison-Wesley. See chapters 4 and 10.

Oracle Corporation. 2019. *Database SQL Language Reference*. Accessed June 2019. https://docs.oracle.com/en/database/oracle/oracle-database/19/sqlrf/index.html

ADVANCED TOPICS

This division of the text covers some advanced topics in database systems that you should be familiar with. The objectives of this division are:

- To introduce you to the salient issues related to database administration
- To introduce you to the theory, advantages, and challenges of distributed databases
- To provide an overview of object databases, and point out the challenges faced by the approach
- To provide an overview of data warehousing
- To provide an overview of Web-accessible databases, and the supporting technologies
- To provide additional elaboration on the fact that database systems are often used to anchor sophisticated software systems

The division includes six chapters, each providing an overview of an area of database systems that could be further explored. The chapters to be covered include the following:

- Chapter 16 — Database Administration
- Chapter 17 — Distributed Database Systems
- Chapter 18 — Object Databases
- Chapter 19 — Data Warehousing
- Chapter 20 — Web-Accessible Databases
- Chapter 21 — Using Database Systems to Anchor Management Support Systems

Please note that for each of these topics, several texts have been written. It will, therefore, not be possible to cover them in detail. Rather, in each case, an overview is provided, outlining the salient issues.

DOI: 10.1201/9781003275725-19

Chapter 16

Database Administration

We have established the importance of a database as a valuable information resource in the organization. This resource must be carefully administered (managed) in order to ensure the continued operation and success of the organization. In this regard, the database administrator is extremely important (database administrators have been among the highest-paid IT professionals).

This chapter provides an overview of database administration. It discusses (from an administrative view point) the following issues:

- Database Installation, Creation, and Configuration
- Database Security
- Database Management
- Database Backup and Recovery
- Database Tuning
- Database Removal
- Summary and Concluding Remarks

Please note that a solitary chapter which overviews database administration will not make you a good database administrator (DBA). To achieve that objective, you will need to apply the knowledge in this course, combined with additional knowledge gained from a special course in database administration. This chapter helps to prepare you for that vocation by providing an overview of database administration issues.

16.1 Database Installation, Creation, and Configuration

Before any work can be done, the database software must be installed. This is usually a straightforward, but time-consuming process. For large, sophisticated products such as Oracle and DB2, installation could get complicated, since decisions have to be made about what components to install, where (in terms of directories or folders) to store certain resources, and what environmental or configuration settings to choose. For simpler products like MySQL and Delphi, the installation process is correspondingly simpler.

DOI: 10.1201/9781003275725-20

Next, the database must be created and configured. Depending on the DBMS suite that is in use, database creation and configuration may be very simple or quite complicated. In some systems (for example, Microsoft Access), database creation is a simple act of creating a directory or folder (depending on the operating system), and then creating database aliases that point to that directory or folder. In others such as MySQL, there is a highly simplified **Create-Database** statement that allows you to create a database within seconds. On the other end of the spectrum, there are systems such as Oracle, for which you must first complete a careful study of the Oracle architecture before you can fully know how to properly create a database (review Sections 11.1–11.3 of Chapter 11). And there are other products in between the two extremes.

Once you have created the database, there are issues that must be determined as part of the database configuration. These issues include the following:

- Location of the database and its related files
- Communication issues for server and client (for client-server databases) in a multi-user environment
- Physical structure of the database
- Logical structure of the database
- Users of the database and their access rights

Like database creation, the complexity of these issues, and the flexibility with which they can be addressed, will depend to a large extent on the products involved, as well as the complexity of the database itself. To obtain a cursory appreciation of this, you are encouraged to spend some time observing the Oracle directory structure on the machine that runs the Oracle DBMS suite. In a corporate environment, you will not have this privilege unless you have been duly authorized. However, if you have the alternative of installing the Oracle DBMS suite on a machine that you have administrative access to, you would then be able to observe the aforementioned directory structure.

16.2 Database Security

Database security is a very important aspect of database administration. Ideally, the security mechanism must be multi-tiered, controlling access to the system, the system resources, and the system data. The DBA must ensure the following:

- Access to the system is controlled.
- Authorized users must be able to access (insert, modify, retrieve, or delete) data that they are authorized to access.
- Authorized users must be restricted to the data and resources that they are duly authorized to access and nothing more.
- Unauthorized users must have absolutely no access.

In order to achieve this, the DBA must be fully conversant with SQL facilities (commands) for managing database users and objects. The information covered in Chapter 13 is particularly relevant here. Typically, DBMS suites are marketed with a GUI, which is superimposed on the basic SQL interface; this facility provides a more user-friendly environment for working, by generating

SQL code (which can be subsequently accessed and modified) from GUI-based instructions. Examples of this in the Oracle suite include Oracle Enterprise Manager (OEM), SQL *Plus (for versions prior to Oracle 12C), and Oracle SQL Developer (OSQLD).

To further reinforce the security mechanism of the database, more sophisticated products provide the facility to encrypt the data stored in database files. In other products, the conventional wisdom is to rely on the encryption feature provided by the underlying operating system.

16.3 Database Management

As mentioned in Chapter 11, once the database has been created, it must be populated with database objects. Database objects include tablespaces, tables, indexes, views, synonyms, procedures, triggers, packages, sequences, users, roles, etc. Most of these database objects are dynamic and will grow or shrink in size, with the passing of time. The database and its objects must therefore be managed.

Database management is a complex matter that covers a wide range of activities, including the following:

- Reorganizing existing database tables and indexes
- Deleting unnecessary indexes or moving other objects
- Making alterations to the database itself
- Making alterations to database components (tablespaces, datafiles, tables, procedures, etc.)
- Creating additional database objects (tablespaces, datafiles, tables, users, indexes, procedures, etc.)
- Training users
- Backup and recovery of database objects
- Database performance tuning

Most of these issues have been discussed in previous chapters. The last two issues — backup and recovery, and database performance tuning — deserve some attention. They will be addressed in the next two sections.

16.4 Database Backup and Recovery

Recall from Section 2.3.1 that among the principal functions of the DBA is the matter of backup and recovery. Two humorous but significant laws emphasized in some DBA training programs are outlined below:

- Law 1: Prepare, prepare, prepare!
- Law 2: Backup, backup, backup!

In general, the term *database backup and recovery* is typically used to refer to the various strategies and procedures involved in protecting a database against data loss, and reconstructing the data should that loss occur. The reconstructing of data is achieved through media recovery, which refers to the various operations involved in restoring, rolling forward, and rolling back a backup of database files. Like database installation, creation, and configuration, backup and recovery can be quite simple or very complex, depending on the database environment, and the desired objectives. For the remainder of this section, we shall consider, as a case study, backup and recovery in the Oracle environment (it does not get any more complicated than this).

16.4.1 Oracle Backups: Basic Concept

A backup is a copy of data. This backup may include important parts of the database such as the control file, datafile(s), or tablespace(s); alternately, it may involve the entire database. A backup is a safeguard against unexpected data loss and application errors. If you lose the original data, then you can reconstruct it by using a backup.

Backups are divided into physical backups and logical backups. Physical backups, which are the primary concern in a backup and recovery strategy, are copies of physical database files. You can make physical backups with either the Oracle Recovery Manager (RMAN) utility or operating system utilities. In contrast, logical backups contain logical data (for example, tables and stored procedures) extracted with the Oracle Export utility and stored in a binary file. You can use logical backups to supplement physical backups.

16.4.2 Oracle Recovery: Basic Concept

Recovery is the opposite of backup. A database recovery is effected from a database backup, so if the backup was not done, the recovery is not an option. Like backup, recovery may involve a component or section of the database (from a control file, datafile(s), or tablespace(s)), or it may involve an entire database.

To restore a physical backup of a datafile or control file is to reconstruct it and make it available to the Oracle database server. The restore only puts the datafile back to the same point in time of when the backup was taken. However, since changes might have been made since the backup, it now requires a recovery process. To recover a restored datafile is to update it by applying archived redo logs, and online redo logs, that is, records of changes made to the database after the backup was taken. If you use RMAN, then you can also recover restored datafiles with incremental backups, which are backups of a datafile that contain only blocks that changed after a previous incremental backup.

After the necessary files are restored, media recovery must be initiated by the user. Media recovery can use both archived redo logs and online redo logs to recover the datafiles. If you use SQL*Plus, then you can run the RECOVER command to perform recovery. If you use RMAN, then you run the RMAN RECOVER command to perform recovery.

16.4.3 Types of Failures

Several circumstances can halt the operation of an Oracle database. The most common types of failure are described in Table 16.1. Oracle provides for complete recovery from all possible types of hardware failures, including disk crashes. Options are provided so that a database can be completely recovered or partially recovered to a specific point in time.

If some datafiles are damaged during a disk failure, but most of the database is intact and operational, the database can remain open while the required tablespaces are individually recovered. Therefore, undamaged portions of a database are available for normal use while damaged portions are being recovered. This is a very desirable feature, especially for large corporate databases that must be up and running "twenty-four by seven."

Table 16.1 Types of Database Failures

Failure Type	*Comment*
User Error	User errors can require a database to be recovered to a point in time before the error occurred. For example, a user may accidentally drop a table. To enable recovery from user errors and accommodate other unique recovery requirements, Oracle provides exact point-in-time recovery. For example, if a user accidentally drops a table, the database can be recovered to the instant in time before the table was dropped.
Statement Failure	**Statement failure** occurs when there is a logical failure in the handling of a statement in an Oracle program (for example, when an update statement attempts to violate a constraint in a table). When statement failure occurs, the effects (if any) of the statement are automatically undone by Oracle (a rollback) and control is returned to the user.
Process Failure	A **process failure** is a failure in a user process accessing Oracle, such as an abnormal disconnection or process termination. The failed user process cannot continue work, although Oracle and other user processes can. The Oracle background process PMON automatically detects the failed user process or is informed of it by SQL*Net. PMON (short for Process Monitor) resolves the problem by rolling back the uncommitted transaction of the user process and releasing any resources that the process was using.
	Common problems such as erroneous SQL statement constructions and aborted user processes should never halt the database system as a whole. Furthermore, Oracle automatically performs necessary recovery from uncommitted transaction changes and locked resources with minimal impact on the system or other users.
Instance Failure	**Instance failure** occurs when a problem arises that prevents an instance from continuing work. Instance failure can result from a hardware problem such as a power outage, or a software problem such as an operating system crash. When an instance failure occurs, the data in the buffers of the system global area is not written to the datafiles.
	Instance failure requires **crash recovery** or **instance recovery**. Crash recovery is automatically performed by Oracle when the instance restarts. In an Oracle Real Application Clusters environment, the SMON process of another instance performs instance recovery. The redo log is used to recover the committed data in the SGA's database buffers that was lost due to the instance failure.
Media (disk) Failure	An error can occur when trying to write or read a file that is required to operate the database. This is called disk failure because there is a physical problem reading or writing physical files on disk. A common example is a disk head crash, which causes the loss of all files on a disk drive.
	Different files can be affected by this type of disk failure, including the datafiles, the redo log-files, and the control files. Also, because the database instance cannot continue to function properly, the data in the database buffers of the system global area cannot be permanently written to the datafiles.

(*Continued*)

Table 16.1 (Continued) Types of Database Failures

Failure Type	*Comment*
	A disk failure requires **media recovery**. Media recovery restores a database's datafiles so the information in them corresponds to the most recent point in time before the disk failure, including the committed data in memory that was lost because of the failure. To complete a recovery from a disk failure, the following are required: backups of the database's datafiles and all online and necessary archived redo log-files.

16.4.4 Database Backups

It is the case that files stored on a typical storage medium can susceptible to physical damage as the result of a disk failure. Media recovery requires the restoration of the damaged files from the most recent backup of a database. There are several ways to back up the files of a database.

Whole Database Backups: A whole database backup is an operating system backup of all datafiles, online redo log-files, and the control file of an Oracle database. A whole database backup is performed when the database is closed and unavailable for use.

Partial Backups: A partial backup is an operating system backup of part of a database. The backup of an individual tablespace's datafiles and the backup of a control file are examples of partial backups. Partial backups are useful only when the database's redo log is operated in ARCHIVELOG mode.

A variety of partial backups can be taken to accommodate any backup strategy. For example, you can back up datafiles and control files when the database is open or closed, or when a specific tablespace is online or offline. When the database is operated in ARCHIVELOG mode, additional backups of the redo log are not necessary. The archived redo log is a backup of filled online redo log-files. In turn, a redo log is a set of two or more pre-allocated files that store all changes made to the database as they occur. As a means of protection against inadvertent instance failure, every Oracle database instance has one or more associated redo log(s).

16.4.5 Basic Recovery Steps

Because of the way the Oracle Database Writer (DBW*n)* writes database buffers to datafiles, at any given time, a datafile might contain some tentative modifications by uncommitted transactions and might not contain some modifications by committed transactions. Therefore, two potential situations can result after a failure:

- Data blocks containing committed modifications were not written to the datafiles, so the changes appear only in the redo log. Therefore, the redo log contains committed data that must be applied to the datafiles.
- Because the redo log can contain data that was not committed, uncommitted transaction changes applied by the redo log during recovery must be erased from the datafiles.

To solve this situation, two separate steps are used by Oracle during recovery from an instance or media failure: rolling forward and rolling back.

Rolling Forward: The first step of recovery is to **roll forward**, which is to reapply to the datafiles all of the changes recorded in the redo log. Rolling forward proceeds through as many redo log-files as necessary to bring the datafiles forward to the required time.

If all necessary redo information is online, Oracle rolls forward automatically when the database starts. After roll forward, the datafiles contain all committed changes as well as any uncommitted changes that were recorded in the redo log.

Rolling Back: The roll forward is only half of recovery. After the roll forward, any changes that were not committed must be undone. After the redo log-files have been applied, then the undo records are used to identify and undo transactions that were never committed yet were recorded in the redo log. This process is called **rolling back**. Oracle completes this step automatically.

16.4.6 Oracle's Backup and Recovery Solutions

There are two methods for performing Oracle backup and recovery: Recovery Manager (RMAN) and user-managed backup and recovery. RMAN is a utility automatically installed with the database that can back up any Oracle8 or later database. RMAN uses server sessions on the database to perform the work of backup and recovery. RMAN has its own syntax and is accessible either through a command-line interface or through the Oracle Enterprise Manager GUI. RMAN comes with an API that allows it to function with a third-party media manager.

One of the principal advantages of RMAN is that it obtains and stores metadata about its operations in the control file of the production database. You can also set up an independent recovery catalog, which is a schema that contains metadata imported from the control file, in a separate recovery catalog database. RMAN performs the necessary record keeping of backups, archived logs, and so forth using the metadata, so restore and recovery is greatly simplified.

An alternative method of performing recovery is to use operating system commands for backups and SQL*Plus for recovery. This method, also called user-managed backup and recovery, is fully supported by Oracle Corporation, although use of RMAN is highly recommended because it is more robust and greatly simplifies administration.

Whether RMAN is used, or user-managed methods, physical backups can be supplemented with logical backups of schema objects made using the Oracle Export utility. The utility writes data from an Oracle database to binary operating system files. You can later use Oracle Import to restore this data into a database.

16.5 Database Tuning

To maintain acceptable database performance, it is necessary to carry out periodic performance tuning. As the database is being used, it will exhibit a natural tendency to degraded performance. The current database system may not be performing acceptably, based on user-defined criteria, due to any of the following:

- Poor database design
- Database growth
- Changing application requirements (possibly including a redefinition of what acceptable performance is)

Note, however, that there may be occasions when database-tuning efforts are not fully effective. When components that are external to the database, yet vital to the entire client-server application performance, fail to perform acceptably, database tuning may not help without the corresponding tuning of these other application infrastructure pieces. The main components external to the backend database are the backend operating system, the network, and the client operating system. Here are three examples of database performance problems that source beyond the immediate database: very weak network clients; network saturation; and saturated or poorly tuned operating systems.

16.5.1 Tuning Goals

There are different ways of determining the goals of a performance tuning effort. A DBA should consider them all. Database systems can be sampled on various quantitative measures; the most important of these are the following:

- **Throughput:** This is the accomplished work per unit of time, as measured by transactions per second (tps); higher is better.
- **Response Time:** This is the time it takes for an application to respond, as measured in milliseconds or seconds, lower is better.
- **Wait Time:** This is the elapsed time a program takes to run; lower is better.

In any system, throughput and response time usually run counter to one another as tuning goals. A high response time is generally not desirable as this will affect the rate at which data can be retrieved; correspondingly, the throughput would be low. Alternately, a high throughput suggests the presence of a low response time, which is desirable.

Common sense helps when sorting out these two conflicting measures. The more users that are concurrently using a system within a certain amount of time, the more likely it is that each user will experience longer delays than normal, but the number of transactions going through the system will be greater. On the other hand, if you decrease the number of concurrent users accessing the system within a certain time window, each user will enjoy faster response time at the expense of fewer overall transactions completing in that duration.

Typically, online transaction processing (OLTP) systems (also called operational databases) want low response time and high throughput, in terms of transactions per second, depending on the application needs. A decision support system (DSS) also wants a low response time. However, a DSS might also want high throughput in terms of data blocks read or written per unit time. This type of throughput is not necessarily counterproductive to high concurrency and low response times. A batch (production) system typically wants lower wait times. For example, everyone likes for the payroll application to complete on time!

Database tuning is essential for the preservation of acceptable database performance. It is important to always consider the following two central tuning goals:

- Maximize your return on investment (ROI). Invest your time and effort wisely by working on the problems most likely to yield the optimum improvement.
- Minimize contention. Bottlenecks are characterized by delays and waits; eliminate or reduce these whenever possible.

Finally, in addition to the central tuning goals, the following general-purpose database tuning goals should be considered:

- Minimize the number of data blocks that need to be accessed; review and rewrite database access code as necessary.

- Use caching, buffering, and queuing whenever possible to compensate for the electromechanical disadvantage (memory is faster than disk).
- Minimize the data transfer rates (the time it takes to read or write data); fast disks, RAID, and parallel operations help do this.
- Schedule concurrent programs that complement instead of compete with each other.

16.5.2 Tuning Methodology

It is best to approach tuning with a structured methodology. After ensuring that the operating system is performing at its peak and that sufficient operating system resources have been allocated to your database system, you should tune following in this order:

- Database application
- Memory management
- I/O management
- Database contention
- Database design

Database Application Tuning: In database application tuning, the concern is on end-user access to the database. During this exercise, the DBA determines ways to better facilitate access to the database by the various applications that need to use it. Among the issues that may be addressed are the following:

- Ascertaining whether existing database application objects (procedures, triggers, etc.) are performing according to expectations.
- Determining whether additional database application objects (procedures, triggers, etc.) are required and where to place them.
- Determining whether adequate database access points (including ODBC connections, database service connections, etc.) are in place and are working acceptably.

Memory and I/O Management Tuning: Memory management tuning is closely related to database design tuning. This is critical because poor database design could lead to poor memory performance which in turn leads to poor database performance. Among the issues that may be addressed are the following:

- Storage allocations for database objects (primarily tablespaces, datafiles, and tables).
- Storage allocations for the database itself (these parameters are set at database creation or database alteration).

Oracle provides a number of utilities for managing memory performance of database tables. However, a full discussion of these is beyond the scope of this course. Suffice it to say that the database fault rate on each table can be monitored. If the fault rate is high, the table needs to be reorganized. A database fault occurs when a data block being requested is not in cache or memory, thus necessitating a fetch from the storage medium where the persistent data for the database are stored. You will no doubt learn more about memory faults in your course on *Operating Systems Design.*

Database Contention: Database contention relates to how the database is handling multi-user access as well as concurrent access. Like memory management, database contention is a complex matter that covers a range of related issues such as memory management, server configuration, and management of certain initialization parameters; a full treatment is beyond the scope of this course, but is normally included in a DBA training course.

Database Design Tuning: Of the stated database-tuning activities, database design tuning is perhaps the most challenging; yet, the database design has the most potent effect on performance of the database. In database design tuning, you are concerned with the physical and logical structure of the database. During this exercise, the DBA examines and takes decisions about refining the logical and physical structure of the database. Among the issues that may be addressed are the following:

- Determining whether critical database components (tablespaces, datafiles, log-files, etc.) need to be redefined and/or relocated (to different directories/folders).
- Determining whether critical database objects (primarily tables and indexes) need to be partitioned (i.e., fragmented into different partitions).
- Determining whether critical database tables need to be restructured.
- Determining whether additional database objects (tables, logical views, etc.) are required, and if so, where they should be placed.

Database design tuning is challenging because the activities have a far-reaching impact on the database, and the existing data must be protected. For instance, if a database table has to be redesigned so as to affect its primary key and/or candidate keys, then such a change would typically involve code conversion from the old format to the new format specified by the redesign. Code conversion of this sort is usually a huge undertaking; this is so because organizations with mission-critical databases are inclined to expect close to 100% conversion of their mission-critical data (see Foster 2021).

16.6 Database Removal

Sometimes, it becomes necessary to remove a database. Often, there is no specific command for this, for obvious reasons. The database exists under the auspices of the host operating system. Database removal is therefore an operating system command.

Depending on the DBMS being used, database removal may be a trivial matter, or one requiring a few steps. For instance, removal of a Delphi or MySQL database involves a single step. On the other hand, removing an Oracle database typically involves several steps of deleting-related folders/directories managed by the DBMS in collaboration with the underlying operating system. These folders/directories were created when the database was created (or altered). Earlier versions of Oracle included a de-install application but the process was still not as seamless as it could have been. With the more contemporary emphasis on cloud-based databases (since Oracle 12C and significantly enhanced in Oracle 19C), database removal has been simplified in the Oracle database environment.

Once a database has been deleted, it is completely gone, and can only be reintroduced via a recovery operation. You should only attempt database removal if the removal target is a test database that is no longer needed, and you are duly authorized to carry out the activity.

16.7 Summary and Concluding Remarks

This chapter covers some database administration issues which would need further exploration if you intend to function as a DBA. Here is a summary of what has been discussed:

- Depending on the DBMS being used, database creation may be complex or simple. Microsoft Access and Oracle are at the two extreme ends of the spectrum — database creation is very simple in Access and very complex in Oracle.
- Database security must ideally be multi-tiered. It must address access to the system, access to the system resources, and access to data.
- Database management must continue after database creation. It must address issues relating to the performance of the database system in the face of growing data collection and changing user needs. Database tuning is an integral part of this.
- Backup must be carefully planned and methodically implemented, in order to minimize or eliminate data loss due to system failures. The recovery procedures must also be reviewed as required.
- Like database creation, depending on the DBMS used, database removal may be trivial or complex.

Database administration is a fascinating field that you are encouraged to explore further. To do so, you would typically pursue a professional course in the area. These courses are usually offered by authorized vendors of the major DBMS suites (Oracle, DB2, MySQL, MS SQL Server, Sybase, etc.). Even if you are not a practicing DBA, or have no intention of pursuing that vocation, you will find the education enlightening.

The upcoming chapter focuses on distributed databases. Your knowledge of database systems is not complete without a solid appreciation of this topic.

16.8 Review Questions

1. What are the main issues to be considered when creating a database?
2. What are the critical issues to be addressed when configuring the security mechanism of a database?
3. What are the main issues to be addressed during the management of a database?
4. Why are backups important? Discuss how backup and recovery are handled in Oracle.
5. Why is performance tuning of a database important? Identify the basic tuning guidelines to be observed.

References and/or Recommended Readings

Coronel, Carlos, & Steven Morris. 2015. *Database Systems: Design, Implementation & Management* 11th ed. Boston, MA: Cengage Learning. See Chapters 11 & 15.

Elmasri, Ramez, & Shamkant B. Navathe. 2011. *Fundamentals of Database Systems* 6th ed. Boston, MA: Pearson. See Chapter 20.

Foster, Elvis C. 2021. *Software Engineering: A Methodical Approach* 2nd ed. New York: CRC Publishing. See Chapters 21 & 22.

Hoffer, Jeffrey A., Ramesh Venkataraman, & Heikki Topi. 2013. *Modern Database Management* 11th ed. Boston, MA: Pearson. See Chapter 11.

Kroenke, David M, & David Auer. 2014. *Database Processing: Fundamentals, Design, and Implementation* 13th ed. Boston, MA: Pearson. See Chapter 8.

Mullins, Craig S. *Database Administration: The Complete Guide to Practices and Procedures* 2nd ed. Reading, MA: Addison-Wesley.

Oracle Corporation. 2019. *Database SQL Language Reference*. Accessed June 2019. https://docs.oracle.com/en/database/oracle/oracle-database/19/sqlrf/index.html

Oracle Corporation. n.d. *Oracle Technology Network*. Accessed June 2015. http://Technet.Oracle.com

Chapter 17

Distributed Database Systems

As wonderful as database systems are, they would not be delivering on their true potential, if they could not be networked in distributed environments. This chapter discusses distributed database systems under the following subheadings:

- Preliminaries
- Advantages of Distributed Database Systems
- Twelve Rules for Distributed Database Systems
- Challenges to Distributed Database Systems
- Database Gateways
- The Future of Distributed Database Systems
- Summary and Concluding Remarks

17.1 Preliminaries

The concept of a distributed system was introduced in Chapter 2. A distributed database system consists of a collection of sites, connected via a communication network in which:

- Each site is a database system in its own right.
- The sites work together (if necessary) so that a user at any given site can access data at any other site as if the data resides on the host (user's) site.

Section 2.7 of Chapter 2 mentioned some connectivity possibilities. Here are a few noteworthy points to remember:

1. The user is given the notion of virtual database systems consisting of data that may reside anywhere in the network.
2. The sites may be distributed over a wide geographical area, or in a local area/building. A distributed database system can therefore be hosted via a LAN (local area network), a MAN (metropolitan area network), or a WAN (wide area network).

DOI: 10.1201/9781003275725-21

3. Distributed database systems are not to be confused with remote access systems, sometimes called distributive processing systems. The latter has been around for some time. In such systems, the user accesses data at remote sites but the operation is not seamless; the user is aware and the consequences may be obvious. In a distributed database system, access across sites is seamless.

The literature [on electronic communications and computer networks] documents several alternate approaches to setting up a distributed database. Four prevalent ones are summarized below:

- Using a LAN as the backbone, set up a federated database system, where the DBMS takes on the responsibility of integrating the autonomous databases in a seamless and transparent manner.
- Using a LAN as the backbone, configure a set of clustered database servers where each server has access to the disks of the other servers, but manages its own memory.
- Using a LAN as the backbone and the internet as the public network, set up a virtual private network (VPN) that uses some tunneling protocol to facilitate distributed access in a relatively secure virtual environment.
- Using cloud technology as the backbone, configure a cloud-based system that provides software as a service (SaaS), data as a service (DaaS), platform as a service (PaaS), and infrastructure as a service (IaaS) to users.

Each option mentioned here is quite involved and beyond the scope of this course. Luckily, there is a prolific reservoir of available resource information to draw from. References Oracle (2002, 2015, 2016b), Stallings (2014), and Tanenbaum & van Steen (2007) should provide you with a useful start. Suffice it to say that Oracle as described earlier (Chapters 10–15 and 22) qualifies as a distributed DBMS. When you install Oracle Server on a node in a network, that node acts as a database server. If you then install Oracle Client on other nodes in the network, your database server can be accessed from anywhere in the network (as well as from other network systems with Web accessibility) in a seamless manner.

17.2 Advantages of Distributed Database Systems

Table 17.1 provides some benefits that distributed database systems provide. The benefits may be summarized in three categories:

- Efficiency and Productivity
- Convenience
- Reliability

When weighed against the challenges posed by distributed databases (discussed later), they outweigh them by a huge margin; we therefore expect the continued proliferation of distributed databases.

Table 17.1 Benefits of Distributed Database Systems

Efficiency and Productivity:	
01.	Improved response time and throughput since the database can be tailored to the needs of the organization(s) served.
02.	Data (from foreign sites) can be replicated (locally) to improve access time.
03.	Easy and fast communication over relatively long distances.
04.	Distributed database systems can ensure that the best resources are brought together and utilized in such systems.
Convenience:	
05.	The systems used (back-ends and front-ends, and at different sites) may be different according to the needs and circumstances of the (users of) respective sites. For instance, at backend, one site may host Oracle, another DB2, another MySQL, and so on. The front-end tools could also vary.
06.	The structure of the database can mirror the structure of the enterprise — local data is stored locally; access is available to other data over the network.
07.	Instead of using one large, expensive database server, it is possible to use smaller, less expensive servers and still achieve what a single powerful server would provide. Due to information sharing, the integrated system could even provide for more data storage, (possibly) increased functionality and features than a single database system would.
08.	The system is more easily designed to facilitate multiple users (this could lead to improved productivity if data are stored closest to respective points of need).
Reliability:	
09.	Reduction of dependence on a central system (also, this might not be practical).
10.	Improved reliability: if one machine fails, the whole system does not fail (there is no reliance on a central system).

17.3 Twelve Rules for Distributed Database Systems

In his classic text, *Introduction to Database Systems*, Christopher Date discusses 12 rules (objectives) for distributed database systems (Date 2004). Let us take a brief look at these rules.

Rule 1 — Local Autonomy: The sites should be autonomous to the maximum possible extent. All operations at a site are governed by that site alone. This is not always entirely possible but is an objective to strive for.

Rule 2 — No Reliance on Central Site: This is a consequence of objective 1. The sites must be treated as equals. There is no reliance on a central site. Reliance on a central site would make the system vulnerable to the central site (bottleneck could occur or the central site could go down). This is undesirable.

Rule 3 — Continuous Operation: The system must be able to run continuously. There should be no need for a planned shut down in order to carry out any function (for instance, backup or tuning).

Rule 4 — Location Independence (Transparency): Users should not need to know where data is physically stored in order to access it; the system should operate as if all the data resided at the local site. The distributed nature of the database must be transparent to the end-users.

Rule 5 — Fragmentation Independence: The system should support data fragmentation — it should be possible to partition a given relation into fragments that are stored at different sites. Thus, data can be stored where it is most frequently used. Network trafficking is therefore reduced. Fragmentation should be transparent to the end uses.

EXAMPLE 17.1: ILLUSTRATING FRAGMENTATION

Objective: Suppose that a large organization has employee records in an Employee relation. The departments are at different localities. Records for each department are stored at different sites in those respective departments. Could fragmentation help?

Action: This scenario can be easily facilitated in Oracle by *partitioning* the **Employee** table, so that different partitions are stored at the locations where they are most relevant. A full discussion of table partitioning is beyond the scope of this course. However, suffice it to say that that this is the technique used by several leading DBMS suites to facilitate fragmentation independence.

Rule 6 — Replication Independence: A given relation (or fragment of a relation) can be replicated at different sites. Replication may improve access time and hence performance. The drawback however, is that when an update is made, all the copies also have to be updated. Replication should be transparent to the end-users.

Rule 7 — Distributed Query Processing: Distributed query processing must be facilitated among different sites. Data is transmitted set (relation) at a time instead of row at a time.

EXAMPLE 17.2: ILLUSTRATING DISTRIBUTED QUERY

Objective: Suppose we have an international company, IBM, say, where employees are stored in the relation Employee, fragmented in various countries, where there are IBM offices. IBM Canada, issues the request: "Find all Jamaican male employees." How would this be facilitated?

Action: A request of this form would require an optimized query operation by the DBMS. The DBMS would collaborate with the underlying operating system and network protocol to ensure that the request is processed efficiently.

a. Suppose that there are *n* records that satisfy this request. If the system is relational, the query will involve two messages — one from Canada to Jamaica and one from Jamaica to Canada. If the system is not relational, but record-at-a-time, the query will involve *2n* messages — *n* from Canada to Jamaica and *n* from Jamaica to Canada.
b. Query optimization of the request occurs before execution (record-at-a-time requests cannot be optimized).

Observe:

The above example should underscore why traditionally speaking, distributive database systems have been predominantly relational. Were they not relational, distributed query processing would not be as feasible and commonplace as it is today.

Rule 8 — Distributed Transaction Management: Each transaction must be atomic — fully committed or fully rolled back. This objective must be met irrespective of the agents (constituent processes) of the transaction. Concurrency control must be ensured (usually by data locking).

Rule 9 — Hardware independence: It should be possible to integrate the system across different hardware platforms. It should therefore be possible to run the DBMS on different hardware systems.

Rule 10 — Operating System Independence: It should be possible to integrate the system across different operating system platforms. It should therefore be possible to run the DBMS on different operating systems.

Rule 11 — Network Independence: The system should be able to support different sites with different hardware and different operating systems and networking protocols.

Rule 12 — DBMS Independence: The DBMS suites used may be different. For instance, DB2 and Oracle both support SQL and open database connectivity (ODBC); it should therefore be possible to link databases running on the two DBMS suites. The same argument should apply for other DBMS suites.

17.4 Challenges to Distributed Database Systems

Distributed databases did not come easy. They are difficult to maintain. Fortunately, the software engineering industry has figured out how to address these challenges. However, improving the algorithms used, and finding new ones are always topical research issues. There are five well-known challenges to distributed database systems. These are:

- Query Optimization
- Catalog Management
- Update Propagation
- Concurrency Control
- Transaction Management

17.4.1 Query Optimization

Query optimizing processing must be distributed in order to minimize network trafficking. To illustrate, consider a query Qa of site A, accessing to relations in a natural join: Rb of site B & Rc of site C. The optimizer must decide on one of the following strategies:

a. Move copies of Rb and Rc to site A
b. Move copy of Rb to site C and process the join there
c. Move copy of Rc to site B and process the join there

The optimizer must be able to calculate what would be the most economical alternative (given the structure and configuration of database as well as the underlying network) and choose that alternative. For example, Oracle implements a *cost-based optimization*. Before executing a query, the query optimizer optimizes the query by converting it to an internal format called an *execution plan* (based on an Oracle algorithm) that will ensure the most efficient execution. For more insight on query optimization in Oracle, please refer to the product documentation (Oracle 2019).

17.4.2 Catalog Management

Catalog management is one of the most complex issues that a distributed database must resolve. This is so, since additional information must be stored for the database objects (e.g., fragmentation, replication, location, etc.). Where and how the catalog should be stored is a complicated issue. Below are some alternatives:

- **Centralized:** The catalog is stored at a centralized location, and is accessible to the other participating sites.
- **Fully Replicated:** The catalog is replicated at each participating site.
- **Partitioned:** Each site maintains its own catalog. The total catalog is the union of each site catalog.
- **Hybrid:** Each site maintains its own catalog; additionally, a central site maintains the global catalog.
- **Cloud-based:** Pursue a cloud-based solution; store the catalog in the cloud.

Each of these approaches has its related advantages and challenges. For instance, one may argue that the centralized option violates Rule 2 for distributed systems. Bear in mind that these rules are meant to be guidelines in pursuit of something bigger and better. The idea is to balance all the related factors in the best interest of the distributed database system. Resolving this balancing challenge is often done with the use of simulation software, and much research into the matter. However, with the emergence and advance of cloud-based solutions, one can expect cloud-based centralized catalogs to become more attractive in the future.

17.4.3 Update Propagation

In the case where data is replicated at different sites, it may not be possible to effectively update all replicas at the desired

time. How is this resolved? The *primary copy* approach is a common method of resolution:

- One replica is deemed the primary copy. As soon as that copy is updated, the update process is deemed completed.
- The site with the primary copy is responsible for updating the other sites as soon as possible.

This somewhat contradicts the objectives of transaction independence and redundancy control. As was emphasized in Chapter 4, once data replication is introduced in a database, with it come various other data integrity problems. Resolution is therefore a matter of tradeoff.

Another strategy that is widely used is the *two-phase commit protocol* (2PC); this is an atomic commitment protocol (ACP) that manages each database transaction by determining whether to commit or rollback the transaction. Here is a summary of each phase of the 2PC:

- The *commit-request phase* (also called the *voting phase* or the *prepare phase*) prepares the transaction for a commit or rollback. As part of this process, the initiating node makes a request to other participating nodes to all agree to either commit or rollback the transaction in question.
- In the *commit-phase*, the actual "voted" action is implemented. The initiating node makes another request to other participating nodes to implement the agreed-upon action. If all participating nodes cannot commit, then the transaction is rolled back.

17.4.4 Concurrency

Concurrency is another issue that must be resolved. To illustrate, consider what might happen if user X tries to retrieve a particular data set for update purposes, but that data set is being updated by another user Y. Consider that in a distributed system, there might be thousands of users, so that this kind of contention could easily develop among several users. Typically, the DBMS handles this problem by *record locking*: a record or data set that is retrieved for update is *locked* to that transaction until the update is completed; it is then released (unlocked) for other users. Requests to release (*unlock*) objects in a distributed database system must be managed. This is a serious but necessary overhead.

In application development where distributed databases are accessed, or where there is multi-user access of a single database, it is good practice to always check for exceptions whenever issuing update(s) or deletion(s) via embedded SQL statements. By doing so, the application program can make a graceful recovery from a record lock situation (normally done by issuing an appropriate message to the end-user and allowing them to defer that particular request until some subsequent time).

17.4.5 Transaction Management

Issues such as when to lock records, and when to commit or rollback, transactions are critical in a distributed database. The application developer must be familiar with the facilities provided by the DBMS and SQL (COMMIT and ROLLBACK) for managing transactions. As explained in the two previous subsections as well as Section 12.4, transaction management is critical matter in any DBMS.

17.5 Database Gateways

Traditionally, a database gateway is a software component that links two different DBMS suites. It could run on either of the two systems running the dissimilar DBMS suites, or on a separate machine for that purpose. The simplest configuration is to run the software on either system as a driver for the other DBMS. Another alternative is to use the software called ODBC. ODBC is marketed with the Windows operating system, and is readily available for other operating systems.

Suppose for instance, that Oracle and DB2 both support SQL (as in fact, they do). It should therefore be possible to link the two DBMS suites, as illustrated in Figure 17.1. In reality, each product includes an ODBC driver that in effect provides some of the functions of the gateway. Following are some desirable functions of the gateway:

- Mapping between the two different protocols (formats)
- Mapping between the two different dialects of SQL (e.g., Oracle's dialect versus DB2's dialect)
- Mapping between the two different system catalogs so that users of one DBMS (Oracle) can access all files from the other DBMS (DB2)
- Mapping between the two different sets of data used
- Resolution of transaction management issues such as data locking, updates, commit, and rollback
- Resolution of semantic mismatches across the two systems (e.g., attribute **Employee.Empno** in a DB2 table may map to **Empl.Emp#** in an Oracle table)
- Resolution of security and accessibility issues across the two different systems

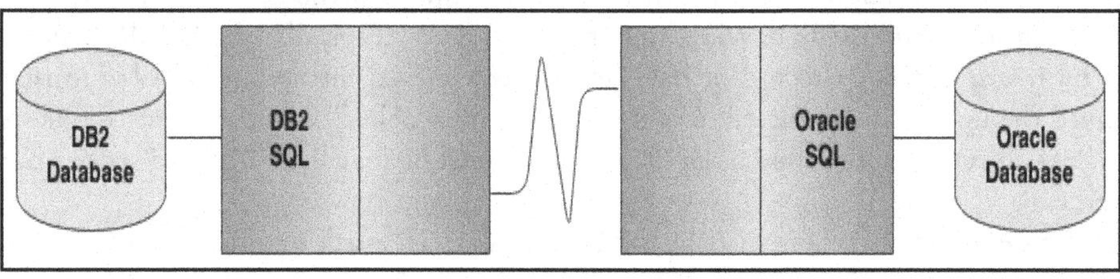

Figure 17.1 ***Illustrating the function of a database gateway.***

If a special database server is used to facilitate the gateway functions, this server could also store metadata to facilitate resolution of issues such as semantic mismatches and security-related issues. By way of example, Oracle operates a very sophisticated gateway system that allows connectivity to other non-Oracle databases either directly through its gateway, or indirectly through ODBC. Among the many features of the Oracle gateway system are the following:

- Location independence and transparency
- Data-type translations between Oracle and non-Oracle systems
- Data dictionary between Oracle and non-Oracle systems
- Read/write access
- Support for large objects (LOBs)
- Support for non-Oracle stored procedures
- Transmission of pass-through SQL between Oracle and non-Oracle system
- Data encryption services

In more recent years, this initiative has grown into the Oracle grid computing which is managed through Oracle Enterprise Manager (OEM) infrastructure (see Oracle 2016a, c). Additionally, as stated earlier, cloud-based database services are on the advance and as expected, Oracle is an integral part of this advancement.

17.6 The Future of Distributed Database Systems

Since the mid-1990s, a number of innovative technologies have significantly influenced the development and direction of distributed databases. Among these technologies are the following:

- Object Technology (OT)
- Computer-based Communication Systems
- Cloud Technology
- Big Data

17.6.1 Object Technology

In a few short years, OT has come to dominate the contemporary software engineering industry. New products are forced by industry demands to support OT in some form. Much work has been done by the Object Management Group (OMG) in establishing the Common Object Request Broker Architecture (CORBA) standards for distributed database systems. The Microsoft equivalents of CORBA are Component Object Model (COM), Distributed COM (DCOM), and more recently, the .NET framework.

CORBA standards span a wide range of specifications from user interface to object–object communication. They are supported by some of the leading software engineering firms in the industry. With the emergence of Java, the software industry has made significant progress in the area of platform independent software components than ever before. As mentioned in Chapter 6, Java (through JDBC) supports both CORBA and ODBC.

17.6.2 Electronic Communication Systems

Complimentary to the advances in OT, the past decade has seen much achievement in the arena of *electronic communication systems* (ECS). Contemporary operating systems are more sophisticated, supporting a wider range of communication protocols. The protocols and their underlying technology have been refined to provide much higher transmission rates. Also, communication protocols provide much more services than previously.

With a refinement of and emphasis on standards, interoperability is now a much more attainable goal than in the previous decade. These advances, when combined with those in OT, will contribute to the proliferation of heterogeneous information models.

17.6.3 Cloud Technology

Cloud technology has emerged as a driving force in contemporary software engineering. The idea is to amass and/or harness a large and complex reservoir of storage and other resources, and to combine these resources with the power of the internet to provide various services for intended end-users. Among the broad categories or services are the following buzz-terms: software as a service (SaaS), data as a service (DaaS), platform as a service (PaaS), and infrastructure as a service (IaaS). These areas also influence and facilitate the provision of database system services, thus providing opportunities for small organizations that would not be able to afford these services by other more traditional means.

17.6.4 Big Data

The term *Big Data* has gradually emerged in database systems, software engineering, and business jargon to represent large, complex datasets. With the ever-expanding need for unstructured and/or and large datasets, it has been found that the relational database model is not always ideal for meeting these demands. The Big Data movement is a response to this challenge. Associated with this movement are methodologies/technologies such as the Hadoop framework, the Apache Spark framework, the NoSQL methodologies, and the Entity-Attributes-Value (EAV) model.

These developments have already made positive impacts on the arena of database systems as we know it, and will do doubt continue to do so for the foreseeable future.

17.7 Summary and Concluding Remarks

Let us summarize what we have covered in this chapter:

- A distributed database system is a conglomeration of database systems in which each system operates as an autonomous system on its own, or collaborates with other systems as required.
- Distributed databases systems provide a number of benefits in the areas of efficiency and productivity, convenience, and reliability.
- Distributed database systems should strive to conform to the standards outlined in Date's 12 rules for such systems.
- Distributed databases face challenges in the areas of query optimization, catalog management, update propagation, concurrency control, and transaction management.
- A database gateway is a software component that links different DBMS suites.
- Distributed databases have been significantly affected by developments in object technology, electronic communications technologies, cloud technology, and big-data methodologies. This trend is expected to continue in the foreseeable future.

Distributed databases have helped to transform our world in a significant way. To fully appreciate the power of distributed databases, just consider for a moment, a world without them: no World Wide Web; minimized reliability on critical company databases; reduced capabilities on operating systems; limited remote access to databases. The next chapter takes a closer look at object databases.

17.8 Review Questions

1. Define a distributed database system. Discuss the advantages of such systems.
2. Outline and clarify the 12 rules for distributed database systems.
3. Discuss the challenges to distributed database system.
4. As an IT professional (perhaps in training), what are your future expectations for distributed database systems?

References and/or Recommended Readings

Chung, P. Emerald, Yennun Huang, Shalini Yajnik, Deron Liang, Joanne C. Shih, Chung-Yih Wang, & Yi-Min Wang. 1997. *DCOM and CORBA Side by Side, Step by Step and Layer by Layer.* Accessed October 2008. http://research.microsoft.com/~ymwang/papers/HTML/DCOMnCORBA/S.html

Connolly, Thomas, & Carolyn Begg. 2015. *Database Systems: A Practical Approach to Design, Implementation and Management* 6th ed. Boston, MA: Pearson. See Chapters 24 and 25.

Coronel, Carlos, & Steven Morris. 2015. *Database Systems: Design, Implementation & Management* 11th ed. Boston, MA: Cengage Learning. See Chapter 12.

Date, Christopher J. 2004. *Introduction to Database Systems* 8th ed. Menlo Park, CA: Addison-Wesley. See Chapter 21.

Elmasri, Ramez, & Shamkant B. Navathe. 2011. *Fundamentals of Database Systems* 6th ed. Boston, MA: Pearson. See Chapter 25.

Hoffer, Jeffrey A., Ramesh Venkataraman, & Heikki Topi. 2013. *Modern Database Management* 11th ed. Boston, MA: Pearson. See Chapter 12.

Oracle Corporation. 2002. "Database Architecture: Federated vs. Clustered.". Accessed May 2016. http://www.oracle.com/technetwork/database/windows/clustercomp-134873.pdf

Oracle Corporation. 2015. "Cloud Computing Comes of Age." *Harvard Business Review*. Accessed May 2016. https://www.oracle.com/webfolder/s/delivery_production/docs/FY15h1/doc16/HBR-Oracle-Report-webview.pdf

Oracle Corporation. 2016a. *Grid Computing*. Accessed May 2016. http://www.oracle.com/technetwork/topics/grid/index.html

Oracle Corporation. 2016b. *Oracle Cloud Computing*. Accessed May 2016. https://www.oracle.com/cloud/index.html

Oracle Corporation. 2016c. *Oracle Database Gateways*. Accessed May 2016. http://www.oracle.com/technetwork/database/gateways/index.html

Oracle Corporation. 2019. *Database SQL Language Reference*. Accessed June 2019. https://docs.oracle.com/en/database/oracle/oracle-database/19/sqlrf/index.html

Silberschatz, Abraham, Henry F. Korth, & S. Sudarshan. 2011. *Database Systems Concepts* 6th ed. Boston, MA: McGraw-Hill. See Chapter 19.

Stallings, William. 2014. *Data and Computer Communications* 10th ed. Boston, MA: Pearson.

Tanenbaum, Andrew S., & Maarten Van Steen. *Distributed Systems: Principles and Paradigms* 2nd ed. Boston, MA: Pearson.

Chapter 18

Object Databases

Since the 1980s, there has been the advancement of the so-called object database management systems (ODBMS) as an alternative to the relational database model for database systems. Indeed, the argument has been made that object databases would be more ideal for scenarios such as multimedia systems and CAD (computer-aided design) systems. This chapter discusses such systems under the following subheadings:

- Overview
- Overview of Object-Oriented Database Management Systems
- Challenges to Object-Oriented Database Management Systems
- Hybrid Approaches
- Summary and Concluding Remarks

18.1 Overview

Object Technology (OT) has been dominating the software engineering industry in recent times. For better or worse, there has been a heightened interest and indulgence in object-oriented methodologies (OOM). Full treatment of OT and OOM is better handled in another course on the subject (for instance, see Lee 2002; Scach 2011). However, for completeness, a brief introduction is made here.

OOM provides obvious advantages to application programming, with benefits of encapsulation, polymorphism and complexity (information) hiding, code reusability, etc. (Table 18.1 summarizes the commonly mentioned advantages). By contrast, an OO approach to database design may or may not bring significant benefits, depending on the situation. Data structure encapsulation may or may not be desirable; besides, the principle of encapsulation often contradicts the principle of data independence in database design.

DOI: 10.1201/9781003275725-22

Table 18.1 Benefits of OOM

01.	**Reusability of Code:** A tested system component can be reused in the design of another component.
02.	**Stability and Reliability:** Software can be constructed from tested components. Organizations can be assured of guaranteed performance of software.
03.	**Increased Sophistication:** More complex systems can be constructed.
04.	**Understandability:** Designer and user think in terms of object and behavior rather than low-level functional details. This results in more realistic modeling that is easier to learn and communicate.
05.	**Faster Design:** Most RAD tools and contemporary CASE tools are object-oriented to some degree. Also, code reusability enhances faster development.
06.	**Higher Quality Design:** New software can be constructed by using tested and proven components.
07.	**Easier Maintenance:** Since systems are broken down into manageable component objects, isolation of system faults is easy.
08.	**Dynamic Lifecycle:** I-OO-CASE tools integrate all stages of the software development life cycle (SDLC).
09.	**Interoperability:** Generic classes may be designed for multiple systems.
10.	**Design Independence:** Classes may be designed to operate and/or communicate across different platforms.
11.	**Clarity:** OT promotes better communication between IS professionals and business people.
12.	**Better CASE Tools:** OT leads to the development of more sophisticated and flexible CASE and RAD tools.
13.	**Better Machine Performance:** OT leads to more efficient use of machine resources.

One would expect an OO database system to embody the established OO principles of inheritance and encapsulation. A third OO principle of polymorphism would probably not feature as much, since this relates to object behavior, which is not relevant to data. In the same way that a relational database is facilitated via a relational DBMS (RDBMS), an OO database is facilitated via an OODBMS. Two examples of such systems are Versant Object Database (VOD), and Object DB. Additionally, products such as Oracle and DB2 have been referred to as universal DBMSs (UDBMSs) in the past to represent their support for the relational database model as well as the OO database model.

It is widely believed that an OO approach to database design and implementation is preferable for certain scenarios such as:

- CAD (computer-aided design) systems
- Multimedia systems and CASE tools
- Geographic information systems (GIS)
- Document Storage and retrieval systems

Which database approach is the better—relational or object-oriented? This is a difficult question to answer. It depends on the situation at hand, but the following is a summary of the alternatives:

- Truly function oriented systems with relational database and procedural application development
- Truly OO systems involving (encapsulation of) both data structure and operation
- Hybrid approach with relational database and an OO user interface

The first approach is a well-known traditional approach that the software engineering industry has long moved away from; for this reason, it will not be discussed further. The latter two approaches are more aligned to current practices in the software engineering industry. They will be discussed in the upcoming sections.

18.2 Overview of Object-Oriented Database Management Systems

You are likely to already be aware of the OO approach to programming and software engineering from earlier undergraduate courses; here is a quick reminder: In a truly object-oriented software system, the following concepts hold (for more details, see Foster 2021; Schach 2011):

- An object type is a concept or thing about which data is to be stored, and a set of operations is to be defined.
- An object is an instance of an object type.
- An operation is a set of actions to be performed on an object.
- A method specifies the way an operation is to be performed.
- Encapsulation is the packaging of data structure and operations, typically into a class. The internal structure of the class is hidden from the outside, and only members (member functions) of the class have access to it.
- A subclass may inherit properties (structure or operations from a subclass. Also, a class may be comprised of several component classes).
- Polymorphism is the phenomenon where a given object or operation may take on a different form, depending on the context of usage.
- Objects communicate by sending messages to each other. These messages are managed via events; an object therefore responds to events.

Since an object-oriented DBMS (OODBMS) is a software system, it stands to reason that these OO concepts and principles would apply to an OODBMS. It turns out that in the context of a database system, pursuing some of these principles to their logical end is somewhat problematic; the challenges are mentioned in the upcoming section.

18.3 Challenges for Object-Oriented Database Management Systems

Several challenges have stood in the way of achieving purely object-oriented DBMSs, some of which have been articulated in Date (2004). Following is a summary of these challenges.

1. **Inheritance:** OOM as proposed by its ardent proponents, encourages generalization and the building of inheritance hierarchies where objects consist of other objects (as in subtype, component, and aggregation relationships). If pursued to their logical conclusion without pragmatic deviations (as discussed in Chapters 3 and 5), the system could break down into a convoluted mess. Another related problem is that hierarchies, as we know them, do not lend themselves to representation of M:M relationships. Consider, for instance, an M:M relationship between **Academic Program** and **Course**. How do we model and implement this in the OO paradigm? Do we assume that academic programs contain courses or vice versa? History has shown that hierarchical based systems (such as CODASYL) are unsuited for complex distributed databases. For this reason, purely OO databases have been criticized as "CODASYL warmed over." Products such as PostGreSQL allow inheritance but there are still efficiency issues to be resolved.
2. **Primary Keys:** The OO model does not encourage the introduction of keys to uniquely identify objects, since object identifiers (OIDs) are generated (as internal address) at the point of instantiation. Note, however, that OIDs do not obliterate the need for user keys, since end-users still need to have a way of identifying objects separate and apart from internal addresses. The problem is further compounded when we have intermediate transient objects, or what in relational terminology is referred to as logical (virtual) objects such as a natural join record. In the OO model, there is no way to negotiate these situations without duplicating data into separate storage variables which must be maintained by the running program, thus increasing system overheads.

   The relational model gracefully handles these situations by allowing the user to define primary keys and surrogates (note that surrogates are not identical to OIDs). Then thanks to data independence, you can define multiple logical views on physical data. It has been argued that encapsulation is to the OO model what data independence is to the relational model. However, when analyzed, encapsulation does not appear to be a perfect replacement for data independence.
3. **Data Sharing:** The OO model does not promote data sharing, a fundamental tenet of the relational model. At best, we could have encapsulated objects sharing their data contents with other objects, but in a distributed environment database (or mere network with one database server and several users), this would significantly increase the system overheads.
4. **Class Implementation:** The matters of class, instance, and collection are particularly difficult to negotiate. In the OO paradigm as we know, a class is essentially a complex data type; an instance is a specific object which belongs to at least one class. Without intense programmatic intervention, it would be difficult to determine a collection of objects belonging to a given class.

   The relational model has no such problem. A relational table defines the type and contains the collection of related data (objects) all in the same place. And as you are aware, this data can be shared and used to construct any number of logical perspectives as required by external end-users.
5. **Encapsulation:** The perfect encapsulation of data structure and operation in a data model is still for the most part, an ideal. We have seen complete achievement of this objective in the programming domain, but seldom in the database domain.

18.4 Hybrid Approach

Due to the above-mentioned challenges, it is unlikely that we will see a proliferation of purely object-oriented DBMS suites involving only encapsulated database objects in the immediate future. The benefits of OOM are, therefore, likely to continue to be more significant in the area of user interface than database (of course, with a few exceptions as mentioned earlier). The relational model, on the other hand, has long proven its worth. The best we can, therefore, expect is a peaceful coexistence of both OO systems and relational databases — a kind of hybrid.

In the *hybrid approach*, a relational database is accessed by an OO user interface. This means that the application development is done via an OO Programming Language (OOPL), CASE tool, or RAD tool. This user interface is then superimposed on top of the relational database.

There is no shortage of object-oriented Programming Languages (OOPLs); some of them are pure OOPLs, others being hybrid OOPLs. The more popular ones include C++, Java, C#, and Object Pascal. Additionally, there are many object-oriented CASE (OO-CASE) tools and RAD tools that support this approach. They include (but are not confined to) products such as Team Developer, Delphi, Oracle JDeveloper, WebSphere, NetBeans, etc.

This hybrid approach is really a case of pragmatism that has turned out to be quite useful. Three advantages of the approach are as follows:

1. The approach provides a way to treat *legacy software systems* (software systems based on old technology), which are prevalent in large organizations.
2. The approach facilitates peaceful coexistence of traditional and more contemporary system approaches.
3. The approach reaps the benefits of a relational database and OO application development environment.

The main disadvantage of the approach is that it does not address the earlier mentioned situations that warrant OO database design. However, the skillful database designer can use techniques mentioned in Chapters 3 and 5 for implementing relationships such as subtype, component, and aggregation.

18.5 Summary and Concluding Remarks

It is now time to summarize what we have discussed in this chapter:

- OOM provides several huge benefits to the software engineering arena.
- An OO approach to database design may or may not bring significant benefits, depending on the situation. In some situations, the OO approach is applicable; however, in a much wider range of scenarios, the relational model appears to be more suitable.
- There are several challenges to object databases that do not exist in a relational database.
- The most pragmatic approach for merging the benefits of the relational model and the OO model is to superimpose an OO user interface on a relational or universal database.

The strength of the relational model lies in its firm mathematical foundations, its huge benefits, and the immense financial outlay that have been placed in relational systems. This third fact is significant: Several large corporations have invested millions of dollars hugely in relational databases. These organizations are not likely to discard these investments. As mentioned in the previous chapter, big-data methodologies have also emerged with significant contributions to the database systems arena. The strengths of these database models lie in the benefits they bring to the scenarios they target — the relational database model for structured, operational databases; the OO database model for multimedia, GIS, and CAD systems; big-data methodologies for large, complex, unstructured datasets. That being the case, we can expect them to continue to peacefully coexist in the foreseeable future, complementing instead of rivaling each other.

18.6 Review Questions

1. Identify the main benefits that object technology brings to the arena of database systems.
2. Describe the main features of the OO DBMS model.
3. Discuss the main challenges to the object-oriented database management systems.
4. Discuss the hybrid approach to database systems.

References and/or Recommended Readings

Connolly, Thomas, & Carolyn Begg. 2015. *Database Systems: A Practical Approach to Design, Implementation and Management* 6th ed. Boston, MA: Pearson. See Chapters 27 & 28.

Date, Christopher J. 2004. *Introduction to Database Systems* 8th ed. Menlo Park, CA: Addison-Wesley. See Chapters 25 & 26.

Foster, Elvis C. 2021. *Software Engineering: A Methodical Approach* 2nd ed. New York: CRC Publishing.

Hoffer, Jeffrey A., Ramesh Venkataraman, & Heikki Topi. 2013. *Modern Database Management* 11th ed. Boston, MA: Pearson. See Chapter 13.

Kifer, Michael, Arthur Bernstein, & Philip M. Lewis. 2006. *Database Systems: An Application-Oriented Approach* 2nd ed. New York: Addison-Wesley. See Chapter 14.

Lee, Richard C., & William M. Tepfenhart. 2002. *Practical Object-Oriented Development With UML and Java*. Upper Saddle River, NJ: Prentice Hall.

Schach, Stephen R. 2011. *Object-Oriented and Classical Software Engineering* 8th ed. Boston, MA: McGraw-Hill.

Silberschatz, Abraham, Henry F. Korth, & S. Sudarshan. 2011. *Database Systems Concepts* 6th ed. Boston, MA: McGraw-Hill. See Chapter 22.

Chapter 19

Data Warehousing

Since the mid-1990s, database technology has expanded into a new area of interest — the development and management of data warehouses. It is a fascinating field of study that continues to expand and therefore deserves some attention. This chapter provides an overview of data warehousing under the following subheadings:

- Overview
- Rationale for Data Warehousing
- Characteristics of a Data Warehouse
- Data Warehouse Architectures
- Extraction, Transformation, and Loading
- Summary and Concluding Remarks

19.1 Overview

The concept of a data warehouse springs from the combination of three main areas of need:

- The business requirement for a global view of information, independent of, and despite its source or underlying structure
- The need of information systems (IS) professionals to manage large volumes of company data in a more effective manner
- The need to reduce the reporting load on operational database servers

The data warehouse has been approached many times and from many directions in the last decade; several implementations exist today. Before proceeding, let us revisit the distinction between data and information; this was mentioned in Chapter 1: Data refers to the raw materials that software systems act on in order to produce useful information to end users. Information is processed and assimilated into data that conveys meaning to its intended users.

This chapter draws significantly from the work of William Inmon (for instance, see Inmon 2002, 2003), regarded by some as the father of data warehousing. However, the work of Ralph

DOI: 10.1201/9781003275725-23

Kimball (as in Kimball & Ross 2013; Kimball 2014) has also been considered. Both Inmon and Kimball are considered as pioneers of data warehousing.

19.1.1 Definitions

Here is a working definition or a data warehouse, based on Inmon's perspective: A *data warehouse* is an *integrated, subject-oriented, time-variant, nonvolatile*, consistent database, constructed from multiple sources, and made available (in the form of read-only access) to support decision-making in a business context. As you will soon see (in Section 19.3), the highlighted terms in this definition are deliberate because of their significance.

Here is an alternate definition that reflects Kimball's perspective: A data warehouse is a database that is typically constructed from multiple *source databases*, and designed for query and analysis rather than transaction processing.

The source databases are typically *transactional databases*, also called *operational databases* or *online transaction processing* (OLTP) databases. The data warehouse usually contains historical data that is derived from transaction data from multiple sources. It separates analysis workload from transaction workload, thus enabling a business to consolidate and standardize data from several sources.

Large and/or complex data warehouses are often made up of *data marts*. A data mart is a miniature data warehouse that stores data for a particular functional area of the business.

A data warehouse environment often consists of an ETL (extract, transformation, and load) solution, an online analytical processing (OLAP) engine, client analysis tools, and other applications that manage the process of gathering data and delivering it to business users. Finally, end users typically require some kind of catalog that describes the data in its business context and acts as a guide to the location and use of this information.

Data mining (also called *knowledge discovery*) is the process of searching through and analyzing large volumes of (often) unstructured data with a view to identifying patterns and relationships, and presenting information in a manner that is consistent with user requirement. Data mining is often applied in the context of data warehouses, so the two terms are closely related. Another related term is *information extraction* (IE) — the extraction of structured information from unstructured text. IE sometimes involves access of data warehouse(s) as either the source or destination of information.

19.1.2 Acquiring a Data Warehouse

If you look closely at the two definitions of a data warehouse in the previous subsection, you will detect a difference between the two. This difference arises from two different perspectives for the data warehouse environment. Here is a brief clarification:

Inmon's Perspective: The Inmon perspective is said to be top-down, proposing that the data warehouse is a central repository, consisting of a comprehensive normalized relational database. This is followed by dimensional data marts created from and supported by the data warehouse. The data warehouse is perceived to be at the center of a so-called *corporate information factory* (CIF), providing a logical framework for the delivery of business intelligence.

Kimball's Perspective: The Kimball perspective is said to be bottom-up, advancing the idea that the data marts are created first, and that the data warehouse is the union (i.e., combination) of existing data marts. These data marts are usually relational databases that have been normalized to at least the third normal form. The data warehouse is denormalized and characterized by star schemas.

So, as you can see, the two approaches are different, and combined with sound judgment, they will affect the level of normalization pursued for the data warehouse. The Inmon approach prescribes a normalized data warehouse; the Kimball approach does not. Nonetheless, the following are some recommended steps for the construction of a data warehouse:

1. Conduct an information infrastructure analysis to determine the required structure of the data warehouse.
2. Identify the source databases that will feed the data warehouse.
3. Design the integrated logical data model and determine the architecture of the data warehouse.
4. Develop and implement a comprehensive metadata methodology.
5. Determine and then implement the physical structure of the data warehouse.
6. Design and implement an integrated staging area for the data warehouse. Depending on the complexity of the data warehousing environment, this step may or may not be required; regardless, it is a good idea.
7. Extract, transform, and load the data (from various sources) into the data warehouse. This usually involves first cleansing the source data of various structural and content errors.
8. Conduct comprehensive postimplementation review(s) to ensure that the data warehouse is performing acceptably.
9. Maintain the data warehouse.

19.2 Rationale for Data Warehousing

Data warehousing is a technology that is fast enhancing more traditional decision support systems (DSS) because of the added flexibility and benefits that the new technology brings. Let us briefly examine the problems that DSS end users tend to have from two perspectives — user constraints and information system (IS) constraints.

User Constraints: In the absence of data warehousing, users of traditional decision support systems developed using the traditional application-driven approach, commonly complained of the following difficulties:

- Difficulty in finding and accessing information needed
- Difficulty in understanding information found
- Information obtained is not as useful as expected

IS Constraints: In the absence of data warehousing, IS personnel also complained of a variety of problems:

- Developing copy programs (for copying data from source database to target database) is often very challenging
- Maintaining copy programs and copy databases presents serious integrity and work scheduling problems
- Data storage volumes tend to grow rapidly
- Database administration also tends to become quite complex

One solution to the above-mentioned problems is the implementation of a data warehouse. A data warehouse provides the decision support benefits that a traditional DSS provides, while providing more flexibility for expansion beyond the confines of the company. This is so for two reasons:

- The data warehouse has the capacity to attract interest in the salient facts about the organization, without providing unnecessary details. For instance, an organization may be more willing to publicize summarized information of total sales over a period of time than it is to publicize the daily internal minutia of sales management.
- While companies may be hesitant about putting their transaction database(s) into the public domain (due to security and confidentiality concerns), they are more likely to be willing to put their data warehouse into the public domain (via the World Wide Web).

19.3 Characteristics of a Data Warehouse

In the definition of a data warehouse, several terms were deliberately highlighted. These terms convey important characteristics about a data warehouse. These will be clarified in the next section. Next, we will examine what kind of data that is typically stored in a data warehouse. We also examine the processing requirements of a data warehouse. Finally, we will review the 12 rules that govern data warehouses.

19.3.1 Definitive Features

In the introduction, it was established that a data warehouse is an integrated, subject-oriented, time-variant, nonvolatile database. Let us briefly examine what these adjectives mean.

Subject-oriented: Data warehouses are designed to aid the analysis of data in order to make decisions. For example, to learn more about your company's sales data, you can build a warehouse that concentrates on sales. Using this warehouse, you can answer questions like "Who was our best customer for this item last year?" This ability to define a data warehouse by subject matter, sales in this case, makes the data warehouse subject-oriented.

Integrated: Integration is closely related to subject orientation. Data warehouses typically contain data from disparate sources into a consistent format. They must resolve such problems as naming conflicts and inconsistencies among units of measure. When a data warehouse achieves this, it is said to be integrated.

Nonvolatile: Nonvolatile means that, once entered into the warehouse, data should rarely change. This is logical because the purpose of a warehouse is to enable you to analyze historical data.

Time-variant: A data warehouse's focus may change over time; also, it could grow (in terms of data volume, data structure, and complexity).

19.3.2 Nature of Data Stored

Of importance also is the nature of data stored in a data warehouse. A data warehouse differs from an operational database in the nature of data stored. An operational database consists of a set of normalized relational tables that store atomic data. A data warehouse, on the other hand, stores decision support data, often in denormalized, aggregated formats. The levels of normalization/de-normalization and aggregation employed in the data warehouse will vary depending on the structure of source databases, the identified objectives of the data warehouse, and the chosen approach (Inmon vs Kimball). Three distinctions can be made between operational data and decision support data:

Time Span: Operational data represent atomic transactions at specific points in time. Decision support data represent (aggregated) data over a period of time.

Granularity: Operational data is atomic; decision support data is often aggregated. Data warehouses contain decision support data that have been aggregated from various sources and transformed to its intended format.

Dimension: Operational data tend to reflect various instances in time; decision support data are often multidimensional, typically involving the dimension of time as well as other factors of concern. Of course, exceptions are possible; for instance, it is conceivable to have multidimensional tables in an operational database (as illustrated in Section 5.6).

19.3.3 Processing Requirements

Data warehouses have very different processing requirements from OLTP systems and/or operational databases, as explained below.

Workload: Data warehouses are designed to accommodate ad hoc queries. The workload of the data warehouse might not be known in advance, so a data warehouse should be optimized to perform well for a wide variety of possible query operations. OLTP systems and operational databases typically support only predefined operations. Your applications might be specifically tuned or designed to support only these operations.

Data Modifications: A data warehouse is updated on a regular basis by the ETL process (run periodically, for example, nightly or weekly) using bulk data modification techniques. The end users of a data warehouse do not directly update the data warehouse. In OLTP systems and operational databases, end users routinely issue individual data modification statements to the database. The OLTP database is always up to date, and reflects the current state of each business transaction.

Schema Design: Data warehouses often embody denormalized relations as well as dimensional models (such as the star schema) to optimize query performance. OLTP systems and operational databases often use fully normalized relations to optimize update/insert/delete performance and to guarantee data consistency.

The star schema was introduced in Chapter 5 (Section 5.6); it is so widely mentioned in database literature, it deserves a bit of attention: A star schema describes a mechanism where there is a "central" table referred to as a *fact table*, and other so-called *dimensional tables* that relate to the fact table via 1:M relationships. The dimensional tables contain characteristic data about details stored in the fact table. While the star schema is widely used in data warehouses, it is also applicable in operational databases as well. Figure 19.1 provides an illustration of a star schema of five relational tables for tracking the graduation statistics from a regional university that operates multiple schools and programs in multiple locations. Tables **Location**, **AcademicProgram**, **TimePeriod**, and **School** qualify as dimensional tables, while table **GraduationSummary** qualifies as the fact table.

Typical Operations: A typical data warehouse query may scan thousands or millions of rows of data. For example, a data warehouse query may involve determining the total sales for all customers in a specific time period. Except for sophisticated queries, a typical OLTP or operational database operation accesses only a small percentage of records, relative to the amount stored. For example, an operational query may involve retrieving the current purchase order for a particular customer, from a table storing hundreds of thousands of purchase orders.

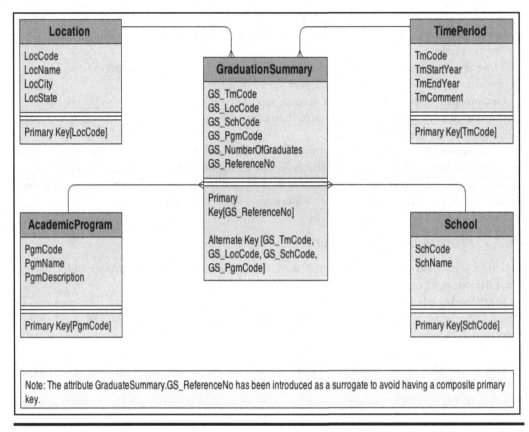

Figure 19.1 Illustration of a star schema for graduations from a regional university.

Figure 19.2 provides a summary of what we have established so far: that a data warehouse is typically constructed for various operational (source) databases and possesses certain characteristics.

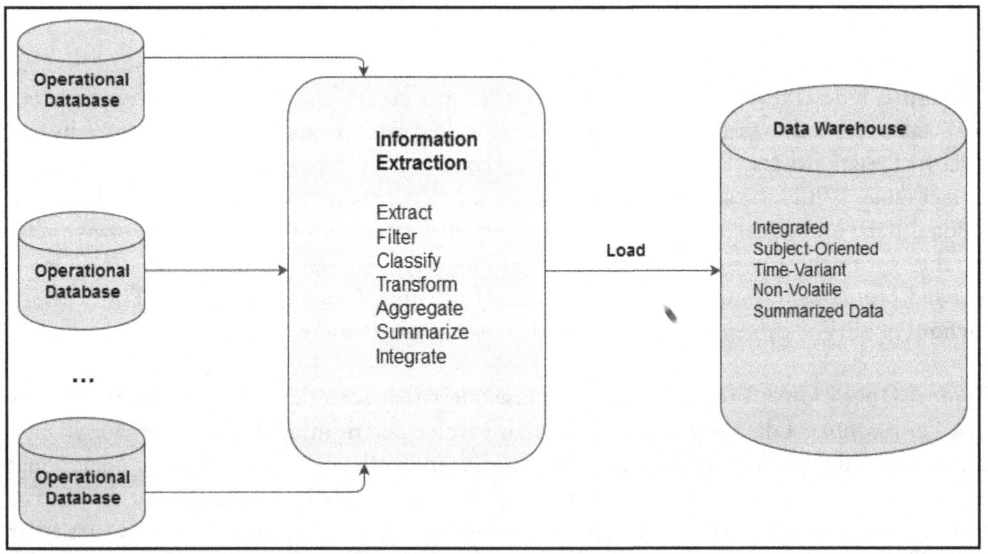

Figure 19.2 Constructing a data warehouse.

19.3.4 Twelve Rules that Govern a Data Warehouse

Data warehouse was first introduced by William Inmon and Ralph Kimball in the 1990s. Like E.F. Codd and C. J. Date of the relational model, Inmon introduced 12 rules for governing data warehouses. These rules aptly summarize the previously mentioned characteristics of the data warehouse. Many of these rules have been subsumed in the foregoing discussions. Nonetheless, for emphasis, they are paraphrased below:

1. **Separation:** The data warehouse and operational database environments should be separated.
2. **Integration:** The data warehouse data are integrated from various operational sources.
3. **Time Horizon:** The data warehouse typically contains historical data with an extended time horizon. This is in contrast to a significantly shorter time horizon for the operational databases that may be used as sources for the data warehouse.
4. **Nature of Data:** The data in a data warehouse represent snapshot captures (from operational data sources) at specific points in time.
5. **Orientation:** The data contained in the warehouse are subject-oriented.
6. **Accessibility:** The data warehouse is a predominantly read-only database, with periodic batch updates from the operational databases that are connected to it. It does not support online interactive updates.
7. **Life Cycle:** The data warehouse development life cycle differs from classical approach to database development in that the data warehouse development is data driven, whereas the classical database approach tends to be process driven.
8. **Levels of Detail:** In a typical data warehouse, there may be several levels of detail. These include current detail, old detail, lightly summarized data, and highly summarized data.
9. **Data Set:** The data warehouse is characterized by read-only transactions on very large data sets. This is in contrast to the operational database, which is characterized by numerous update transactions on a more narrowly defined data set.
10. **Relevance:** The data warehouse environment has a system that keeps track of all data sources, transformations and storage. This is essential if the data warehouse is to maintain its relevance.
11. **Metadata:** The data warehouse's metadata forms a critical component of its environment. The metadata provides the following functions: definitions of data elements in the warehouse; identification of data source, transformation, integration, storage, usage, relationships, and history of each data element.
12. **Resource Usage:** The data warehouse typically enforces optimal usage of the data by enforcing some form of chargeback mechanism for resource usage.

19.4 Data Warehouse Architecture

The architecture of a data warehouse varies depending on the specifics of an organization's situation. Three common architectures have been identified:

- Basic Data Warehouse
- Data Warehouse with a Staging Area
- Data Warehouse with Staging Area and Data Marts

19.4.1 Basic Data Warehouse Architecture

Figure 19.3 shows a simple architecture for a data warehouse. End users directly access data derived from several source systems through the data warehouse.

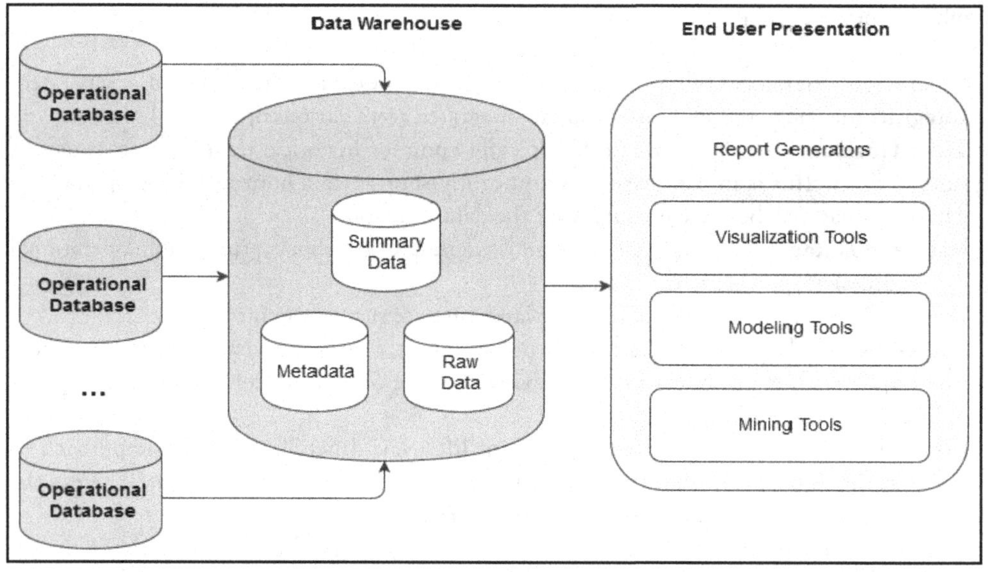

Figure 19.3 Basic data warehouse.

The data warehouse consists of raw data (from operational databases), summary data, and metadata. Summaries are very valuable in data warehouses because they pre-compute long operations in advance. For example, a typical data warehouse query may be to retrieve something like aggregate sales for a specific period. In Oracle, this may be implemented as a materialized view or snapshot relation set up for that purpose.

19.4.2 Data Warehouse Architecture with a Staging Area

In the basic data warehouse, you need to clean and process your operational data before putting it into the warehouse. You can do this programmatically, although most data warehouses use a staging area instead. A staging area simplifies building summaries and general warehouse management. Figure 19.4 illustrates this typical architecture.

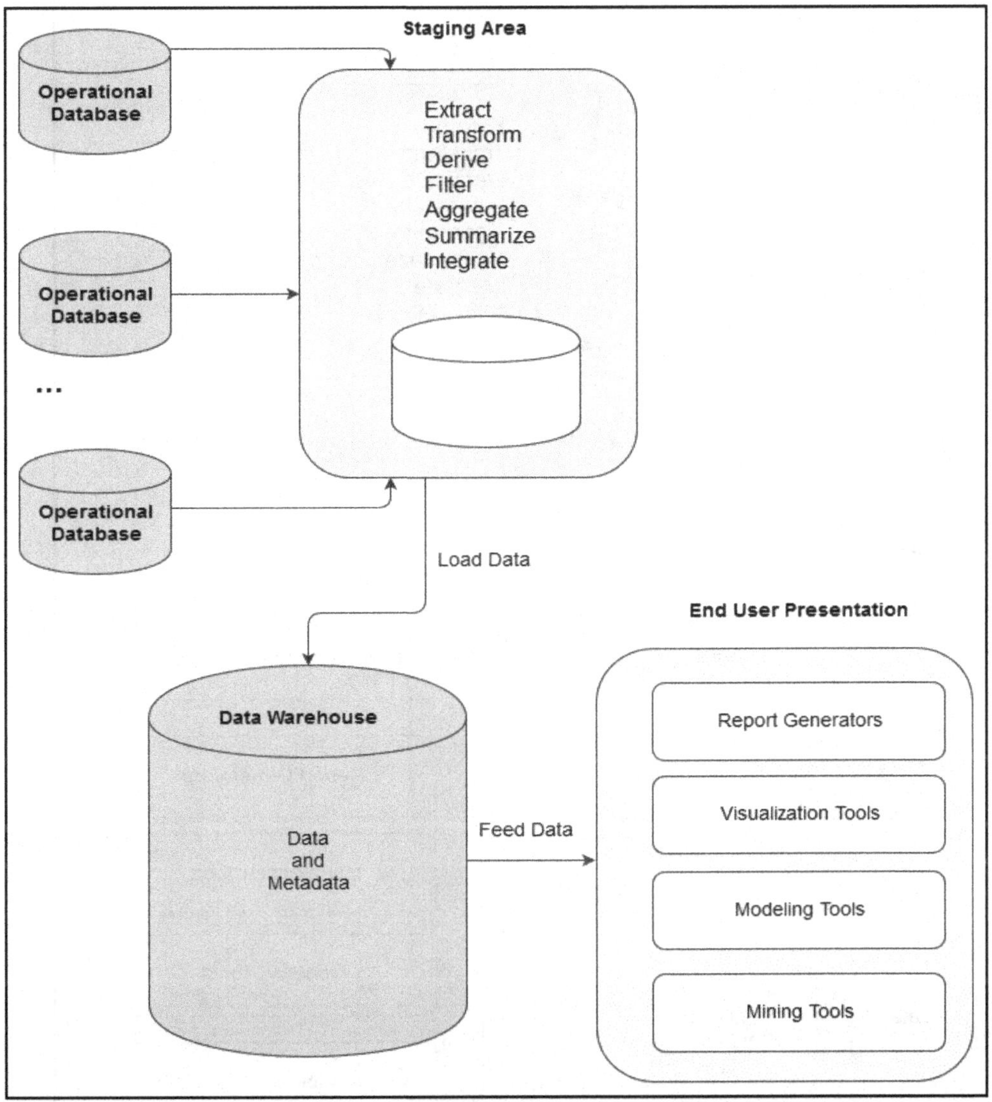

Figure 19.4 Data warehouse with staging area.

19.4.3 Data Warehouse Architecture with a Staging Area and Data Marts

Although the data warehouse with a staging area is quite common, you may want to customize your warehouse's architecture for different groups within your organization or across different related organizations. You can do this by adding data marts that focus on specific functional areas of the business. Figure 19.5 summarizes the approach for three data marts. For example, the data marts may, respectively, represent information relating to purchasing, sales, and inventory for an organization. A financial analyst would then be able to conduct separate analyses on purchasing, sales, and inventory, and then a global analysis on the three aspects combined.

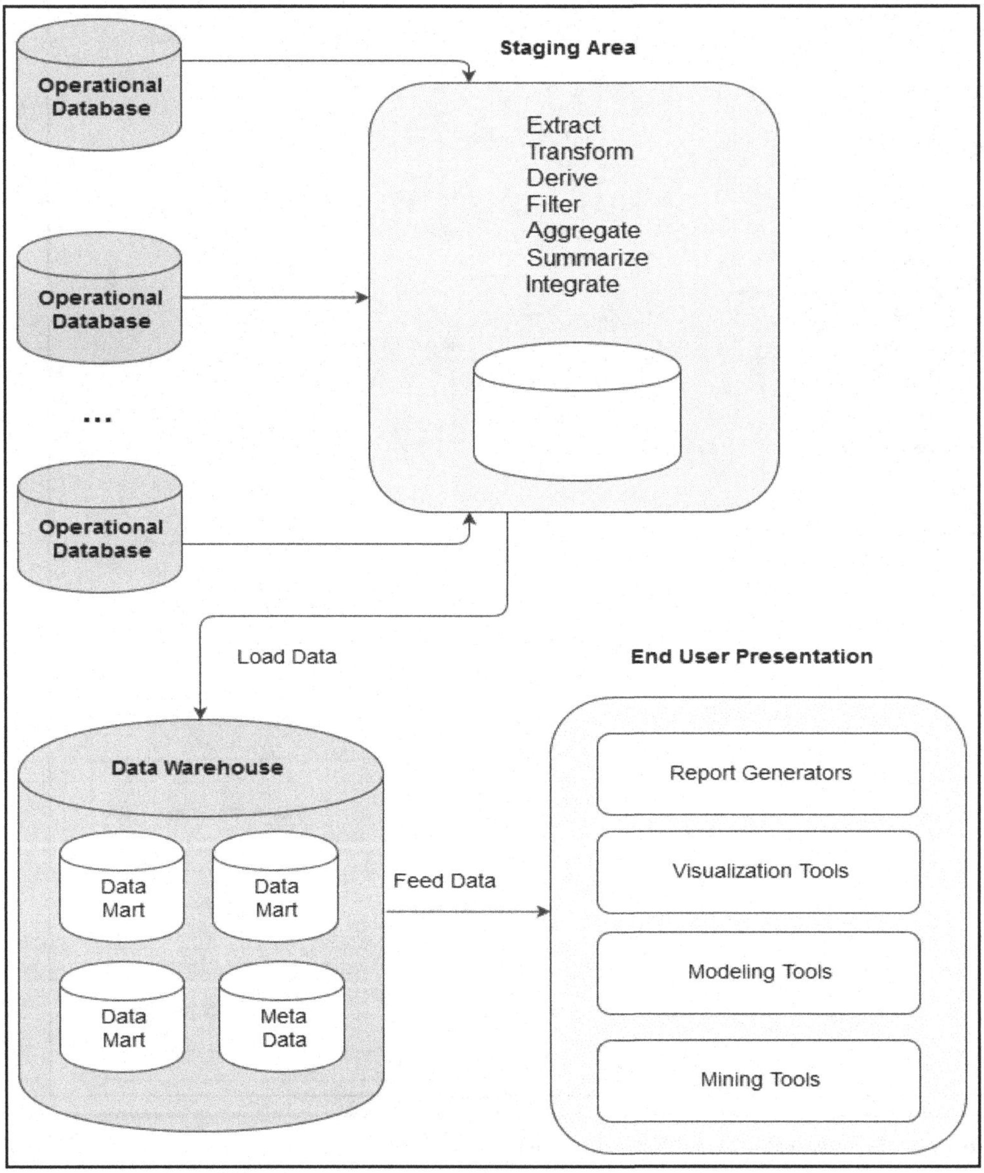

Figure 19.5 Data warehouse with a staging area and data marts.

19.5 Extraction, Transformation, and Loading

The data warehouse must be loaded regularly so that it can serve its purpose of facilitating business analysis. To do this, data from one or more operational databases needs to be extracted, transformed (where necessary), and loaded into the warehouse. The acronym ETL is often used to refer to this extraction, transformation, and loading of data. The acronym is perhaps too simplistic, since it omits the transportation phase and implies that each of the other phases of the process is distinct. We refer to the entire process, including data loading, as ETL. However, you should understand that ETL refers to a broad process and not three well-defined steps.

The methodologies and tasks of ETL have been well known for many years, and are not necessarily unique to data warehouse environments. A wide variety of proprietary applications and database systems are used as the IT backbone business enterprises. Data are shared between applications or systems, trying to integrate them, giving at least two applications the same picture of the world. This data sharing was mostly addressed by mechanisms similar to what we now call ETL. Data warehouse environments face the same challenge with the additional burden that they not only have to exchange, but to integrate, rearrange and consolidate data over many operational systems, thereby providing a new unified information base for business intelligence. Additionally, the data volume in data warehouse environments tends to be very large.

19.5.1 What Happens During the ETL Process

During extraction, the desired data is identified and extracted from many different sources, including database systems and applications. Very often, it is not possible to identify the specific subset of interest; therefore, more data than necessary has to be extracted, so the identification of the relevant data will be done at a later point in time.

Depending on the source system's capabilities (for example, operating system resources), some transformations may take place during this extraction process. The size of the extracted data varies from hundreds of kilobytes up to gigabytes, depending on the source system and the business situation. The same is true for the time difference between two (logically) identical extractions: the time span may vary between days/hours and minutes to near real-time. Web server log files, for example, can easily become hundreds of megabytes in a very short period of time. After extracting data, it has to be physically transported to the target system or an intermediate system for further processing. Depending on the chosen mode of transportation, some transformations can be done during this process, too. For example, an SQL statement that directly accesses a remote target through a gateway can concatenate two or more columns as part of the **Select** statement.

After transportation to the target system, the data may undergo transformation into the desired formats for the target system. Once this process is completed, the data is loaded into the data warehouse.

19.5.2 ETL Tools

Designing and maintaining the ETL process is often considered one of the most difficult and resource-intensive portions of a data warehouse project. Many data warehousing projects use ETL tools to manage this process.

19.5.2.1 Oracle Tools

The Oracle database management system (DBMS) suite facilitates data warehousing via its Oracle Data Integrator (ODI) component. This ODI provides ETL capabilities and takes advantage of inherent database abilities. Other data warehouse builders create their own ETL tools and processes, either inside or outside the database.

Besides the support of extraction, transformation, and loading, there are some other tasks that are important for a successful ETL implementation as part of the daily operations of the data warehouse and its support for further enhancements. The ODI is quite a sophisticated component that facilitates the construction of simple as well as complex data warehouses, the population of them via the ETL process, and the management of them.

19.5.2.2 IBM Db2 Tools

The Db2 DBMS suite facilitates data warehousing of various levels of complexity via two main families or related products:

- IBM InfoSphere Data Architect
- IBM Data Warehousing in Db2

Each of these product lines consists of related components that allow for the creation, population (via ETL transactions), and management of data warehouses, according to organizational requirements.

19.5.3 Daily Operations and Expansion of the Data Warehouse

Successive ETL transactions to the data warehouse must be scheduled and processed in a specific order. Depending on the success or failure of the operation or parts of it, the result must be tracked and subsequent, alternative processes might be started. The control of the progress and the definition of a business workflow of the operations are typically addressed by special ETL tools provided by the DBMS suite.

As the data warehouse is an active information system, data sources and targets are not beyond the prospect of change. These changes must be maintained and tracked through the life span of the system without overwriting or deleting the old ETL process flow information. To build and keep a level of trust about the information in the warehouse, the process flow of each individual record in the warehouse can be reconstructed at any point in time in the future. With time, the data warehouse could therefore expand into something larger and different.

19.6 Summary and Concluding Remarks

Here is a summary of what we have discussed in this chapter:

- Based on Inmon's perspective, a data warehouse is an integrated, subject-oriented, time-variant, nonvolatile, consistent database, obtained from a variety of sources and made available to support decision-making in a business context. The data warehouse is constructed in a top-down manner that feeds into dimensional data marts.
- Based on Kimball's perspective, a data warehouse is a relational database that is designed for query and analysis rather than transaction processing. The data warehouse is constructed in a bottom-up approach as the conglomeration (i.e., union) of related data marts.
- The data warehouse is updated via the ETL process.
- A data warehouse provides the decision support benefits that a traditional DSS provides, while providing more flexibility for meeting additional needs of the organization.
- Data warehouses often use denormalized schemas or alternate storage schemas (for instance, star schemas) to optimize query performance.
- Data warehouses should conform to Inmon's 12 rules for data warehouses.
- Three common data warehouse architectures are the basic data warehouse, data warehouse with a staging area, and data warehouse with a staging area and data marts.

The field of data warehousing is a fascinating breakthrough and is the subject of many contemporary researches. Since its introduction in the 1990s, data warehousing has flourished into a growing field that impacts other related areas such as business intelligence and data analytics; vast opportunities for data warehouse architecture, and ETL transaction optimization still abound. Data warehousing is studied as an advanced course in several undergraduate degree programs as well as in graduate programs. The supporting technologies are provided via products from the three leading software engineering companies — IBM's DB2, Oracle (from Oracle Corporation), and Microsoft's SQL Server. Additionally, these companies provide readily available documentation on the topic.

With the advancement of the World Wide Web and Web-accessible databases, it is anticipated that the fascination in data warehousing will continue into the foreseeable future. And speaking of Web-accessible databases, the next chapter discusses this topic.

19.7 Review Questions

1. Give the definition of a data warehouse. Clearly outline what data warehousing entails.
2. Provide a rationale for data warehousing.
3. Describe a scenario that would warrant the use of a data warehouse.
4. Discuss the main characteristics of a data warehouse in terms of:
   - Definitive features
   - Nature of data stored
   - Processing requirements
   - Rules that govern the data warehouse
5. State the three architectural approaches to data warehouse design. For each approach, describe the basic architecture and provide a scenario that would warrant such an approach.
6. Clearly explain the ETL process. Give examples of ETL tools.

References and/or Recommended Readings

Adelman, Sid, & Larissa Terpeluk Moss. 2000. *Data Warehouse Project Management*. Boston, MA: Addison-Wesley.

Connolly, Thomas, & Carolyn Begg. 2015. *Database Systems: A Practical Approach to Design, Implementation and Management* 6th ed. Boston, MA: Pearson. See chapters 31–34.

Date, Christopher J. 2004. *Introduction to Database Systems* 8th ed. Menlo Park, CA: Addison-Wesley. See chapter 22.

Elmasri, Ramez, & Shamkant B. Navathe. 2011. *Fundamentals of Database Systems* 6th ed. Boston, MA: Pearson. See chapters 28 & 29.

Inmon Associates Inc. 2003. Accessed June 2008. http://www.inmongif.com/inmon.html

Inmon, William. 2002. *Building the Data Warehouse* 3rd ed. New York: Wiley.

Kimball, Ralph, & Margy Ross. 2013. *Data Warehouse Toolkit Classics: The Definitive Guide to Dimensional Modeling* 3rd ed. New York: Wiley.

Kimball, Ralph, & Margy Ross. 2014. *Data Warehouse Toolkit Classics: 3 Volume Set*. New York: Wiley.

Oracle Corporation. 2013. "Oracle Database 12c for Data Warehousing and Big Data." Accessed December 2015. http://www.oracle.com/technetwork/database/bi-datawarehousing/data-warehousing-wp-12c-1896097.pdf

Oracle Corporation. 2015. *Oracle Database for Data Warehousing and Big Data*. Accessed December 2015. http://www.oracle.com/technetwork/database/bi-datawarehousing/index.html

Chapter 20

Web-Accessible Databases

Another area of database systems that has become widespread since the 1990s is Web-accessible databases. The proliferation of these databases is strongly correlated to the growth of the World Wide Web (WWW or W3); both phenomena have become part and parcel of the 21st-century lifestyle, and both promise to be an integral part of life in the foreseeable future. This chapter provides an overview of Web-accessible databases. The chapter proceeds under the following subheadings:

- Introduction
- Web-Accessible Database Architecture
- Supporting Technologies
- Implementation with Selected DBMS Suites
- Generic Implementation via Front-end and Back-end Tools
- Summary and Concluding Remarks

20.1 Introduction

A Web-accessible database is simply a database that is accessible via the WWW. The rationale for Web-accessible databases can be easily appreciated when one considers the huge benefits that they bring to the business community. Table 20.1 provides a brief summary of some of the significant benefits. Because of these benefits, Web accessibility has become the norm (de facto standard) for many contemporary database systems. Moreover, this feature is provided by all the major DBMS suites.

DOI: 10.1201/9781003275725-24

Table 20.1 Significant Benefits of Web-Accessible Databases

Benefit Area	*Brief Comment*
Electronic commerce	Through electronic commerce (e-commerce), organizations are able to market their products and services via consumer-to-business (C2B) services as well as business-to-business (B2B) services.
Organizational scope	The online market is not constrained by space, time, or geographic region. Organizations that trade in this market have virtually joined a global community in which resources and services are only seconds away.
Convenience	Organizations that use Web-accessible databases afford themselves easy access to mission-critical company information (preferably without adversely sacrificing security) in very cost-effective way. These conveniences would have much more expensive (if at all possible) if pursued via more traditional methods.
Improved productivity	By using Web-accessible databases, organizations often improve their productivity by making use of resources that otherwise would have been more expensive (financially and in terms of time).
Improved competitiveness	By using Web-accessible databases, organizations often improve their competitive advantage in the marketplace.
Distributed systems	In many cases, Web-accessible databases involve the use of distributed databases. In such cases, the benefits of distributed databases (review Section 17.2) also apply.

20.2 Web-Accessible Database Architecture

In Sections 2.6 and 2.7, we discussed the idea of separating a database system into front-end and back-end systems. This principle is commonly used in implementing Web-accessible databases. Two approaches are common: the two-tiered approach as represented in Figure 20.1 and the three-tiered approach as represented in Figure 20.2. In the two-tiered approach, client applications send requests to the DBMS, which is running on a database server. These requests are processed according to some scheduling algorithm. In the three-tiered approach, additional sophistication is provided by an intersecting application server, which services client requests from various (heterogeneous) applications and filters them to the DBMS for processing. This provides additional flexibility and functionality to the system.

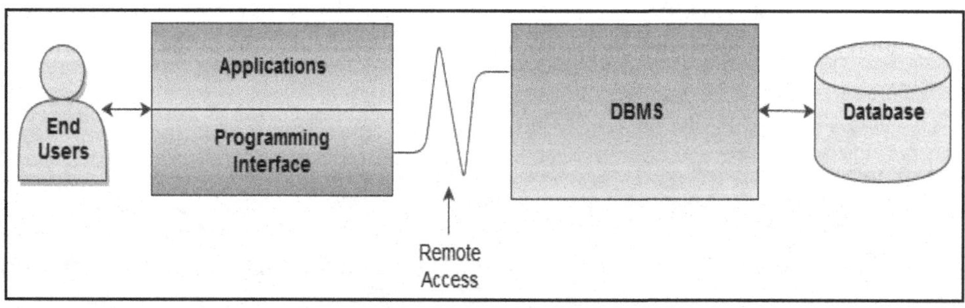

Figure 20.1 Two-tiered approach to Web-accessible database.

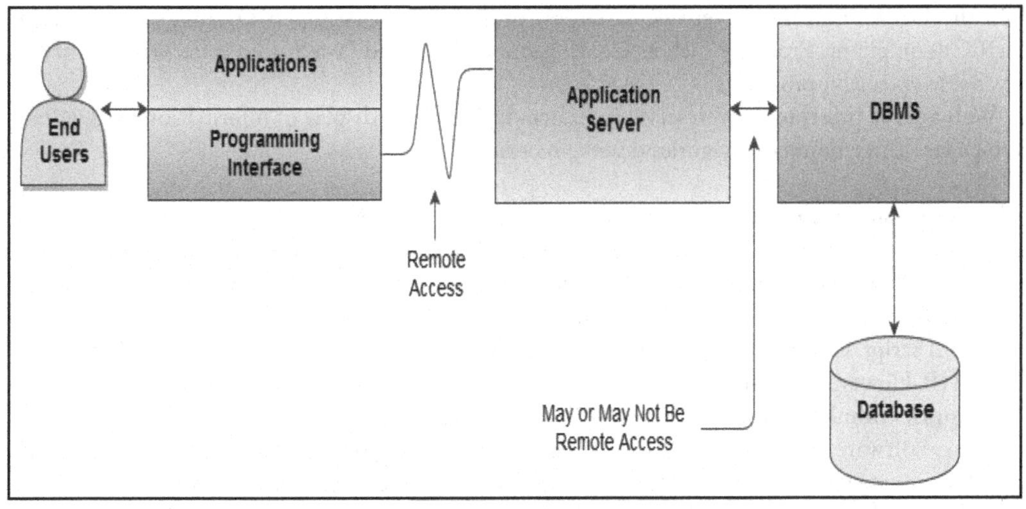

Figure 20.2 Three-tiered approach to Web-accessible database.

As you view these two figures, bear in mind that users (in the figures) include end users as well as businesses making electronic requests from other Web-accessible databases. This is so because there are two types of Web-accessible database systems that are prevalent:

- Consumer-to-business (C2B) systems facilitate individual users accessing a company database.
- Business-to-business (B2B) systems facilitate businesses accessing other Web-accessible databases of other businesses in a manner that is often transparent to the end user. This market provides huge opportunities for companies to improve their efficiency and productivity by concentrating on what they do best and outsourcing nonessential functions that other businesses provide more efficiently. This gives them the opportunity to also forge powerful alliances for more effective operation.

20.3 Supporting Technologies

The usefulness of separating front-end and back-end subsystems was illustrated in the previous section. This strategy allows us to design and secure the database as a separate activity from developing applications that access that database. Web applications could be one of the many different applications that access the database. Following are some supporting technologies for Web-accessible databases.

Web Servers: A Web server is a sophisticated software system that allows a computer to provide various services to multiple client requests made via the WWW. The server runs on an operating system (which runs on a computer). However, loosely speaking, we usually refer to the entire package of machine (physical or virtual) and software as the Web server. Popular Web server software products include Apache, Microsoft's Internet Information Service (IIS), CERN server, NCSA (National Center for Supercomputing Applications) server, Spinner server, Plexus server, Perl server, and Tomcat server.

Server-side Extensions: A server-side extension is a software that communicates with a Web server to handle assorted client requests. Often, the server-side extension program acts as an intermediary between the Web server and the database, farming out all SQL requests to the DBMS.

Both the DBMS and the server-side extension program must be Open Database Connectivity (ODBC)-compliant. Products such as ColdFusion, Delphi, and Java Studio Enterprise qualify as server-side extension programs.

Web Server Interfaces: Web server interfaces facilitate the display of information on dynamic Web pages. Three popular categories deserve mention here:

- **Common Gateway Interface (CGI)** is a standard set of methodologies that facilitates communication between Web servers and executable programs on servers that dynamically generate Web pages. The rules specify how parameters are passed between client programs and Web servers. A client program that can run on a Web server is called a script, hence the term CGI script. Common scripting languages include JavaScript, Active Server Pages (ASP), and PHP. However, high-level languages such as C++, Java, and Perl also qualify.
- **Application Program Interface (API)** is a set of routines, protocols, and tools that facilitate easy software construction. Since APIs are typically shared code that is resident in memory (in the case of Web technology, they reside on the Web server), they tend to be more efficient than CGI scripts.
- **Model-View-Controller (MVC) Framework:** Many of the newer scripting languages do not enforce data types on variables but support inline definitions with HTML; this makes it more difficult to write easily maintainable code. To address this problem, PHP and other languages started creating frameworks that enforce the MVC paradigm. Two variations of this paradigm are the Model-View-Adapter framework and the Model-View-Presenter framework. Common examples of the MVC framework are Cake (PHP), Ruby on Rails (Ruby), Code Igniter (PHP), Django (Python), etc. Typically, these frameworks create models and provide functions for the programmer to interact with, while ensuring that modularity and good coding practices are enforced.

Extensible Markup Language (XML): XML is a meta-language that was designed specifically to facilitate the representation, manipulation, and transmission of structured data over the WWW. The language was first published by the WWW Consortium (W3C) in 1998 and, to no surprise, has become a de facto data exchange standard for e-commerce, thus circumventing the preexisting problem of interoperability among different Web servers (see W3W 2015).

XML was developed from an earlier standard called the Standard Generalized Markup Language (SGML). And as expected, other XML-based languages are emerging. Three examples are Extensible Business Reporting Language (XBRL), Structured Product Labeling (SPL), and Extensible Style Language (XSL).

JSON: JavaScript Object Notation (JSON) is a minimalistic format for representing data; this is normally favored over XML due to its conservation in required data packets needed for transmission. JSON is also considered easier to implement than XML — in terms of machine parsing as well as readability.

Hypertext Transport Protocol (HTTP): HTTP is the protocol that acts as the foundation of data communication over the WWW. It is characterized by a simple request–response structure that represents interactions between a client (Web browser) and a Web server. It is assumed that you are familiar with the Internet, basic Web page construction, Uniform Resource Locators (URLs), domain names, and other related issues.

Client-side Extensions: These add functionality to the Web browsers. Client-side extensions are available in various forms. Following is a summary of the most commonly encountered ones:

- **Plug-ins:** A plug-in is an external application program that is automatically invoked by the Web browser when required. For example, a Web browser, upon receiving a portable document format (PDF) document may invoke the available PDF reader on the host operating system.
- **Java:** As you are no doubt aware, Java is a platform-independent programming language. Most operating systems have a Java Virtual Machine (JVM), which allows Java code to execute on the local machine (if there isn't a JVM in the operating system, this is readily available from the Sun Microsystems Web site). Calls to Java routines are often embedded in HTML pages. When a Web browser encounters this, it invokes the local JVM to execute the code.
- **JavaScript:** JavaScript is a Java-like scripting language (developed by Netscape), but is much simpler. JavaScript code is often embedded in Web pages. It is downloaded whenever the Web page is activated, as well as on certain specific events (such as loading of a specific page from the server, or a mouse-click). Whenever the browser encounters this, it invokes the JavaScript plug-in to execute this code.
- **VBScript:** VBScript is another Microsoft product that is often used to add functionality to the Web browser. Like JavaScript, VBScript code is often embedded in Web pages. Invocation and execution are also similar to JavaScript.
- **AJAX:** Asynchronous JavaScript and XML (AJAX) is a technology that facilitates communication between XML-HTTP-Request objects and server-side scripts. Data can be sent or received via a number of formats, including JSON, XML, HTML, and even text files. AJAX allows a webpage to update a portion of its content without reloading the entire page. This facility gives the Web developer the capacity to create faster, more dynamic Web applications. This technology can be used to request information from a database without freezing the website.
- **Cookies:** Cookies are used to expedite user requests when the user visits a site more than once. When the site is first visited, the Web server creates a cookie with basic user information (such as email) and sends it to the browser. When the browser is subsequently used to visit that site, the cookie, upon recognition by the Web server, is used to expedite the user request.

20.4 Implementation with Selected DBMS Suites

Web accessibility is supported in the leading DBMS suites that are currently available. This section summarizes how this objective is achieved via Oracle, DB2, and MySQL.

20.4.1 Implementation in Oracle

Oracle implements a database that is by definition Web-accessible. This is achieved primarily through the products Oracle Enterprise Manager (OEM), which runs on the database server, and SQL Developer (OSQLD), which may run on the server or client side:

- When you install the Oracle DBMS suite (including OEM) on a node in your company or home network, the machine is automatically configured as a database server.
- Oracle automatically installs and configures a database listener on database server to respond to incoming requests from client nodes in the network, on the Internet or an extranet.
- Typically, your database server should have at least one database. This database can be created during DBMS installation; alternately, it may be created subsequent to DBMS

installation via the Oracle Database Configuration Assistant (DBCA) component or the OEM component.

- You can access your database server from any machine that has an Internet connection, through OEM via the URL **http://<Machine.Domain>:5500/em** (you supply the machine name and domain name for your network). Of course, you will need a valid user account, password, and appropriate privileges.
- Through Oracle JDeveloper (OJD) component, Oracle allows Web-accessible Java-based applications to be constructed. These applications may access local or remote Oracle (or heterogeneous) databases. **JDeveloper** is an *integrated development environment* (IDE) facilitating software development in Java, XML, SQL and PL/SQL, HTML, JavaScript, BPEL, and PHP.
- Similar to Oracle JDeveloper, the NetBeans IDE facilitates the development of Web-accessible Java-based applications. Through Java database connectivity (JDBC), a NetBeans application may access an Oracle (or heterogeneous) database. NetBeans may be extended to facilitate PHP, C, C++, or HTML coding.
- Oracle SQL Developer may be configured to access multiple Oracle databases from any computer with an Internet connection. With a user-friendly GUI-type interface, this Java-based application is very easy to learn and use. You may use it to connect to a single database instance or multiple database instances at multiple locations.
- As explained in Sections 11.1 and 11.7, the latest versions of Oracle support cloud-based database hosting, where all aspects of the database are hosted in an Oracle cloud and accessible via the W3. This option hides a significant portion of the database complexities from the end user, providing seamless access to the database services.

This development represents a huge step forward for Oracle, and has no doubt fueled its invigorated claim of being the leading software engineering firm for Web-accessible databases.

20.4.2 Implementation with DB2

Like Oracle, DB2 implements a Web-accessible database via various alternatives. This is achieved through products such as Aqua Data Studio, WebSphere, and SQL Developer in the following ways:

- If you configure a node in your corporate or home network to be a DB2 server and create your DB2 database on it, you can access the database over an intranet or the Internet by running a product called Aqua Data Studio from your client machine. As in the case of the Oracle database, you will need to know the machine name, the domain name, and port number that the database server listens on, in order to connect to it. Aqua Data Studio very closely mirrors the Oracle product called SQL Developer, exhibiting similar features and providing similar services to the user.
- Through WebSphere, IBM allows Web-accessible Java-based applications to be constructed. These applications may access local or remote DB2 (or heterogeneous) databases. There is also a WebSphere Everyplace version for cell phones and other pervasive devices.
- The Oracle product SQL Developer may also be used to connect to a DB2 database for queries; however, the product does not support database writes to a DB2 database.

Like Oracle, IBM also claims to be the leading software engineering firm for Web-accessible databases. The truth is that both companies have been archrivals at the top of the database systems market for several years.

20.4.3 Implementation in MySQL

In the area of Web-accessible relational databases, MySQL is crown-prince if not king. The MySQL DBMS suite is most widely used open-source DBMS; moreover, MySQL ranks among the top seven DBMS suites in the marketplace (see ITCS 2015; DB Engines 2015). One of the most attractive features of the product is its seamless support for Web accessibility.

As you perhaps know (or will find out from courses such as *Internet Programming* and *Electronic Commerce*), MySQL is often packaged with the PHP programming language. Following are some common bundles for MySQL:

- WAMP — Windows operating system, Apache Web server, MySQL, and PHP programming language
- LAMP — Linux operating system, Apache Web server, MySQL, and PHP programming language
- XAMPP — Any operating system, operating system, Apache Web server, MySQL, PHP programming language, and Perl programming language
- WIMP — Windows operating system, Internet Information Services (IIS), MySQL, and PHP programming language
- MAMP — Mac operating system, Apache Web server, MySQL, and PHP programming language
- SAMP — Solaris operating system, Apache Web server, MySQL, and PHP programming language
- FAMP — FreeBSD operating system, Apache Web server, MySQL, and PHP programming language
- iAMP — iSeries operating system, Apache Web server, MySQL, and PHP programming language

Whenever you install any of the above software bundles (depending on your operating system platform), that typically creates a Web-accessible database on your machine. One popular front-end application that is used to access the MySQL database in the back-end is **phpMyAdmin**. This application allows access of the MySQL database via the Web.

20.5 Generic Implementation via Front-End and Back-End Tools

This section describes a generic approach to constructing a Web-accessible database system, based on front-end and back-end tools. The operation can be summarized in two steps:

1. Create the database using an appropriate back-end tool. The tool used must support ODBC, JDBC, or both. Of course, it is assumed that appropriate planning and design as discussed in earlier chapters have taken place.
2. Create the Web application using an appropriate front-end tool that incorporates the requisite Web-supporting technologies as discussed in Section 20.3. The tool must support ODBC and/or JDBC and must facilitate code in at least one of the accepted scripting languages (JavaScript, PHP, ASP, etc.). Again, the basic assumption is that sound principles of software engineering will be used to design the user interface as a prerequisite to this activity.

Table 20.2 provides a list of commonly used tools. They have been listed in three categories: front-end tools, back-end tools, and programming languages. For each tool, it is recommended that you use the most current version of the stated software product.

Table 20.2 Commonly Used Tools for Web-Accessible Databases

Product	*Parent Company*
Front-end RAD Tools that Support Web-Accessible Databases	
Delphi	Embarcadero Technologies
WebSphere	IBM
NetBeans	Sun Microsystems
ColdFusion	Adobe
Oracle JDeveloper	Oracle
Visual Studio	Microsoft
Relational and/or Universal DBMS Suites	
DB2	IBM
Oracle	Oracle
SQL Server	Microsoft
Informix	IBM
MySQL	Oracle
Programming/Scripting Languages	
Java, C++, C#, Object Pascal, JavaScript, PHP, XML, ASP, VBScript, Perl	

20.6 Challenges and Opportunities

Due to the rapid advance of Internet-related technologies and market demand, all major DBMS suites have made the transition to Web accessibility (some with more features than others). This wave of innovation has spawned a number of new areas of challenges as well as opportunities. Three such areas are cloud computing, big data, and cybersecurity.

20.6.1 Cloud Computing and Big Data

Both cloud computing and big data were mentioned in Chapter 17 (Sections 17.6.3 and 17.6.4). Cloud technology allows organizations to store and/or access vast repositories of storage and other resources as services. The emerging practice is for leading software engineering firms that market DBMS suites to include various cloud-based features in their products. To illustrate, Oracle has named the latest version of its DBMS Oracle 12C, where the "C" represents cloud computing. Through cloud computing, Oracle provides the following cloud-based services to its customers: software as a service (SaaS), data as a service (DaaS), platform as a service (PaaS), infrastructure as a service (IaaS), private cloud services, and managed cloud services.

The impact of cloud computing on business is quite significant. Because of this technology, smaller organizations can purchase the aforementioned services from the host company for a fraction of what it would cost to set up and manage their own information services facilities. Additionally, software engineering and/or information technology firms can set up their own cloud infrastructures and then market cloud-based services to other organizations and/or individuals who need them. The end result is a wave of new business opportunities, expanded business services, increased level of flexibility, and improved efficiency — all for lower costs than would be possible without the (cloud) technology.

20.6.2 Cybersecurity

Another area of increased threats, challenges, and opportunities is that of cybersecurity. It is the case that thanks to web-accessible database systems, organizations can conduct electronic commerce (C2B as well as B2B), communicate with a large and complex marketplace beyond the proximity of their immediate communities, can create and explore new business opportunities, and explore several other benefits/opportunities. It is also the case that in opening up to these opportunities, participants (individuals as well as organizations) are exposed to more vulnerabilities. For instance, in making an online purchase, one faces the risk of having his/her credit card information being hacked by malicious users of the internet.

The term *cybersecurity* refers to methodologies, technologies, and strategies designed to protect computers, computer networks, software systems, and data from damage, unauthorized access, or misuse. With the advancement of the Internet and the WWW, cybersecurity has emerged as a subfield of computer science (CS) that is being actively studied at the undergraduate level as well as graduate level.

Among the various known forms of cyberattacks are the following: malware — malicious code carrying virus(s) end up on someone's computer system; phishing — sensitive information is stolen from unsuspecting email user(s); password compromise; denial of service (DoS) — the disrupting of network services; man in the middle (MITM) — impersonation of someone after intercepting (and stealing) the individual's information; drive-by downloads — unintended download of malicious code by an unsuspecting visitor to a website; mal-advertising (called malvertising) — malicious codes hidden in online ads end up being downloaded by an unsuspecting user; SQL injection — the injection of nefarious SQL code to distort originally intended SQL code; and rogue software — malware masqueraded as the legitimate resource. Another area of heightened concern is the compromise of public systems such as government information systems, public telephone systems, and the electric grid. Obviously, a full discussion of the subject matter is beyond the scope of this course. Suffice it to say, cybersecurity attempts to prevent and/or recover from cyberattacks.

While the threats of cyberattacks are real, the opportunities for significant breakthroughs for the public good are great. This realization has fueled research in the area at various levels.

20.7 Summary and Concluding Remarks

Let us summarize what we have covered in this chapter:

- A Web-accessible database is simply a database that can be accessed via the WWW.
- Web-accessible databases provide a number of benefits to companies and individuals that implement and/or use them. Among the benefits are facilitation of e-commerce, broadening of the company's scope of operation and market reach, a wide range of conveniences, improved productivity, and improved competitive advantage.
- A Web-accessible database system may be implemented as a two-tiered system or a three-tiered system. Additionally, they may be C2B or B2B.

- The supporting technologies for Web-accessible databases include Web servers, server-side extensions, server interfaces, XML, JSON, HTTP, and client extensions.
- As a matter of policy, Web accessibility is provided by leading DBMS suites including Oracle, DB2, and MySQL.
- Implementation of a Web-accessible database can be summarized into two simple but profound steps: creating the database and creating the Web-accessible user interface. The tools used must meet minimum industry standards.
- The expanse of Web-accessible database systems has given rise to significant challenges and opportunities in other areas such as cloud computing, big data, and cybersecurity.

Web-accessible databases are among the technology-related phenomena that have transformed life in the 21st century. They are expected to continue to be an integral part of life in the foreseeable future.

20.8 Review Questions

1. Give the definition of a web-accessible database. Provide justification for their existence.
2. Discuss two examples of the usefulness of Web-accessible databases.
3. Discuss the two-tiered approach to implementing Web-accessible database. When should you use such an approach?
4. Discuss the three-tiered approach to implementing Web-accessible database. When should you use such an approach?
5. Briefly describe the main supporting technologies for Web-accessible databases.
6. Describe how Oracle implements Web-accessible databases.
7. Describe how DB2 implements Web-accessible databases.
8. Describe a generic approach for implementing a Web-accessible database. What precautions must be taken?

References and/or Recommended Readings

Coronel, Carlos, & Steven Morris. 2015. *Database Systems: Design, Implementation & Management* 11th ed. Boston, MA: Cengage Learning. See chapter 14.

Date, Christopher J. 2004. *Introduction to Database Systems* 8th ed. Menlo Park, CA: Addison-Wesley. See chapter 27.

DB Engines. 2015. "DB-Engines Ranking of Relational DBMS." Accessed December 2015. http://db-engines.com/en/ranking/relational+dbms

Elmasri, Ramez, & Shamkant B. Navathe. 2011. *Fundamentals of Database Systems* 6th ed. Boston, MA: Pearson. See chapter 27.

IT Career Success. 2015. "Top 10 Database Software Systems." Accessed December 2015. http://www.itcareersuccess.com/tech/database.htm

Kifer, Michael, Arthur Bernstein, & Philip M. Lewis. 2006. *Database Systems: An Application-Oriented Approach* 2nd ed. New York: Addison-Wesley. See chapters 15 & 25.

Oracle Corporation. 2013. "Big Data Analytics: Advanced Analytics in Oracle Database." Accessed May 2016. http://www.oracle.com/technetwork/database/options/advanced-analytics/bigdataanalyticswpoaa-1930891.pdf

Oracle Corporation. 2016a. *Grid Computing.* Accessed May 2016. http://www.oracle.com/technetwork/topics/grid/index.html

Oracle Corporation. 2016b. *Oracle Cloud Computing.* Accessed May 2016. https://www.oracle.com/cloud/index.html

Riccardi, Greg. 2003. *Database Management with Website Development Applications.* Boston, MA: Addison-Wesley. See chapters 12–15.

World Wide Web Consortium. 2015. *Extensible Markup Language (XML).* Accessed December 2015. http://www.w3.org/XML/

Chapter 21

Using Database Systems to Anchor Management Support Systems

At the beginning of the course (Section 1.1), the relevance and importance of database systems to software engineering was mentioned. At this point in the course, this nexus between the two related disciplines of computer science should be very clear to you. This chapter reinforces that understanding even more by introducing the concept of *management support systems* (MSSs). Next, the chapter revisits a concept that was raised in Chapter 13 (toward the end of Section 13.2.3) — the idea that for many software systems requiring stringent security mechanisms, the option of incorporating such requirements into the underlying database design is a viable alternative. This will be followed by a discussion of a case study and some insights into selected MSS project ideas. The chapter advances through the following captions:

- Overview of Management Support Systems
- Building System Security through Database Design
- Case Study: Dynamic Menu Interface Designer
- Selected MSS Project Ideas
- Summary and Concluding Remarks

21.1 Overview of Management Support Systems

As pointed out at the outset of the course, databases typically exist to provide storage requirements for software systems that use them. Put another way, there are many software systems that rely on underlying databases for information support. Additionally, for many software systems, it can be observed that the user interface and system security are intricately linked. One family of software systems that exhibits these two features is the group referred to as *management support systems* (MSSs).

DOI: 10.1201/9781003275725-25

Drawing from a previous research recorded in Foster (2015b), management support systems (MSSs) refer to a family of software systems that are geared toward the promotion and facilitation of efficient and effective management and decision-making in the organization. Included among MSSs are the following categories: *strategic information systems* (SISs), *decision support systems* (DSSs), *executive information systems* (EISs), *expert systems* (ESs), *knowledge management systems* (KMSs), *business intelligence systems* (BISs), and *enterprise resource planning systems* (ERPSs). Following is a brief clarification on each member of the MSS family.

- A *strategic information system* (SIS) is defined in Wiseman (1988) as a software system that is designed to be aligned with the corporate and strategic vision of an organization or group of related organizations, thus giving strategic and competitive advantages to the host organization(s).
- A *decision support system* (DSS) has been defined by Keen and Morton (1978) as a software system that provides information on enabling managers and executives to make informed decisions.
- An *executive information system* (SIS) is a special DSS that focuses exclusively on information reaching the business executive (Rockart & DeLong 1988).
- An *expert system* (ES) is a software system that emulates a human expert in a particular problem domain. The classic text, *Introduction to Expert Systems*, by Peter Jackson, represents a significant work in this area (see Jackson 1999).
- A recent addition to the MSS family is *knowledge management system* (KMS). This was recognized by Thomas Clark and colleagues in 2007, when they defined MSS to include DSS, EIS, BIS, and KMS (see Clark et al. 2007). These systems have emerged out of the need for organizations to have access to much larger volumes of (often unstructured) information than those at any point in the past.
- A *business intelligence system* (BIS) is a software system that incorporates a set of technologies that allow a business to operate on relevant information that is made available to its decision-makers (see Mulcahy 2007; Oracle 2016).
- An *enterprise resource planning system* (ERPS) is a comprehensive software system that facilitates strategic management in all principal functional areas of operation of a business enterprise. The ERPS typically includes several interrelated subsystems, each of which may qualify as a software system in its own right (see APTEAN 2015; Foster 2015b).

In addition to their intricate connection to the decision-making process in the organization, MSSs are characterized by their data-centeredness, i.e., they rely on data pulled from underlying database(s) to drive the information they provide for decision support. Additionally, MSSs typically require stringent security mechanisms, allowing only authorized access to the data that they manage. These two observations will become clearer in the upcoming discussions; hopefully, you will also develop a more profound understanding of the importance of carefully designing the underlying database for each of these software systems.

21.2 Building System Security through Database Design

Contemporary software engineering is typically influenced by quality factors including (but not limited to) development speed, precision, interoperability, user-friendliness, usefulness, and reusability. Software consumers have become quite impatient and reluctant about persisting with products that do not meet their expectations. Moreover, software developers are expected to deliver projects on schedule or face the wrath of disgruntled consumer(s).

As you are aware from your knowledge of software engineering (see Foster 2021; Schach 2011), the user interface is an integral part of a software system. Additionally, planning, constructing, and testing the user interface for a software system is a time-consuming and labor-intensive effort, which often takes relatively more time than other aspects of the software design (Kivisto 2000). Despite this, system flaws due to a lack of thorough testing are quite common. Moreover, there appears to be a preponderance of poorly designed user interfaces for software systems that are heavily used in corporate organizations. Given the importance of the user interface, taking meticulous steps to get it right is of paramount importance.

In the foregoing discussions (review Sections 1.2.1 and 13.2), the importance of system security has been emphasized. It turns out that in many cases, user interface and system security are intricately related. This is so because end users typically access system resources via the user interface. It therefore stands to reason that if we could build into the software design the capacity for integrating these two areas, such an achievement would significantly reduce the time spent on software design as well as software development. One way to achieve this end is through the development and application of a software application for that purpose; let's call it a *dynamic menu interface designer* (DMID).

The idea of a DMID is creation of a generic software system that facilitates the specification of the operational components and constraints of other software systems being constructed as well as the user access matrices for those software systems. This information is then used to dynamically generate the user interface for each end user who attempts to access the target software system(s). The generated user interface contains only resources to which the user has been given the authority to access. Moreover, multiple software systems can be facilitated, and end-user access to these systems is facilitated only through the DMID. In the upcoming section, the basic architecture of the DMID is described.

The DMID as proposed offers a number of advantages to the software engineering experience, particularly in the area of information system (IS) development. Included among the proposed advantages are the following:

- User interface specification is significantly simplified by transforming the problem to mere data entry. By providing the facilities for menus to be defined (via data entry) and loaded dynamically, based on the user's access rights, the software engineer is spared the responsibility of major user interface planning and design. The time gained here could be used in other aspects of the project. This, in practice, should significantly shorten the software development life cycle (SDLC).
- The shortening of the SDLC could result in a noticeable improvement in software engineering productivity, particularly for large projects. The hours gained in not having to program a user interface could be spent on other aspects of the project.
- The DMID addresses not only menu design but also user accessibility. It is constructed in such a way as to enable the following two constraints: Firstly, only users who are defined to the DMID can gain access to the software system(s) employing it. Secondly, through logical views, each user gets a picture of the system that is based on the user's authorization log. Only those resources, to which the user is authorized, will show up on the user's display. So apart from not being able to access other resources, the user is given the illusion that they do not exist. Hence, the user's perspective of the system is as narrow or broad as his/her span of authorization.
- The DMID provides support for future changes in the operational environment of the software system, without forcing a corresponding change in the underlying code. This will become clear as we proceed.

21.3 Case Study: Dynamic Menu Interface Designer

This section provides excerpts from other resources previously developed for the DMID project (see Foster 2011, 2012). We will examine the database requirements followed by the user interface requirements.

21.3.1 Database Requirements of the DMID

The basic architecture of the DMID calls for the use of a few relational tables, which are described in Figure 21.1. Notice that to aid subsequent referencing, each table is assigned a reference code (indicated in parentheses); attributes of each table are also assigned reference codes; all reference codes are indicated in square brackets; foreign key references are indicated in curly braces.

The relationship details of Figure 21.1 are clarified in the ERD of Figure 21.2. As the figure conveys, there are 10 entities and 12 relationships connecting the entities. Also observe that entities **E6 Subsystem (Menu) Constituents**, **E7 User Authorization to Operations**, **E8 User Authorization to Menu**, **E9 User Authorization to Systems**, and **E10 Organization – System Mapping** are associative entities (i.e., intersecting relations) that resolve M:M relationship between the respective connected entities.

Through these relational tables, one can accurately and comprehensively define the constituents, structure, and security constraints of the user interface for any software system that requires the use of user menu(s). Specifically, here is a summary of information that could be specified for each software system:

- Definitional details for the software system(s) employing the DMID
- Basic information about the participating organization(s)
- Basic information about users who will access their respective software system(s) via the DMID
- Definitional details for the operations that comprise each participating software system
- Definitional details for the menus and submenus for each participating software system
- The structure and operational constituents of each menu/submenu comprising each participating software system
- User authorization matrix for subsystems comprising each participating software system
- User authorization matrix for operations comprising each participating software system
- User authorization matrix for participating software systems
- Mapping of participating system(s) for each participating organization (particularly relevant if there are multiple participating organizations and multiple systems)

To facilitate the users having different perspectives of the software system(s), based on their authorization schedule, a number of logical views are required; the salient ones are described in Figure 21.3. With these logical views in place, the next step is to superimpose on the database, a user interface that provides the user with the operational objectives mentioned in Section 21.2, but reiterated here for reinforcement:

- Specification of operational components of software system(s) being constructed
- Specification of user access matrices for end users of the software system(s)
- User access to related software system(s) via the DMID
- Automatic generation of user menu(s) for each user based on his/her access matrix

E01. System Definitions: for storage of internal identifications of all information systems that use the DMID. Each system is assigned a unique identifier. Essential attributes include:

- System Code [SysCode]
- System Name [SysName]
- System Abbreviation [SysAbbr]
- Home Path [SysHome]

The primary key: {SysCode}

E02. Participating Organizations: for storage of internal identification(s) of the organization(s) that have access to the software system(s) of the host organization. Essential attributes include:

- Organization Code [OrgCode]
- Organization Name [OrgName]
- Organization Abbreviation [OrgAbbr]

The primary key: {OrgCode}

E03. System Users: for identification of all legitimate users of the system. Each user must belong to an organization that is recognized by the system. Essential attributes include:

- User Identification Code [UsrCode]
- User Login Name [UsrName]
- User First Name [UsrFName]
- User Last Name [UsrLName]
- User' Organization [UsrOrgCode] {**Refers to E02**}
- User Classification [UsrClass]
- User Password [UsrPssWrd]
- User Password Change Ceiling in days [UsrPssCeil]
- Date of Last Password Change [UsrPssChgD]

The primary key: {UsrCode}

E04. System Operations: for definition of all user operations (options) used. Essential attributes include:

- Operation Code [OpCode]
- Operation Implementation Name [OpIName]
- Operation Descriptive Name [OpDName]
- Operation Description [OpDscr]
- Operation Home Path [OprHome]

The primary key: {OpCode}

E05. Subsystem (Menu) Definitions: for definition of all major menus and sub-menus used in the (possibly different) software system(s). Each menu is assigned a unique identifier, and is attached to an information system. Essential attributes include:

- Menu Code [MnuCode]
- Menu Implementation Name [MnuIName]
- Menu Descriptive Name [MnuDName]
- Menu Description [MnuDscr]
- Menu's System Code [MnuSysCode] {**Refers to E01**}
- Menu's Home Path [MnuHome]

The primary key: {MnuCode}

Figure 21.1 Normalized relations for the DMID. (Continued)

E06. Subsystem (Menu) Constituents: the implementation of a M:M relationship between System Menu Definitions (E05) and System Operations (E04). Essential attributes include:

- Menu Code [MC_MnuCode] {**Refers to E05**}
- Menu Sequence Number [MC_MnuSeqN]
- Constituent Operation Code [MC_OpCode] {**Refers to E04**}

Candidate keys: {MC_MnuCode, MC_MnuSeqN}; {MC_MnuCode, MC_OpCode}

E07. User Authorization to Operations: the implementation of a M:M relationship between System Users (E03) and System Operations (E04). Essential attributes include:

- User Identification Code [AO_UsrCode] {**References E03**}
- Authorized Operation Code [AO_OpCode] {**References E04**}

The primary key: {AO_UsrCode, AO_OpCode}

E08. User Authorization to Menus: the implementation of a M:M relationship between System Users (E03) and System Menu Definitions (E05). Essential attributes include:

- User Identification Code [UM_UsrCode] {**Refers to E03**}
- Authorized Menu Code [UM_MnuCode] {**Refers to E05**}
- User Menu Sequence Number [UM_MnuSeqN]

Candidate keys: {UM_UsrCode, UM_MnuCode}; {UM_UsrCode, UM_MnuSeqN}

E09. User Authorization to Systems: the implementation of a M:M relationship between System Users (E03) and System Definitions (E01). Essential attributes include:

- User Identification Code [US_UsrCode] {**Refers to E03**}
- Authorized System Code [US_SysCode] {**Refers to E01**}
- User System Sequence Number [US_SysSeqN]

Candidate keys: {US_UsrCode, US_SysCode}; {US_UsrCode, US_SysSeqN}

E10. Organization – System Mapping: the implementation of a M:M relationship between Participating Organizations (E02) and System Definitions (E01). Essential attributes include:

- Organization Code [OS_OrgCode] {**Refers to E02**}
- System Code [OS_SysCode] {**Refers to E01**}
- System Sequence Number [OS_SysSeqN]

Candidate keys: {OS_OrgCode, OS_SysCode}; {OS_OrgCode, OS_SysSeqN}

Figure 21.1 (Continued) Normalized relations for the DMID.

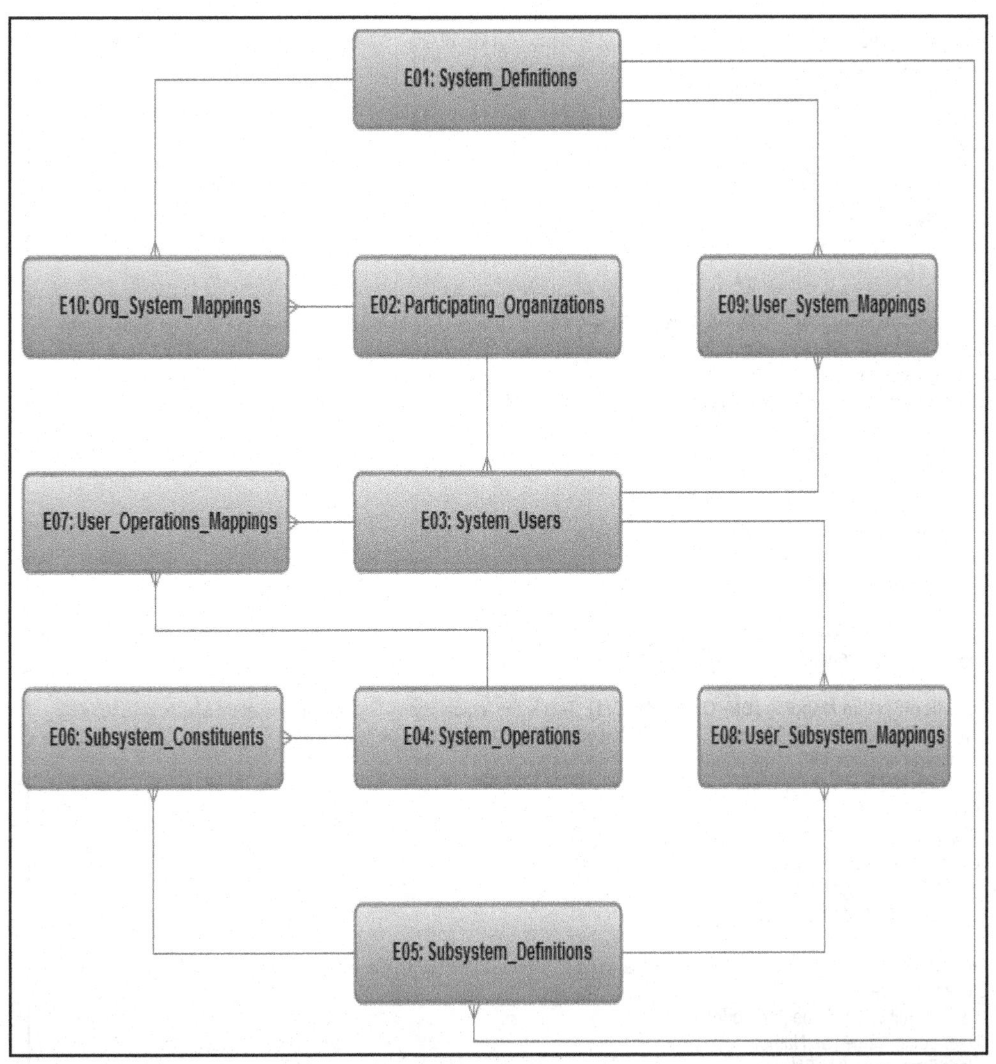

Figure 21.2 ERD for the DMID.

User's System Overview [DM_UsrSysA_LV1]: This is the logical join of User-System Authorizations (E09) with System Definitions (E01), and System Users (E03). Attributes will be read-only:

- User Identification Code [US_UsrCode]
- User Login Name [UsrName]
- User First Name [UsrFName]
- User Last Name [UsrLName]
- System Code [US-SysCode]
- User System Sequence Number [US_SysSeqN]
- System Name [SysName]
- System Abbreviation [SysAbbr]

User Menus Summary [DM_UsrMnuA_LV1]: This is the logical join of User-Menu Authorizations (E08) with Subsystem (Menu) Definitions (E05) and System Users (E03), and Subsystem (Menu) Definitions (E05) with System Definitions (E01). Attributes will be read-only:

- User Identification Code (UM_UsrCode)
- User Login Name [UsrName]
- User First Name [UsrFName]
- User Last Name [UsrLName]
- Menu Code (UM_MnuCode)
- Menu Implementation Name [MnuIName]
- Menu Descriptive Name [MnuDName]
- Menu's System Code [MnuSysCode]
- System Name [SysName]
- System Abbreviation [SysAbbr]
- User Menu Sequence Number (UM_MnuSeqN)

Organization-System Mapping [DM_OrgSysM_LV1]: This is the logical join of Organization-System Mapping (E10) with Participating Organizations (E02), and System Definitions (E01). Attributes will be read-only:

- Organization Code [OS_OrgCode]
- Organization Name [OrgName]
- Organization Abbreviation [OrgAbbr]
- System Sequence Number [OS_SysSeqN]
- System Code [OS_SysCode]
- System Name [SysName]
- System Abbreviation [SysAbbr]

System Users [DM_User_LV1]: This is the logical join of System Users (E03) with Participating Organizations (E02). Attributes will be read-only:

- User Identification Code [UsrCode]
- User Login Name [UsrName]
- User First Name [UsrFName]
- User Last Name [UsrLName]
- User' Organization [UsrOrgCode]
- Organization Name [OrgName]
- Organization Abbreviation [OrgAbbr]
- User Classification [UsrClass]
- User Password [UsrPssWrd]
- User Password Change Ceiling in days [UsrPssCeil]
- Date of Last Password Change [UsrPssChgD]

Figure 21.3 Important logical views for the DMID. (Continued)

User Operations Summary [DM_UsrOprA_LV1]: This is the logical join of User-Operation Authorization (E07) with Subsystem (Menu) Constituents (E06), System Operations (E04), System Users (E03), and Menu Definitions (E05). Attributes will be read-only:

- User Identification Code [UO_UsrCode]
- User Login Name [UsrName]
- User First Name [UsrFName]
- User Last Name [UsrLName]
- Authorized Operation Code [UO_OpCode]
- Operation Implementation Name [OpIName]
- Operation Descriptive Name [OpDName]
- Menu Code [MC_MnuCode]
- Menu Sequence Number [MC_MnuSeqN]
- Constituent Operation Code [MC_OpCode]
- Menu Implementation Name [MnuIName]
- Menu Descriptive Name [MnuDName]

System Menu Definitions [DM_MenuD_LV1]: This is the logical join of Subsystem (Menu) Definitions (E05) and System Definitions (E01). Attributes will be read-only:

- Menu Code [MnuCode]
- Menu Implementation Name [MnuIName]
- Menu Descriptive Name [MnuDName]
- Menu Description [MnuDscr]
- Menu's System Code [MnuSysCode]
- System Name [SysName]
- System Abbreviation [SysAbbr]

System Menu Constituents [DM_MenuC_LV1]: This is the logical join of Subsystem (Menu) Constituents (E06), Subsystem (Menu) Definitions (E05), System Definitions (E01), and System Operations (E04). Attributes will be read-only:

- Menu Code [MC_MnuCode]
- Menu Implementation Name [MnuIName]
- Menu Descriptive Name [MnuDName
- Menu's System Code [MnuSysCode]
- System Name [SysName]
- System Abbreviation [SysAbbr]
- Menu Sequence Number [MC_MnuSeqN]
- Constituent Operation Code [MC_OpCode]
- Operation Implementation Name [OpIName]
- Operation Descriptive Name [OpDName]

Key:

- Attributes in this blue-green color are taken from the reverencing relation
- Attributes in this rust-brown color are taken from the referenced relation(s)

Figure 21.3 (Continued) Important logical views for the DMID.

21.3.2 Overview of the DMID's User Interface Requirements

Having described the database requirements, let us now examine the user interface requirements for the DMID. This user interface will be superimposed on top of the relational database. Through the DMID's user interface, it must be possible to define and configure the user interface requirements for other software systems. In its first incarnation, the DMID's user interface itself is subject to future enhancements; however, it is fully operational.

The operational objectives of the DMID were stated in the previous subsection. From an end-user perspective, the DMID should also meet the following end-user objectives:

- **Usefulness:** Software engineers must be able to use the DMID as a means of shortening the development time on software systems that they have under construction.
- **Interoperability:** The DMID must be applicable across various software systems and operating systems platforms.
- **User-friendliness:** The system must be user friendly and easy to use. The interface must be intuitive so that there is a very short learning curve.
- **Reusability:** It must be possible to use the DMID on various independent software systems, as well as a conglomeration of software systems operating as part of a larger integrated system.
- **Flexibility:** The DMID must provide users with the flexibility of specifying the relative order of menu options comprising the software system being constructed.

The user's first interaction with the DMID is via logging on. In logging on, the user specifies the following: user identification code or name; organization the user represents; and password (not displayed in the interest of security). This information is checked against the internal representations stored in the database. If a match is found, the user is taken to the next stage; otherwise, the user is given an appropriate error message and allowed to try logging on again (the number of allowable attempts may be appropriately restricted). The current incarnation does not encrypt the password, but this is an enhancement that could be pursued.

In its preliminary configuration, two classes of users are facilitated — end users and system administrators (user classification is specified when the user account is defined). System administrators have access to the Administrative Specification Management Subsystem (ASMS). This subsystem provides select, update, delete and insert (SUDI) privileges to all the data contained in the DMID's internal database. This means the administrator can carry out functions of defining the operational requirements and constraints of software systems, subsystems, and users (as described in the upcoming section). End users have access to the End-user Access Control Subsystem (EACS), which will be elaborated on shortly.

21.3.3 Management of System Constraints via the DMID

The ASMS provides facilities for defining, reviewing, and modifying the operational scope of participating software systems using the DMID, as well as the operational constraints for end users of the participating software system(s). Among the related activities for this feature are the following:

- Addition of new system(s) and subsystems
- Deletion of (obsolete) system component(s)
- Addition of new operations and/or menu options
- Deletion of menu options
- Configuring/restructuring of system menus
- Setting user authorization schedules in respect of access to systems, subsystems, and operations
- Change of user authorization schedules in respect of access to systems, subsystems, and operations
- Addition of new users
- Deletion of users
- Setting the organization–system mapping
- Changing the organization–system mapping
- Reviewing (querying) system constraints information

Figure 21.4 shows a screen capture from the ASMS. In this illustration, a test user called Lambert is working with system definitions. At first entry into this option, an initial list of all software systems being managed through the DMID is provided. As these system definitions are modified, or if items are removed during the session, the list is updated. New entries can also be made. Finally, notice that there is a **Search** button to the bottom right of the panel. If this is clicked, a related operation is invoked as illustrated in Figure 21.5. This will allow the user (in this case Lambert) to peruse through the system definitions using any combination of the search criteria provided.

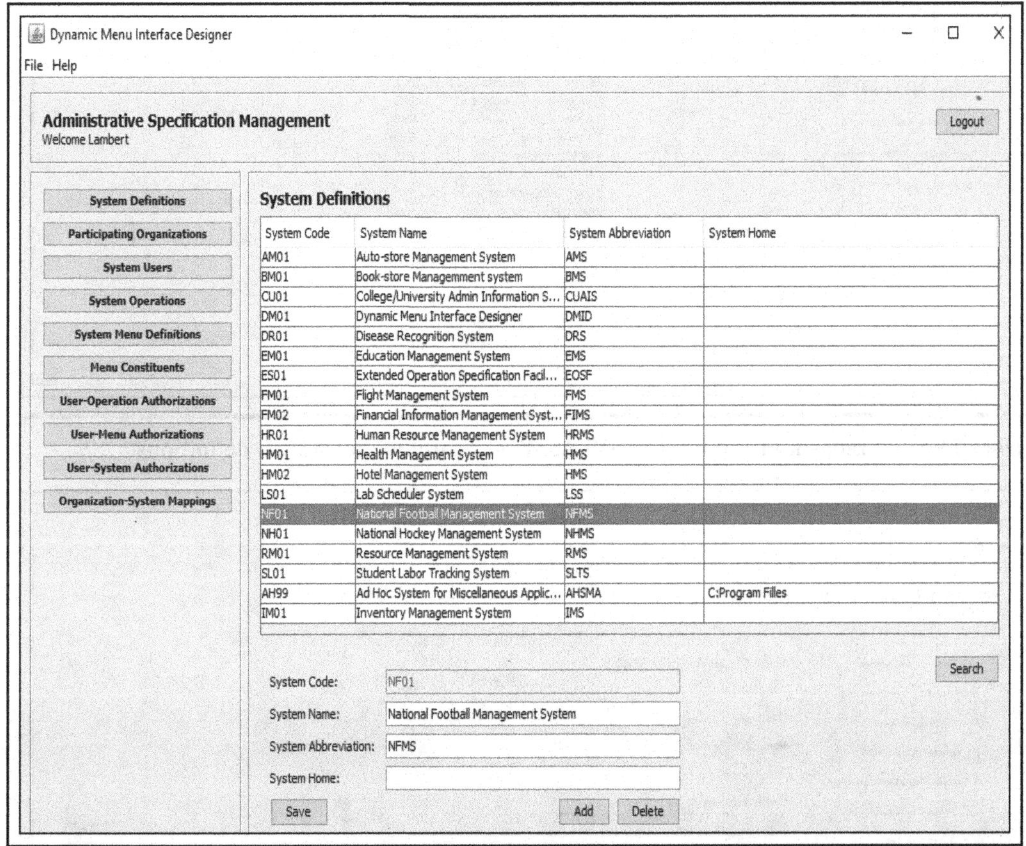

Figure 21.4 Screenshot from the DMID's ASMS —managing system definitions.

It should be noted that the illustrations of Figures 21.4 and 21.5 represent the most basic operational features of the ASMS. There are other more sophisticated operations involving access of the logical views of Section 21.2.1 (revisit Figure 21.3). Figure 21.6 provides an example. Here, user-menu authorizations are being managed; specifically, user Lambert is perusing a logical view that joins multiple tables (review the spec for **User Menus Summary [DM_UsrMnuA_LV1]** in Figure 21.3). He has the option of listing all user-menu combinations stored, or narrowing the search by specifying data for any of the six search criteria shown. Please note that each operation follows a similar design concept.

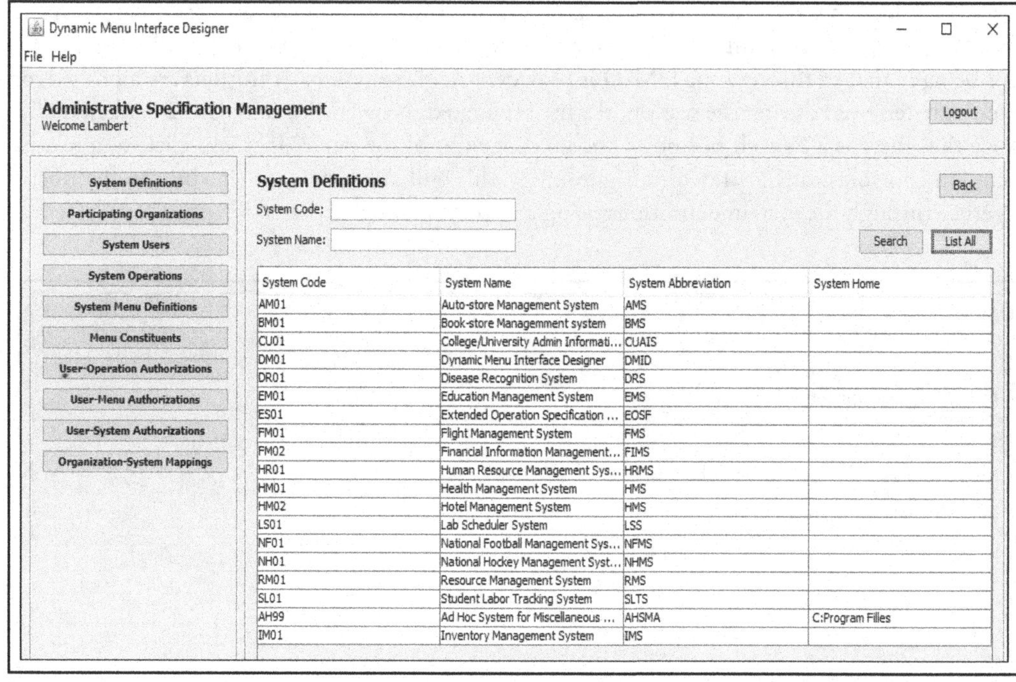

Figure 21.5 Screenshot from the DMID's ASMS — searching on system definitions.

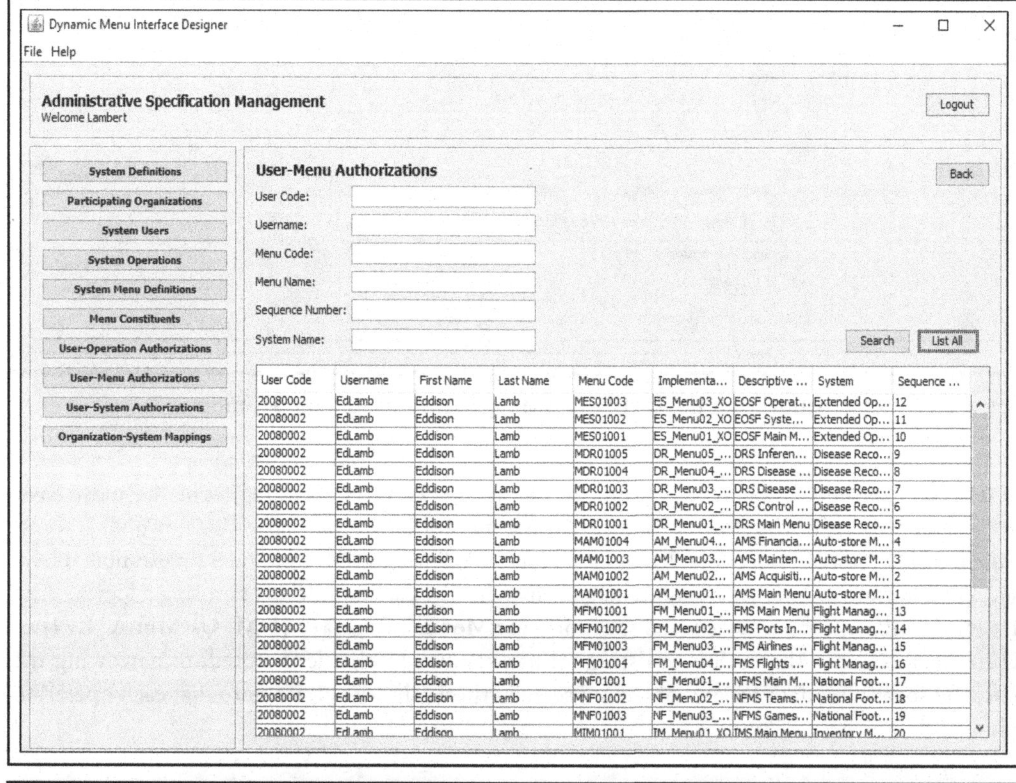

Figure 21.6 Screenshot from the DMID's ASMS — searching user-menu authorizations.

Only individuals who are duly authorized to carry out the functions of software system configuration and management will have access to the ASMS; they must have the *administrator* classification. Database administrators (DBAs) and software engineers would typically fall in this category.

21.3.4 Access to System Resources

Let us now revisit the End-user Access Control Subsystem (EACS) and provide some additional clarifications. Through this subsystem, an end user can only perform functions defined by his/her authorization schedule; in fact, only these capabilities will appear on that user's menu. End users will be oblivious to system resources that they are not authorized to access; and if by a stroke of luck or unauthorized means they attempt to access such resources, those attempted access will be denied.

Assuming successful logon, the user gets a dynamic system-generated menu, depending on his/her authorization schedule that is stored in the underlying DMID database. Three mutually exclusive scenarios are likely:

a. Being an end user, the user gets a menu with the software system(s) to which he/she has access rights.
b. The end user is provided with a blank menu, representing zero access to resources.
c. The user, being a system administrator, gains access to the ASMS menu. Administrative users have access to all resources.

From here on, the user's display panel varies according to what system resources he/she is authorized to access; only resources to which the user is authorized are shown on his/her display panel. Figures 26.7 and 26.8 illustrate two extremes: one in which a user has access to all the resources managed through the DMID, and another in which the user has negligible access to system resources.

In Figure 21.7a, the main menu for a user called Edison is shown. Notice that Edison has access to all the software systems managed via the current instance of the DMID; you can tell this is so by comparing the screenshot portrayed in Figure 21.7a with the one displayed in Figure 21.5. Also observed from Figure 21.7a that user Edison has selected the *Auto-store Management System* for the session depicted. In Figure 21.7b, the user (in this case Edison) is provided with a submenu of subsystems of the previously selected system, to which he/she has access privileges. Then, in Figure 21.7c, after selecting the *Acquisitions and Sales* subsystem from the previous screen, another submenu of related operations to which the user is authorized is provided. At this point, the user may select any desired operation for execution. At each level, the user's display panel is filled with options so that he/she is able to scroll and select the option of choice (by highlighting and clicking the **Select** push button). The end user is never shown a resource that he/she is not authorized to access.

In Figures 21.8a, 21.8b, and 21.8c, the user (in this case Ann Marie) has more limited access to systems, subsystems, and operations managed by the DMID. In the most extreme scenario, a user may have no access to any resource (system, subsystem, or operation), in which case a blank screen would appear, and the user would not be able to do anything.

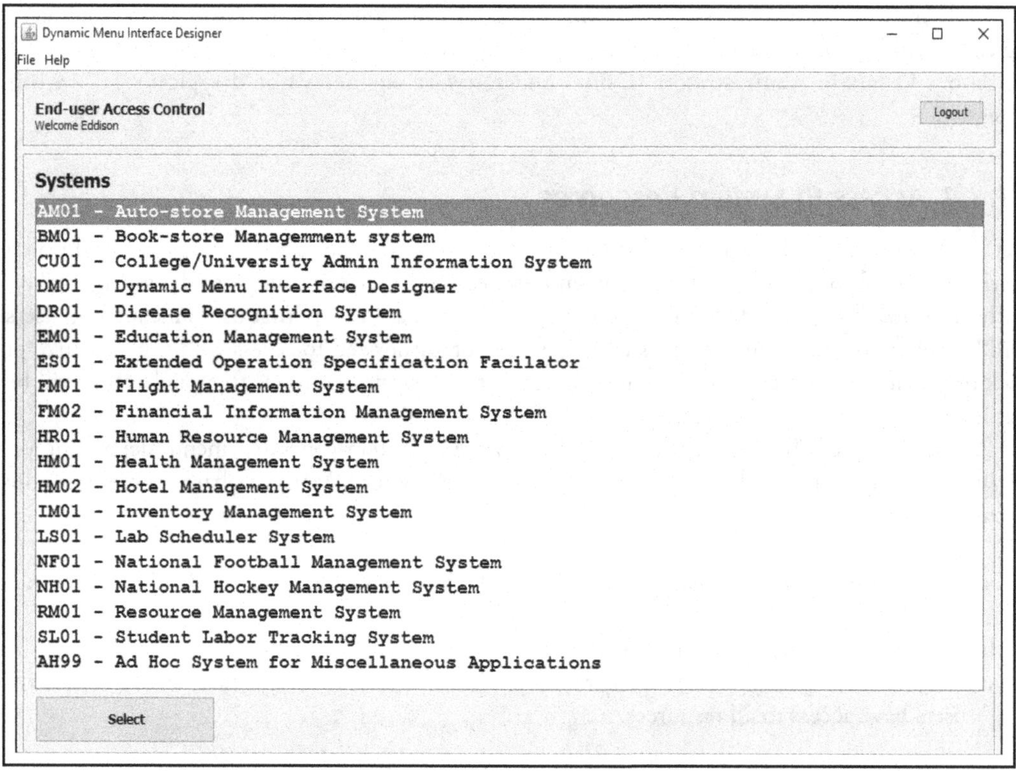

Figure 21.7a Screenshot from the DMID's EACS — main menu.

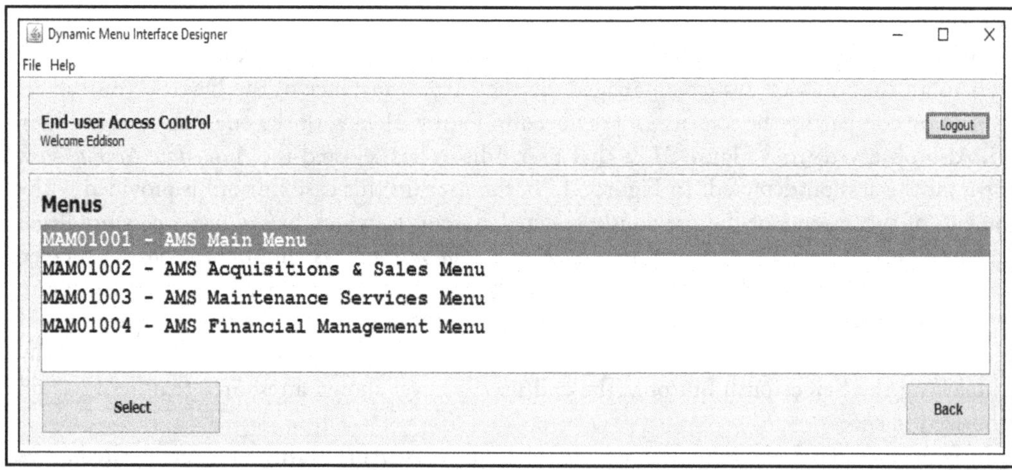

Figure 21.7b Screenshot from the DMID's EACS — subsystems menu for a specific software system.

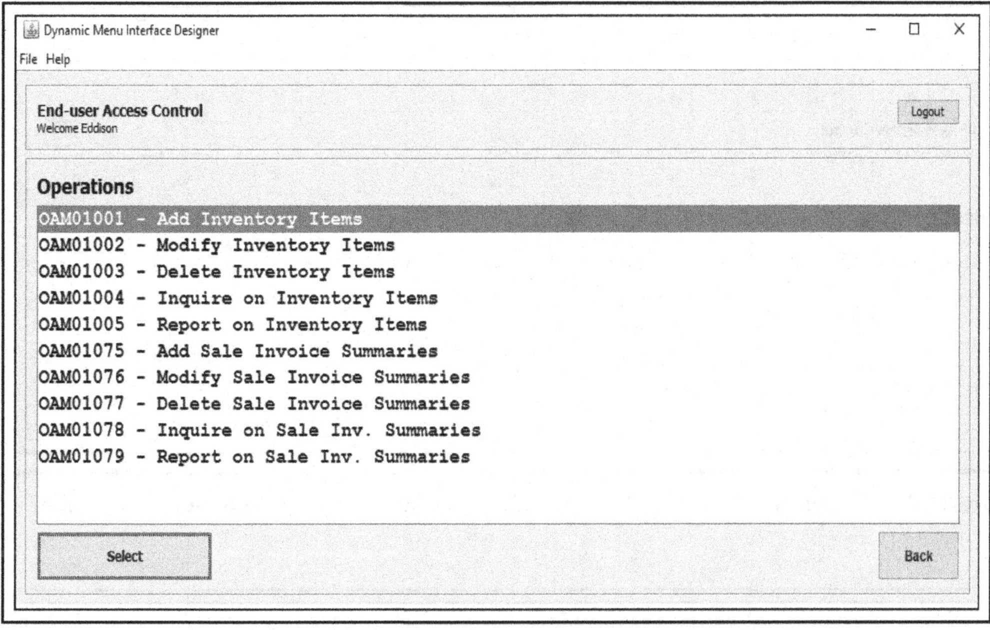

Figure 21.7c Screenshot from the DMID's EACS — operations menu for a specific subsystem.

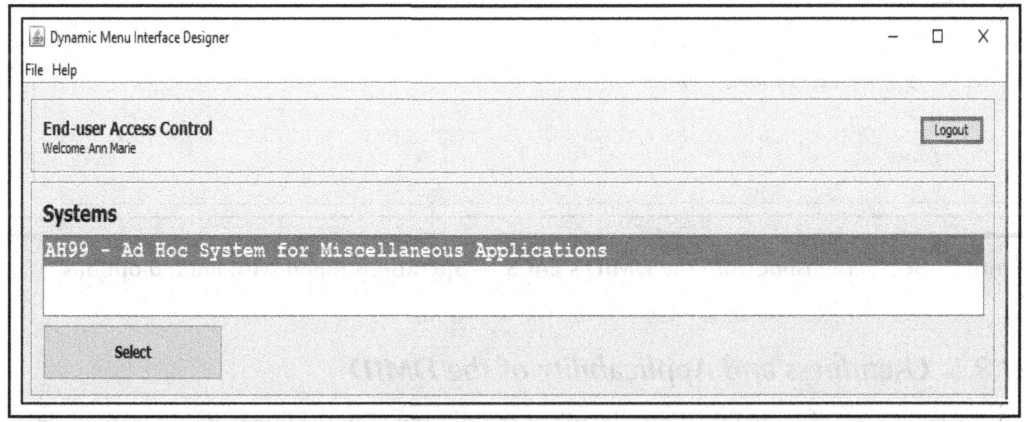

Figure 21.8a Screenshot from the DMID's EACS — main menu with only one option.

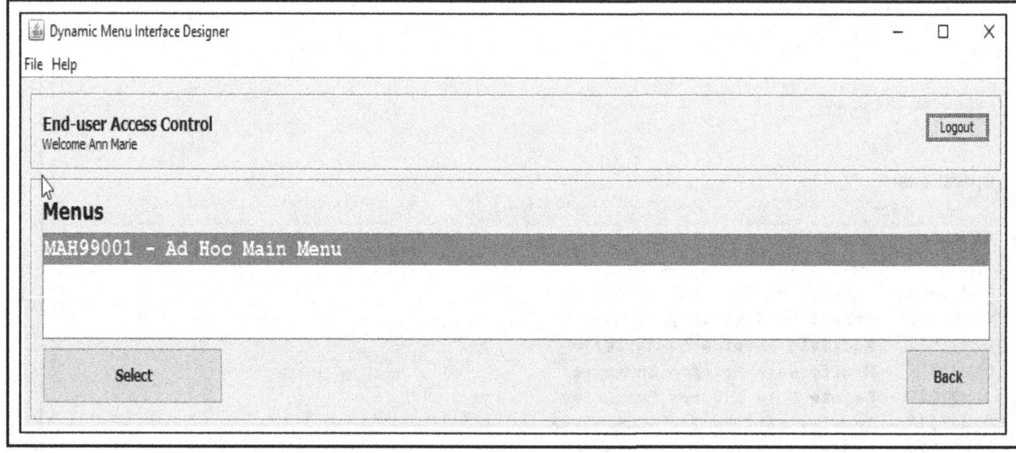

Figure 21.8b Screenshot from the DMID's EACS — subsystems menu with only one option.

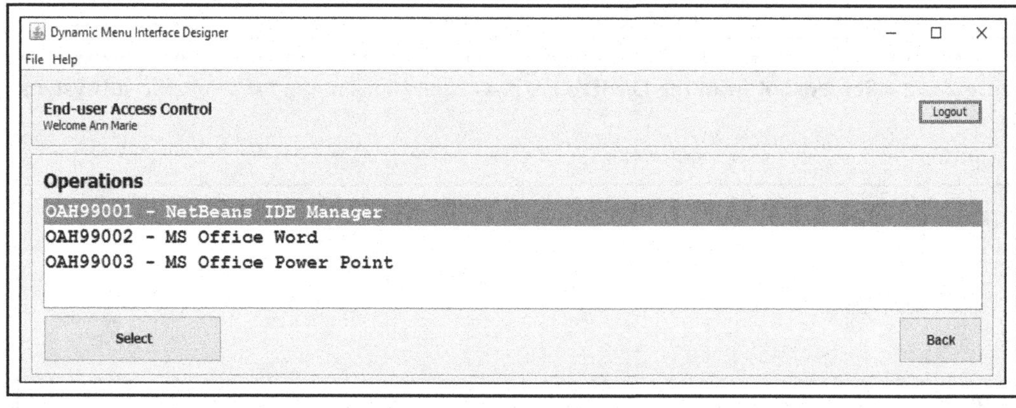

Figure 21.8c Screenshot from the DMID's EACS — operations menu with limited options.

21.3.5 Usefulness and Applicability of the DMID

The design concepts for the DMID, as described and illustrated in the foregoing subsections, may be replicated for any set of one or more complex software systems. Moreover, the DMID prototype that has been described above can be easily adopted and used for any software engineering endeavor requiring an elaborate user interface and stringent security arrangements for various end users. If used in this context, the DMID could significantly reduce the timeframe for the software development life cycle (SDLC) on specific software engineering projects by transforming previously arduous activities of time-consuming user interface design and development to mere data entry. When one considers the push toward more ubiquitous software components (for instance, see Niemelä & Latvakoski 2004), this design approach offers some promise.

21.4 Selected MSS Project Ideas

Management support systems appear in various aspects of life, including (but not confined to) engineering, management, education, health care. Following are the selected examples of MSSs and the kinds of information they manage; these examples are drawn from the author's ongoing software engineering research projects with undergraduate students of computer science. For each case presented, it would be necessary to carefully design the identified information entities to conform to relational database norms as discussed in Chapters 3–5. Also remember that for each project, only main entities are mentioned; it is anticipated that other related entities would emerge as part of the database design process. Finally, for each of the projects mentioned, the concepts of a DMID (as described and illustrated earlier) would be applicable.

21.4.1 Electoral Management System

An *Electoral Management System* (EMS) is an example of a SIS (as defined in Section 21.1) that could be used to support local as well as national elections. Among the main information entities that would be tracked are the following:

- Country employing the EMS (stored to avoid hard-coding)
- Provinces/territories in the country
- Electoral districts in each province
- Cities/towns in each electoral district
- Voter information for each eligible voter
- Candidate information for each candidate
- Election definition for each election managed by the EMS
- Election schedule for each election managed by the EMS
- Election referendum issues being tested
- Electoral vote
- Referendum vote
- Election results
- Referendum results
- Election summarized outcome — a summarized READ-ONLY-derived view that anonymizes each voter in the final result

While the list of information entities for this EMS appears very small, the following must be noted:

1. Data volume on some of these entities would be very large, depending on the population participating in the election in question.
2. The system could be configured to facilitate multiple local as well as national elections.
3. Stringent data security mechanisms would be required to ensure the integrity of the voting and vote-counting.
4. Depending on the country and/or territory, the use of special voting technology may be required. However, it is possible to develop a generic solution that uses readily available technologies and methodologies.

21.4.2 Health Information Management System

Another example of a SIS is a *Health Information Management System* (HIMS). This HIMS could be used to facilitate management of the health information for a conglomerate of health organizations (hospitals and clinics), a local town, a county, a province/state, or a whole nation. Following are some of the main information entities that would be managed in such a system:

- Country employing the HIMS (stored to avoid hard-coding)
- Provinces/territories in the country
- Patron information (for individuals requiring health care)
- Medical professional information (for medical professionals)
- Health institutions
- Medical examinations of patrons
- Medical prescriptions for patrons
- Medical records
- Health insurance providers
- Health insurance programs
- Common medical cases
- Answers to frequently-asked questions

At first glance, this list of information entities for the HIMS appears very small; however, bear in mind the following:

1. Similar to the previous case, data volume on some of these entities would be very large, depending on the population being served.
2. Medical records often involve the use of high-precision graphical information (for example X-rays and other similar records); the system must be designed to efficiently and effectively handle such data.
3. Stringent data security mechanisms would be required to ensure the integrity of all information stored and accessed, with a zero tolerance for unauthorized access. This security mechanism should include a strategy for encrypting the data.
4. The system would be tailored for the application of various technologies used in modern medicine.

21.4.3 Strategic Education Management System

Most governments have a vested interest in their education system at the local, provincial, and national levels. A *Strategic Education Management System* (SEMS) is a SIS that would be geared toward facilitation of the management of an education system at various levels all the way up to the national level and for various categories of educational institutions. The core information entities that would be managed in a SEMS are the following:

- Country implementing the SEMS (stored to avoid hard-coding)
- Provinces/territories in the country
- Educational districts in each province
- Cities/towns in each province
- Categories of educational institutions (elementary school, high school, college, university, vocational)
- Educational institutions in various categories
- Performance criteria for institutions in various categories
- Performance criteria for students in various categories
- Evaluation of institutional performance at various levels against established institutional performance criteria

- Evaluation of the student performance at various levels against established student performance criteria
- Tracking of outputs (i.e., graduates) from the various institutions of learning
- Forecasting future outputs and demands based on historical trends

As in the previous two examples, the list of information entities for this HIMS may appear to be small; however, the following must be noted:

1. Similar to the previous case, data volume on some of these entities would be very large depending on the population being served.
2. Educational records often involve the use of high-precision graphical information (for example, X-rays and other similar records); the system must be designed to efficiently and effectively handle such data.
3. Stringent data security mechanisms would be required to ensure the integrity of all information stored and accessed, with a zero tolerance for unauthorized access. This security mechanism should include a strategy for encrypting the data.
4. The system would be tailored for the application of various technologies (for library management, financial management, learning management, etc.) used in modern education.

21.4.4 Flight Information Management System

The definition for a DSS is provided in Section 21.1. A *Flight Information Management System* (FLIMS) qualifies as DSS that facilitates effective management of flights at an airport. Moreover, the FLIMS could be designed to be customizable for a group of airports for an entire country. Following are some of the main information entities that would be managed in a FLIMS:

- Countries of interest
- Ports of interest in various countries
- Airlines with which the port does business
- Aircraft information on each aircraft for each airline that uses the airport
- Information on each runway (including terminals and gates)
- Information on all pilots and flight attendants
- Standard flight schedule of all flights to and from the airport
- Log of actual arrivals/departures for flights to and from the airport

Depending on the airport where implementation takes place, FLIMS could be a data-intensive system. For instance, at a busy airport such as Logan International, there are over 20,000 flights per month on average. At a less busy airport, this average may be down to a few hundred flights per month. Additionally, security and data integrity would be extremely important objectives, with zero tolerance for unauthorized access or incorrect data.

21.4.5 Financial Information Management System

Though this example has become an almost trivial solution in many organizations due to the abundance of existing solutions; it still remains a rather challenging problem for undergraduate students of computer science to appreciate, hence its inclusion: A *Financial Information Management*

System (FIMS) is an example of an EIS as defined in Section 21.1. The FIMS qualifies as an EIS because it provides useful decision support for the financial manager in any organization. Among the core information entities managed are the following:

- Country implementing the SEMS (stored to avoid hard-coding)
- Provinces/territories in the country
- Organization in which the FIMS is being implemented (stored to avoid hard-coding)
- Chart of accounts — a list containing all valid account numbers and clarifications about them
- Account balances
- Financial transactions
- Transaction classifications
- Purchase payment plans for payments to suppliers
- Payments made to suppliers
- Financial institutions
- Sales payment plans for payments from customers
- Payments received from customers
- Investments log
- Departments of Interest within the organization
- Customers (representing sources of revenue)
- Suppliers (representing sources for expenditure)
- Inventory/resource Items (representing items used to conduct business)
- Purchase orders generated and sent out to suppliers
- Purchase invoices received from suppliers
- Purchase returns made to suppliers
- Sale orders generated and sent to customers
- Sale invoices generated and sent to customers
- Sale returns received from customers

This list of entities gets even longer after normalization (for more insight, see Appendix 3 of Foster 2021). Here are some additional insights:

1. Depending on the size, scope, and complexity of the organization in which implementation takes place, data volume on some of these entities would be very large.
2. As in the previous cases, stringent data security mechanisms would be required to ensure the integrity of all information stored and accessed, with a zero tolerance for unauthorized access.
3. The system would be tailored for the application of various technologies used in financial management.
4. Depending on the context of implementation, entities from Departments of Interest to the end of the list may already have internal representations in other existing software systems within the organization. Should that be the case, integration (or at minimum, collaboration) with such system(s) would be necessary.

21.4.6 Disease Recognition System

Expert systems (ESs as defined in Section 21.1) appear frequently in various aspects of the 21st-century lifestyle. Among the areas that such systems frequent are engineering, business, manufacturing, health care, etc. One example of an ES is a *Disease Recognition System* (DRS), tailored to facilitate preliminary diagnosis of certain diseases in specified problem domains. In this regard, the DRS fulfills the role of a medical expert. The required database to support this DRS would include the following core entities:

- Disease Categories
- Disease Definitions
- Disease Symptoms — listing various symptoms for the diseases of interest
- Disease Symptoms Matrix — for mapping each disease to its various symptoms
- Countries of Interest — for addressing purposes
- Provinces/Territories within each country of interest — for addressing purposes
- Vising Patrons — for tracking individuals that use the DRS
- Standard Prescriptions — stored for various diseases
- Diagnosis Summaries — for summarizing each diagnosis
- Diagnosis Symptoms — for noting the observed symptoms of the diagnosed disease
- Diagnosis Prescriptions — for storing information on prescriptions issued
- Medical Professionals — for storing information about the participating physicians

Since the number of known diseases is extremely large, it is impractical to think that such a system covers all possibilities. More realistically, the system would be configured to cover a specific set of diseases. Following are some additional insights:

1. Depending on the set of diseases studied as well as the targeted number of participants, data volume on some of these entities could be very large or relatively small.
2. As in the previous cases, stringent data security mechanisms would be required to ensure the integrity of all information stored and accessed, with a zero tolerance for unauthorized access. This security mechanism should include a strategy for encrypting the data.
3. For the system to work, it will be necessary to observe several software engineering procedures: constructing an information-gathering instrument for specific diseases and their symptoms; codification of that information and entry into the database; construction of an inference engine; testing and refinement. The intent here is to highlight the role of the underlying database for such a system.
4. In recent years, IBM's supercomputer Watson, which started out with emulation of the game Jeopardy, is taking on the role of a DRS as described above (for more information on this, see IBM 2015). This DRS initiative started long before Watson, and no doubt, there will be other initiatives after Watson. Nonetheless, the Watson initiative offers huge potential benefits to the fields of expert systems as well as health.

21.4.7 Cognitive Leadership Analysis System

Another ES project worthy of mention is a *Cognitive Leadership Analysis System* (CLAS). The CLAS project involves the design, construction, and testing of an expert system to identify and encourage the development of leadership styles and qualities in individuals and organizations. The idea is to identify certain leadership styles and qualities, allow respondents to fill out a questionnaire to measure disposition to those leadership styles and qualities, and then generate a statistical profile of the respondent's leadership dispositions based on the responses made. Among the Core information entities included in the underlying database are the following:

- Leadership Styles of interest
- Leadership Qualities of interest
- Leadership Style Questions
- Leadership Quality Questions
- Leadership Analysis Rules
- Individual Evaluation Instruments
- Participating Individuals
- Individual Responses for Style
- Individual Responses for Quality
- Individual Response Analysis for Style
- Individual Response Analysis for Quality
- Participating Organizations
- Organizational Evaluation Instruments
- Organizational Responses for Style
- Organizational Responses for Quality
- Organizational Response Analysis for Style
- Organizational Response Analysis for Quality

Observe that the various leadership styles and qualities are not hard-coded, but treated as data entry. The same is true for the questions geared toward measuring these styles and qualities. The questions are connected to their related styles and/or qualities through leadership analysis rules. These strategies make the system very flexible and customizable for different scenarios of interest. This flexibility is enabled and facilitated by thoughtful database design. Following are some additional insights on the project:

1. The proposed system can be used by individuals or by an organization. Depending on the context of usage as well as the number of participants, data volume on some of these entities could be very large or relatively small.
2. For the system to work, the following software engineering procedures will be necessary: constructing an information-gathering instrument for specific leadership styles and qualities, as well as the criteria (questions) for measuring them; codification of that information and entry into the database; codification of the leadership analysis rules; construction of an inference engine to evaluate the leadership capacity of the respondents; testing and refinement. The intent here is to highlight the role of the underlying database for such a system.

21.4.8 Financial Status Assessment System

Here is another example also qualifies as an ES: A *Financial Status Assessment System* (FSAS) to evaluate the financial health of an individual or organization based on known information about the financial dealings and obligations of the past for that person or organization. This is similar to what banks and other financial institutions do in order to determine one's credit worthiness. Indeed, the current credit score system that is administered by the three leading credit evaluation agencies (Experian, Equifax, and TransUnion) fulfill this role for the American public. The idea is to identify financial performance criteria, evaluate individual and/or organizational performances based on those criteria, and then generate a statistical profile of the respondent's financial profile. However, rather than exacting real penalties and/or rewards on its users, the purpose of such a software system would be educational — working with what-if scenarios and test data, to teach the consequences of financial decisions.

Among the core information entities that may be included in the underlying database for this FSAS are the following:

- Country implementing the SEMS (stored to avoid hard-coding)
- Provinces/territories in the country
- Individual identity information
- Sources of Periodic Income/Expense — for identifying all sources of income/expense
- Inflow/Outflow Categories — for analyzing inflow/outflow of funds
- Financial Institutions — for storing data on various financial institutions
- Financial Obligations in loans
- Financial Obligations in regular bills excluding loans
- Cash Inflows — for storing periodic inflow of funds
- Cash Outflows — for storing periodic outflow of funds
- Financial Investments — for storing data on investments made by the individual
- Investment Categories — for analyzing investments
- Benchmarks and Evaluation Rules — for storing formulas and/or performance benchmarks
- Personal Evaluation Summaries — for storing the summaries of financial evaluations
- Personal Evaluation Summaries — for storing the details of financial evaluations

While the specific formulas used by credit agencies are not known (to this course), a software system like this enhances a more profound understanding of what these agencies do, and what financiers look for in customers that they sponsor. Here are some additional insights on the project:

1. As in the case of the previous case, the proposed FSAS project can be used by individuals or by an organization. Depending on the context of usage as well as the number of participants, data volume on some of these entities could be very large or relatively small.
2. As in the previous cases, stringent data security mechanisms would be required to ensure the integrity of all information stored and accessed, with a zero tolerance for unauthorized access.
3. For the system to work, the following software engineering procedures will be necessary: constructing an information-gathering instrument for specific the financial dealings of individuals/organizations, as well as the criteria for measuring them; codification of that information and entry into the database; construction of an inference engine to evaluate the financial status of the respondents; testing and refinement. Again, the intent here is to highlight the role of the underlying database for such a system.

21.4.9 College/University Administrative Information System

Recall from Section 21.1 that an ERPS is a software system that addresses all the principal functional areas of an organization. This final example qualifies as an ERPS: a College/University Administrative Information System (CUAIS). This is a huge software system that facilitates effective management of a college environment. The project encompasses over 100 information entities; full discussion is beyond the scope of this course.

For the purpose of initial discussion, Figure 21.9 shows a partial *information topology chart* (ITC) for the CUAIS project. The ITC is a diagramming methodology described in Foster (2015a); it presents the main information entities for a software engineering project, organized in the logical areas (subsystems or modules) where they will likely be managed. Also notice from the diagram that eight subsystems are included — Infrastructure and Resource Management System (IRMS), Public Relations and Alumni Management System (PRAMS), Curriculum and Academic

Management System (CAMS), Human Resource Management System (HRMS), Student Affairs Management System (SAMS), Financial Information Management System (FIMS), Library Management System (LMS), and Cafeteria Services Management System (CSMS).

Then, Figure 21.10 shows an object flow diagram (OFD) for the CUAIS project. In this illustration, you will notice the eight aforementioned subsystems and a ninth one for the central database that is accessed by each subsystem. The bidirectional arrows convey that each subsystem has read-write access to the database.

Referring again to Figure 21.9, note that for each subsystem only a partial list of data entities is provided. After applying database design principles of Chapters 3–5, the number of entities would significantly increase. Here are some additional insights:

1. Depending on the size of the institution and the time period over which the underlying database is used, the data volume on some of these entities would be expected to get very large.
2. As in the previous cases, stringent data security mechanisms would be required to ensure the integrity of all information stored and accessed, with a zero tolerance for unauthorized access.
3. The system would be tailored for the application of various technologies used in higher education.
4. Similar to the discussion of Section 21.4.3, educational records often involve the use of high-precision graphical information; the system design would need provisions to efficiently and effectively handle such data.

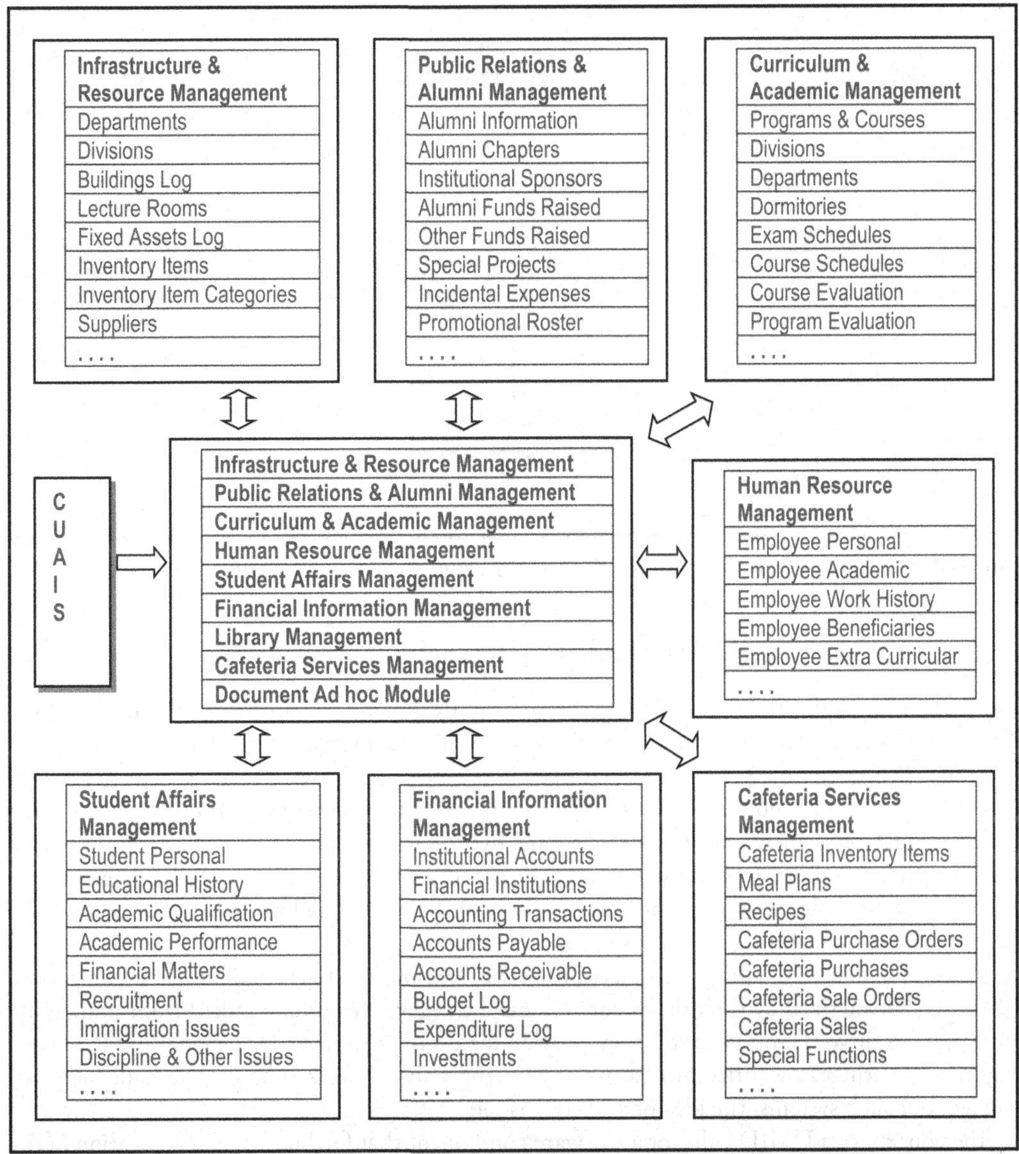

Figure 21.9 Partial information topology chart for the CUAIS project.

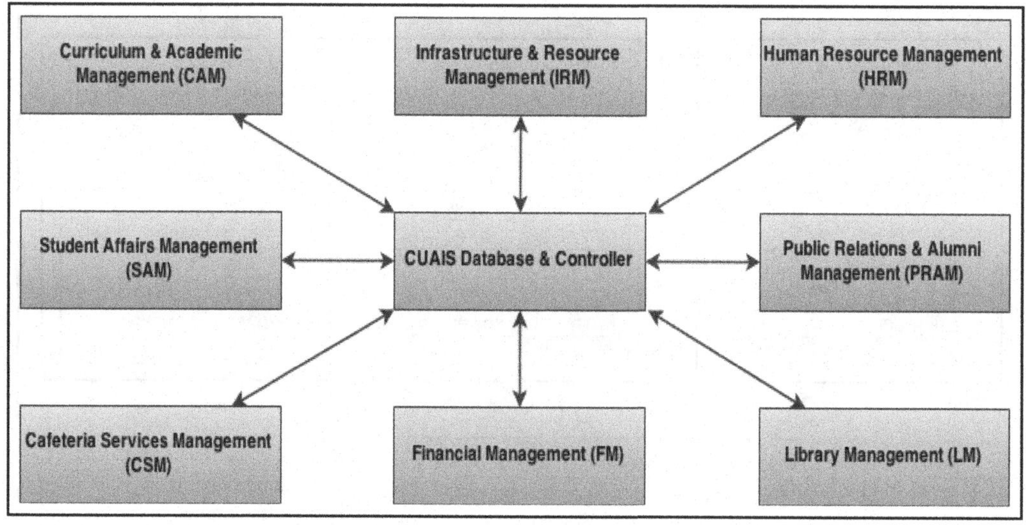

Figure 21.10 Object flow diagram for the CUAIS project.

21.5 Summary and Concluding Remarks

This chapter highlights the dominant role that a carefully designed database can play in the design of a software system. Here is a summary of the salient points covered in the chapter:

Management support systems (MSSs) refer to a family of software systems that are geared toward the promotion and facilitation of efficient and effective management and decision-making in the organization. Included among MSSs are the following categories: strategic information systems (SISs), decision support systems (DSSs), executive information systems (EISs), expert systems (ESs), knowledge management systems (KMSs), business intelligence systems (BISs), and enterprise resource planning systems (ERPSs).

In addition to having intricate connections to the decision-making process in the organization, MSSs are characterized by their data-centeredness, i.e., they rely on data pulled from underlying database(s) to drive the information they provide for decision support. Additionally, MSSs typically require stringent security mechanisms, allowing only authorized access the data that they manage. For such systems, the idea of a DMID is relevant.

The concept of a DMID calls for a software component that facilitates the specification of the operational components and constraints of another software system being constructed, as well as the user access matrices for that software system. This information is then used to dynamically generate the user interface for each end user who attempts to access the target software system.

The DMID discussed in this chapter embodies a database that includes 10 normalized relations and 12 relationships connecting them. Through logical views, these relations are combined to create a virtual environment that facilitates easy enforcement of desirable security constraints. Superimposed on top of this database design is a user interface that facilitates specification of aforementioned constraints, while achieving objectives of usefulness, interoperability, user-friendliness, reusability, and flexibility.

The DMID proposed in this chapter has two main subsystems — the Administrative Specification Management Subsystem (ASMS) and the End-user Access Control Subsystem (EACS). The ASMS is accessed only by system administrators; it provides SUDI privileges to all the data contained in the DMID's internal database. This means the administrator can carry out functions of defining the operational requirements and constraints of software systems, subsystems, and users (as described in the upcoming section). End users have access to the End-user Access Control Subsystem (EACS), which dynamically generates their menus at login time. The EACS also controls user access by including on the user's menu-only system resources to which they have been granted access privileges.

The design concepts for the DMID may be replicated for any set of one or more complex software systems. Moreover, the DMID prototype that has been described above can be easily adopted and used for any software engineering endeavor requiring an elaborate user interface and stringent security arrangements for various end users. If used in this context, the DMID could significantly reduce the timeframe for the software development life cycle (SDLC) on specific software engineering projects by transforming previously arduous and time-consuming activities of user interface design and development to mere data entry.

MSSs appear in various aspects of life, including (but not confined to) engineering, management, education, and health care. Among the examples of MSSs discussed are the following:

- Electoral Management System (EMS) — for facilitating local as well as national elections
- Health Information Management System (HIMS) — for facilitating management of the health information for a conglomerate of health organizations (hospitals and clinics), a local town, a county, a province/state, or a whole nation
- Strategic Education Management System (SEMS) — for enhancing the management of an education system at various levels all the way up to the national level, and for various categories of educational institutions
- Flight Information Management System (FLIMS) — for empowering effective management of flights at an airport
- Financial Information Management System (FIMS) — for providing useful decision support for the financial manager in any organization
- Disease Recognition System (DRS) — for facilitating preliminary diagnosis of certain diseases in specified problem domains
- Cognitive Leadership Analysis System (CLAS) — for identifying and encouraging the development of leadership styles and qualities in individuals and organizations
- Financial Status Assessment System (FSAS) — for evaluating the financial health of an individual or organization based on known information about the financial dealings and obligations of the past for that person or organization
- College/University Administrative Information System (CUAIS) — for effective management of a college environment

In your course on software engineering, you will no doubt learn about legacy software systems (see Chapter 22 of Foster 2021). As you begin to practice database systems and/or software engineering, you will learn that at the heart of most problematic legacy systems is a poorly designed database system; you will also learn and understand that replacing a poorly designed database after it has been placed in production is often a very daunting task. It therefore behooves every practicing software engineer and/or database expert to invest much time in getting the database properly designed at the outset; in so doing, you will save yourself and/or others much grief later on.

This takes us to the end of division D of the text, and the end of the critical mass of DBS knowledge that you need to master. If you understand most of the material covered, and now have a desire to delve more deeply into some aspects of database systems, then the course has succeeded in its intent. If you find that you now have a strong desire to make database systems one of your areas of expertise, then welcome to the community! You will find it a wonderfully rewarding and progressive field. The next division of the course provides you with a cursory overview of selected database management system (DBMS) suites.

21.6 Review Questions

1. What two aspects of many software systems are usually intricately linked?
2. What do you understand by the acronym DMID? Briefly explain how a DMID as described in this chapter allows you to seamlessly tie the aforementioned two aspects of a software system together.
3. Describe the essential database components of a DMID.
4. Describe the essential user interface components of a DMID.
5. What does the acronym MSS mean? What is an MSS? Identify the different categories of software systems that comprise the MSS family.
6. Identify two dominant characteristics of MSSs.
7. Identify four examples of MSSs. For each, briefly outline the database requirements.

References and/or Recommended Readings

APTEAN. 2015. *ERP Solutions.* Accessed June 1, 2016. http://www.aptean.com/solutions/application/erp-solutions

Clark, J., Thomas D., Jones, M. C., & Armstrong, C. P. 2007. "The Dynamic Structure of Management Support Systems: Theory Development, Research Focus, and Direction." *MIS Quarterly* vol. 31, no. 3, pp. 579–615. Accessed February 1, 2013. http://ezproxy.library.capella.edu/login?url=http://search.ebscohost.com.library.capella.edu/login.aspx?direct=true&db=bth&AN=25980839&site=ehost-live&scope=site

Foster, Elvis C. 2011. *Design Specification for the Dynamic Menu Interface Designer.* Accessed January 2016. www.elcfos.com

Foster, Elvis C. 2012. "Dynamic Menu Interface Designer." In *Athens Institute for Education and Research, Enterprise Management Information Systems.* Selected Papers from the 7th and 8th International Conferences on Information Technology and Computer Science, June 13–16, 2011, May 21–24, 2012. Athens, Greece. Chapter 23.

Foster, Elvis C. 2015a. "Three Innovative Software Engineering Methodologies." In 2nd Global Online Conference on Information & Computer Technology, November 4–6 2015. Sullivan University, Louisville, Kentucky.

Foster, Elvis C. 2015b. "Towards Measuring the Impact of Management Support Systems on Contemporary Management." In ATINER 11th Annual International Conference on Information Technology & Computer Science, May 18–21, 2015. Athens, Greece.

Foster, Elvis C. 2021. *Software Engineering: A Methodical Approach* 2nd ed. New York: CRC Publishing.

IBM. 2015. *IBM Watson Health.* Accessed January 2015. http://www.ibm.com/smarterplanet/us/en/ibm-watson/health/

Jackson, P. 1999. *Introduction to Expert Systems* 3rd ed. New York: Addison-Wesley.

Keen, P. G. W., & M. S. Scott Morton. 1978. *Decision Support Systems: an Organizational Perspective.* Reading, MA: Addison-Wesley.

Kivisto, Kari. 2000. *A Third Generation Object-Oriented Process Model.* Department of Information Processing Science, University of Oulu. Accessed January 2016. http://herkules.oulu.fi/isbn9514258371/isbn9514258371.pdf (July 2010). See chapter 3.

Mulcahy, Ryan. 2007. "Business Intelligence Definition and Solutions." *CIO*, March 2007. Accessed June 1, 2016. http://www.cio.com/article/2439504/business-intelligence/business-intelligence-definition-and-solutions.html

Niemelä, E., & J. Latvakoski. 2004. "Survey of Requirements and Solutions for Ubiquitous Software." In Proceedings of the 3rd International Conference on Mobile and Ubiquitous Multimedia MUM 04 (2004). ACM Press, 71–78.

Oracle. 2016. *Business Intelligence and Data Warehousing*. Accessed June 1, 2016. http://www.oracle.com/technetwork/topics/bi/whatsnew/index.html

Rockart, J., & D. DeLong. 1988. *Executive Support Systems*. Homewood, IL: Dow Jones-Irwin.

Schach, Stephen R. 2011. *Object-Oriented and Classical Software Engineering* 8th ed. Boston, MA: McGraw-Hill.

Wiseman, C. 1988. *Strategic Information Systems*. Homewood, IL: Irwin.

OVERVIEW OF SELECTED DBMS SUITES AND TECHNOLOGIES

E

This division of the text is dedicated to providing an overview of selected database management system (DBMS) suites and emerging technologies. The intent is not to make you experts in these products; that would be impossible. Rather, it is hoped that the material provided, though cursory, will help you develop an appreciation for the products, and a desire to want to delve deeper into these products. For each product, the discussion proceeds as follows: introduction; identification of the main components; discussion of selected components; identification of some shortcomings; summary and concluding remarks.

The division consists of five chapters, each focused on a particular DBMS suite. The chapters to be covered include the following:

- Chapter 22 — Overview of Oracle
- Chapter 23 — Overview of DB2
- Chapter 24 — Overview of MySQL
- Chapter 25 — Overview of Microsoft SQL Server
- Chapter 26 — Other Emerging Database Methodologies

Please note that for each product, several volumes of documentation have been written. It will therefore not be possible to cover them in detail. Rather, in each case, an overview is provided, based on these captions.

DOI: 10.1201/9781003275725-26

Chapter 22

Overview of Oracle

This chapter provides an overview of the Oracle database management system (DBMS). The truth is, you have been exposed to Oracle throughout division C of the text, since the implementation of Structured Query Language (SQL) that we have studied is Oracle-based. The chapter proceeds under the following subheadings.

- Introduction
- Main Components of the Oracle Suite
- Shortcomings of Oracle
- Summary and Concluding Remarks

22.1 Introduction

Oracle is widely regarded as one of the world's leading relational DBMS (RDBMS) suites. The product was developed by a company that bears the same name as the product. The product, unlike many of its competitors, has benefited from an extremely focused corporate mission. The Oracle DBMS was first introduced in the 1980s. By the early 1990s, Oracle Corporation was the world's leading software engineering company. Today, the company consistently ranks among the world's top ten largest and most successful software engineering companies and first in the areas of database management systems and Internet applications.

The Oracle DBMS suite is a comprehensive conglomeration of software development tools for development and/or management of any information system with an underlying Oracle database. The product has been through several stages of revision and upgrade. For instance, the product Oracle 12C was widely known and appropriately named for its support of cloud computing, as well as no upper limit on database size. As a follow-up, the product Oracle 19C was known and named primarily for its support of autonomous technology. In addition to its support of autonomous technology, Oracle 19C provides the reinforcement of the multitenant architecture in cloud-based settings, enhanced performance via various strategies including Automatic Indexing and Database In-Memory for storing frequently accessed table data in memory, high availability strategies for minimizing database downtime and enhancing database availability, enhanced security features, and many more automatic features.

DOI: 10.1201/9781003275725-27

The latest release, Oracle 21C, is regarded by the company as the *innovation release*—the ultimate long-term release, supporting the 12C, 18C, and 19C families. Oracle 21C introduces several new features and enhancements that further improve productivity for developers, analysts, and data scientists; the release also increases query performance. Among the more widely touted highlights of Oracle 21C are the following:

- **Blockchain Tables:** These tables operate like any normal heap table, but with some important differences. The most notable difference is that rows are cryptographically hashed as they are inserted into the table, ensuring that the row can no longer be changed later.
- **Native JSON Datatype:** JSON support is further enhanced by offering a native data type, "JSON." This means that instead of having to parse JSON on read or update operations, the parse only happens on an insert and the JSON is then held in an internal binary format that makes access much faster. This can result in read and update operations being five times faster and updates to very large JSON documents being 20 to 30 times faster. Oracle also added a new JSON function JSON_TRANSFORM, which makes it much simpler to update and remove multiple attributes in a document in a single operation.
- **Executing JavaScript inside Oracle Database:** Developers can now execute JavaScript code snippets inside the database where the data resides. This provision allows them to execute short computational tasks easily expressed in JavaScript, without having to move the data to a mid-tier or browser. The Multilingual Engine (MLE) in Oracle Database 21C, powered by GraalVM, automatically maps JavaScript data types to Oracle Database data types and vice versa so that developers don't have to deal with data-type conversion themselves.
- **SQL Macros:** Allow SQL expressions and table functions to be replaced by calls to stored procedures that return a string literal to be inserted in the SQL we want to execute.
- **In-Memory Enhancements:** These include Database In-Memory Vector Joins, Self-Managing In-Memory Column Store, and In-Memory Hybrid Columnar Scans.

The Oracle suite is marketed for major modern operating systems including Windows, AIX, Linux, and Solaris. The Oracle 21C product editions are Personal Edition, Express Edition, Standard Edition One (SE1), Standard Edition (SE), and Enterprise Edition (EE). These editions all come with various components and features to service different needs. Typically, when you purchase Oracle, you specify what edition you are interested in. You will be allowed to install the appropriate edition when you access the Oracle website.

Oracle provides several benefits to its users. Some of the noteworthy advantages of the Oracle DBMS suite are as follows:

- The product is recognized as one of the leading DBMS suites in the industry.
- Oracle provides a comprehensive and at times innovative implementation of SQL.
- The product supports rapid application development (RAD, though limited).
- The Oracle DBMS handles large databases very effectively.
- Oracle handles distributed databases quite well.
- Oracle also handles object databases quite well.
- Oracle facilitates the construction and management of small, medium size, and large data warehouses.
- Oracle facilitates communication with other databases.
- The Oracle DBMS hosts a comprehensive system catalog, thus allowing it to effectively handle complex databases consisting of different types of objects.

- Oracle provides a user interface that encourages Oracle experts, while facilitating novices.
- Oracle provides availability and scalability with grid computing, industry-leading security, and lower maintenance costs with its self-managing database.
- Oracle provides vertical integration across different versions (for example, 19C–21C) as well as different editions within each version. This is so because all editions share the same core code.
- Based on a new multitenant architecture, Oracle Database 21C makes it easier for customers to standardize, consolidate, and automate database services through the incorporation of cloud technology.
- Enhanced cloud services, performance, availability, and security as mentioned above.
- Enhanced client services via SQL Developer and APEX, as described in Section 11.7 of Chapter 11.

22.2 Main Components of the Oracle Suite

The Oracle DBMS suite is a large and complex aggregation of software components. The more visible ones are listed in Table 22.1. In addition, there are several other add-ons as well as less visible but significant components in the Oracle suite. A brief discussion (overview) of a few of these components follows. For more detailed clarifications, see the Oracle product documentation (Oracle 2015, 2016a–c, 2019).

22.2.1 Oracle Server

The Oracle Server is the central database engine of the Oracle Suite. It hosts a fairly comprehensive implementation of SQL, supporting all major SQL statements, with enhancements of its own. Traditionally, the software developer or database administrator (DBA) typically writes SQL statements in an SQL*Plus environment (as described in the upcoming section). However, the more contemporary tendency is to facilitate alternate development environments.

The Oracle Server supports a very comprehensive system catalog. There are system tables for all the database object types supported. These include (but are not confined to): databases; tablespaces; datafiles; tables; sequences; table columns; constraints; synonyms; logical views; materialized views; indexes; privileges; table comments; and column comments. These catalog tables are typically accessed indirectly via logical views prefixed by "DBA." "User," or "All" (e.g., we have DBA_Tables, User_Tables, and All_Tables).

*22.2.2 Oracle PL/SQL and SQL*Plus*

Oracle has its own programming language, PL/SQL. This high-level language is what is often used to develop Oracle database applications such as functions, stored procedures, and triggers. PL/SQL was specifically developed for Oracle applications and is portable on Oracle Servers. It exhibits all the main features of a classical high-level language (HLL), but avoids ambiguities (for instance about the **If-Statement**). PL/SQL is a limited HLL; it focuses solely on database application development. It has limited treatment of arrays and pointers.

SQL*Plus is a working environment that is included in every Oracle installation. Oracle provides a standard SQL Editor through SQL*Plus. SQL*Plus provides a basic line editor that allows you to enter and run SQL statements. You were introduced to basic SQL*Plus commands in

Section 11.1. Due to its limitations, many Oracle users tend to use other text editors to specify SQL instructions and integrate them into SQL*Plus. This process is seamless and is handled quite well by SQL*Plus. While SQL*Plus is still widely used, alternate products such as Oracle SQL Developer make it easier for writing, managing, and executing SQL statements. SQL Developer is also rich with many other features to help the developer and DBA.

Table 22.1 Prominent Oracle Components

Prominent Oracle 12C Components	*Prominent Oracle 21C Related Products*
• Oracle Server Core • Oracle PL/SQL • Oracle Developer Suite • Oracle Database Configuration Assistant (DBCA) • Oracle Network Services • Oracle Admin Assistant • Oracle SQL Developer • Oracle Enterprise Manager (OEM) Database Control • Oracle Enterprise Manager Grid Control • Oracle Enterprise Manager Configuration Assistant • Oracle HTTP Server • Oracle Transparent Gateways • Oracle JDBC/OCI Interfaces • Oracle Real Application Clusters • Oracle Enterprise Integration Gateways • Oracle Application Server • Oracle Integration • Oracle Internet File System • Oracle Recovery Manager • Oracle Data Integrator • Oracle Online Analytic Processing (OLAP) Client • Oracle Active Data Guard • Oracle Advanced Analytics • Oracle Advanced Compression • Oracle Advanced Security • Oracle Label Security • Oracle Vault • Oracle Multitenant • Oracle Spatial and Graph • Oracle Partitioning • Oracle Cloud Management Pack • Oracle Database Lifecycle Management Pack • Oracle Diagnostic Pack • Oracle Tuning Pack	• Oracle Application Express • Oracle SQL Developer Data Modeler • Oracle SQL Developer Command Line (SQLcl) • Oracle REST Data Services • Simple Oracle Document Access (SODA) • Oracle Audit Vault and Database Firewall • Oracle Database Migration Assistant for Unicode • Zero Downtime Migration • Industry Data Models for Utilities, Airlines, Retail, and Communications • Oracle Machine Learning • Oracle Data Safe • Oracle Key Vault • Oracle Database XBRL Extension • Oracle Data Integrator • Oracle Business Intelligence Foundation Suite

22.2.3 Oracle Developer Suite

Oracle Developer is a sophisticated suite component that facilitates rapid application development (RAD). Like most RAD tools, Oracle Developer provides several facilities in a graphical user interface (GUI) environment. Following is a summary of the prominent subsystems:

- **Oracle Business Intelligence Beans:** This subsystem provides a set of standard JavaBeans components to build *business intelligence* (BI) applications.
- **Oracle Designer:** This subsystem provides a complete toolset to be employed in modeling the requirements and design of enterprise applications.
- **Oracle Forms Developer:** This subsystem is a PL/SQL-based development environment for building GUI-based applications that may be designed to be *Web-accessible* (see Chapter 20).
- **Oracle JDeveloper:** This subsystem is a Java-based alternative to the Forms Developer, facilitating the development of a GUI-based application that may be designed to be Web-accessible. It provides an Interactive Development Environment (IDE) that mirrors that of the Sun Microsystems equivalent product called NetBeans. With the recent acquisition of Sun Microsystems, we can expect a further improvement of the related services and capabilities in this area. This is a huge acquisition for the Oracle Corporation.
- **Oracle Reports Developer:** This subsystem provides report-building tools for designing attractive reports by pulling information from an Oracle database.
- **Oracle Software Configuration Manager:** This subsystem facilitates management of structured and unstructured data and all types of files throughout the software development life cycle.
- **Oracle Application Express (APEX):** This is a low-code development platform that enables you to build stunning, scalable, and secure apps with world-class features that can be deployed anywhere.
- **Oracle Application Development Framework (ADF):** Oracle ADF is an end-to-end Java EE framework that simplifies application development by providing out-of-the-box infrastructure services and a visual and declarative development experience.
- **Oracle XML Developer Kit (XDK):** This is a set of components, tools and utilities in Java, C, and C++ bundled with Oracle Database 12c that eases the task of building and deploying XML-enabled applications with a commercial redistribution license.

22.2.4 Oracle Enterprise Manager Database Control and SQL Developer

The Oracle Enterprise Manager (OEM) Database Control is a component that provides a user-friendly GUI environment for the DBA to perform administrative work on the database. Through OEM, the DBA can manage a given database or a group of databases in a cluster. Moreover, OEM is Web-accessible; through it, you can access your database from any computer that has an Internet connection via the URL address **http://<Machine.Domain>:5500/em** (you supply the machine name and domain name for your network).

When you access a database through OEM, all its component objects (schemas, users, tablespaces, tables, procedures, triggers, indexes, constraints, etc.) are available in a hierarchical manner. Each of these database components can be accessed, and their properties changed. OEM then generates the required SQL code or database scripts, in order to effect these changes, and executes them. OEM can also be used to create database objects (user accounts, tables, views, tablespaces, indexes, constraints, etc.). The productivity of the DBA is therefore greatly enhanced. Of course, you will need a valid user account, password, and appropriate privileges.

Oracle also provides a GUI-based component called Oracle SQL Developer (OSQLD). OSQLD provides functionalities similar to OEM. Students as well as practitioners will find this component very effective in providing a shield from the slightly gorier rigor of SQL syntax (but the truth is SQL syntax is by no means a tragedy; we could live with it).

22.2.5 Oracle Enterprise Manager Grid Control

Oracle Enterprise Manager Grid Control is an HTML-based interface that provides complete monitoring across the entire Oracle technology stack — business applications, application servers, databases, and the E-Business Suite — as well as non-Oracle components. The components of Grid Control include:

- Oracle Management Service (OMS)
- Oracle Management Agents
- Oracle Management Repository

These components communicate with each other through HTTP and we can achieve secure communications between tiers within firewall-protected environments by enabling the secure socket layer (SSL) protocol. Using Grid Control, an administrator can view alerts, overall system status, and performance metrics, and be alerted when failure occurs.

Starting with Oracle Grid Infrastructure 19C, the Grid Infrastructure Management Repository (GIMR) is optional for new installations of Oracle Standalone Cluster. Oracle Domain Services Clusters still require the installation of a GIMR as a service component.

22.2.6 Oracle Database Configuration Assistant

Unfortunately, creating, configuring, and managing Oracle databases can be quite complex. The Oracle Database Configuration Assistant (DBCA) helps to take some complexity out of this by automating the process. The user is shielded from gory syntactical details via a user interface, which generates the requisite SQL and database scripts, based on the user responses to friendly prompts.

DBCA and OEM are complementary products; the former is particularly useful for database creation and initial configuration; the latter is extremely helpful during database administration and monitoring.

22.2.7 Oracle Data Integrator

The Oracle Data Integrator (ODI) replaces the Oracle Warehouse Builder (OWB) in earlier versions (12C). The ODI comprises several components that facilitate the construction and management of a data warehouse environment. For more about data warehousing, please review Chapter 19. Recall that a data warehouse is a database consisting of read-only information obtained by extracting, aggregating, and possibly reformatting data from multiple source databases. With ODI, you can construct a simple data warehouse for a set of departmental databases, as well as a complex data warehouse for an entire enterprise or group of enterprises.

In addition to expected conventional data warehousing services (mentioned in Chapter 19), ODI is a data integration platform focused on fast bulk data movement and complex data transformations. ODI is mature, proven, flexible, and rich in features and capabilities. It boasts a vast range of pre-built connectors to databases, object stores, big data, and applications (both standalone and SaaS), as well as the ability to customize existing connectors or build new ones from scratch.

22.3 Shortcomings of Oracle

Oracle, despite its apparent monopoly on the industry, has a few significant shortcomings, primarily in the area of its user interface:

SQL Environment: Over the past few years, we have witnessed an avalanche of wide-ranging improvements in the Oracle DBMS suite. While products such as DBCA, OEM, and Oracle SQL Developer (OSQLD), and APEX are GUI based, there remains room for improvement (for instance, in the area of automatic code generation). If the recent past is prolog, we expect to see these refinements in the near future.

PL/SQL Support: Like SQL support, Oracle's support of its own PL/SQL is somewhat lacking: no GUI or interactive command prompt. The developer is forced to memorize PL/SQL command syntax, and there are no context-sensitive prompts to help.

System Integration: As stated earlier, the Oracle suite consists of several components. Here are two observed problems related to integration of the components:

- In familiarizing with the Oracle DBMS suite, one easily gets the impression that there are too many splinter components, many of which could be merged.
- Some of the components are not all gracefully integrated.

The reason for this jerky integration is that Oracle shows signs of after-thought evolution, rather than purposeful planning. This problem becomes evident when one considers, for example, components OEM and DBCA. These two components could have been easily merged with a third — Recovery Manager (RMAN) — into one integrated component, providing comprehensive coverage of database administration. This problem of integration becomes evident when full installation of Oracle is attempted. After installation, if you invoke the Oracle main menu, you will notice that not all the related components are accessible from the menu; you will likely see a menu link for DBCA, but will not see one for OSQLD; nor are you likely to see a link to the OEM component; you need to take separate actions to access these components and you will need to have prior knowledge of them.

As the Oracle DBMS continues to improve, we have noticed significant integration improvements in the areas such as database accessibility and data warehousing. It is expected that this trend will continue.

Code Generation: Earlier versions of Oracle provided very negligible automatic code generation. For instance, early versions of Form Builder provided a useful environment for application development but did not generate much PL/SQL code. The developer was still required developers to memorize SQL and PL/SQL syntax, and there was more reliance on the SQL*Plus component. It must be said, however, that components such as OEM, OSQLD, and DBCA do generate a considerable amount of code. It is hoped that this trend will continue in the future.

Database Management: Most DBMS suites have a simple, straightforward way of dealing with creating a database and populating it with database objects. Oracle does not. To achieve this objective, you need to observe the following procedure:

a. Create an Oracle database.
b. Create one or more pluggable databases (PDBs) — applicable to Oracle 12C and beyond.
c. Create one or more tablespaces. Each tablespace will consist of at least one datafile.
d. Create user account(s) and grant appropriate quotas to the tablespace(s).
e. Grant appropriate privileges to the user account(s).
f. Populate the tablespace(s) with database objects.

The DBA or someone with DBA privileges typically does the first five steps. Each of these steps is multifaceted, involving several subservient steps. By using DBCA and OEM, these steps have been greatly simplified, but they can still be demanding. The final step can be done by users with appropriate privileges.

Affordability: Traditionally, Oracle solutions have been prohibitively expensive for small- and medium-sized companies. In recognition of this, the Oracle Corporation provides a special educational program for colleges and universities. This program is called the Oracle Academy; under this initiative, enrolled institutions have free access to Oracle products, which they are authorized to use for education and research purposes.

Combined Effect: When we combine the effect of all the flaws mentioned, the result is that Oracle remains a product that is relatively difficult to learn and use. On the other hand, if the rate of improvements continues, the product is expected to soon qualify as one of the leading "killer applications" (Downes 1998) of our lifetime.

22.4 Summary and Concluding Remarks

Let us summarize what we have covered in this chapter:

- Oracle is widely regarded as one of the world's leading RDBMS suites. The product runs on major operating system platforms (Windows, AIX, Linux, and Solaris) and is marketed under two editions — Standard Edition (SE) and Enterprise Edition (EE).
- The Oracle suite includes a number of components. These include (but are not confined to) Oracle Server, Oracle PL/SQL, Oracle JDeveloper, Oracle Database Configuration Assistant, Oracle Enterprise Manager, Oracle SQL Developer, and Oracle SQL*Plus.
- Oracle has some shortcomings relating to the component integration, limited code generation, complex database creation, and affordability.

Notwithstanding the shortcomings, Oracle is an excellent DBMS suite, and for this reason, it is expected to continue to dominate the database systems arena well into the foreseeable future. Oracle's recent acquisition of Sun Microsystems represents a significant possession for the company. This means that we can expect to see greater integration of Java-based interfaces in the Oracle products (something Oracle has been working on for some time), as well as increased platform independence. This transformation has already begun.

22.5 Review Questions

1. What are the main editions of Oracle that are marketed?
2. What are the main components of an Oracle suite? Briefly discuss how these components are related.
3. Identify some benefits and drawbacks of the Oracle suite.

References and Recommended Readings

Downes, Larry, & Chunka Mui. 1998. *Unleashing the Killer App: Digital Strategies For Market Dominance.* Boston, MA: Harvard Business School Press.

Oracle Corporation. 2015. "Oracle Database 12C Product Family." Accessed June 2016. http://www.oracle.com/technetwork/database/oracle-database-editions-wp-12c-1896124.pdf

Oracle Corporation. 2016a. *Oracle Cloud Computing.* Accessed May 2016. https://www.oracle.com/cloud/index.html

Oracle Corporation. 2016b. *Oracle Database Online Documentation 12c.* Accessed June 2016. http://docs.oracle.com/database/121/index.htm

Oracle Corporation. 2016c. *Oracle Multitenant.* Accessed June 2016. https://www.oracle.com/database/multitenant/index.html

Oracle Corporation. 2019a. *Oracle 19C Introduction and Overview.* Accessed June 2019. https://www.oracle.com/middleeast/a/tech/docs/database19c-wp.pdf.

Oracle Corporation. 2019b. *Oracle Data Integrator.* https://docs.oracle.com/en/middleware/fusion-middleware/data-integrator/index.html

Shah, Nilesh. 2002. *Database Systems Using Oracle: A Simplified Guide to SQL and PL/SQL.* Upper Saddle River, NJ: Prentice Hall.

Thomas, Biju, Gavin Powell, Robert Freeman, & Charles Pack. 2014. *Oracle Certified Professional on 12C Certification Kit.* Indianapolis, IN: Wiley.

Chapter 23

Overview of Db2

This chapter provides an overview of the Db2 database management system (DBMS). The chapter proceeds under the following captions:

- Introduction
- Main Components of the Db2 Suite
- Shortcomings of Db2
- Summary and Concluding Remarks

23.1 Introduction

In 2017, IBM re-branded its DB2 and dashDB product offerings into the new name "Db2." Db2 is another leading regional DBMS (RDBMS) suite in the software engineering industry (Oracle being its main archrival). Developed by IBM Corporation, the Db2 RDBMS was first introduced for mainframes MVS/370 and MVS/XA in 1983 and for MVS/ESA in 1988. In 1996, the Db2 Universal Database, the industry's first fully scalable, Web-accessible database management system was announced by IBM. It is called "universal" because of its ability to sort and query alphanumeric data as well as text documents, images, audio, video, and other complex objects. Also, Db2 supports the two dominant database models — the relational model and the object-oriented model, hence the term universal DBMS (UDBMS). Like Oracle, the Db2 suite is quite comprehensive.

Db2 Universal Database (UDB) offers many database and information management enhancements. The most recent version of the product is Db2 UDB version 11.5.5. It is the database of choice for the development and deployment of critical solutions in areas such as:

- E-business
- Business intelligence
- Content management
- Enterprise Resource Planning
- Customer Relationship Management
- Data Warehousing

DOI: 10.1201/9781003275725-28

The latest version of Db2 goes by the code name Kepler and runs on operating systems Linux, Unix, and Windows. With this latest release, IBM has combined the functionality and tools offered in the prior generations of Db2 and InfoSphere Warehouse on Linux, UNIX, and Windows to create a single multi-workload DBMS. Included in this version are a number of new capabilities, three of which are mentioned here:

- AI-ready: Db2 is the AI (Artificial Intelligence) database. Db2 databases are powered by AI in order to optimize and improve the data management process with machine learning query optimization, confidence-based querying, and adaptive workload management.
- Developer inclusivity: Db2 is for developers. The Db2 team has been actively working to improve all aspects of developer and database administrator interactions.
- Resilience and enterprise grade: This allows customers to have unprecedented flexibility to deploy on the infrastructure of their choice, whether it is on-premises, dedicated appliance, or an IaaS platform like IBM Cloud, AWS, Azure, or Google Cloud.

Being a leading DBMS product, Db2 provides a wide range of advantages. The following are some of the more widely acclaimed advantages of the Db2 UDB suite:

- The product is recognized as one of the leading DBMS suites in the industry.
- Db2 provides a comprehensive and at times innovative implementation of SQL.
- The Db2 suite includes several front-end RAD tools that all seamlessly support Db2.
- The Db2 UDB handles large databases very effectively.
- Db2 UDB handles distributed databases and object databases quite well.
- Db2 UDB facilitates the construction and management of small, medium, and large data warehouses.
- Db2 UDB facilitates communication with other databases.
- The Db2 UDB hosts a comprehensive system catalog, thus allowing it to effectively handle complex databases consisting of different types of objects.
- Db2 UDB provides a user interface that encourages experts, while facilitating novices.
- Db2 facilitates seamless integration of various products.
- Db2 facilitates mobile application development for various mobile application platforms.

The Db2 suite is marketed for major modern operating systems including i5/OS, Windows, AIX (IBM's implementation of Unix), Solaris, Linux, HP-UX, and Mac OS-X; additionally, there appears to be a marketing focus on Linux, Unix, and Windows (referred to as LUW). The following are the Db2 product editions currently marketed (see IBM 2020c):

- Community edition (free, unsupported, cannot be used in production systems)
- Standard edition
- Advanced edition

These editions all come with various components and features to service different needs. For a long time, IBM used the mantra "Db2 is Db2 is Db2" to mean that all Db2 database applications are platform independent across the above-mentioned editions and operating systems (IBM 2015b).

According to IBM, "Db2 12 for z/OS is a leading AI enterprise database for mission-critical data. This solution handles rapidly changing, diverse and unpredictable workloads while optimizing resource utilization and investment. It is among the most scalable, reliable and cost-effective data servers available" (see IBM 2020b). This DBMS suite provides the following additional benefits to organizations:

- Provision of analytics to give more value from enterprise data
- Extending mobile and AI support for organizations
- Providing IBM Cloud services to reduce complexity and cost

23.2 Main Components of the Db2 Suite

The Db2 product family includes the following main components, starting with the Db2 Database Core; arguably, these may be more appropriately described as complimentary tools (IBM 2020e):

- Db2 Database Core
- IBM Data Management Console
- IBM Advanced Recovery Feature
- IBM Db2 Connect
- IBM Data Studio
- IBM InfoSphere Data Architect
- IBM Data Warehousing in Db2

Depending on the edition of the Db2 server, various other components may be included. Additionally, IBM provides several complimentary tools that enhance the Db2 environment. Some of these Db2 tools are summarized in Table 23.1. This list is not comprehensive. However, it should convey the reality that the Db2 suite is quite comprehensive.

Following is a summary of the main components mentioned above. As the Db2 product documentation is a much more comprehensive source than this chapter, you are advised to check that source (see IBM 2015a, b) for additional information.

Table 23.1 Prominent Db2 Enhancement Tools

Db2 Universal Database Core: This is the core product for creating and managing a relation/object database.
Db2 High-Performance Upload (Db2 HPU): Enables the DBA to manage large quantities of data, and efficiently manage functions such as backup, data upload, recovery, and data migration.
Db2 Performance Expert for LUW: Useful in optimizing the performance of the database environment.
Data Encryption Expert: Facilitates encryption and compression of data for both online and offline environments.
InfoSphere Data Architect: Facilitates the modeling, designing, and standardization of various distributed data resources.
InfoSphere DataStage: Facilitates the extraction and transformation of data from multiple sources, and loading them into a staging area.
InfoSphere Federation Server: Facilitates the amalgamation of data from multiple heterogeneous sources into a single virtual view for end users.
InfoSphere Information Analyzer: For analyzing the structure, content, and quality of the information from heterogeneous data sources.
InfoSphere Connectivity Software: Provides the necessary protocol for efficient cross-platform connections to multiple sources.
InfoSphere Metadata Workbench: Allows for tracing of relationships and attributes from source databases to a target database where they are used (even across editions and platforms).
InfoSphere QualityStage: Facilitates analysis of data from source databases, based on established business rules, in order to determine the data quality; also includes data cleansing features.
InfoSphere Replication Server: Enables data consolidation and/or replication in support of data availability.
Optim Database Administrator for Linux, Unix, and Windows (LUW): Used for various administrative activities such as data migration, structural database changes, disaster recovery, etc.
Optim Database Relationship: Used for managing groups of related tables in support of a set of business operation(s).
Optim Development Studio: A development environment for Oracle, Db2, and Informix, supporting SQL, XQuery, stored procedures, Web services, and Java data access.
Optim pureQuery Runtime for LUW: Used for building high-performance database applications.
Optim Query Tuner for LUW: Provides expert advice for writing high-performance queries and improving database design.

(*Continued*)

Table 23.1 (Continued) Prominent Db2 Enhancement Tools

Optim Test Data Management Solution for Custom and Packaged Application: Facilitates the thorough testing of database applications for data quality.
InfoSphere Warehouse Departmental Edition: For constructing and managing a data warehouse, typically at the departmental level.
InfoSphere Warehouse Enterprise Edition: For constructing and managing a data warehouse at the enterprise level.

23.2.1 IBM Data Management Console

IBM Db2 Data Management Console is a new browser-based console that helps you administer, monitor, manage, and optimize the performance of IBM Db2 for Linux®, UNIX, and Windows databases. Db2 Data Management Console is packaged with all IBM Db2 editions at no extra charge. The key features of IBM Data Management Console are:

- Real-time and historical monitoring
- Alerts and notifications
- Fully integrated SQL editor
- Database object exploration and management
- Team collaboration with composable user interface
- Support for RESTful services APIs

23.2.2 IBM Advanced Recovery Feature

IBM Db2 Advanced Recovery Feature combines three Db2 tools for advanced database backup, recovery, and data extraction: Db2 Merge Backup for Linux, UNIX and Windows; Db2 Recovery Expert for Linux, UNIX and Windows; and IBM InfoSphere Optim High-Performance Unload for Db2 for Linux, UNIX, and Windows. These tools help improve data availability, mitigate risk, and accelerate crucial administrative tasks. Db2 Advanced Recovery Feature can be purchased separately and used with various Db2 editions. These tools help improve data availability, accelerate crucial administrative tasks, and mitigate the risk of downtime, which can be very costly.

23.2.3 IBM Db2 Connect

IBM Db2 Connect facilitates connection among different applications in business enterprise to your mainframe. It provides application enablement and a communication infrastructure that lets you connect Web, Microsoft Windows, UNIX, Linux, and mobile applications to IBM z/OS, AS/400, iSeries, and System I data. The software offers data integration, a secure application environment and centralized management for data servers and clients. It is available in six editions to meet different application, development, and scalability needs. In each case, license charges are not affected by number of users, Db2 Connect server size, or size of IBM System Z or I database servers (IBM 2020c). The main features of IBM Db2 Connect are:

- IBM Data Studio
- Query Tuner

- Runtime Environment
- Configuration Manager
- Autonomic Capabilities
- SQL Technology
- Scalability

23.2.4 IBM Data Studio

IBM Data Studio provides an integrated, modular environment to facilitate database development and IBM Db2 administration. It also offers improved collaboration through an open-source integrated environment and database development tools for Db2 for z/OS, IBM Informix, and Db2 Big SQL.

The Data Studio spans the entire life cycle (design, development, implementation, deployment, and management) for all Db2 applications, irrespective of editions and OS platform. Among the services provided are the following:

- Accelerated solution delivery
- Integrated database administration
- Data growth management
- Optimized performance
- Data privacy
- Streamlined data test management
- Streamlined upgrades and migration

Under these services, several components have been bundled. Table 23.2 provides a list of some of the more prominent ones.

Table 23.2 Prominent Constituent Components of the Data Studio

• InfoSphere Data Architect • Optim Database Administrator • Optim Database Relationship • Optim Development Studio • Optim pureQuery Runtime for LUW • Optim Query Tuner for LUW • Optim Database Administrator for LUW • Optim Test Data Management Solution for Custom and Packaged Applications • Db2 High-Performance Upload • Db2 Performance Expert for LUW • Data Encryption Expert

23.2.5 IBM InfoSphere Data Architect

IBM InfoSphere Data Architect is a collaborative enterprise data modeling and design solution that can simplify and accelerate integration design for initiatives related to business intelligence, master data management, and service-oriented architecture. InfoSphere Data Architect enables database professionals to work with end users at every step of the data design process from project management to application design to data design. The tool helps to align processes, services, applications, and data architectures. Important features of InfoSphere Data Architect solution are:

- Native data querying
- Logical and physical data modeling
- Import and export of constant mappings
- Source control management
- Integration with related products

23.2.6 IBM Data Warehousing Products

IBM provides a rich suite of data warehousing resources to support data warehouses of any kind and configuration (review Chapter 19). The following three options are supported (IBM 2020a):

a. IBM Netezza Performance Server for IBM Cloud Pak for Data is an advanced data warehouse and analytics platform available both on premises and on cloud. With enhancements to in-database analytics capabilities, this next generation of Netezza enables you to do data science and machine learning with data volumes scaling into the petabytes.
b. IBM Db2 Warehouse on Cloud is a fully managed, elastic cloud data warehouse that delivers independent scaling of storage and compute. An award-winning dashboard interface (link resides outside IBM) makes it easier to see and manage your data. A highly optimized columnar data store, actionable compression, and in-memory processing help supercharge your analytics and machine learning workloads.
c. IBM Db2 Warehouse: The AI-ready data warehouse for operational analytics, a client-managed, highly flexible operational data warehouse for private clouds and containerized deployments.

23.3 Shortcomings of Db2

Db2 represents a huge effort by IBM to develop and market a top-quality product. To a large extent, the company has succeeded. There have not been many serious complaints about the product. Nonetheless, as the saying goes, there is no perfect software. A few complaints have been made against the product, as summarized below.

Support of Domains: Some users would like to see Db2 support domain-based calculus. It is felt that by adding domains to the DBMS, stronger data integrity constraints can be achieved.

Affordability: Like Oracle, Db2 is prohibitively expensive for small and medium size companies. In recognition of this, IBM provides unsupported free Community edition, which may be used for educational purposes but cannot be used in production systems for commercial purposes. Additionally, IBM provides the public with the flexibility of choosing from three different editions according to prevailing needs and financial constraints.

Documentation: IBM provides a Web-accessible information center for the Db2 product family (see IBM 2015b). It must be stated that compared to other similar resources, this site could benefit from some improvements.

Combined Effect: Db2 is on the verge of becoming a truly superb product. Contrasted with the Oracle DBMS that shows signs of after-thought evolution, the Db2 DBMS shows signs of careful and deliberate planning and design. It is hoped that IBM can address the few problem areas for the product.

23.4 Summary and Concluding Remarks

It is now time to summarize what we have covered in this chapter:

- Db2 is widely regarded as one of the world's leading RDBMS suites. The product runs on all the major operating system platforms and is marketed under three editions: Community Edition, Standard Edition, and Advanced Edition.
- The Db2 suite includes a number of components/tools. These include (but are not confined to) Db2 Database Core, IBM Data Management Console, IBM Advanced Recovery Feature, IBM Db2 Connect, IBM Data Studio, IBM InfoSphere Data Architect, and IBM Data Warehousing in Db2.
- Db2 has some shortcomings primarily relating to affordability and documentation.

Notwithstanding the shortcomings, Db2 is an excellent DBMS suite and is arguably the one with the most comprehensive and coherent design. For these and other reasons, it is expected that the product will continue to dominate the database systems arena well into the foreseeable future.

23.5 Review Questions

1. What are the main editions of Db2 that are marketed?
2. What are the main components of the Db2 suite? Briefly discuss how these components are related.
3. Identify some benefits and drawbacks of the Db2 suite.

References and/or Recommended Readings

IBM Corporation. 2015a. *Big Data and Analytics*. Accessed January 2016. http://www.ibm.com/developerworks/analytics/

IBM Corporation. 2015b. *IBM Db2 Database for Linux, UNIX, and Windows*. Accessed January 2016. http://www-01.ibm.com/support/knowledgecenter/SSEPGG/welcome.

IBM Corporation. 2015c. *InfoSphere Warehouse Components by Edition*. Accessed January 2016. http://www-01.ibm.com/support/docview.wss?uid=swg21585911

IBM Corporation. 2020a. *Data Warehouse Solutions*. https://www.ibm.com/analytics/data-warehouse

IBM Corporation. 2020b. *IBM Db2 12 for z/OS* https://www.ibm.com/products/Db2-12-for-zos

IBM Corporation. 2020c. *IBM Db2 Connect*. https://www.ibm.com/products/db2-connect

IBM Corporation. 2020d. *IBM Db2 Database*. https://www.ibm.com/products/Db2-database/pricing

IBM Corporation. 2020e. *IBM Db2 Tools*. https://ibm.com/analytics/Db2/tools

Chapter 24

Overview of MySQL

This chapter provides an overview of MySQL. The chapter proceeds under the following captions:

- Introduction to MySQL
- Main Features of MySQL
- Main Components of MySQL
- Alternate Storage Engines
- Shortcomings of MySQL
- Summary and Concluding Remarks

24.1 Introduction to MySQL

MySQL has become the most popular open-source database management system (DBMS) and the fastest growing DBMS in the industry. The product is an attractive alternative to higher cost and more complex DBMS suites. Its award-winning speed, scalability, and reliability, combined with the fact that it is free, are some of the reasons for the product's increasing popularity. The most recent production release series is MySQL 8.0.

Table 24.1 provides the salient differences between these editions, in terms of their characteristic services and features. The editions and features are clarified on the MySQL website (Oracle 2021d). MySQL is currently marketed in three editions:

- MySQL Standard Edition
- MySQL Enterprise Edition
- MySQL Cluster Edition

The original creator of MySQL was a Swedish company called MySQL AB. The company had been in operation for over ten years (since 1995–2008), before being acquired by Sun Microsystems, which has been acquired by Oracle. In the early stages, MySQL was used primarily for internal purposes. Over the past ten years, the product's ascendancy to international acclaim has been quite noticeable. MySQL runs on multiple platforms, including Unix, Linux, Windows, Solaris,

DOI: 10.1201/9781003275725-29

and MacOS. The effective maximum table size for MySQL databases is usually determined by operating system constraints on file sizes, not by MySQL internal limits (Oracle 2015a).

Table 24.1 Prominent Services and Features of the MySQL Editions

	MySQL Editions		
Characteristic Services and Features	*Standard*	*Enterprise*	*Cluster*
Continuous Support	Yes	Yes	Yes
Maintenance, Updates, and Patches	Yes	Yes	Yes
Knowledge Base	Yes	Yes	Yes
MySQL Core Database	Yes	Yes	Yes
MySQL Document Store	No	Yes	Yes
MySQL Connections (via ODBC, JDBC, and .Net)	Yes	Yes	Yes
MySQL Replication	Yes	Yes	Yes
MySQL Router	No	Yes	Yes
MySQL Partitioning	No	Yes	Yes
MySQL Workbench	Yes	Yes	Yes
MyISAM Storage Engine	Yes	Yes	Yes
InnoDB Storage Engine	Yes	Yes	Yes
NDB Storage Engine	No	No	Yes
MySQL Enterprise Monitor (for MySQL Servers)	No	Yes	Yes
MySQL Enterprise Backup	No	Yes	Yes
MySQL Enterprise Security	No	Yes	Yes
MySQL Cluster Manager	No	No	Yes
MySQL Enterprise Scalability	No	Yes	Yes
MySQL Enterprise High-Availability	No	Yes	Yes
MySQL Cluster Manager	No	No	Yes
MySQL Cluster Geo-Replication	No	No	Yes

MySQL brings a number of advantages to the database arena. The main advantages are summarized below:

- **Reliability and Performance:** In its relatively short existence, the product has established itself as being fairly reliable. Also, due to their relatively small size, MySQL databases tend to be relatively high on performance, when compared to larger, more complex products.
- **MySQL Software is Open Source:** Because MySQL is open source, it is free, and it brings all the benefits of open-source products. For this reason, the product is enjoying increased popularity in the academic community as well as among small businesses.
- **Platform Independence**: MySQL runs on multiple operating system platforms. This provides users with flexibility in terms of project development and implementation.
- **Ease of Use and Deployment:** Because of the above-mentioned benefits, MySQL is very easy to deploy. The product is also easy to learn and use.
- **Oracle Support:** Oracle's 2010 acquisition of Sun Microsystems means that products such as MySQL and NetBeans are now marketed under the Oracle umbrella. This bodes well for MySQL (and other related products) since Oracle remains one of the world's leading software engineering firms, and has maintained the first or second rank of the world's database systems market share for over two decades.
- **Flexibility:** MySQL offers huge flexibility to developers who are seeking to have a nimble DBMS on their laptops or personal desktops as they pursue their development objectives, without any concerns about subsequent migration of the database to an enterprise version.

24.2 Main Features of MySQL

In order to fully appreciate MySQL, it is important to take note of the underlying features of the product. These are summarized in Table 24.2. These features were adopted from one of the leading websites on the product (Oracle 2015a, b; 2021d); minor modifications have been added to improve readability. From the list of features provided, it is clear that MySQL is maturing into a comprehensive DBMS.

Table 24.2 Main Underlying Features of MySQL

Internals and Portability:

- MySQL was written in C and C++.
- The product has been tested with a broad range of different compilers.
- MySQL runs on many different operating system platforms.
- Includes APIs for C, C++, Eiffel, Java, Perl, PHP, Python, Ruby, and Tcl are available.
- Implements a fully multi-threaded using kernel threads. It can easily use multiple CPUs if they are available.
- Provides transactional and non-transactional storage engines.
- Uses very fast B-tree disk tables with index compression.
- It is relatively easy to add other storage engines. This is useful if you want to add an SQL interface to an in-house database.
- Implements a very fast thread-based memory allocation system.
- Implements very fast joins using an optimized one-sweep multi-join.
- Employs in-memory hash tables, which are used as temporary tables.
- SQL functions are implemented using a highly optimized class library and should be as fast as possible. Usually there is no memory allocation at all after query initialization.
- The MySQL code is tested with Purify (a commercial memory leakage detector) as well as with Valgrind, a GPL tool (*http://developer.kde.org/~sewardj/*).
- The server is available as a separate program for use in a client/server networked environment. It is also available as a library that can be embedded (linked) into standalone applications.

Data Types:

- Numeric Types: INTEGER, INT, SMALLINT, TINYINT, MEDIUMINT, BIGINT
- Floating Point: FLOAT, DOUBLE, DECIMAL, NUMERIC
- Bit-value: BIT
- Date and Time: DATE, DATETIME, TIMESTAMP, TIME, YEAR
- String Types: CHAR, VARCHAR, BINARY, VARBINARY, BLOB, TEXT, ENUM, SET

Statements and Functions:

- MySQL supports the standard DDL and DML statements of SQL.
- The MySQL-specific **SHOW** command can be used to retrieve information about databases, database engines, tables, and indexes.
- The **EXPLAIN** command can be used to determine how the optimizer resolves a query.
- Function names do not clash with table or column names. For example, ABS is a valid column name. The only restriction is that for a function call, no spaces are allowed between the function name and the '(' that follows it.
- You can mix tables from different databases in the same query.

Security:

- A stringent but flexible privilege-based security mechanism that allows host-based verification. Passwords traffic is encrypted when you connect to a server.

Scalability and Limits:

- Handles small as well as large databases. There is no limit on the number of databases or the number of tables in any given database.
- Table sizes vary with the underlying operating system. Typical upper boundaries for file-size limits are 2TB for Windows, 2GB for Linux, 16 TB for Solaris, and 2 TB for Mac OS. However, there are ways to go around these limits and create very large tables and tablespaces.

(*Continued*)

Table 24.2 (Continued) Main Underlying Features of MySQL

Connectivity: • Clients can connect to the MySQL server using TCP/IP sockets on any platform. On Windows systems family, clients can connect using named pipes. On Unix systems, clients can connect using Unix domain socket files. • The Connector/ODBC (MyODBC) interface provides MySQL support for client programs that use ODBC connections. For example, you can use MS Access to connect to your MySQL server. • The Connector/J interface provides MySQL support for Java client programs that use JDBC connections. Clients can be run on Windows or Unix. Connector/J source is available. • MySQL Connector/NET enables developers to easily create .NET applications that require secure, high-performance data connectivity with MySQL. It implements the required ADO.NET interfaces and integrates into ADO.NET aware tools. Developers can build applications using.NET languages. MySQL Connector/NET is a fully managed ADO.NET driver written in 100% pure C#.
Geographic Localization: • Even prior to acquisitions by Sun Microsystems and then Oracle, MySQL was widely used around the world. Under Oracle, this widespread reach has continued and is expected for the foreseeable future.
Clients and Tools: • MySQL Server has built-in support for SQL statements to check, optimize, and repair tables. These statements are available from the command line through the **mysqlcheck** client. MySQL also includes **myisamchk**, a very fast command-line utility for performing these operations on **MyISAM** tables. • All MySQL programs can be invoked with the **--help** or **-?** option to obtain online assistance.
Alternate Storage Engines: • Support for multiple table types
MySQL 8.0 highlights include [Oracle 2021b]: • MySQL Document Store • Transactional Data Dictionary • SQL Roles • Default to utf8mb4 • Common Table Expressions and Window Functions

24.3 Main Components of MySQL

Compared to other DBMS suites on the market, MySQL is relatively simple and therefore does not include a sophisticated (or convoluted) list of components. Rather, there is a list of important programs that make up the MySQL core. Some of the most important programs are summarized in Table 24.3. The information provided here was adopted from the MySQL website with minor changes to improve readability and clarity (for more information, see Oracle 2015a).

Table 24.3 Important Core MySQL Component Programs

mysqld: This is the SQL daemon (that is, the MySQL server). To use client programs, **mysqld** must be running, because clients gain access to databases by connecting to the server.
mysqld_safe: This is a server startup script. **mysqld_safe** attempts to start **mysqld. mysqld_safe** is the recommended way to start a **mysqld** server on Unix and NetWare. **mysqld_safe** adds some safety features such as restarting the server when an error occurs and logging runtime information to an error log file.
mysql.server: This is a server startup script. This script is used on systems that use System V-style run directories containing scripts that start system services for particular run levels (Linux, Solaris and Mac OS X). It invokes **mysqld_safe** to start the MySQL server. The **mysql.serv er** component can be found in the **support-files** directory under your MySQL installation directory or in a MySQL source distribution.
mysqld_multi: This is a server startup script that can start or stop multiple servers installed on the system. **mysqld_multi** is designed to manage several **mysqld** processes that listen for connections on different Unix socket files and TCP/IP ports. It can start or stop servers, or report their current status.
mysqlmanager: This is an alternative to the **mysql_multi** program; it is called the MySQL Instance Manager (IM). This program is a daemon running on a TCP/IP port that serves to monitor and manage MySQL Database Server instances. MySQL Instance Manager is available for Unix-like operating systems, as well as Windows. MySQL Instance Manager can be used in place of the mysqld_safe script to start and stop the MySQL Server, *even from a remote host.* MySQL Instance Manager also implements the functionality (and most of the syntax) of the **mysqld_multi** script.
mysql_install_db: This script creates the MySQL database and initializes the grant tables with default privileges. It is usually executed only once, when first installing MySQL on a system. Installation is usually seamless on Windows platforms. After installing MySQL on Unix, you need to initialize the grant tables, start the server, and make sure that the server works satisfactorily. On Unix, the grant tables are set up by the **mysql_install_db** program. For some installations, this program is run for you automatically:
mysql_secure_installation: This program provides facilities to secure the MySQL installation.
mysql_upgrade: This program is used after a MySQL upgrade operation. It checks tables for incompatibilities, repairs them if necessary, and updates the grant tables with any changes that have been made in newer versions of MySQL. The **mysql_upgrade** program should be executed each time you upgrade MySQL. It checks all tables in all databases for incompatibilities with the current version of MySQL Server. If a table is found to have a possible incompatibility, it is checked. If any problems are found, the table is repaired. **mysql_upgrade** also upgrades the system tables so that you can take advantage of new privileges or capabilities that might have been added. All checked and repaired tables are marked with the current MySQL version number. This ensures that next time you run **mysql_upgrade** with the same version of the server, it can tell whether there is any need to check or repair the table again. **mysql_upgrade** also saves the MySQL version number in a file named **mysql_upgrade.info** in the data directory. This is used to quickly check if all tables have been checked for this release so that table-checking can be skipped.
mysqladmin: This program facilitates various administrative activities such as creation or deletion of databases, flushing tables to the hard drive, managing log files, among other related activities.
mysqlcheck: This program systematically checks, repairs, analyzes, and optimizes MySQL tables.

24.4 Alternate Storage Engines

MySQL supports *storage engines* — components designed to handle SQL operations for different table types. MySQL storage engines facilitate *transaction-safe tables* as well as *non-transaction-safe tables*.

- A transaction-safe table (TST) preserves the data stored, even after a system crash; this is so because of the presence of a transaction log. TSTs are typically implemented in **InnoDB** or **BerkeleyDB** (BDB) table type.
- A non-transaction-safe table (NTST) has fewer overheads (since there are no transaction logs) and is therefore faster on processing. However, data lost during a failed transaction is not preserved in the face of the related system failure; data will remain what it was prior to the failure. NTSTs are typically implemented in **Memory**, **Merge**, or **MyISAM** table type.

MySQL employs a *pluggable-storage-engine* architecture. This means that different storage engines may be loaded into or unloaded from the running MySQL Server. The default storage engine is **InnoDB**. At any point in time, the MySQL database user may issue the **Show Engines** command to obtain a list of the storage engines running on the database server. Other alternative storage engines are **Archive**, **CSV**, **Blackhole**, **Federated**, and **Example**. For additional insight, please refer to the MySQL documentation (Oracle 2015a).

MySQL 8.0 has added one more alternate storage engine called NDB (also known as NDB CLUSTER): This clustered database engine is particularly suited for critical applications that require the highest degree of uptime and availability.

24.5 MySQL Database Service

The MySQL Database Service is a fully managed database service that empowers developers to quickly develop and deploy secure, cloud native applications using the open-source database. MySQL Database Service is the only MySQL cloud service with an integrated, high-performance analytics engine called HeatWave. This engine enables customers to run sophisticated analytics directly against their operational MySQL databases, thus eliminating the need for more complex, time-consuming, and expensive data movement and integration with a separate analytics database. MySQL Database Service is 100% built, managed, and supported by the Oracle Cloud Infrastructure (OCI) and the MySQL engineering teams (Oracle 2021c).

The MySQL Database Service represents a significant enhancement of the MySQL suite, offering the following advantages:

- Ease of use
- Cheaper than Amazon RDS
- Security and compliance
- Optimized for cloud infrastructure
- Unique, integrated high-performance analytics engine (HeatWave).

24.6 Shortcomings of MySQL

Like all software products, MySQL does have some limitations in its current production release. What is most impressive about the MySQL venture is that the developers readily admit the limitations and document them (see Oracle 2015a). Following is a summary on some of these limitations:

24.6.1 Limitations on Logical Views

Limitation on Joins: The maximum number of tables that can be referenced in a single join is 61. This also applies to the number of tables that can be referenced in the definition of a view. This is a reasonable threshold that should not cause anyone to lose sleep.

Limitations on View Processing: View processing is not optimized in MySQL. Two noteworthy restrictions are mentioned here:

1. You are not allowed to create indexes on views. This is significant, so MySQL would do well to remove it in the near future.
2. Indexes can be used for views processed using MySQL's **merge** algorithm. However, a view that is processed with the **temptable** algorithm is unable to take advantage of indexes on its underlying tables (although indexes can be used during generation of the temporary tables). This is also a significant restriction that should be lifted in the future.

To briefly clarify, the MySQL **Create-View** statement allows the user to optionally specify the algorithm to be used in creating the view; this is done via the **Algorithm-Clause**, which allows for one of three options — UNDEFINED, MERGE, or TEMPTABLE. The MERGE option is the most efficient but the default is UNDEFINED (forcing MySQL to choose between the other two options).

24.6.2 Limitations on Subqueries

MySQL developers list a number of restrictions on subqueries as summarized below (for more details, see Oracle 2015a). Three such limitations are summarized here:

1. You are not allowed to modify a table and select from that said table in the same sub-query.
2. Subquery optimization for the IN (<sub-query>) construct is not as effective as for the equal (=) operator or for IN (<value-list>) constructs. A typical case for poor performance of the IN (<sub-query>) construct is when the subquery returns a small number of rows but the outer query returns a large number of rows to be compared to the subquery result. The problem is that, for a statement that uses an IN (<sub-query>) construct, the optimizer rewrites it as a correlated subquery.

EXAMPLE 24.1: ILLUSTRATING AN UNDESIRABLE CASE OF QUERY OPTIMIZATION IN MYSQL

The following SQL statements illustrate the conversion of an IN (<sub-query>) construct to a correlated subquery.

```
SELECT ... FROM Table1 t1 WHERE t1.a IN (SELECT b FROM Table2);
// The optimizer rewrites the statement to a correlated subquery:
SELECT ... FROM Table1 t1 WHERE EXISTS (SELECT b FROM Table2) t2 WHERE
t2.b = t1.a);
```

If the inner and outer queries return **M** and **N** rows, respectively, the execution time becomes on the order of **f(M×N)**, rather than **f(M+N)** as it would be for an uncorrelated subquery. An implication is that query construct with IN (<sub-query>) can be much slower than a query written using an IN(<value-list>) construct that lists the same values that the subquery would return. Obviously, this is a very undesirable situation.

3. The optimizer is more mature for joins than for subqueries, so in many cases a statement that uses a subquery can be executed more efficiently if you rewrite it as a join. An exception to this occurs for the case where an IN (<sub-query>) construct can be rewritten as a SELECT DISTINCT join construct.

EXAMPLE 24.2: ILLUSTRATING ANOTHER CASE OF QUERY OPTIMIZATION IN MYSQL

The following SQL statements illustrate conversion of a subquery to a join query:

```
SELECT <ColumnsList> FROM <Table1> WHERE
<ID_col> IN (SELECT <ID_col2>
FROM <Table2> WHERE <Condition>);
// The above statement format can be rewritten as follows:
SELECT DISTINCT <ColumnsList> FROM <Table1> t1, <Table2> t2
WHERE t1.<ID_col> = t2.<ID_col2> AND <Condition>);
```

Note, however, that in this case, the join requires an extra DISTINCT operation and is not necessarily more efficient than the subquery. If you are not looking for distinct values in the result set, then the **Select** statement with the join is preferred to the **Select** statement with a subquery.

24.6.3 Limitation on Server-Side Cursors

Server-side cursors are implemented in the C programming language via the **mysql_stmt_attr_set()** function. The same implementation is used for cursors in stored routines. A server-side cursor allows a result set to be generated on the server side only; the entire result set is not necessarily transferred to the client, but only for the rows requested by the client. For example, if a client executes a query but is only interested in a few rows, the remaining rows are not transferred. In MySQL, a server-side cursor is materialized into a temporary table. Initially, this is a memory table, but is converted to disk table if its size reaches the value of the **max_heap_table_size** system variable. One limitation of the implementation is that for a large result set, retrieving its rows through a cursor might be slow.

24.6.4 Other Limitations

The MySQL website discusses other product limitations related (but not limited) to issues such as stored procedures, condition handling, performance schema, and extended architecture (XA) transactions. The term *XA transaction* is often used to describe transactions involving different systems or databases. For instance, a client application program acting as a transaction manager in one system may request information from a database server acting as a resource manager in another system.

The MySQL manual lists a number of additional limitations, some of which are mentioned below (Oracle 2021a):

- Identifier Length Limits
- Grant Table Scope Column Properties
- Limits on Number of Databases and Tables
- Limits on Table Size
- Limits on Table Column Count and Row Size

24.7 Summary and Concluding Remarks

Let us summarize what has been discussed in this chapter:

- MySQL is the most popular open-source DBMS. It runs on various operating system platforms (including Unix, Linux, Windows, Solaris, and MacOS). The product is marketed under three editions: Standard Edition, Enterprise Edition, and Cluster Edition. The MySQL suite consists of a number of important component programs, including but not confined to: **mysqld**, **mysqld_safe**, **mysql.server**, **mysqlmanager**, **mysql_multi**, etc.
- MySQL supports *storage engines* — components designed to handle SQL operations for different types of tables.
- The official MySQL website reports a number of shortcomings that are being addressed. These include but are not confined to limitations on joins, limitations on views, limitations on subqueries, limitations on server-side cursors, etc.

MySQL has made an impressive entry into the software engineering industry, and for this reason, it is expected that the product will be around for the foreseeable future.

24.8 Review Questions

1. Briefly account for the history of MySQL.
2. Outline the characteristic features of MySQL.
3. Explain the use of alternate storage engines in MySQL.
4. What are the main components of the MySQL suite? Briefly discuss the functional responsibilities of each component.
5. Briefly describe MySQL Database Service.
6. Discuss the benefits and shortcomings of MySQL.

References and Recommended Readings

Dubois, Paul. 2005. *MySQL* 3rd ed. Indianapolis, IN: Sams.

Kofler, Michael. 2005. *The Definitive Guide to MySQL 5* 3rd ed. Berkeley, CA: Apress Publishing.

Oracle. 2015a. *MySQL 5.7 Reference Manual*. Accessed January 2016. https://docs.oracle.com/cd/E17952_01/refman-5.7-en/index.html

Oracle. 2015b. *MySQL Editions*. Retrieved January 2016. https://www.mysql.com/products/

Oracle. 2021a. *Limits in MySQL*. https://dev.mysql.com/doc/mysql-reslimits-excerpt/8.0/en/limits.html
Oracle. 2021b. *MySQL Alternative Storage Engines*. https://dev.mysql.com/doc/refman/8.0/en/storage-engines.html
Oracle. 2021c. *MySQL Database Service*. https://www.oracle.com/mysql/
Oracle. 2021d. *MySQL Editions*. Retrieved March 2021. https://www.mysql.com/products/
Oracle. 2021e. *What's new in MySQL 8.0*. https://www.mysql.com/why-mysql/white-papers/whats-new-mysql-8-0/

Chapter 25

Overview of Microsoft SQL Server

This chapter provides an overview of the Microsoft SQL Server DBMS. Developed and marketed for the Microsoft Windows operating system platform, the product has emerged as one of the leading RRDBMS suites in the marketplace. The chapter proceeds under the following captions:

- Introduction
- Main Features of SQL Server
- Main Components of the SQL Server Suite
- Shortcomings of SQL Server
- Summary and Concluding Remarks

25.1 Introduction

Microsoft SQL Server is Microsoft's flagship relational database engine product. The product was developed by Microsoft for its Windows operation system platform, and has been through several stages of revision.

25.1.1 Brief History

Microsoft SQL Server uses a version called Transact-SQL (T-SQL). Microsoft initially developed SQL Server (a database product that implements the SQL language) with Sybase Corporation for use on the IBM OS/2 platform; the first release was made in 1989. After the IBM–Microsoft collaboration broke down, Microsoft abandoned OS/2 in favor of its then new network operating system, Windows NT Advanced Server. At that point, Microsoft decided to further develop the SQL Server engine for Windows NT by itself. The resulting product was Microsoft SQL Server 4.2, which was updated to version 4.2.1.

After Microsoft and Sybase parted ways, Sybase further developed its database engine to run on Windows NT (currently known as Sybase Adaptive Server Enterprise), and Microsoft developed SQL Server 6.0, then SQL Server 6.5, which also ran on top of Windows NT. SQL Server

DOI: 10.1201/9781003275725-30

7.0 introduced the capability to run on Windows NT as well as on Windows 95 and Windows 98. With SQL Server 7.0, Microsoft dramatically rewrote and modified the Sybase code. The company subsequently redesigned the core database engine and introduced a sophisticated query optimizer and an advanced database storage engine.

Microsoft SQL Server versions are currently named by the year of introduction in the marketplace. Recent versions are SQL Server 2016, SQL Server 2017, and SQL Server 2019; moreover, SQL Server 2021 is expected to be released ahead of the publication of this volume.

25.1.2 Operating Environment

As mentioned earlier, Microsoft SQL Server has been specifically designed, developed, and tailored to operate in and maximize the use of the features of the Windows operating system. The DBMS runs as a Windows service. As you are no doubt aware, a service is an application that Windows can start either automatically when booting up or manually on demand. Services on Windows have a generic application programming interface (API) that can be controlled programmatically. Services facilitate the running of applications such as Microsoft SQL Server without requiring that a user be logged in to the server computer. SQL Server 2019 runs on Linux as well as Microsoft Windows. To find out more about this release, see "What's new in SQL Server 2019 for Linux" (Microsoft 2021c).

Technically speaking, Microsoft SQL Server is a back-end system. However, being developed and marketed by Microsoft, all the front-end Microsoft RAD tools are designed to integrate with SQL Server. The effect is that when SQL Server is implemented in a Microsoft Windows environment (as it must be), one has a choice from several front-end tools that will seamlessly integrate with the SQL Server database.

25.1.3 Microsoft Server Editions

SQL Server is marketed through the several editions, each tailored to a specific set of needs (Microsoft 2021b). Following are the main editions:

- Free Editions include then Express Edition, Developer Edition, and the Evaluation Edition
- Paid Editions include the Enterprise Edition, Standard Edition, and Azure SQL Database
- Specialized Editions include Web Edition, Business Intelligence Edition, Analytics Platform System (APS) — formerly Parallel Data Warehouse PDW), and Big Data Clusters (BDC)

Across these editions, the storage capacity ranges from 10 GB to 524 PB. These editions are further clarified through the product documentation (see Microsoft 2016).

25.2 Main Features of Microsoft SQL Server

According to Microsoft, SQL Server provides a number of significant features; some of the more noteworthy ones are mentioned below:

Database Support: Like other DBMS suites, the SQL Server database engine provides facilities for supporting relational databases as well as unstructured data (via XML). It also includes security features and other related features required to create and support complex Web-accessible databases.

Replication Services: The SQL Server relational database engine supports the features required to support demanding data processing environments. The database engine protects data integrity while minimizing the overhead of managing thousands of users who may be concurrently modifying the database. SQL Server distributed queries facilitate referencing of data from multiple sources as if the data all resided in the local SQL Server database. At the same time, the distributed transaction support protects the integrity of any updates of the distributed data. Replication facilitates maintenance of multiple copies of data, while ensuring that the separate copies remain synchronized.

Ease of Installation and Usage: SQL Server includes a set of administrative and development tools that improve upon the process of installing, deploying, managing, and using SQL Server across several sites. SQL Server also supports a standard-based programming model integrated with the Microsoft Distributed interNet Applications (DNA) architecture, allowing for easy integration with the World Wide Web (WWW). These features allow software engineers to rapidly deliver SQL Server applications that customers can implement with a minimum of installation and administrative overhead.

Interoperability: SQL Server includes facilities for communicating with heterogeneous databases. Interoperability has become a standard requirement for contemporary DBMS suites.

Integration Services: These services include data extraction, transformation, and loading (ETL) capabilities for data warehousing and enterprise-wide data integration.

Analysis Services: These services include online analytical processing (OLAP) capabilities for the rapid and sophisticated analysis of large and complex data sets using multidimensional storage.

Reporting Services: These services provide a comprehensive solution for creating, managing, and delivering both traditional, paper-oriented reports and interactive, Web-based reports from preexisting databases.

Management Tools: SQL Server includes integrated management tools for administering and tuning the database. Two significant products provided are Microsoft Operations Manager (MOM) and SQL Server Management Studio (SSMS).

Development Tools: SQL Server ships with various integrated development tools for the database engine, data extraction, transformation, loading, data mining, OLAP, and application development through Microsoft Visual Studio.

SQL Server 2019 (15.x) introduces Big Data Clusters for SQL Server. It also provides additional capability and improvements for the SQL Server database engine, SQL Server Analysis Services, SQL Server Machine Learning Services, SQL Server on Linux, and SQL Server Master Data Services. For details on SQL Server 2019 new features, please refer to Microsoft documentation (Microsoft 2021d).

Part of the Azure SQL family, Azure SQL Database, is an intelligent, scalable, relational database service built for the cloud, which offers optimized performance and durability with automated, AI-powered features that are always up to date. With serverless compute and Hyperscale storage options that automatically scale resources on demand, users are free to focus on building new applications without worrying about storage size or resource management (Microsoft 2021a).

25.3 Main Components of Microsoft SQL Server Suite

The SQL Server suite is a complex system consisting of various component systems. Two primary component systems in the SQL Server suite are the Server Components system and the

Management Tools system. In addition, the following components have been carried forward (with appropriate enhancements) from earlier versions: Development Tools, Client Connectivity, Code Samples, and Optional Components.

25.3.1 Server Components

The main server components are as follows:

- SQL Server Database Engine concerns itself primarily with storing, securing, and processing data; it includes the DBMS core, tools for managing the stored data, and Data Quality Services (DQS) server.
- Analysis Services include tools for supporting OLAP and data mining services.
- Reporting Services include components for the creation, management, and deployment of tabular, matrix, graphical, and free format reports.
- Integration Services cover GUI-based tools for moving, copying, and transforming data; DQS components are also included.
- Master Data Services (MDS) cover services for data management, including hierarchies, security transactions, data versioning, business rules, and add-ins for Excel.

25.3.2 Management Tools

The management tools are enabled by default; they consist of the following:

- SQL Server Management Studio (SSMS) is an integrated environment for access, configuration, management, administration, and deployment of a database environment.
- SQL Server Configuration Manager facilitates the configuration management for services, protocols, and client aliases.
- SQL Server Extended Events is an event-handling system that provides a uniform way of handling events in the DBMS. One spinoff from this arrangement is that users are able to monitor active processes in the system. Moreover, where necessary, this component interacts with the Event Tracing for Windows (ETW) component of the Windows operating system.
- Database Engine Tuning Advisor facilitates the creation of optimal sets of indexes, indexed views, and partitions of the database.
- Data Quality Client Provides is an intuitive GUI to connect to the DQS server and perform data cleansing operations. It facilitates central monitoring of various activities performed during the data cleansing operation.
- SQL Server Data Tool (SSDT) provides an IDE for building solutions for the Business Intelligence components: Analysis Services, Reporting Services, and Integration Services. The SSDT also includes an integrated environment that database developers can use to conduct their database design work for any SQL Server platform within Visual Studio.

25.3.3 Development Tools

The main development tools are described below:

- Headers and Libraries are the C++ files needed for development of SQL Server programs.
- Microsoft Data Access Components (MDAC) system development kits (SDKs) are the Software Development Kits for XML and the Microsoft Data Access Components.

These SDKs allow enable and support the development of programs using XML and MDAC.

- Backup/Restore API includes a sample program, necessary C/C++ files, and documentation on how to build backup and restore programs.
- Microsoft Visual Studio and SSMS (mentioned above) are two components that are commonly used by software developers who need to access SQL Server databases.

25.3.4 Client Connectivity

Client Connectivity is a set of components that facilitates communication with the SQL Server. This collection includes components such as ActiveX Data Objects (ADO), Open Database Connectivity (ODBC), and Object Linking and Embedding Database (OLE DB). Collectively, they are known as the MDAC (Microsoft Data Access Components).

25.3.5 Code Samples

None of the code samples is installed by default. However, depending on your choice, their inclusion could serve to enrich your SQL Server environment. The following are the main options:

- ADO includes programming examples for ActiveX Data Objects (ADO).
- DBLIB includes programming examples for the DB-Library API. DB-Library was the native Database Application Programming Interface (API) of SQL Server in earlier releases, and is supported in SQL Server for backward compatibility.
- Desktop includes code samples on setting up unattended install operations for the Microsoft SQL Server Desktop Engine (MSDE).
- ESQLC includes the programming examples for Embedded SQL for the C programming language.
- MSDTC includes the programming examples for the Microsoft Distributed Transaction Coordinator.
- ODBC includes the programming examples for the open database connectivity programming API in SQL Server.
- ODS includes the programming examples for the open data services (ODS) API for SQL Server.
- OLE Automation includes the programming examples to support OLE Automation for SQL Server.
- Replication includes the programming examples for SQL Server replication.
- SQLDMO includes programming examples for the SQL-Distributed Management Objects administrative programming interface.
- SQLNS includes programming examples for the SQL Name Space administrative programming interface.

25.4 Shortcomings of Microsoft SQL Server

SQL Server represents a huge effort by Microsoft to develop and market a top-quality product. The indications are that this effort has begun to yield huge benefits for the company as well as users of the product. There are, however, a few areas of concern as summarized below:

Poor Performance and Configuration Out of the Box: Earlier versions of SQL Server drew widespread criticisms from database systems experts on its poor performance. The main area of contention was that the DBMS did not promote separate disk storage locations for the program, data, and transaction log files. This resulted in poor performance. Microsoft argues that the situation is significantly improved in SQL Server 2012 (see Microsoft 2012a; SQL Sentry 2016). This has changed over the past few years and the latest version of SQL Server is very stable and performs well in various kinds of environments.

Reading Transaction Log Files: In earlier versions there was no capability with SQL Server's own tools to read the transaction logs. This meant that database experts were forced to either buy third-party tools, or take wild guesses at restoration points for database recovery. Transaction logs are used only during recovery of database, if the database is corrupt. This has also been addressed in SQL Server 2012 (see Microsoft 2012b), but complaints still persist.

Combined Effect: Despite the shortcomings, Microsoft SQL Server has made a reasonably impressive entrance to the DBMS market place. SQL Server 2019 provides tremendous improvements in performance and scalability, high availability, security, temporal tables, hybrid cloud solutions, and easy access to data.

Given the fact that for better or for worse, the Windows operating system has become a household name, and Microsoft SQL Server seamlessly integrates with that operating system (both products produced and marketed by Microsoft), the future of Microsoft SQL Server is secure. Moreover, with the DBMS now also running on the Linux platform, its future is bolstered even more.

25.5 Summary and Concluding Remarks

Here is a summary of what we have covered in this chapter:

- Microsoft SQL Server is an emerging DBMS suite that is seeking to increase market share in the database systems market. The product runs on Windows and Linux operating systems and is currently marketed under the following editions: Enterprise Edition, Business Intelligence Edition, Standard Edition, Web Edition, Azure SQL Database, Express Edition, Developer Edition, and Parallel Data Warehouse Appliance.
- The Microsoft SQL Server suite includes a number of components. These components have been classified under Server Components, Database Management Tools, Client Connectivity Tools, Development Tools, Code Samples, and Optional Components.

Microsoft SQL Server has made a fairly successful entry into a very competitive market. If not for any other reason, we expect the product to be around for the foreseeable future, due to its availability on Windows and Linux platforms, and competitive pricing for small- and medium-sized customers or organizations. Plus, Azure SQL Database is a fully managed platform as a service (PaaS) database engine that handles most of the database management functions such as upgrading, patching, backups, and monitoring without user involvement.

25.6 Review Questions

1. What are the main editions of Microsoft SQL Server that are marketed?
2. What are the main components of the Microsoft SQL Server suite? Briefly discuss how these components are related.

3. Briefly discuss the main features of the Microsoft SQL Server suite.
4. Identify some benefits and drawbacks of the Microsoft SQL Server suite.

References and/or Recommended Readings

Microsoft. 2012a. "SQL Server 2012 Performance White Paper." Accessed June 2013. http://www.bteksoftware.com/docs/SQL_Server_2012_performance_White_Paper_FINAL.pdf

Microsoft. 2012b. *The Transaction Log (SQL Server)*. Accessed June 2013. http://msdn.microsoft.com/en-us/library/ms190925.aspx

Microsoft. 2016. *SQL Server*. Accessed January 2016. https://msdn.microsoft.com/en-us/library/mt590198%28v=sql.1%29.aspx

Microsoft. 2021a. *Azure SQL Database*. https://azure.microsoft.com/en-us/services/sql-database/

Microsoft. 2021b. *Microsoft SQL Server Editions*. https://sqlserverbuilds.blogspot.com/2019/02/sql-server-editions.html

Microsoft. 2021c. *What's New for SQL Server 2019 on Linux*. https://docs.microsoft.com/en-us/sql/linux/sql-server-linux-overview?view=sql-server-ver15

Microsoft. 2021d. *What's New in SQL Server 2019 (15.x)*. https://docs.microsoft.com/en-us/sql/sql-server/what-s-new-in-sql-server-ver15?view=sql-server-ver15

SQL Sentry. 2016. "Performance Baselines and Benchmarks for Microsoft SQL Server." Accessed September 2016. https://www.sqlsentry.com/white-papers/performance-baselines-and-benchmarks-for-microsoft-sql-server

Chapter 26

Other Emerging Database Methodologies

Throughout the course, we have placed an emphasis on the relational database model; and in Chapter 18, we discussed object databases. This chapter provides an overview of some emerging database approaches that are worthy of mention. The chapter proceeds under the following captions:

- Introduction
- The Entity–Attributes–Value (EAV) Model
- Database-Supporting Frameworks
- NoSQL
- Other Burgeoning Trends
- Summary and Concluding Remarks

26.1 Introduction

In addition to the dominant contemporary database approaches, there are some emerging database approaches that are worthy of discussion. They are summarized below:

- **Entity–Attributes–Value (EAV) Model:** This approach reduces a database to three principal storage entities — an **entities-table** for defining other entities; an **attributes-table** for defining the properties (attributes) of entities; a **values-table** that connects the other two entities and stored values for entity–attribute combinations (see Wikipedia 2016).
- **Database-Supporting Frameworks:** In recent years, software frameworks that support database systems have become commonplace. One prominent example is Hadoop — a framework for handling distributed processing of large data sets (see Apache 2014). However, there are several others.
- **NoSQL:** This approach refers to a family of non-relational database approaches that are designed for managing large data sets, while providing benefits such as flexibility, scalability,

DOI: 10.1201/9781003275725-31

availability, lower cost, and special capabilities. Four related methodologies are key-value stores, graph stores, column stores, and document stores (see IBM 2015).

- **Other Burgeoning Trends** include multimedia databases, digital libraries, mobile databases, and spatial databases.

While a detailed discussion is beyond the scope of this course, these issues will be briefly introduced and summarized in the upcoming sections.

26.2 Entity–Attributes–Value Model

The EAV model attempts to meet the needs of a database by being compact and flexible. The idea is to pack the logical database requirements into three core tables — the **entities-table** for storing details about entities comprising the system; the **attributes-table** for storing details about attributes of system entities; and the **values-table** for storing the actual data. Alternate names for the EAV model are the object–attribute–value (OAV) model, the vertical database model, and the open schema model.

If you give this some thought, you should begin to see that this is a brilliant model for covering the requirements for almost any database. By pursuing this approach, one could reduce a significant aspect of logical database design to data entry. A similar approach has been described in Chapter 21 (Section 21.3) for the Dynamic Menu Interface Designer. The approach is also applied in other ongoing software engineering research projects conducted by the author. Finally, it must be stated that the EAV model is comparable to the extended relational model of Chapter 5 (Sections 5.4 and 5.5), which is used to influence the design and management of the system catalog for relational databases (review Chapter 14).

26.2.1 Rationale for the EAV Model

The main attraction of the EAV model is its flexibility in dealing with different design scenarios. For instance, this model gracefully handles the situation where certain entities may require additional attributes after the database is already in usage. In the more conventional relational model, specifying additional attributes for an entity involves a table alteration.

The EAV model also has the propensity to be more compact, especially if the DBMS being used supports variable-length data. However, in the interest of simplicity, multiple values-tables may be employed (this will be further elucidated shortly).

26.2.2 Challenges of the EAV Model

The EAV model is more abstract than the conventional relational model. This additional abstraction, while it provides additional flexibility, tends to result in slower database performance, particularly on complex queries over large databases. The reason for this is that increased abstraction will necessitate additional joins across multiple tables in order to support queries.

Another challenge of the EAV model is that unlike the relational model, it is counter-intuitive and therefore requires time to learn and practice. If you struggle with learning the relational model, then learning the EAV model poses even more challenge since mastery is more likely if you have a firm grasp of the relational model.

26.2.3 Illustrating Application of the EAV Model

Let us take an example that should reinforce how the EAV model may be applied to a database scenario. Recall the situation in Chapter 19 (Section 19.3.3) where a star schema was used to represent the database needs for a regional educational institution that operated in different locations. The star schema that was employed covered four dimensional tables, namely **Location**, **AcademicProgram**, **TimePeriod**, and **School**; there was also a fact table, **GraduationSummary**. For ease of reference, the ERD for the schema (Figure 19.1) is repeated here as Figure 26.1.

Let us now try to construct an alternate schema based on the EAV model. From Figure 26.1, we know that at least five entities must be represented. These would be covered in an **Entities** table (as data). Next, we would need an **Attributes** table to facilitate specification of attributes for each entity. Finally, how do we store attribute values? We may attempt to use one **Values** table; however, in the interest of simplicity, it is preferable to use multiple **Values** tables based on the attribute types supported by the host DBMS and the requirements of the candidate database. For the problem at hand, let us consider the data types Decimal, Char, and VarChar2. Based on this arrangement, we would arrive at a schema for five tables as summarized in Figure 26.2.

From this simple illustration, it should be clear to you that the EAV model, while being more flexible, it tends to result in a more abstract schema for the target database. Notice that the original entities of Figure 26.1 are represented indirectly as data in the EAV model of Figure 26.2. In order to obtain information for any of the original entities, it will be necessary to construct queries that join all five tables, and with search criteria to select out the data of interest. For these reasons, it is reasonable to anticipate that the performance of queries on an EAV-based database will be less favorable than queries on a conventional relational database.

There are several online resources that provide clarification and support for the EAV model. One example that you may find helpful is Magestore (2020).

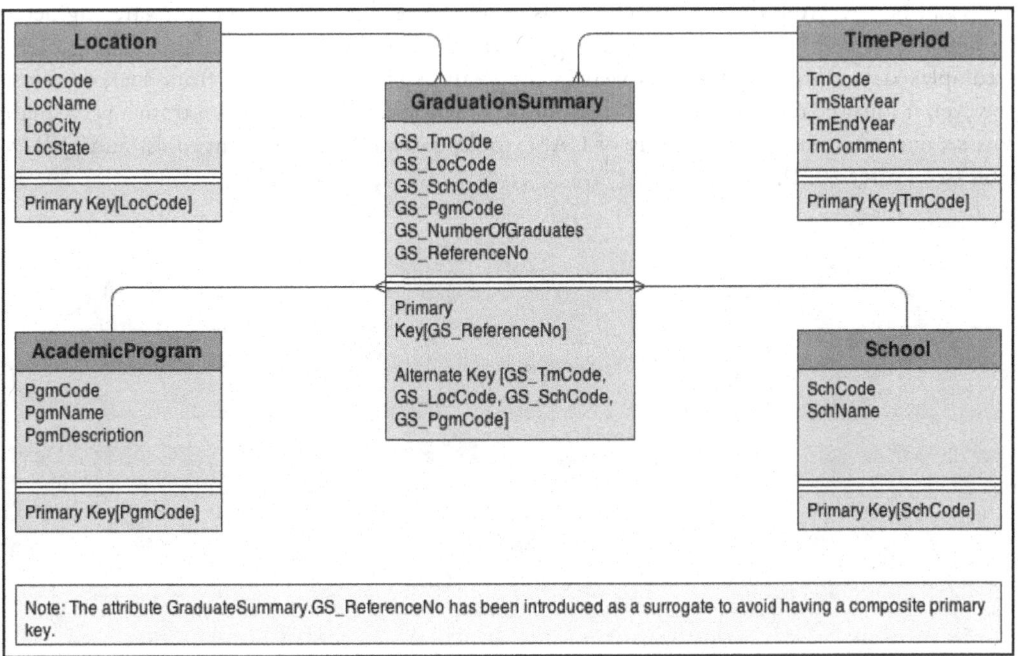

Figure 26.1 Illustration of a star schema for graduations from a regional university.

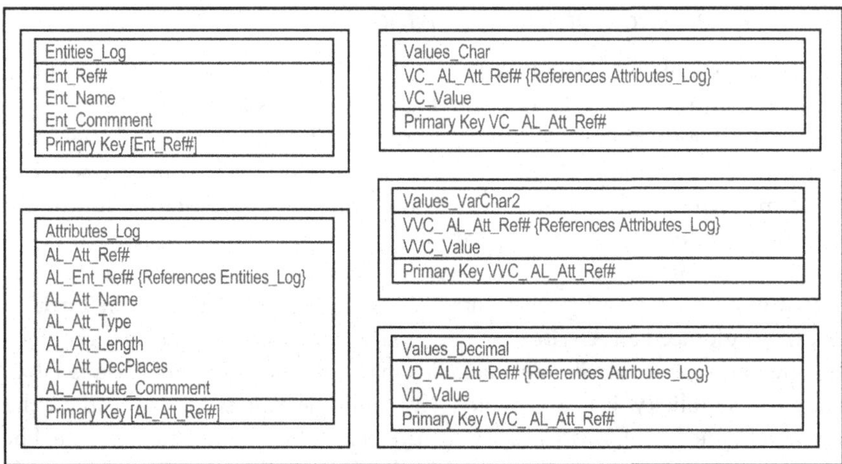

Figure 26.2 Alternate schema for graduations from a regional university based on the EAV model.

26.3 Database-Supporting Frameworks

Over the past two decades, we have seen a growing emergence of database-supporting frameworks, i.e., software systems that facilitate the easy creation and management of sophisticated cloud-based or local database systems. While this text does not endorse any product, Table 26.1 provides a list of prominent alternatives in the marketplace (for more information, see Ilyukha n.d).

As can be seen from the list, Apache tends to dominate this niche of the software engineering industry. These frameworks provide significant benefits in the creation and management of complex data storage systems, as well as the analysis of data retrieved from such systems. However, it must be remembered that the products mentioned here are drawn from a very small cross section of a much larger family of CASE tools, software development tools, and DBMS suites (see Foster 2021).

Table 26.1 Prominent Database-Supporting Frameworks

Product	*Host Company*	*Purpose*
Apache Hadoop	Apache	A framework that facilitates storage and processing of large data sets over multiple platforms.
MapReduce	Google	A framework that started out as part of Hadoop but has morphed into its own; uses functional programming principles to support parallel processing big data across multiple platforms.
Apache Spark	Apache	Another framework that is similar to Hadoop but with a more flexible (and sometimes more efficient) data retrieval algorithm.
Apache Hive	Facebook	A widely used big-data analytics framework; integrates well with other alternatives.
Apache Storm	Twitter	A framework that works well with large real-time data flow; integrates well with other complimentary components such as Bolt and Topology.
Apache Samza	Apache and Kafka	A stateful stream processing big-data framework that is intended for quick single-stage processing.
Apache Flink	Apache	A robust big-data processing framework for both stream and batch processing; also useful in ETL activities.
Apache Heron	Twitter	A framework intended to replace Apache Storm; highly recommended for real-time spam detection, ETL tasks, and trend analytics.
Apache Kudu	Apache	An SQL-like storage component intended to complicated pipelines in the Hadoop ecosystem.
Presto	Facebook	A fast, flexible query tool for multi-tenant data environments with different storage types; intended to be an alternative to Apache Hive for smaller tasks.

26.4 NoSQL

The acronym "NoSQL" is sometimes interpreted as non-SQL; however, it more accurately means "not only SQL." In short, NoSQL databases are non-traditional (and sometimes non-relational) databases that facilitate big data, cloud, high-volume Web, and mobile applications. Moreover, NoSQL databases support SQL-like queries; in fact, it is not unusual for a conventional relational database and a NoSQL database to converge in a single software system (see IBM 2019).

While the requirements and virtues of the relational database model are well established and applauded, pragmatism dictates recognition that there are situations in which some elements of the model may be relaxed in favor of accessibility and efficiency (even if only short term). It turns out that NoSQL databases are well suited for such circumstances. Two examples of such scenarios are document-retrieval systems and multimedia systems.

NoSQL databases are useful in scenarios where conventional relational databases may struggle. Among those scenarios are the following:

- There is a need for the storage of large volumes of unstructured data.
- Rapid and/or agile software development is required.
- A hybrid of conventional relational database and other creative database approaches are required.

26.4.1 NoSQL Databases versus SQL Databases

While there are gray areas of overlap between NoSQL databases and SQL databases, there are clear areas of distinction. Following is clarification on some of these areas of demarcation.

Definitional Differences: The term *SQL database* is used to refer to a conventional relational database that is subject to the structural, integrity, and manipulation constraints of most of the text and crystallized in Chapter 9. In contrast, the term *NoSQL database* refers to a database that is not necessarily tethered to the constraints of the relational database model.

Applicational Differences: SQL databases store and manage structured data through the auspices of a DBMS. NoSQL databases typically store unstructured data but can also handle structured data; they are not necessarily reliant on a DBMS, though there is usually an SQL-like retrieval system in place.

Support for Big Data: Through data warehousing methodologies as well as data types such as BLOB and BFile, SQL databases can be configured to handle big data; however, special processing measures are required. In contrast, NoSQL databases are designed specifically for treatment of big data and do not rely on SQL to analyze data.

Complexity of Design: As evidenced by the content of this course, the relational database model is complex, requiring a significant investment of time and effort in order to gain mastery. The NoSQL database model is simpler, more flexible, and less demanding on one's time and effort.

26.4.2 Emerging NoSQL Data Models

NoSQL databases are relatively new to the software engineering arena. Currently, four data models have emerged; each is briefly described here.

Key-Value Storage: In this approach, the prevailing file structure consists of two attributes — a key consisting of an appropriate string of characters, and a value which may consist of any combination of primitive data (integer, string, or array) and/or more complex data object(s). Typically, no query language is required; rather, the standard practice is to have a system for efficiently storing, retrieving, and updating data.

Document-based Storage: This approach is similar to the key-value storage model. The main difference here is that all objects are stored as documents, each with a key for accessibility purposes. Obviously, documents may be of variable lengths and are easily accessed for manipulation.

Column-based Storage: For this strategy (also called wide-column), data is stored via row key, column, and cell timestamp. Columns are not tethered to rows as in the relational database model. Rather, for each row, columns may be added or removed as required, thus providing a flexible arrangement for the database.

Graph-based Storage: In this approach, the database is organized and managed as a graph, consisting of nodes (vertices or data) and edges (connections). The principles of graph theory as covered in your course in Data Structures and Algorithms would apply here.

26.4.3 Examples of NoSQL Databases

Many of the extant NoSQL database solutions were developed by technology and/or software engineering companies to facilitate easy access to their online resources and services. Among the prominent solutions are the following:

- **Apache CouchDB:** A JSON document-based, open-source database system that employs JavaScript as its query language.
- **Apache Casandra:** An open-source, column-based database that has been designed to handle large data sets across multiple servers.
- **Redis:** An open-source, key-value database that is typically used for session caching, message queueing, stream processing, traffic shaping, and other in-memory type applications.
- **MongoDB:** An open-source, document-based database that is comparable to Apache CouchDB with its JSON-like document schema.
- **Elasticsearch:** An open-source, document-based database with a full-text search engine.
- **JanusGraph:** An open-source, scalable graph-based database that is suited to handle graphs of varying size and complexity, and dispersed across multiple machines.

26.5 Other Burgeoning Trends

We have touched on some exciting trends in contemporary database systems, but we are not quite finished. In addition to the foregoing discussion of the chapter, four other burgeoning trends deserve attention (even if brief) — multimedia databases, digital libraries, mobile databases, and spatial databases.

Multimedia Databases: As the name suggests, a multimedia database stores data of different formats and storage media. The data is typically unstructured, covering documents, images, text, video, and audio. While multimedia databases may be implemented using the conventional relational database model, they seem ideal for NoSQL databases (see previous section).

Digital Libraries: A digital library is a computerized online repository that provides traditional library services electronically. Over the past two decades, we have seen the proliferation of digital libraries for institutions of higher learning, national libraries (for example, Library of Congress), publishing organizations, retail companies providing E-commerce (such as Walmart, Amazon), professional associations (such as ACM, IEEE), and academic journals. Most digital libraries are supported by relational databases.

Mobile Databases: The exponential advances in mobile technology (hence mobile devices) have spawned a corresponding advance in mobile databases to support the various mobile devices. For instance, a typical mobile telephone is supported by a multimedia database storing contact information, notes, documents, texts, images, etc. Mobile databases have become commonplace in business as well as in social connections.

Spatial Databases: As the term suggests, a spatial database is a database system that stores and supports spatial data. Spatial databases are widely used in for complex problems with deep geographically and/or mathematical underpinnings. For instance, global positioning systems (GPS), meteorological systems, and complex geometrical systems rely on sophisticated spatial databases. Many spatial databases rely on graph theory for implementation.

26.6 Summary and Concluding Remarks

This brief chapter has taken a look at the database systems landscape, focusing on the emerging methodologies. It's time to summarize what has been covered in the chapter:

- The EAV database model reduces a database to three principal storage entities — an entity for defining other entities; an entity for defining properties (attributes) of entities; an entity that connects the other two entities and stores values for entity–attribute combinations.
- In recent years, software frameworks that support database systems have become commonplace. One prominent example is Hadoop; however, there are several others.
- NoSQL databases comprise a family of non-relational database approaches that are designed for managing large data sets, while providing benefits such as flexibility, scalability, availability, lower cost, and special capabilities. Four related methodologies are key-value stores, graph stores, column stores, and document stores.
- Other burgeoning trends include multimedia databases, digital libraries, mobile databases, and spatial databases.

The field of database systems continues to be an exciting area of computer science. As methodologies and supporting technologies continue their emergence and/or refinement, one thing is as certain as sunrise and sunset: databases are here to stay.

26.7 Review Questions

1. Describe the EAV database model. Identify two favorable features of the model, followed by two challenges.
2. What is the primary purpose of a database-supporting framework? State and summarize four such frameworks.
3. What are the main differences between an SQL database and a NoSQL database?
4. Describe four prominent NoSQL data models. For each model, provide an example of a software system that supports that model.
5. Describe four burgeoning trends in contemporary database systems.

References and/or Recommended Readings

Apace Software Foundation. 2014. "Hadoop." Accessed February 2016. http://hadoop.apache.org/

Foster, Elvis C. 2021. *Software Engineering: A Methodical Approach* 2nd ed. New York: CRC Press. See chapters 2, 25, & 26.

IBM Corporation. 2015. "Analytics White Paper." Accessed February, 2016. https://cloudant.com/wp-content/uploads/Why_NoSQL_IBM_Cloudant.pdf

IBM Corporation. 2019. "NoSQL Databases." Accessed July 2021. https://www.ibm.com/cloud/learn/nosql-databases

Ilyukha, Vitaliy. n.d. "10 Best Big Data Tools for 2021." Accessed July 2021. https://jelvix.com/blog/top-5-big-data-frameworks

Magestore. 2020. "Entity Attribute Value in Magento." Accessed June 2021. https://blog.magestore.com/entity-attribute-value-in-magento/

Wikipedia. 2016. "Entity-Attribute-Value Model." Accessed February 2016. https://en.wikipedia.org/wiki/Entity%E2%80%93attribute%E2%80%93value_model

APPENDICES F

This final division contains a number of related topics that you may find interesting. First, there are two review topics from your *data structures and algorithms* course — review of trees and hashing. This is followed by a review from your *software engineering* course — review of information gathering strategies. Next, there is a chapter providing Backus–Naur Form (BNF) notations for selected Structured Query Language (SQL) statements covered in the course. Finally, Appendix 5 provides some sample assignments. The chapters are presented in the following order:

- Appendix 1: Review of Trees
- Appendix 2: Review of Hashing
- Appendix 3: Review of Information Gathering Strategies
- Appendix 4: BNF Syntax for Selected SQL Statements
- Appendix 5: Sample Exercises and Examination Questions

DOI: 10.1201/9781003275725-32

Appendix 1: Review of Trees

This appendix provides a brief review of trees. You should pay specific attention to the section on B-trees, since most database management system (DBMS) suites implement them by default. Note: The appendix is not meant to replace a full course (and text) in data structures. It should therefore be regarded as an overview, not a final authority on the subject matter.

This review covers the following sub-topics:

- Introduction to Trees
- Binary Trees
- Threaded Binary Trees
- Binary Search Trees
- Height-Balanced Trees
- Heaps
- M-way Search Trees and B-trees
- Summary and Concluding Remarks

A1.1 Introduction to Trees

The main difference between $O(N^2)$ sorting algorithms and O(N-log-N) sorting algorithms is that the latter repeatedly reduce (by approximately one half) the number of keys remaining to be compared with each other, while the former does not. Trees are excellent sources of f (N-log-N) sorts.

As you are no doubt aware, in computer science, we use trees (and graphs) to represent and implement complex data structures. Here is a working definition of a tree: A tree T is a finite set of nodes (V_1, V_2 ...V_n) such that:

a. There is one designated node called the root.
b. The remaining nodes are partitioned into $M \geq 0$ disjoint sets T_1, T_2 ...T_n such that each T_i is itself a tree.
c. Except for the root, each node has a parent node.

Figure A1.1 shows the graphic representation of a general tree. Observe that the figure embodies each of the three properties that the definition articulates. Also, you will readily recall from your earlier computer science (CS) courses that CS trees are typically represented in an upside-down manner.

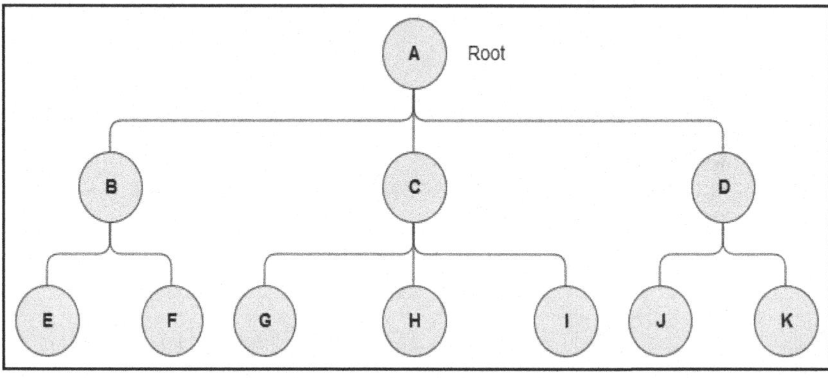

Figure A1.1 An example of a tree.

Nodes below the root of a tree are said to be sub-trees; for instance, nodes B, E, and F form a sub-tree. Any node that does not itself have one or more sub-nodes (called children) connected to it is called a leaf. In Figure A1.1, the root of the tree is node A; leaves are nodes E, F, G, H, I, J, and K. Here are a few additional conventions about trees:

- The *leaves* are the terminal nodes.
- Parent nodes are said to have *siblings*.
- A forest is a combination (abundance) of trees.

The root is at level zero (0). If we denote *k* to represent the highest level of the tree, then the following observations apply:

- The height is determined by the formula: $H = k + 1$.
- Weight of a node is the value of the node multiplied by [its level plus 1], assuming that the root is at level zero. This is summarized by the following formula: *Weight [i] = Value [i] * (1 + Level [i]).*

- The weight of the tree is the summation of the weight of the nodes. We may summarize this by the following formula: *Tree Weight* = *Σ[0 … k] Weight [i]* = *Σ[0 … k] {Value [i]* * *(1 + Level [i])}*.

A1.2 Binary Trees

While general trees are interesting, they are a bit too unstructured for a focused study. In computer science, we are more interested in trees for which we can observe patterns of behavior, thus making predictions or formulating reasonable expectations. Fundamental to this study is an understanding of binary trees.

A1.2.1 Overview of Binary Trees

A binary tree is a tree in which each node is either empty or consists of two disjoint binary trees: the left *sub-tree* and the right *sub-tree*. Figure A1.2 illustrates this concept.

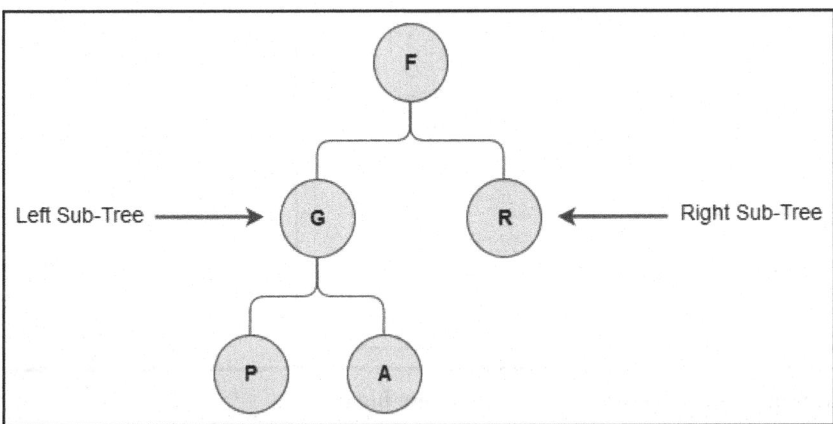

Figure A1.2 Illustrating a binary tree.

An understanding of binary trees is fundamental to understanding more complex data structures that are often applied in software engineering and database systems. Here are a few additional notes:

- No parent has more than two children.
- A *full* (also called *perfect*) binary tree is a binary tree that contains the maximum number of nodes possible. Figure A1.3 shows a complete binary tree.
- A *complete* binary tree is a binary tree that is full down to the penultimate level, with nodes at the final level filled in from the left to the right.

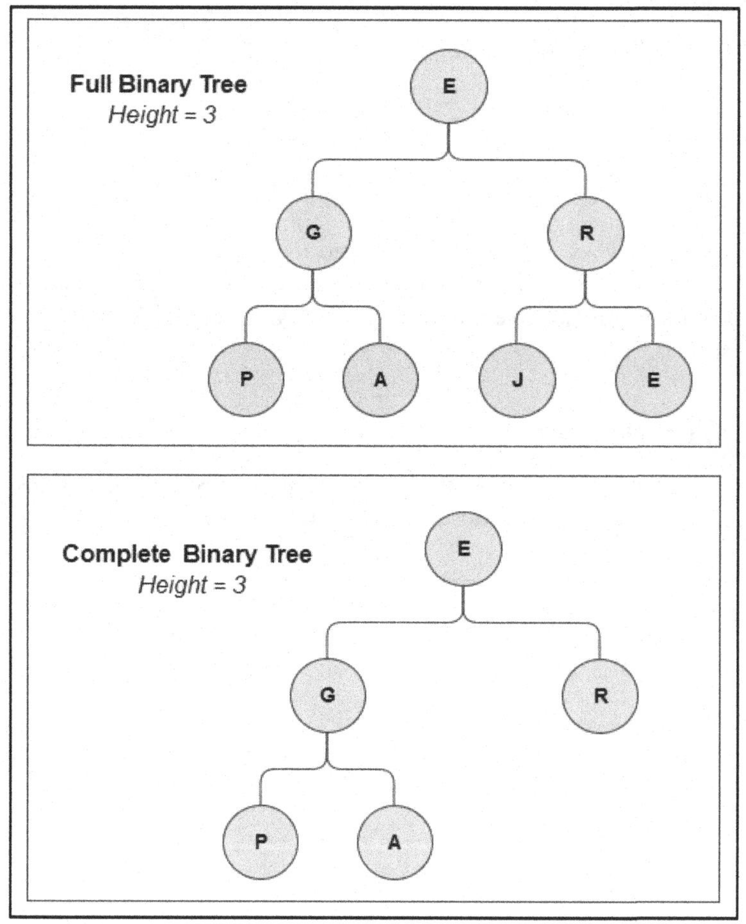

Figure A1.3 Illustration of a full binary tree and complete binary tree.

The maximum number of nodes of a binary tree may be determined by the following formulas:

- *Maximum number of nodes at the ith level is* 2^i.
- *Maximum number of nodes =* Σ *[0 … k] of* 2^i.

Observe that that the latter formula describes a geometric series of first term 1 and common ratio 2 (review your discrete mathematics). The sum is therefore $2^k - 1$. So, for a binary tree of height k,

- *Maximum number nodes =* $2^k - 1$

To illustrate, in the complete binary tree above of Figure A1.3, the maximum number of nodes is $2^3 - 1 = 7$. So once you know the height of the tree, you can determine its maximum size (in terms of number of nodes).

A1.2.2 Representation of Binary Trees

Binary trees may be represented by arrays or preferably linked lists. Figure A1.4a illustrates a tabular representation of a binary tree while Figure A1.4b illustrates the graphic representation.

Location	Information	Left	Right
1	B	4	6
2	G	0	0
3	F	0	0
4	D	0	0
5	A	1	7
6	E	0	0
7	C	3	2

Note:
1. The root is node A. It has no pointer to it.
2. Leaves do not have left or right sub-trees. From the table, leaves are D, E, F, G.

Figure A1.4a Tabular representation of a binary tree.

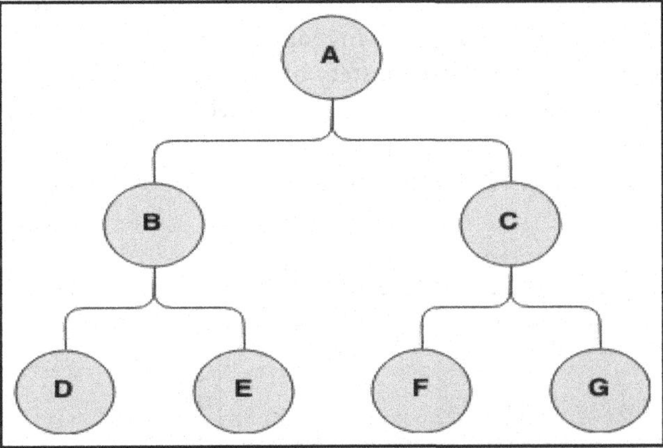

Figure A1.4b Graphic representation of a binary tree.

A1.2.3 Application of Binary Trees

Binary trees are typically applied in the following ways:

- Binary trees (and extensions of binary trees) are used extensively in database management system (DBMS) and operating system (OS) construction to effectively manage indexes of data files.
- Binary trees are used in calculation of expressions during compilation.
- Binary trees are used in data compressions, for example, Huffman Coding Tree.

A1.2.4 Operations on Binary Trees

The following are the main operations that are normally defined on a binary tree:

- **Creation:** At creation the tree either has no nodes or a dedicated root node.
- **Insertion:** We may allow insertion at the root, after terminal nodes only, or we may allow insertion anywhere in tree.
- **Deletion:** Again, we may allow deletion of terminal nodes only, or we may allow deletion of nodes from anywhere in tree.
- **Clear:** Remove all nodes from the binary tree, thus leaving it empty.
- **Check-Size:** Return the size of the tree.
- **Check-Empty:** Check if the tree is empty.
- **In-order:** In-order traversal of the tree.
- **Pre-order:** Pre-order traversal of the tree.
- **Post-order:** Post-order traversal of the tree.

A1.2.5 Implementation of Binary Trees

Suppose that you desire a binary tree of **LibraryPatron** objects. The **LibraryPatron** class includes data items including patron number, name, major, and status; it also includes methods for manipulating instances of the class. You can implement this tree using an array, array-list, or linked-list. The preference is for the linked-list implementation, since it provides more flexibility. Figure A1.5 shows the Unified Modeling Language (UML) diagrams for implementing this binary tree. It is assumed that you have mastery of fundamental programming principles; therefore, elaboration on these diagrams is not considered necessary. Note also that Figure A1.5c lists a number of methods that you may choose to implement. Except for traversal methods (the last three), you should be able to figure out the required logic for these methods on your own.

LibraryPatron
protected int PatronNumber protected String Name protected String Major private String Status
LibraryPatron() LibraryPatron(int ThisNumber) void Modify(LibraryPatron ThisPatron) void InputData(int x) String PrintMe() int GetPatronNumber()

Figure A1.5a The UML diagram of the LibraryPatron class.

PatronNode
protected LibraryPatron Info protected PatronNode Left, Right
PatronNode() void Modify(PatronNode ThisNode) void InputData(int x) String PrintMe()

Figure A1.5b The UML diagram of the PatronNode class.

PatronsBinaryTree
protected PatronNode Root protected int Size, TravRef LibraryPatron[] Traversal // May also be implemented as an array-list or string; stores the traversal result

public PatronsBinaryTree()
public void addRoot(LibraryPatron ThisPatron) // Inserts at the root, assuming previously empty tree
public void addLeftLeaf(PatronNode ThisLeaf, LibraryPatron ThisPatron) // Inserts at the specified leaf
public void addRightLeaf(PatronNode ThisLeaf, LibraryPatron ThisPatron) // Inserts at the specified leaf
public void addLeftSubtree(PatronNode ThisNode, PatronNode NewNode) // Inserts at the specified node
public void addRightSubtree(PatronNode ThisNode, PatronNode NewNode) // Inserts at the specified node
public void removeSubtree(PatronNode ThisNode)
public void Modify(PatronNode ThisNode, LibraryPatron ThisPatron)
public void clearTree()
public LibraryPatron getInfo(PatronNode ThisNode)
public PatronNode getNode(PatronNode ThisNode)
public int getSize()
public boolean isEmpty()
public void inOrderTraversal(PatronNode ThisNode)
public void preOrderTraversal(PatronNode ThisNode)
public void postOrderTraversal(PatronNode ThisNode)

Figure A1.5c The UML diagram of the PatronsBinaryTree class.

PatronsBinaryTreeMonitor
public static PatronsBinaryTree PatronsTree public static final String HEADING = "Library Patrons Tree" public static final int DEFAULT_NUMBER = 0
public static void main(String[] args) public static void InputPatrons() public static void TraverseTree() public static void RemovePatrons() public static void CheckSize() public static void InitializeTree() public static void Empty()

Figure A1.5d The UML diagram of the PatronsBinaryTreeMonitor class.

Figure A1.6a provides the pseudo-code for the methods identified in Figure A1.5c. Immediately following, Figure A1.6b provides an illustration of how selected letters of the alphabet would be inserted in a binary tree of characters. The last three methods of Figure A1.5c relate to traversal of the binary tree. This matter will be discussed in the next section.

A1.2.6 Binary Tree Traversals

There are three commonly used traversal algorithms for binary trees: Pre-order, In-order, and Post-order. The algorithm for each traversal is recursive; each algorithm produces a different result.

Pre-order traversal (also called prefix walk or polish notation) obeys the algorithm shown in Figures A1.7a and A1.7b. These are followed by an illustration of the algorithm at work in Figure A1.7c.

In-order traversal (also called symmetric order or infix notation) obeys the algorithm shown in Figures A1.8a and A1.8b. Figure A1.8c provides an example of the application of the in-order algorithm.

Post-order traversal (also called suffix walk or reverse polish notation) obeys the algorithm shown in Figures A1.9a and A1.9b. These are followed by an example of the application of the post-order algorithm in Figure A1.9c.

Assume that Root always points to the root of the tree and that each node consists of the following:

- Info // Euphemism for details of the node
- Left // Pointer to the left sub-tree
- Right // Pointer to the right sub-tree
- Size // The number of nodes in the tree

```
void addRoot(LibraryPatron ThisPatron)
START
 If (Root = NULL)
  Root.Info := ThisPatron;
  Set Root.Left and Root.Right to NULL;
  Add 1 to Size;
 End-If;
STOP

void addLeftLeaf(PatronNode ThisLeaf, LibraryPatron ThisPatron)
START
 Let NewNode be a PatronNode;
 Instantiate NewNode;
 NewNode.Info := ThisPatron;
 Set NewNode.Left and NewNode.Right to NULL;
 ThisLeaf.Left := NewNode; // No longer a leaf, but now a sub-tree
 Add 1 to Size;
STOP

void addRightLeaf(PatronNode ThisLeaf, LibraryPatron ThisPatron)
START
 Let NewNode be a PatronNode;
 Instantiate NewNode;
 NewNode.Info := ThisPatron;
 Set NewNode.Left and NewNode.Right to NULL;
 ThisLeaf.Right := NewNode; // No longer a leaf, but now a sub-tree
 Add 1 to Size;
STOP

void addLeftSubtree(PatronNode ThisNode, PatronNode NewNode)
START
 Let Temp, LeftEnd be PatronNode references;
 Temp := ThisNode.Left;
 LeftEnd := NewNode;
 While (LeftEnd.Left <> NULL) LeftEnd := LeftEnd.Left; End-While;
 ThisNode.Left := NewNode;
 LeftEnd.Left := Temp;
 Increase Size by the size of the sub-tree;
STOP

void addRightSubtree(PatronNode ThisNode, PatronNode NewNode)
START
 Let Temp, RightEnd be PatronNode references;
 Temp := ThisNode.Right;
 RightEnd := NewNode;
 While (RightEnd.Right <> NULL) RightEnd := RightEnd.Right; End-While;
 ThisNode.Right := NewNode;
 RightEnd.Right := Temp;
 Increase Size by the size of the sub-tree;
STOP
```

Figure A1.6a Binary tree algorithms. (Continued)

```
void removeSubtree(PatronNode ThisNode)
START
 If (ThisNode.Left = ThisNode.Right = NULL) // a leaf
        Kill(ThisNode); Subtract 1 from Size;
 End-If;
 Else
        If (ThisNode.Left <> NULL) removeSubtree(ThisNode.Left); End-If;
        If (ThisNode.Right <> NULL) removeSubtree(ThisNode.Right); End-If;
        Kill(ThisNode);
 End-Else;
STOP

void Modify(PatronNode ThisNode, LibraryPatron ThisPatron)
START
 ThisNode.Info := ThisPatron;
STOP

void clearTree()
START
 removeSubtree(Root);
STOP

LibraryPatron getInfo(PatronNode ThisNode)
START
 Return ThisNode.Info;
STOP

PatronNode getNode(PatronNode ThisNode)
START
 Return ThisNode;
STOP

int getSize()
START
 setSize (Root, 0);
Return Size;
STOP

boolean isEmpty()
START
 Return whether Size is 0 or not;
STOP
```

Figure A1.6a (Continued) Binary tree algorithms. (Continued)

```
void setSize (PatronNode ThisNode, Size)
START
 // Assume that this method is first called with ThisNode pointing to Root
 Let Current, HoldLeft, and HoldRight be PatronNode instances;
 Current := ThisNode;
 If (Current <> NULL)
        Add 1 to Size;
        Holdeft := Current.Left;
        HoldRight := Current.Right;

      If (HoldLeft <> NULL)
               Current := HoldLeft;
               setSize(Current, Size);
        End-If;
        If (HoldRight <> NULL)
               Current := HoldRight;
               setSize(Current, Size);
        End-If;
  End-If;
STOP
```

Figure A1.6a (Continued) Binary tree algorithms.

The following tree is the result of these insertions (in the order shown):

1. addRoot("M")
2. addLeftLeaf(Root, "F") and addRightLeaf(Root, "S")
3. addLeftSubtree(nodeF,D-C-G) // assume D is the root of the subtree, C is the left node and G is the right node
4. addRightSubtree(nodeS, T-X-W) // assume T is the root of the subtree, X is the left node and W is the right node
5. addLeftSubtree(nodeD, H-A-B) // assume H is the root of the subtree, A is the left node and B is the right node

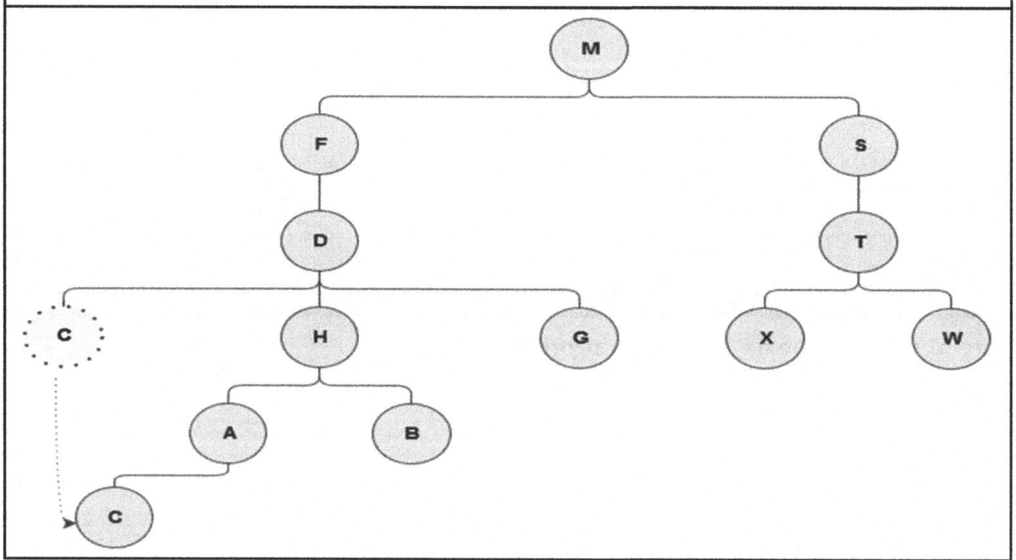

Figure A1.6b Illustrating insertions in binary tree.

```
Visit the root;
Traverse the left sub-tree in Pre-order;
Traverse the right sub-tree in Pre-order;
```

Figure A1.7a Summary of the Pre-order Traversal algorithm.

```
void preOrderTraversal(PatronNode ThisNode)
START // Assume the UML diagram of figure A1-5
 // Assume that Traversal is a global array of Size LibraryPatron objects;
 If (ThisNode is not NULL)
   Append ThisNode.Info to Traversal;
   preOrderTraversal(ThisNode.Left);
   preOrderTraversal(ThisNode.Right);
 End-If;
STOP
```

Figure A1.7b Detailed Pre-order Traversal algorithm.

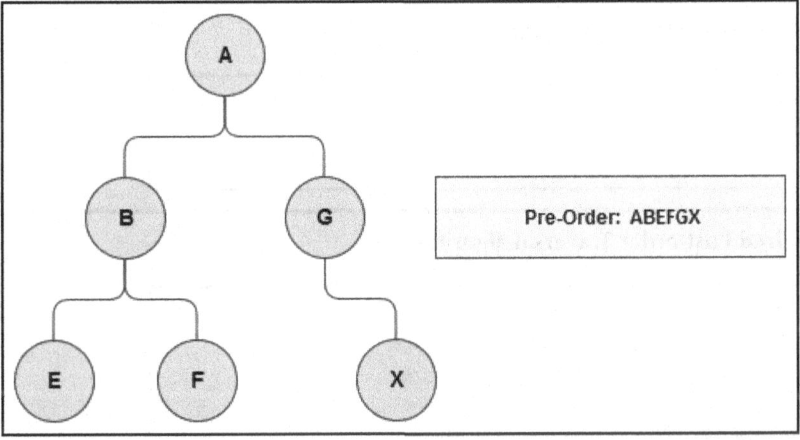

Figure A1.7c Example of Pre-order Traversal algorithm.

```
Traverse the left sub-tree in In-order;
Visit the root;
Traverse the right sub-tree in In-order;
```

Figure A1.8a Summarized In-order Traversal algorithm.

```
void inOrderTraversal(PatronNode ThisNode)
START // Assume the UML diagram of figure A1-5
 // Assume that Traversal is a global array of Size LibraryPatron objects;
 If (ThisNode is not NULL)
   inOrderTraversal(ThisNode.Left);
   Append ThisNode.Info to Traversal;
   inOrderTraversal(ThisNode.Right);
 End-If;
STOP
```

Figure A1.8b Detailed In-order Traversal algorithm.

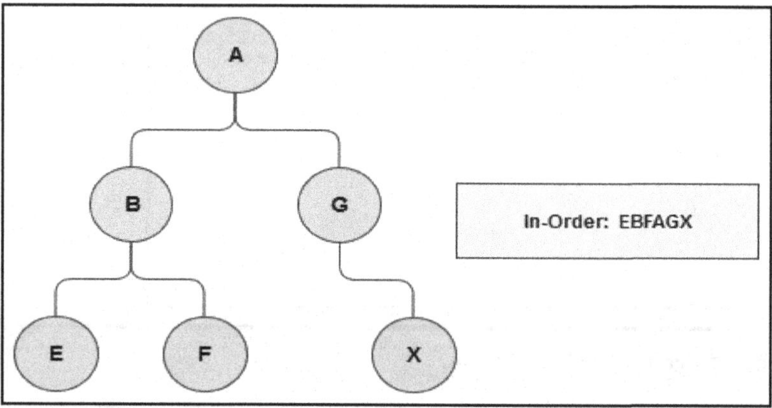

Figure A1.8c Example of In-order Traversal algorithm.

```
Traverse the left sub-tree in Post-order;
Traverse the right sub-tree in Post-order;
Visit the root;
```

Figure A1.9a Summarized Post-order Traversal algorithm.

```
void postOrderTraversal(PatronNode ThisNode)
START // Assume the UML diagram of figure A1-5
  // Assume that Traversal is a global array of Size LibraryPatron objects;
  If (ThisNode is not NULL)
    postOrderTraversal(ThisNode.Left);
    postOrderTraversal(ThisNode.Right);
    Append ThisNode.Info to Traversal;
  End-If;
STOP
```

Figure A1.9b Detailed Post-order Traversal algorithm.

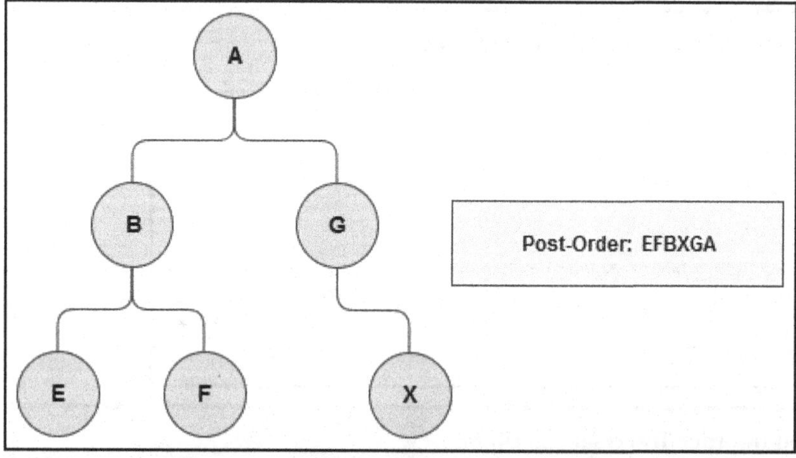

Figure A1.9c Example of Post-order Traversal algorithm.

A1.2.7 Using Binary Tree to Evaluate Expressions

We can use binary trees to evaluate expressions by following a simple convention: All operands are leaves of the tree and the operators are parent nodes. To load an expression to tree, repeatedly select the operator that divides the expression into two, as the root. In the end, leaves are operands and roots are operators.

> **EXAMPLE A1.1: USING BINARY TREE TO REPRESENT ARITHMETIC EXPRESSIONS**
>
> The expression (a - b) * c is represented on a binary tree as shown in Figure A1.10a.

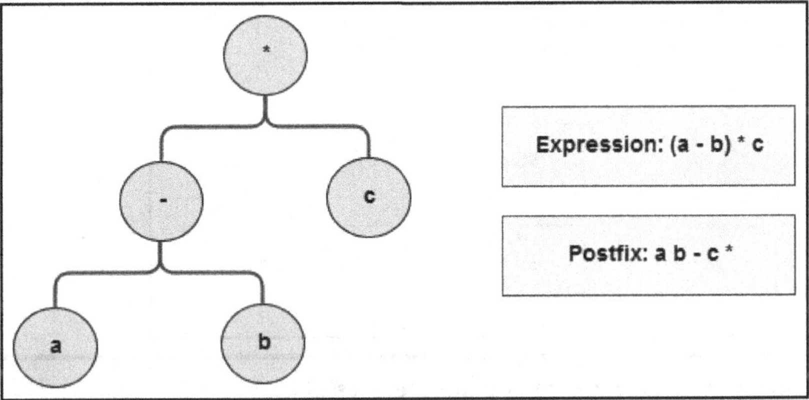

Figure A1.10a Using binary tree to represent arithmetic expressions.

> **EXAMPLE A1.2: USING BINARY TREE TO REPRESENT ARITHMETIC EXPRESSIONS**
>
> The expression [(((a+b)*c)/d) + e^f]/g is represented on a binary tree as shown in Figure A1.10b. Note that in the figure, the symbol ↑ is used to denote the exponent.

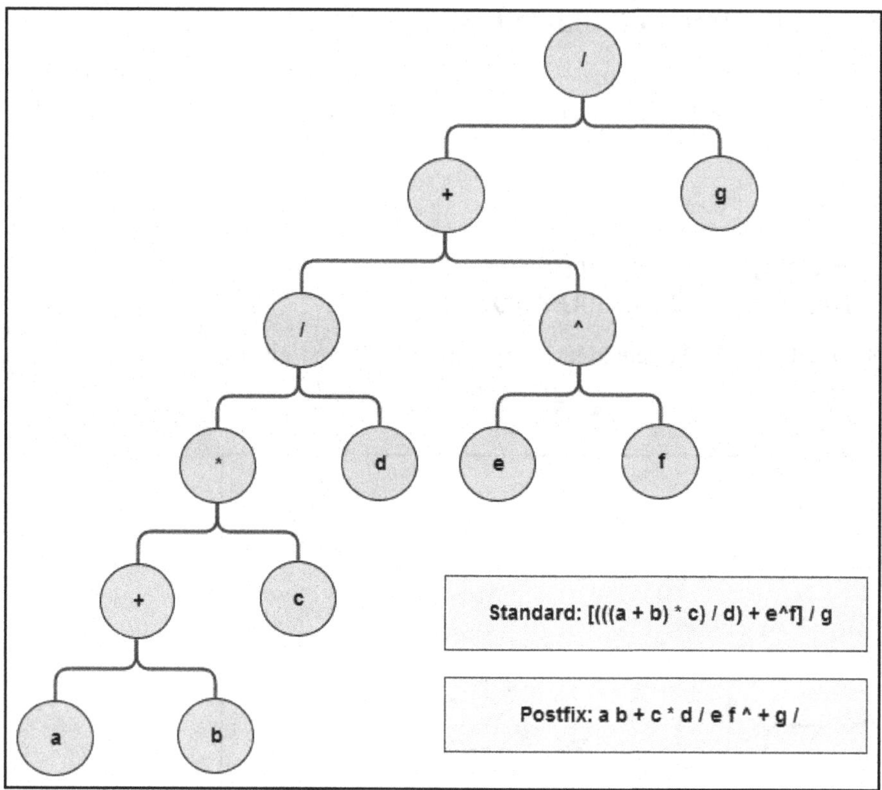

Figure A1.10b Using binary tree to represent arithmetic expressions.

Observe:

You should be able to convince yourself that a post-order traversal of a binary tree for arithmetic expressions of this sort produces the expression in postfix notation. Referring to the two trees above, the postfix notation for each expression tree is indicated in the bottom-right corner. As an exercise, try loading various arithmetic expressions into binary trees and then obtaining the postfix notation.

A1.3 Threaded Binary Trees

In order to facilitate easy traversal of a binary tree, the tree may be threaded with pointers that explicitly show a traversal ordering. The threads link the nodes in the sequence of the traversal method. A tree can only be threaded according to one traversal method at a time. Traditionally, two types of threads were used:

- Right thread links a node to its successor.
- Left thread links a node to its predecessor.

To illustrate if a binary tree is threaded for In-order processing, the following conventions are observed:

- Left threads (except for leftmost leaf) point to predecessor nodes in in-order.
- Left thread from leftmost leaf points to the root.
- Right threads (except for rightmost leaf) point to successor nodes in in-order.
- Right thread of rightmost leaf is NULL.

A1.4 Binary Search Trees

A binary search tree (BST) is a binary tree in which the following properties hold:

a. All nodes in the left sub-tree of R_i precede (by way of ordering) R_i, so that
 If R_j = R_i.Left, then R_j.Info < R_i.Info
b. All nodes in the right sub-tree of R_i succeed (by way of ordering) R_i, so that
 If R_j = R_i.Right, then R_j.Info >= R_i.Info

Observe:

In-order traversal of a BST produces a sorted list (in ascending values). Moreover, it can be shown that this sort algorithm is an O(N-Log-N) algorithm on the average.

Mastery of BSTs is a fundamental steppingstone in the journey to more advanced software development. This is so because the BST principles are often applied to more complex programming scenarios. The BST provides the following advantages:

- The BST facilitates an O(N-Log-N) sort, which is significantly more efficient sort than N^2 sort algorithms.
- The BST facilitates faster (binary) search than searching a linear linked-list.

Despite their advantages, BSTs also exhibit some shortcomings. The main disadvantages associated with a BST are as follows:

- The BST takes up slightly more space than linear linked-list.
- The order in which nodes are inserted affects the height and shape of the BST. In the worst-case scenario, if an attempt is made to insert a sorted list into a BST, the BST degenerates into a linked-list with a set of wasted pointers on either the left or the right.
- As nodes are added to the BST, there is no control on the height (hence structure) of the tree. This is often undesirable.

EXAMPLE A1.3: INSERTING NODES INTO A BST

Suppose that we wanted to load the string HBXAM to a BST. Assuming that the string is traversed from left to right, we would obtain a BST as shown in Figure A1.11.

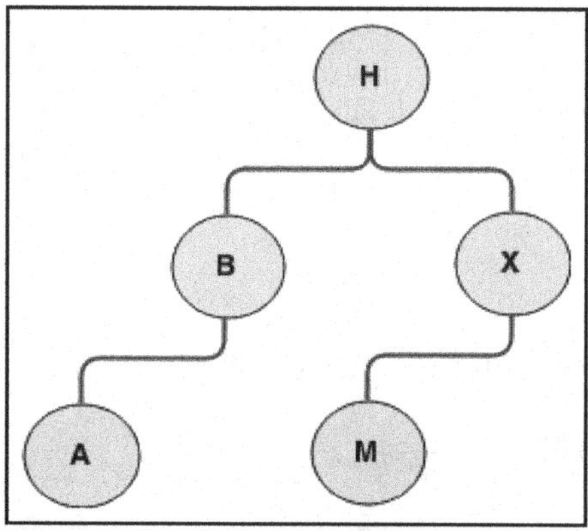

Figure A1.11 Example of inserted BST nodes.

Assuming the conventions of Figure A1.5, the algorithms of Figure A1.6 would still be applicable but with two extensions: We need to introduce an algorithm for inserting a node in the tree while preserving its properties. Secondly, we need to be able to search for a particular value in the tree. Figures A1.12 and A1.13 provide the insertion algorithm, and the search algorithm, respectively. Note that the order in which nodes are added to the BST affects the structure of the tree. In fact, if a sorted list is added to the BST, it degrades to a linear linked-list.

```
void Insert (PatronNode ThisNode, LibraryPatron ThisPatron)
START
  // ThisNode represents the point in the BST where an assessment for insertion begins.
  // ThisPatron represents the information to be inserted in the BST.
  // Typically, this algorithm will be called with the argument for ThisNode being Root.
  // However, it could called with any node as the starting point.

Let NewNode and Current be instances of PatronNode;
If (ThisNode = NULL) // BST is empty
   Instantiate NewNode; NewNode.Info := ThisPatron;
   Set NewNode.Left and NewNode.Right to NULL;
   Root := NewNode;
End-If;
Else
   // Find the insertion point and insert the node there
   Current := FindInsertionPoint(ThisNode, ThisPatron);
   If (ThisPatron < Current.Info) addLeftLeaf(Current, ThisPatron); End-If; // See figure 7.6
   Else addRightLeaf(Current, ThisPatron);  End-Else;
End-Else;
  // Add 1 to Size; // This is already done
STOP
```

Figure A1.12a Algorithm to insert a node in the BST.

```
PatronNode FindInsertionPoint(PatronNode ThisNode, LibraryPatron ThisPatron)
START
 Let Point be a PatronNode;
 Point := ThisNode;
 If (ThisPatron < Point.Info)
  If (Point.Left <> NULL)
    Point := FindInsertionPoint(Point.Left, ThisPatron);
  End-If;
  Else // ThisPatron >= Point.Info
  If (Point.Right <> NULL)
   Point := FindInsertionPoint(Point.Right, ThisPatron);
  End-If;
 End-If;
 Return Point;
STOP
```

Figure A1.12b Algorithm to find the insertion point in the BST.

```
PatronNode Search(LibraryPatron SearchValue)
START
 Let Current and Sought be PatronNode instances;
 Current := Root; Sought := NULL;
 While (Current.Info <> SearchValue) AND (Current.Info <> NULL) do the following:
        If (SearchValue < Current.Info)
               Current  := Current.Left;
        End-If
        Else      Current :=  Current.Right; End-Else;
 End-While;
 If (Current.Info = SearchValue) Sought := Current; End-If;
 Return Sought;
STOP
```

Figure A1.13 Algorithm to search the BST.

EXAMPLE A1.4: INSERTING A SORTED LIST INTO A BST

Inserting the sorted list A B C D E F G to a BST results in a lopsided tree as illustrated in Figure A1.14. Searching for a node in this BST would be reduced to searching a linear linked-list! Solving this problem is not trivial. It is generally referred to as balancing the tree. Several algorithms have been proposed for balancing a BST.

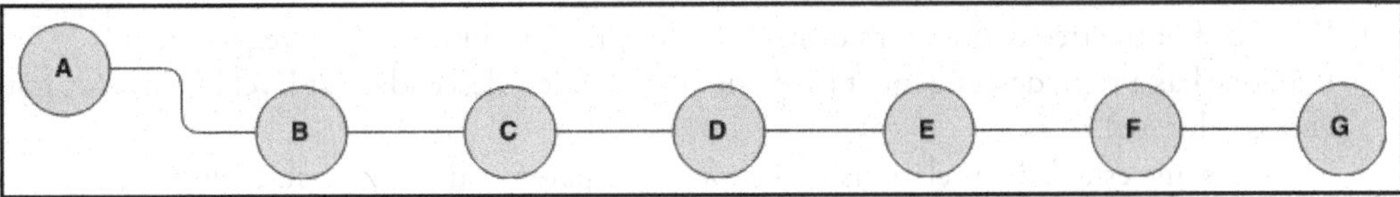

Figure A1.14 Inserting a sorted list into a BST.

A1.5 Height-Balanced Trees

A height-balanced k-tree (denoted (HB (k)) is a BST in which the maximum allowable difference in height between any two sub-trees sharing a common root (not necessarily a parent) is k. Put another way, the maximum possible difference in height between leaves of the tree is k.

An AVL tree (named after Adelson-Velskii and Landis, the Russian founders) is an HB (1) tree, that is, the maximum possible difference in height between any two sub-trees sharing a common root is 1. Put another way, leaves are either at level m or level m + 1. Figure A1.15 provides some illustrations. Heaps (discussed next) are also examples of HB trees.

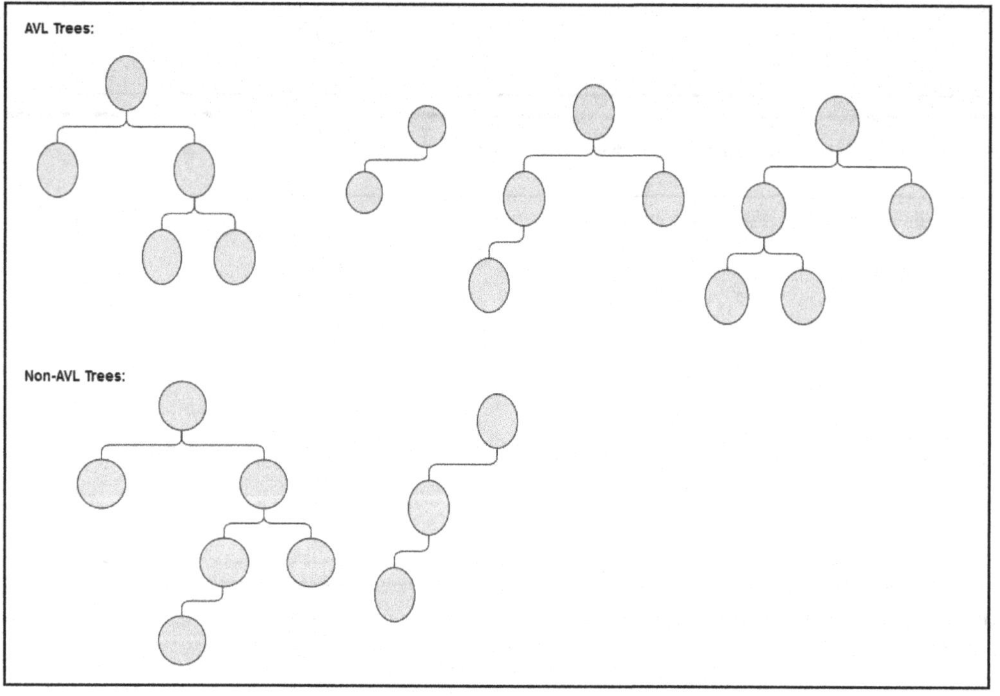

Figure A1.15 Illustrating AVL and non-AVL trees.

A1.6 Heaps

Like BSTs, heaps are a source of N-log-N sort algorithms. A heap is an almost complete binary tree upon which the following properties have been imposed:

1. Every leaf of the tree is at level m or m + 1, where m is an integer (AVL requirement).
2. If a node has a right descendant at level l, it also has a left descendant at level l (converse, not necessarily true).
3. There is some established relationship between the parent-value and each child-value.

Figure A1.16 provides some illustrations. From this definition of a heap, observe that there are two types of heaps: *max-heaps* and *min-heaps*:

- **Max-heap:** Every node stores a value that is greater than or equal to the value of either of its children. The root therefore stores the maximum value of all nodes in the tree.
- **Min-heap:** Every node stores a value that is less than or equal to the value of either of its children. The root therefore stores the minimum value of all nodes in the tree.

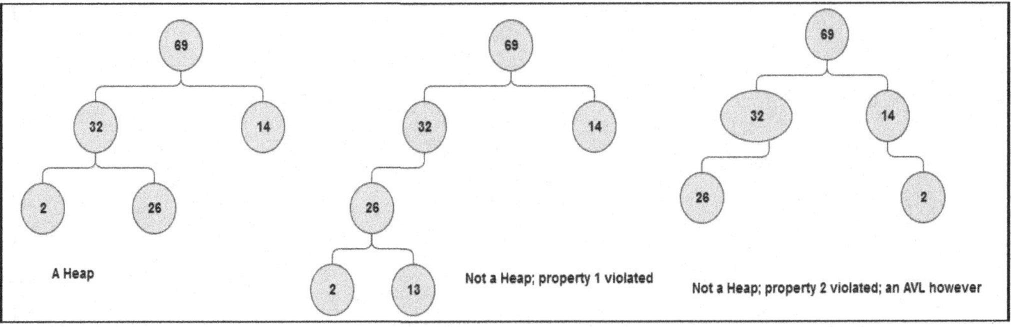

Figure A1.16 Illustrations of heap.

The heapsort algorithm may use a max-heap or a min-heap. For the purpose of illustration, let us assume a max-heap. Heapsort involves two phases:

- Creating (building) the heap
- Processing the heap

A1.6.1 Building the Heap

Figure A1.17 provides a summarized form of the algorithm to construct a heap. A more detailed form of the algorithm would be similar (but of course not identical) to the algorithm provided in Figure A1.12; this is left as an exercise for you.

1. Start reading the input stream from left to right and gradually build the heap;
2. For each node, introduce left sub-tree whenever possible before introducing a right sub-tree at each level; introduce a right subtree only if introduction of a left subtree would violate a heap principle;
3. If addition of a node results in any heap principle being violated, rotate node(s) until the anomaly is resolved;

Figure A1.17 Overview of the heap construction algorithm.

EXAMPLE A1.5: ILLUSTRATING HEAP CONSTRUCTION

Figure A1.18 shows how a max-heap is constructed with the following nodes:
6 25 7 2 14 10 9 22 3 16.

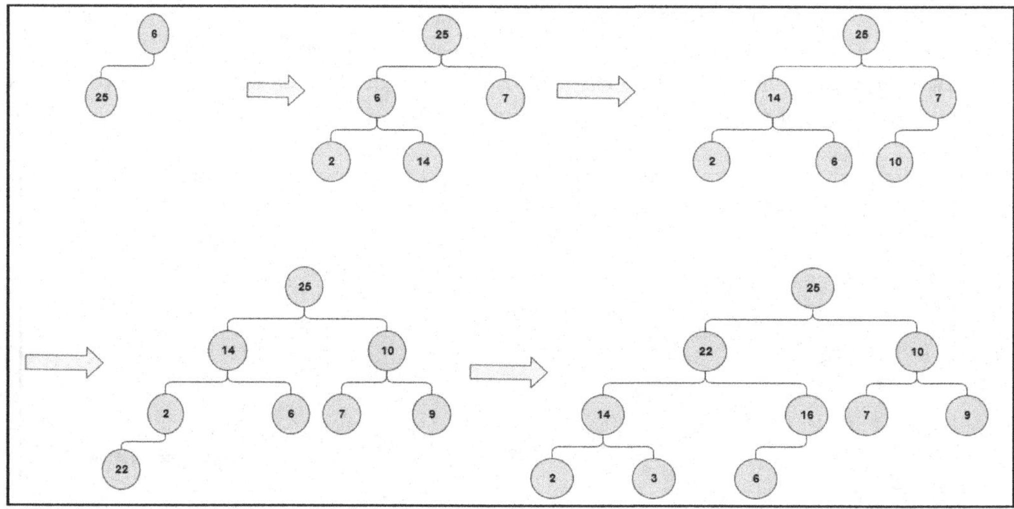

Figure A1.18 Illustration of heap construction.

A1.6.2 Processing the Heap (Heapsort)

The algorithm for processing the heap (also called heapsort because it produces a sorted list) is shown in Figure A1.19. The algorithm shown assumes a max-heap; it progressively removes the root of the heap until it is empty. Heapsort consistently performs as an O(N-log-N) algorithm, even in the worst-case scenario. This compares favorably with quick-sort, which performs as an O(N-Log-N) algorithm on the average and best-case scenario, and an O(N^2) algorithm in the worst-case scenario.

1. Repeatedly remove the maximum value from the heap (that, is the root);
2. Place this at end of a growing list (growing backwards);
3. Re-arrange the tree so that the heap properties are maintained;
4. Continue until the heap is empty.

Figure A1.19 Overview of the heap-sort algorithm.

EXAMPLE A1.6: ILLUSTRATING HEAPSORT ALGORITHM

Figure A1.20 illustrates how the heap of Example A1.5 would be processed to yield a sorted list.

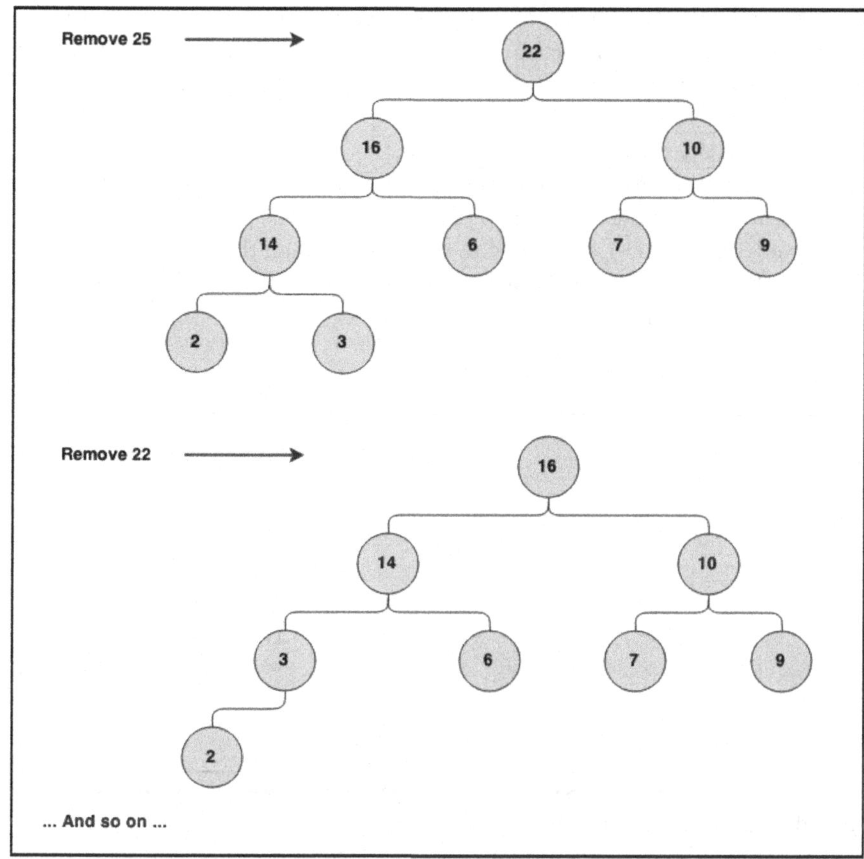

Figure A1.20 Illustrating heap-sort.

A1.7 M-Way Search Trees and B-Trees

An m-way search tree is a tree in which each node has out-degree ≤ m. The out-degree of a node is the number of sub-trees that it has. Figure A1.21 illustrates an m-way search tree in which the root has out-degree 2, the left node has out-degree 3 and the right node has out-degree 3. The leaves (always) have out-degree 0. Figure A1.22 illustrates a 3-way search tree.

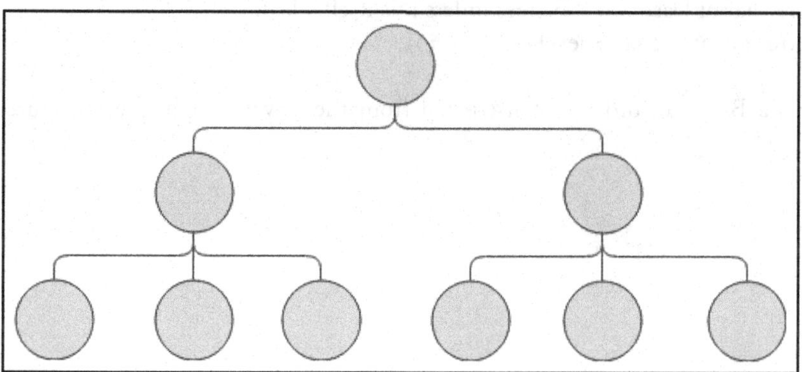

Figure A1.21 Illustrating an M-way search tree.

As you will soon see, understanding m-way search trees is essential to understanding B-trees, which form the core of contemporary database management system (DBMS). The M-way tree has the following properties (which are illustrated in Figure A1.22):

a. Each node is of the form $P_0, K_0, P_1, K_1, \ldots P_{n-1}, K_{n-1}, P_n$
   where $P_o, P_1 \ldots P_n$ are pointers to successor nodes and $K_o, K_1 \ldots K_{n-1}$ are key values.
b. The key values in a node are in the ascending order so that $K_i \leq K_{i+1}$.
c. All key values in nodes of the sub-tree pointed to by P_i are less than the key value K_i.
d. All key values in nodes of the sub-tree pointed to by P_i are greater than the key value k_{i-1}.
e. The sub-tree pointed to by P_i are also m-way search trees.

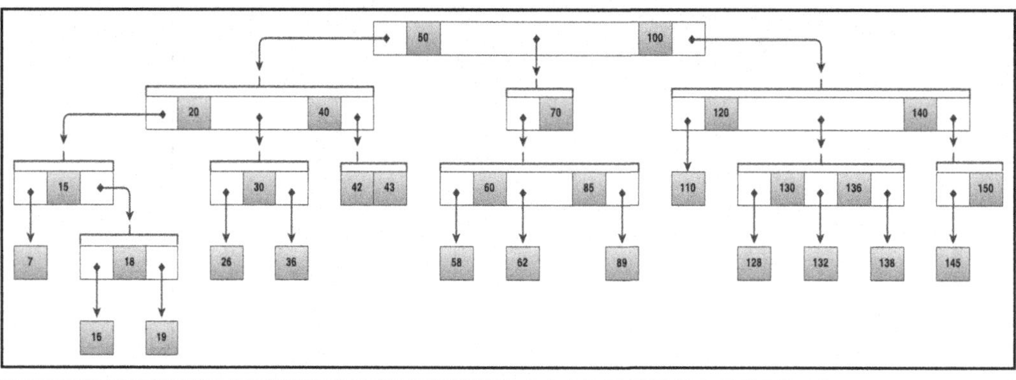

Figure A1.22 Illustrating a three-way search tree.

A1.7.1 Definition of B-tree

We come now to the B-tree; here is the definition: A B-tree of order m is an m-way search tree with the following properties:

a. Except for the root and leaves, each node of the tree has at least [m/2] sub-trees and no more than m sub-trees so that [m/2] ≤ number of sub-trees ≤ m. (Note: [x] means, the smallest integer greater than x; e.g. [1.5] = 2).
b. The root of the tree has at least two sub-trees unless it is itself a leaf.
c. All leaves of the tree are at the same level.

Figure A1.23 illustrates a B-tree of order 3, constructed from the 3-way search tree of Figure A1.22.

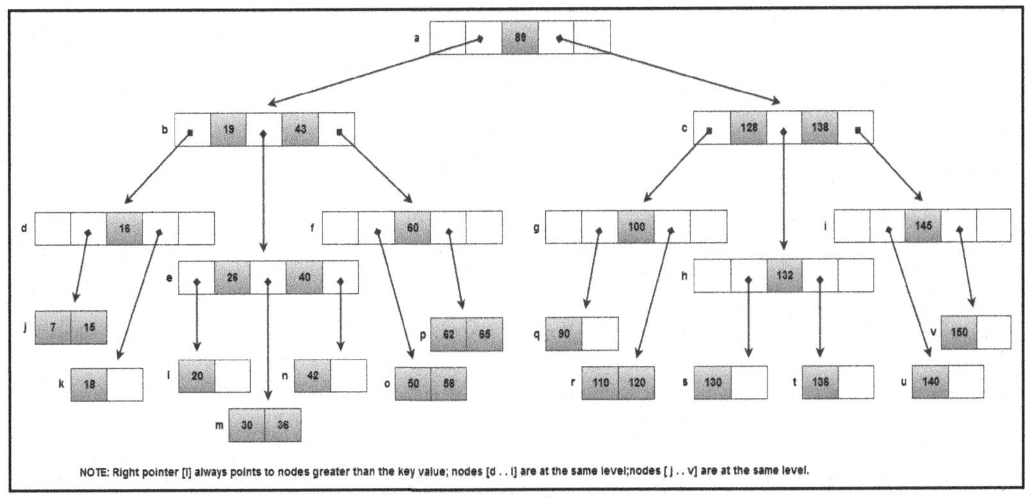

Figure A1.23 B-tree of order 3 (corresponding to 3-way search tree of Figure A1.22).

A1.7.2 Implementation of the B-tree

The main operations to be considered for the B-tree include Creation; Direct Search; Sequential Search; Insertion; and Deletion.

A possible implementation of a B-tree of **LibraryPatron** objects is shown in Figure A1.24 (to refresh your memory on the composition of the **LibraryPatron** class, review Figure A1.5). As you may well imagine, maintaining such a tree is much more challenging than a BST or heap, but by no means insurmountable. We will look more closely at some of these algorithms shortly.

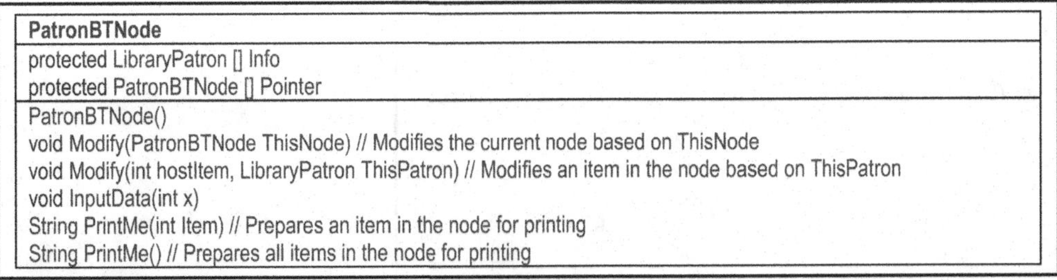

PatronBTNode
protected LibraryPatron [] Info protected PatronBTNode [] Pointer

PatronBTNode()
void Modify(PatronBTNode ThisNode) // Modifies the current node based on ThisNode
void Modify(int hostItem, LibraryPatron ThisPatron) // Modifies an item in the node based on ThisPatron
void InputData(int x)
String PrintMe(int Item) // Prepares an item in the node for printing
String PrintMe() // Prepares all items in the node for printing

Figure A1.24a The UML diagram of the PatronBTNode class.

PatronsBTree
protected PatronBTNode Root protected int Size, TravRef LibraryPatron[] Traversal

public PatronsBTree()
public void addRoot(LibraryPatron ThisPatron) // Inserts an item at the root
public void addItem(LibraryPatron ThisPatron) // Inserts an item
public void addNode(PatronBTNode ThisNode) // Inserts a node
public LibraryPatron removeItem(PatronBTNode ThisNode, int x) // Deletes an item
public PatronBTNode removeNode(PatronBTNode ThisNode) // Deletes a node
public void Modify(PatronBTNode ThisNode, int x, LibraryPatron ThisPatron) // Modifies an item based on **ThisPatron**
public void clearTree()
public LibraryPatron getInfo(PatronBTNode ThisNode, int x)
public PatronBTNode getNode(PatronBTNode ThisNode)
public int getSize()
public boolean isEmpty()
public void inOrderTraversal()

Figure A1.24b The UML diagram of the PatronsBTree class.

Sequential search of the B-tree is achieved by an In-order traversal, which also produces a sorted list. Several values may be sought simultaneously. Note, however, that internal nodes will be visited more than once since they contain several keys. Performance is therefore not optimal.

The direct search algorithm is shown in Figure A1.25. It is used to find a specific node in the tree. It is a very efficient algorithm. For instance, if you consider a simple case involving a B-tree of integers, on the average, searching for an item among 1,000,000 items takes roughly about 20 comparisons! B-trees are also excellent for storing large volumes of data without deteriorating. For these reasons, B-trees are widely used as the file systems for compilers, database management systems, and operating systems.

```
PatronBTNode diretSearch(LibraryPatron SearchValue)
START // Assume that the maximum number of items per node is N
 Current := Root;
 i :=      0;
 While (SearchValue <> Current.Info[i]) do the following:
           If (SearchValue < Current.Info[i]) OR (i = N)
                     Current := Current.Pointer[i];  i := 0;
         End-If;
           Else  If ((i + 1) <= N)  i :=   i + 1; End-If; End-Else;
 End-While;

// Current now points to the correct node
 Return Current;
STOP
```

Figure A1.25 Direct search algorithm.

Figure A1.26 provides the summarized form of the B-tree insertion algorithm. This algorithm is easier described than implemented. Fortunately, you do not need to implement it in a typical data structures course. However, you do need to be able to demonstrate an understanding of the algorithm, so let us take an example:

1. Search for and locate appropriate insertion node;
2. If the node is not full, insert the data item;
3. If the node is full, node splitting occurs as follows:
   - Introduce a new node;
   - Place half of the keys in the new node, and half of the keys in the original node;
   - Move the remaining value up to the parent node;

Note In the worst case, node splitting continues up to the root and the tree height increases by one.

Figure A1.26 Summary of B-tree insertion algorithm.

EXAMPLE A1.7: ILLUSTRATING INSERTION OF ITEMS INTO THE B-TREE

Referring to the B-tree of Figure A1.23, consider inserting the following items: 22, 41, 59, 57, 54. Figure A1.27 illustrates how these items would be added.

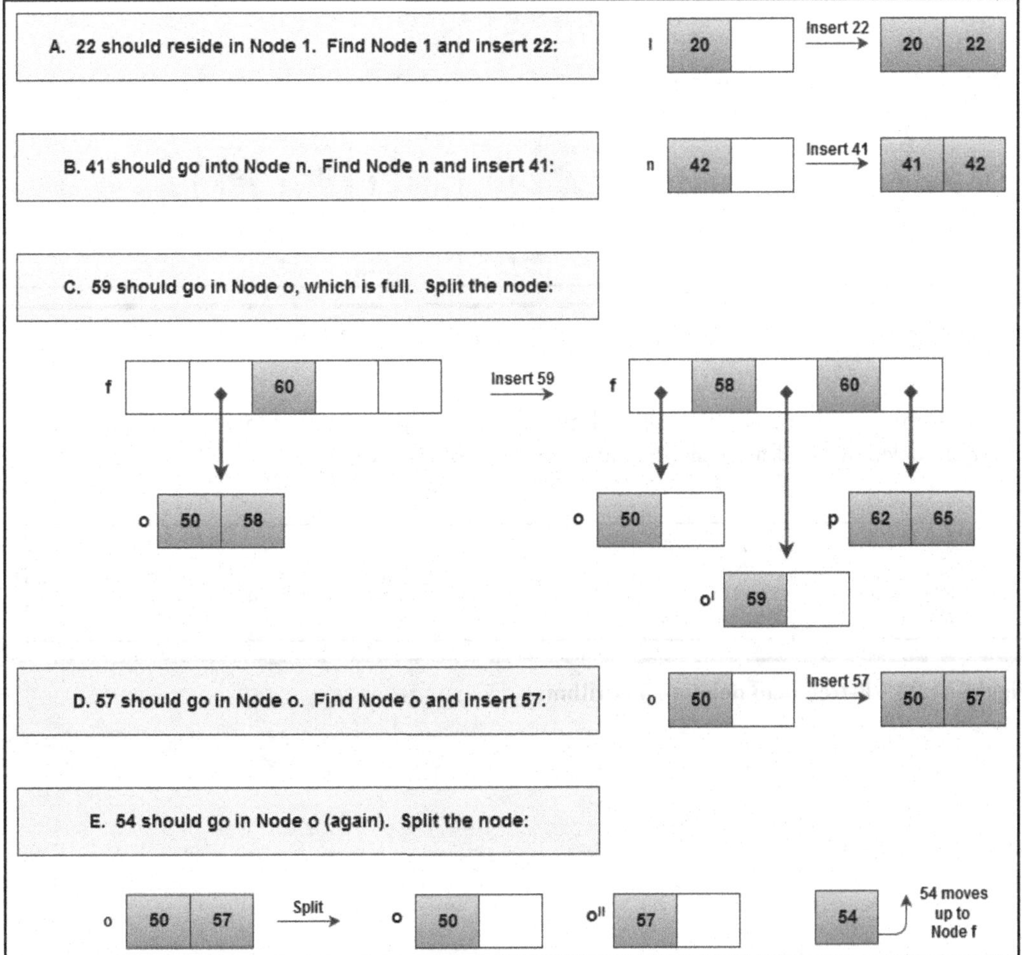

Figure A1.27 Illustrating addition of items to the B-tree. (Continued)

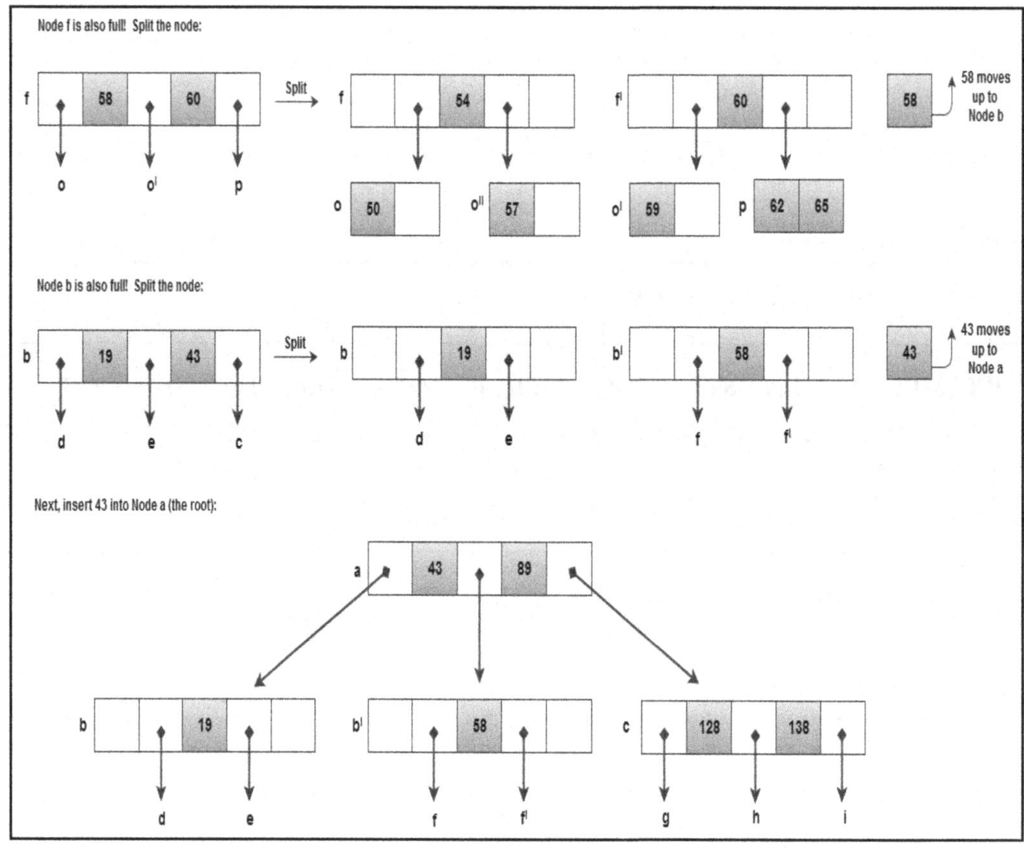

Figure A1.27 (Continued) Illustrating addition of items to the B-tree.

The summarized item deletion algorithm is shown in Figure A1.28. Note that as in the case of insertion, deletion may lead to significant adjustment of the tree.

1. Search for and locate the appropriate node;
2. Remove the item;
3. If resultant node is now empty, adjust the tree, by moving up a key value from one of its sub-tree giving preference to the leftmost sub-tree. The worst-case scenario is that the tree loses a leaf.

Figure A1.28 B-tree item deletion algorithm.

EXAMPLE A1.8: ILLUSTRATING DELETION FROM B-TREE

Referring to the B-tree of Figure A1.23, consider deleting the following items:

65	7	16

a) 65 is in node **p**, remove it: p

62				

b) 7 is in node **j**, remove it: j

15				

c) 16 is in node **d**, remove it. Move up 15 from node **j** and place a dummy (null) value at node **j**.

A1.8 Summary and Concluding Remarks

Here is a summary of what has been discussed in this chapter:

- A tree has one root and zero or more other sub-trees connected to it. Except for the root, each node has a parent node.
- A binary tree (BST) is a tree in which each node is either empty or consists of two disjoint binary trees: the left sub-tree and the right sub-tree. The BST may be traversed in-order, pre-order, or post-order.
- A binary search tree is a binary tree in which all nodes to the left of a node are less than or equal to that node; all nodes to the right of a given node are greater than or equal to that node.
- A height-balanced k-tree (denoted (HB (k)) is a BST in which the maximum allowable difference in height between any two sub-trees sharing a common root is k. For AVL trees, k = 1.
- A heap is an almost complete binary tree, such that: Every leaf of the tree is at level m or m + 1, where m is an integer; if a node has a right descendant at level l, it also has a left descendant at level l; there is some established relationship between the parent-value and its two children-values.
- An m-way search tree is a tree of out-degree less than or equal to m, i.e., a maximum of m pointers point from a node. The tree has the following properties: Each node has a maximum of m pointers and m – 1 key values; in any node, the key values are in ascending order; the pointer to the left of key value k[i] points to a node with values less than or equal to k[i]; the pointer to the right of key value k[i] points to a node with values greater than or equal to k[i].
- A B-tree is a special type of m-way search tree with the following properties: Except for the root and leaves, each node of the tree has at least m/2 sub-trees and no more than m sub-trees; the root of the tree has at least two sub-trees, unless it is itself a leaf; all leaves of the tree are at the same level.

Trees (and particularly BSTs, heaps, and B-trees) are widely used in software engineering to solve various programming problems related to the organization and retrieval of data.

References and/or Recommended Readings

Carrano, Frank. 2012. *Data Abstraction & Abstractions with Java* 3rd ed. Boston, MA: Pearson Education. See chapters 23–27.

Drake, Peter. 2006. *Data Structures and Algorithms in Java.* Upper Saddle River, NJ: Prentice Hall. See chapter 10.

Folk, Michael, Bill Zoellick, & Greg Riccardi. 1998. *File Structures: An Object-Oriented Approach with C++* 3rd ed. Reading, MA: Addison-Wesley. See chapters 8 & 9.

Ford, William H., & William R. Topp. 2005. *Data Structures with Java.* Upper Saddle River, NJ: Prentice Hall. See chapters 16 – 18, 22.

Goodrich, Michael T., & Roberto Tamassia. 2010. *Data Structures & Algorithms* 5th ed. New York: Wiley. See chapters 7, 8, & 10.

Knuth, Donald. 1973. *The Art of Computer Programming* Vol. 3. Reading, MA: Addison-Wesley.

Kruse, Robert L., & Alex Ryba. 1999. *Data Structures and Program Design in C++.* Eaglewood Cliffs, NJ: Prentice Hall. See chapters 10 & 11.

Langsam, Yedidya, Moshe Augenstein, & Aaron M. Tanenbaum. 2003. *Data Structures Using Java.* Upper Saddle River, NJ: Prentice Hall. See chapter 5.

Main, Michael. 2012. *Data Abstraction & Other Objects Using Java* 4th ed. Boston, MA: Pearson Education. See chapters 9 and 10.

Prichard, Janet J., & Frank Carrano. 2011. *Data Abstraction & Problem Solving with Java* 3rd ed. Boston, MA: Pearson Education. See chapter 11.

Venugopal, Sesh. 2007. *Data Structures Outside In with Java.* Upper Saddle River, NJ: Prentice Hall. See chapters 9–11.

Weiss, Mark Allen. 2010. *Data Structures and Problem Solving Using Java* 4th ed. Boston, MA: Addison-Wesley. See chapters 18, 19, 21, & 22.

Weiss, Mark Allen. 2012. *Data Structures and Algorithm Analysis in Java* 3rd ed. Boston, MA: Pearson Education. See chapter 4.

Appendix 2: Review of Hashing

Hashing is a technique for mapping data from a large set to limited space in a much smaller set. This appendix provides an overview of the subject. The appendix covers the fundamentals of hashing under the following captions:

- Introduction
- Hash Functions
- Collision Resolution
- Hashing in Java
- Summary and Concluding Remarks

A2.1 Introduction

In computer science, it is always important to establish mapping functions between (what is called) the conceptual user view of data and actual physical storage reality in the computer system. The former is what the end user sees; the latter is what is actually stored on the storage media. For reasons that will become clear in your courses in software engineering, database systems, and operating systems, the two perspectives are seldom identical.

Another equally important issue is how to find data that is stored. For example, consider a transaction file for a financial institution, with millions of records. As you can well imagine (based on earlier discussions), sequential access of records in this file would be suitable for a non-interactive program running in batch mode. However, it would not be suitable for a program that needs to provide interactive responses to the end user based on different customers. You certainly would not want to process this file relying solely on an array, linked-list, queue, or stack. You could use a BST, a heap, or better yet, a B-tree! But how would you determine where on the storage media to actually store records of your file so they can be easily retrieved when needed? Hashing is one technique for addressing this problem.

Hashing is a technique that addresses the problems of where (on a physical storage medium) to store data and how to retrieve the data stored. In hashing, there is a predictable relationship between a key value used to identify a record, and the record's location on a storage medium.

A hash function takes a key value and applies an algorithm to determine its location. Mathematically, we represent this as follows (Figure A2.1).

The hash function maps key values from a relatively large domain to a relatively smaller range of address locations. Because of this, collision may occur and must be resolved. Collision occurs when two or more keys map to the same address, as illustrated in Figure A2.2.

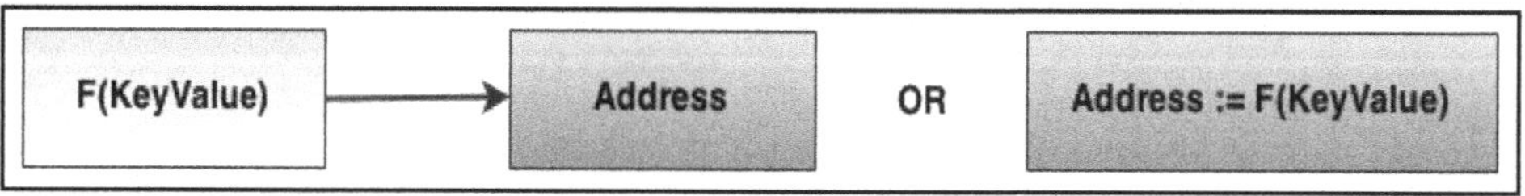

Figure A2.1 Basic concept of hashing.

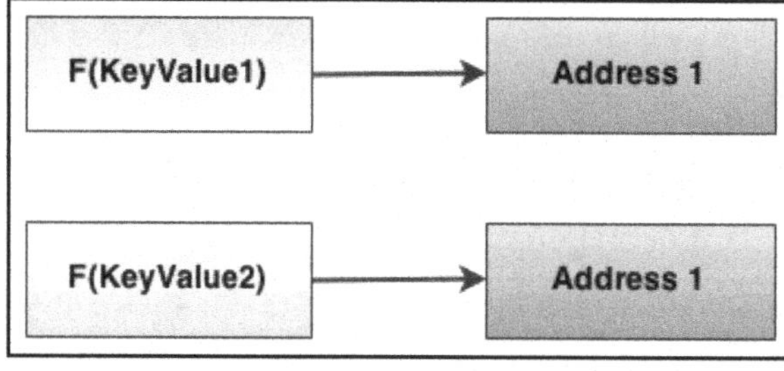

Figure A2.2 Illustrating collision.

Since two records cannot be stored in the same location, the collision must be resolved. Various collision resolution techniques have been proposed. These will be discussed shortly.

A2.2 Hash Functions

A good hash function should exhibit two important features: Firstly, it should be easy and fast to compute. Secondly, it must attempt to scatter the key values (random or non-random) evenly over the address space. Various hash functions have been proposed and tested with varying results. Among the commonly discussed hash functions are the following: absolute addressing, direct table lookup, division-remainder, mid-square, folding, and truncation. It should be noted that due to the randomness of data in a typical database, even the best hash function is likely to produce less than desirable results. We will briefly discuss each technique.

A2.2.1 Absolute Addressing

In absolute addressing, the address is the key value. Thus,

```
Address := KeyValue
```

This is a simplistic approach to hashing; it will only work for trivial cases. In most practical cases, this approach will be inadequate, since storage will be limited.

A2.2.2 Direct Table Lookup

Technically speaking, a direct table lookup is not a hashing function; rather, it's a hashing strategy. In direct table lookup, the following steps are taken:

- The addresses are initially generated via some method. The method used to generate these addresses could be any of the techniques discussed in this section, or some other technique.
- The keys and addresses are stored as a separate index (directory).
- To access a record, the index is first consulted to determine its address.

Figure A2.3 provides an example of an index lookup table for an employee file. The assumptions here are as follows:

- The file is keyed on the employee's identification number.
- Each employee record occupies 128 bytes of space.
- A means of determining the address for each record is in place.
- The notation TnSn is used to denote the track number and sector number on a hard drive (for instance, T1S1 means track 1, sector 1).
- Each sector can store up to 4096 bytes of data.

Key Value (ID_Number)	Address on Storage Medium
2001000	T1S1-0000
2001010	T1S1-0128
2001020	T1S1-0256
2001030	T1S1-0384
2001040	T1S1-0512
2001050	T1S1-0640
2001060	T1S1-0768
2001070	T1S1-0896
2001080	T1S1-1024
...	...

Note: These are fictitious numbers just for illustration purpose

Figure A2.3 Example of Table Lookup for employee records.

The technique is applied as follows:

- When a record is first written, its address is logged into the index (lookup) table.
- When a record is to be retrieved, the index is consulted for its exact location.

The technique has three significant advantages: Table lookup is a very efficient way of storing small to medium sized files. Additionally, the technique is simple and easy to implement. Finally, the technique avoids the problem of collision (by presumably addressing it up front). For these reasons, the technique is widely used in database management system (DBMS) suites and operating systems.

The approach has two problems: Firstly, storing a data file requires management of two separate physical files on the storage medium. Secondly, as the size of the data file grows, so does the size of the index (table lookup) file.

A2.2.3 Division-Remainder

The division-remainder technique prescribes that you divide the key value by an appropriate number, then use the remainder of the division as the address for the record. In computer science, we refer to the operation that yields the remainder of an integer division as the modulus (commonly abbreviated as mod). Thus,

Address := KeyValue Mod Divisor

You will find the mod function in several high-level languages (HLLs). Additionally, observe the following about the division-remainder technique:

- If the divisor is N, then address space must be of minimum size N (that is 0 to N−1).
- To minimize the number of collisions, the divisor is typically a large number that does not contain any prime factors less than 20. For example, 997 or 1011 is preferred to 1024.

The technique is applied as follows: When a record is to be stored or retrieved, its address is first determined via the hash function. The key value used to generate the hash address is the numeric value or representation of the record's primary key. To illustrate with a simple example, if the key value is 7895 and the divisor is 1011, then the address could be determined as follows:

Address := 7895 mod 1011 = 818

The division-remainder technique, implemented in many programming languages via the modulus function/operation, has three significant advantages:

- It is simple and easy to implement.
- It is very efficient for small files.
- It avoids using two physical files for a single data file.

The technique has three significant disadvantages that discourage widespread use as a primary hash function:

- It offers no guarantee against collision (though it can be minimized by carefully choosing the divisor).
- It does not perform very well in attempting to distribute keys evenly across the address space.
- As the size of the data set increases, so does the likelihood of collision.

A2.2.4 Mid-Square

In the mid-square technique, the key value is squared, and then a specific number of digits is extracted from the "middle" of the result to yield the hash address. If an address space of N digits is desired, then digits are truncated at both ends of the squared key value, leaving N digits from the middle.

As an example, suppose that we need a 4-digit address, and we have a key value of 7895. The address could be determined as follows:

Address := Mid-square(7895) = ~~62~~3310~~25~~ = 3310

Like the division-remainder technique, when a record is to be stored or retrieved, its address is first determined via the hash function. The technique is a slight improvement over the division-remainder technique, providing the following advantages:

- It is simple and easy to implement.
- It is very efficient for small files as well as large files.
- It performs reasonably well in attempting to distribute the keys evenly over the address space.
- It avoids using two physical files for a single data file.

The mid-square technique has one significant disadvantage: It offers no guarantee against collision (but the performance is better than the division-remainder technique).

A2.2.5 Folding

The folding technique involves partitioning the key into several parts and combining the parts in a convenient way (via multiplication, addition, or subtraction) to obtain a hash address.

To illustrate this technique, suppose that the folding method is addition and the address space is for four digits. Then, we may derive a hash address from the key value of 625149 via one of the following means:

Address := 625 + 149 = 0774	Or	Address := 6251 + 49 = 6300

The advantages and disadvantages of folding are similar to those of the mid-square technique. Figure A2.4 shows a comparison of how both techniques form on data set of key values. For this particular data set, folding appears to distribute the key values more gracefully over the address space. On the other hand, the mid-square technique appears to distribute the key values more randomly over the address space. Of course, this simple experiment is not enough for us to make conclusion as to how the techniques will perform on other data sets.

Key-Values:	2001010	2001020	2001030	2001040	2001050	2001060	2001070	2001080
Folding Address:	2011	2021	2031	2041	2051	2061	2071	2081
Mid-square Address:	4102	8104	2106	6108	0110	4112	8114	2116

Note: The folding spread for this data set tends to be rhythmically distributed over the address space. The mid-square spread displays a more random distribution over the address space.

Figure A2.4 Comparing folding with mid-square on a data set.

A2.2.6 Truncation

In the truncation technique, part of the key is ignored, and the other part is used as the hash address.

To illustrate this technique, suppose that the address space is for four digits. Then, we may derive a hash address from the key value of 625149 via one of the following means:

Address := 6251~~49~~ = 6251	Or	Address := ~~62~~5149 = 5149

Truncation, like absolute addressing is too simplistic for practical purposes. It could result in repeated collisions, and it does not come close to distributing the keys evenly over the address space.

A2.2.7 Treating Alphanumeric Key Values

In many cases, key values will be alphanumeric instead of numeric. For these situations, you will need to convert the alphanumeric data to numeric form. There are several ways to do this, so there is no need to panic. Figure A2.5 provides a summary of two commonly used techniques that have been proposed. Of course, you can come up with your own.

1. **Simple Character Sequencing**
   - Assign a unique number for each possible character that can be used as part of a key-value (a good place to start is characters on the keyboard). For example, A-Z could be assigned 1-26; a-z could be assigned 27-52; digits could remain 0-9; and so on.
   - For each alphanumeric key-value, convert each character to an integer, using the above-mentioned convention, and then add the individual numeric pieces to obtain a final numeric value for that key-value.
   - **Observe:** This approach would not guarantee that each key-value has a unique numeric value. For example, "care" and "race" would have the same numeric value.

2. **Use the ASCII/Unicode Character Sequencing**
   - Instead of developing your own sequencing, use the numeric value of each ASCII/Unicode character in the key-value. In Java, this can be easily achieved by invoking any of the following two static methods of the **Character** class (method signatures given):

     static int digit(char ch, int radix) // Returns the numeric value of character **ch** in the specified radix.
     static int getNumericValue(char ch) // Returns the numeric value of the specified Unicode character.
   - For each alphanumeric key-value, convert each character to an integer, and then add the individual numeric pieces to obtain a final numeric value for that key-value.
   - **Observe:** Again, this approach would not guarantee that each key-value has a unique numeric value.

Figure A2.5 Converting alphanumeric keys to numeric.

A2.3 Collision Resolution

As mentioned earlier, when hash functions are applied, collisions are likely to occur. Three collision resolution techniques are prevalent: *linear probing*, *synonym chaining*, and *rehashing*.

A2.3.1 Linear Probing

The linear probing strategy prescribes that whenever a collision occurs, the empty location that is closest to the hash address is found and used. If the end of the table is reached, wrap-around is allowed. Figure A2.6 illustrates this. In the figure, collision has been resolved for nodes D, E, and G.

Key Value	Hash Address	Storage Address
A	1	1
B	2	2
C	3	3
D	2	4
E	4	5
F	6	6
G	4	7

Note: Collision has been resolved for nodes D, E, and G.

Figure A2.6 Illustrating linear probing.

The *load factor* **f** is defined as the number of records (to be) stored in the table to the actual size of the table. Thus,

f = n/s where **n** is the number of records (to be) stored and **s** is the size of the table.

As items are added to the table, blocks of occupied cells start forming. This is referred to as *primary clustering*. The result is an increased likelihood of collision and lengthened probes for subsequent items to be inserted. As the table fills up, these searches become longer.

For insertions and unsuccessful searches, the expected number of probes is $\frac{1}{2}[1 + 1/(1 - f)^2]$. For successful searches, the expected number of probes is $\frac{1}{2}[1 + 1/(1 - f)]$. The derivation of these formulae is beyond the scope of this course.

From a programming perspective, linear probing is easy to implement. The technique has the following advantages:

- The technique is conceptually simple and easy to implement.
- The amount of calculation done at each stage is very small.

Disadvantages of linear probing include the following:

- Primary clustering — the hash-table fills up in clusters, rather than in even distribution over the address space.
- Primary clustering leads to longer search and increased likelihood of collision, particularly as the file becomes full.

Linear probing sometimes described as an *open addressing* strategy. Another commonly discussed open addressing strategy is *quadratic probing*. Since its performance is not much different from linear probing, it will not be discussed any further. *Rehashing* is another open addressing strategy; it will be discussed shortly.

A2.3.2 Synonym Chaining

The synonym chaining technique uses the hash address, not as the actual storage location for the record, but as the index into a list (implemented as a linked-list, vector, or array-list). This way we can facilitate multiple records with the same hash address! Figure A2.7 provides a graphic representation of this approach.

Programmatically, the hash-table may be implemented via any of the following strategies (you are encouraged to try implementing this on your own):

- An array of linked lists or queues
- An array-list of linked lists or queues
- An array-list/vector of array-lists/vectors
- A vector of linked lists or queues

Turning to the load factor, please observe the following:

- The table size **s** is the number of chains (buckets), not the number of nodes.
- The average length of each chain is n/s where **n** is the number of records (nodes). Moreover, from the previous subsection, n/s is also the load factor, f.
- The load factor **s** is therefore the average length of each chain.
- An unsuccessful search will involve the examination of **f** items; a successful search will require an average of 1 + f/2 examinations. So as expected, the performance is significantly better than linear probing.

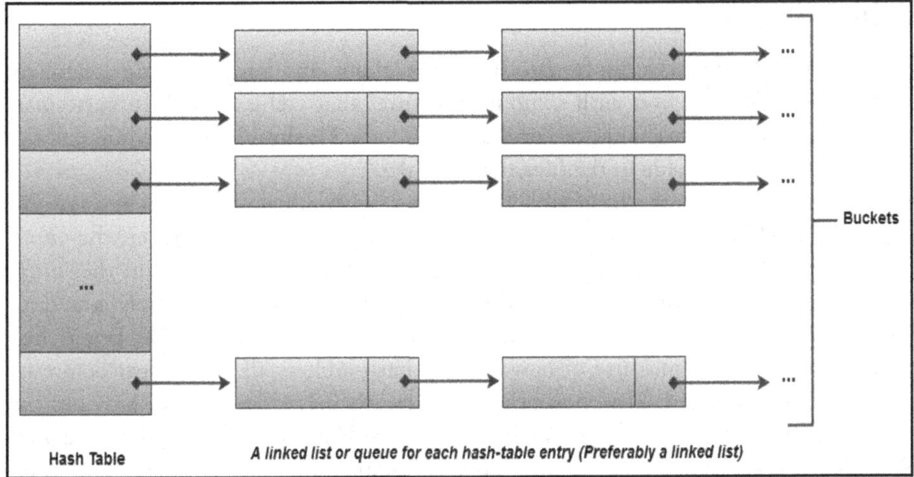

Figure A2.7 Graphic representation of synonym chaining.

Synonym chaining has the following advantages:

- The technique reduces access time for records with collision addresses.
- It is relatively easy to insert nodes in a given chain.
- There is even distribution of keys (nodes) over address space.

The technique has three main disadvantages:

- Additional space is required for the linked lists. However, since linked lists are dynamic structures, this is not significant.
- Linked lists cannot be accessed randomly; they are sequential access structures. That would be a slight setback if the buckets are implemented as linked lists. However, the buckets could be implemented as vectors or array-lists.
- Determining the appropriate number of buckets can be challenging. Programmatically, this can be resolved by allowing the user to specify an initial size and desired load factor when the hash-table is being created. Additionally, the hash table can be (and is typically) implemented to self-adjust its size over time in order to maintain its load factor.

A2.3.3 Rehashing

Rehashing is another open addressing strategy that significantly reduces clustering. The technique involves the application of a second hash function to resolve any collision that might have occurred from the first hash function. In principle, rehashing can continue to other levels until collision is resolved. However, in practice, if you have to hash more than twice to resolve collision, you probably need to change your hash function(s).

The main advantage of rehashing is that it tends to produce an even spread of key values over the address space. One significant disadvantage of the technique is that records may be moved some distance from their home addresses.

A2.4 Hashing in Java

You can develop and implement hash functions in just about any programming language. However, as usual, Java provides you with some nice features that facilitate this. In particular, there are three Java classes that you should be familiar with — the **Hashtable**, **HashMap**, and the **TreeMap** classes. These classes reside in the **java.util** package.

The **Hashtable** class implements a hash-table that maps keys to values. A key can be any valid object. Each instance of **Hashtable** class has two properties that affect its performance: the *initial capacity* and *load factor*. The *capacity* is the number of *buckets* in the hash-table, and the *initial capacity* is simply the capacity at the time the hash-table is created. A bucket is simply a holding area for key values that have the same address (synonymous to a chain); the bucket is searched sequentially. The *load factor* is a measure of how full the hash-table is allowed to get before its capacity is automatically increased. The conventional wisdom is 0.75. Figure A2.8 provides the Unified Modeling Language (UML) diagrams for the class. The **HashMap** class is roughly equivalent to **Hashtable**, except that it is unsynchronized and permits nulls. Additionally, there is no guarantee that the order of the hash-table will remain constant over time.

The **TreeMap** class implements a non-synchronized map (similar to a hash-table), in which keys are sorted. Figure A2.9 provides the UML diagrams for the class.

Hashtable <K, V> /* Implements a hash-table which maps keys to values as specified in the **Map** interface. Note: If K and V are not specified at instantiation, they revert to Object. */
int capacity, loadFactor

// Constructors
public Hashtable() // Constructs an empty **Hashtable** instance with a default initial capacity of 11 and load factor of 0.75.
public Hashtable(int initCap) // Constructs an empty **Hashtable** instance with initial capacity of **initCap** and load factor of 0.75.
public Hashtable(int initCap, float loadFactor) /* Constructs an empty **Hashtable** instance with initial capacity of **initCap** and load factor of **loadFactor**. */
 	
// Other methods of interest
public void clear() // Clears the hash-table
public boolean contains(Object value) // Checks whether the specified value is contained in the hash-table.
public boolean containsValue(Object value) // As for **contains**().
public boolean containsKey(Object key) // Checks whether the specified key is contained in the hash-table.
public boolean equals(Object obj) // Checks whether the current hash-table (map) is equal to the specified object.
public V get(Object key) // Returns the value that the specified key maps to, or null.
public int hashCode() // Returns the hash-code value of this hash-table (map).
public boolean isEmpty() // Checks of the hash-table is empty.
public V put(K key, V value) // Inserts a new key-value mapping into the hash-table
protected void rehash() // Increases the capacity and internally reorganizes the hash-table.
public V remove(Object key) // Removes the specified key and corresponding value from the hash-table.
protected int size() // returns the size (number of keys) of the hash-table.
protected String toString() // Returns a string representation of the hash-table.

Figure A2.8 UML diagram showing main properties of the Hashtable class.

TreeMap <K, V> /* Implements a map which maps keys to values as specified in the **Map** interface. Note: If K and V are not specified at instantiation, they revert to Object. */

// Constructor
public TreeMap() // Constructs an empty **TreeMap** instance, using the natural ordering of its keys.
 	
// Other methods of interest
public K ceilingKey(K thisKey) // Returns the least key that is >= **thisKey**, or null.
public void clear() // Clears the map.
public boolean containsKey(Object value) // Checks whether the specified value is contained in the map.
public boolean containsKey(Object key) // Checks whether the specified key is contained in the map.
public boolean equals(Object obj) // Checks whether the current map is equal to the specified object.
public K firstKey() // Returns the first (i.e. lowest) key-value in the map.
public K floorKey(K thisKey) // Returns the greatest key that is <= **thisKey**, or null.
public V get(Object key) // Returns the value that the specified key maps to, or null.
public int hashCode() // Returns the hash-code value of this hash-table (map).
public K higherKey(K thisKey) // Returns the least key that is > **thisKey**, or null.
public V put(K key, V value) // Inserts a new key-value mapping into the map.
protected void rehash() // Increases the capacity and internally reorganizes the map.
public boolean isEmpty() // Checks of the hash-table is empty.
public K lastKey() // Returns the last (i.e. greatest) key-value in the map.
public K lowerKey(K thisKey) // Returns the greatest key that is < **thisKey**, or null.
public V remove(Object key) // Removes the specified key and corresponding value from the map.
protected int size() // returns the size (number of keys) of the hash-table.
protected String toString() // Returns a string representation of the map.

Figure A2.9 UML diagram showing main properties of the TreeMap class.

Let us now look at a programming example. Assume the presence of the **LibraryPatron** class of earlier discussions (see Section A1.2.5), suppose that it is desirous to build a hash-table of **LibraryPatron** objects (representing users of a library). To do this effectively, while providing a reasonable amount of flexibility, it will be necessary to have a hash-table, as well as an array-list to hold key values. The UML diagram for the **LibraryPatron** class is repeated in Figure A2.10a for ease of reference. The driver class will be called **PatronsHashMonitor**; its UML is shown in Figure A2.10b. The program listing for this class is provided in Figure A2.10c. Here is a summary of each method:

- The **main(...)** method provides a menu of all the services provided by the program and invokes the other methods based on the user's choice.
- The **InputPatrons()** method facilitates entry of data into the hash-table called **PatronsList**. It also allows corresponding entry into the array-list called **KeyValues**.
- The **QueryPatrons( )** method facilitates searching for specific items in the hash-table via their keys.
- The **ListPatrons( )** method provides a list of all items in the hash-table.
- The **RemovePatrons( )** method facilitates searching for specific items in the hash-table via their keys, and removing them.
- The **CheckSize( )** method displays the number of items in the hash-table.
- The **Empty( )** method removes all items from the hash-table and the corresponding array-list.
- The **InitializeList( )** method instantiates both the hash-table and the corresponding array-list.

LibraryPatron
protected int PatronNumber protected String Name protected String Major private String Status
LibraryPatron() LibraryPatron (int ThisNumber) void Modify(LibraryPatron ThisPatron) void InputData(int x) String PrintMe() int GetPatronNumber()

Figure A2.10a UML diagram of the LibraryPatron class.

PatronsCrudeHashMonitor
public static Hashtable <Integer, LibraryPatron> PatronsList public static ArrayList KeyValues public static final String HEADING = "Library Patrons Hash Table" public static final int DEFAULT_NUMBER = 0
public static void main(String[] args) public static void InputPatrons() public static void QueryPatron() public static void ListPatrons( ) public static void RemovePatrons() public static void CheckSize( ) public static void InitializeList() public static void Empty()

Figure A2.10b UML diagram of the PatronsHashMonitor class.

```
// ********************************************************************************************************** **********
// Class Name: PatronsHashMonitor
// Purpose: Implements hash table of LibraryPatron objects
// Author: Elvis Foster
// ********************************************************************************************************** **********

package javaapplication11;
import javax.swing.JOptionPane; // This package facilitates dialog boxes, etc.
import java.util.Hashtable;
import java.util.ArrayList;

public class PatronsHashMonitor
{
  // Global Data Items
  public static Hashtable <Integer, LibraryPatron> PatronsList;  // The hash table
  public static ArrayList KeyValues; // Used to list the contents of the hash table
  public static final String HEADING = "Library Patrons Hash Table";
  public static final int DEFAULT_NUMBER = 0;
  // public static int PatronsListSize;

// main method
public static void main(String[] args)
{
  // Declare main variables
  String PromptString = "1. Add Library Patrons \n" + "2. Query a Library Patron \n" + "3. List Patrons \n" +
  "4. Remove Library Patrons \n" + "5. Check Size of Hash Table \n" + "6. Empty the Hash Table \n" + "0. Quit Processing";
  boolean ExitTime = false;
  int NextUserAction, Option;
  // Object[] PatronsArray = null;

InitializeList(); // Initialize the queue
while (!ExitTime) // While user wishes to continue
{
  // Present menu and process user's request
  // PatronsListSize = PatronsList.size();
  Option = Integer.parseInt(JOptionPane.showInputDialog(null, PromptString, HEADING,  +
    JOptionPane.QUESTION_MESSAGE));

  switch (Option)
  {
    case 0: {ExitTime = true; break;}
    case 1: {InputPatrons(); break;}
    case 2: {if (!PatronsList.isEmpty()) QueryPatron(); break;}
    case 3: {if (!PatronsList.isEmpty()) ListPatrons(); break;}
    case 4: {if (!PatronsList.isEmpty()) RemovePatrons(); break;}
    case 5: {CheckSize(); break;}
    case 6: {if (!PatronsList.isEmpty()) Empty(); break;}
  }

  } // End While
} // End of main method

// Continued on the following page
```

Figure A2.10c Essential Java Code for the PatronsHashMonitor class. (Continued)

```
// The InputPatrons Method
public static void InputPatrons()
{
 int NumberOfPatrons, x, PatronsListSize = PatronsList.size();
 LibraryPatron CurrentPatron;
 NumberOfPatrons = Integer.parseInt(JOptionPane.showInputDialog(null, "Number of Patrons: ", HEADING, +
   JOptionPane.QUESTION_MESSAGE));

 KeyValues.ensureCapacity(PatronsListSize + NumberOfPatrons); // Ensure correct size of list for key values
 for (x =1; x <= NumberOfPatrons; x++)
 {
  CurrentPatron = new LibraryPatron();
  CurrentPatron.InputData(x); // Prompt For and Accept LibraryPatron Data
  PatronsList.put(CurrentPatron.GetPatronNumber(), CurrentPatron);
  KeyValues.add(x-1, CurrentPatron.GetPatronNumber()); // Store the key values for subsequent listing of the hash table
 }; // End For
} // End of InputPatrons Method

// The QueryPatron Method
public static void QueryPatron()
{
 // Declarations
 String OutString;
 int NextUserAction, SearchKey;
 LibraryPatron SoughtPatron;
 String Heading = "Library Patron Hash Table Query";
 boolean ExitNow = false;

 // Prompt for the then check it it's in the list
 while (!ExitNow) // While more processing required
 {
  SearchKey = Integer.parseInt(JOptionPane.showInputDialog(null, "Patron Key Value: ", Heading, +
        JOptionPane.QUESTION_MESSAGE));
  SoughtPatron = new LibraryPatron();

  // Output the result of the search
  if (PatronsList.containsKey(SearchKey))
  {
   SoughtPatron.Modify(PatronsList.get(SearchKey));
   OutString = SoughtPatron.PrintMe();
  }
  else OutString = "Library patron specified is not in the list.";
  JOptionPane.showMessageDialog(null, OutString, Heading, JOptionPane.INFORMATION_MESSAGE);

  // Check whether user wishes to continue
  NextUserAction = JOptionPane.showConfirmDialog(null, "Click Yes to query another. Click No or Cancel to exit.");
  if ((NextUserAction == JOptionPane.CANCEL_OPTION) || (NextUserAction == JOptionPane.NO_OPTION)) ExitNow = true;
 } // End-While more processing required

} // End QueryPatron

// Continued on next page
```

Figure A2.10c (Continued) Essential Java Code for the PatronsHashMonitor class. (Continued)

```
// The ListPatrons Method
public static void ListPatrons()
{
  int x, PatronsListSize = PatronsList.size();
  String OutString = "The hash table contains the following: \n";

  // Use the KeyValues list to retrieve and print items in the hash table
  for (x = 1; x <= PatronsListSize; x++)
  { OutString += PatronsList.get(KeyValues.get(x-1)).PrintMe() + "\n"; }

  JOptionPane.showMessageDialog(null, OutString, HEADING, JOptionPane.INFORMATION_MESSAGE);
} // End of ListPatrons Method

// The RemovePatrons Method
public static void RemovePatrons()
{
  String RemovalPrompt, Heading = "Removal of Items from the Hash Table";
  int x, RemovalKey, NextUserAction;
  boolean ExitNow = false;

  while (!ExitNow) // While the user wishes to continue
  {
    // Prompt for number of items to be popped
    RemovalKey = Integer.parseInt(JOptionPane.showInputDialog(null,"Specify the key value of the patron you wish to remove: ", +
            Heading, JOptionPane.QUESTION_MESSAGE));
    while (!PatronsList.containsKey(RemovalKey)) // While invalid key value
    {
      JOptionPane.showMessageDialog(null, "The hash table does not contain the specified key value", Heading, +
            JOptionPane.ERROR_MESSAGE);
      RemovalKey = Integer.parseInt(JOptionPane.showInputDialog(null, "Specify the key value of the patron that you +
                 wish to remove: ", Heading, JOptionPane.QUESTION_MESSAGE));
    } // End of While invalid key value

    // Allow user to confirm removal of the specified key value
    // Note: The item is removed from both the hash table and the key-values list
    RemovalPrompt = "Patron " + RemovalKey + " is about to be removed from the hash table.\n" + "Click Yes to remove the items. +
            Click No or Cancel to exit.";
    NextUserAction = JOptionPane.showConfirmDialog(null, RemovalPrompt);
    if (NextUserAction == JOptionPane.YES_OPTION)
    {PatronsList.remove(RemovalKey); KeyValues.remove(KeyValues.indexOf(RemovalKey));}

    // Check whether user wishes to continue
    NextUserAction = JOptionPane.showConfirmDialog(null, "Click Yes to query another. Click No or Cancel to exit.");
    if ((NextUserAction == JOptionPane.CANCEL_OPTION) || (NextUserAction == JOptionPane.NO_OPTION)) ExitNow = true;

  } // End while the user wishes to continue
} // End of RemovePatrons Method

// The CheckSize Method
public static void CheckSize()
{
  JOptionPane.showMessageDialog(null, "There are " + PatronsList.size() + " patrons in the hash table", HEADING, +
      JOptionPane.INFORMATION_MESSAGE);
} // End of CheckSize Method
// Continued on next page
```

Figure A2.10c (Continued) Essential Java Code for the PatronsHashMonitor class. (Continued)

```
// Initialize Method
public static void InitializeList() // Creates an empty hash table and array-list
{
  PatronsList = new Hashtable <Integer, LibraryPatron>();
  KeyValues = new ArrayList();
}

// The Empty Method
public static void Empty()
{
  int x, NextUserAction;
  String RemovalPrompt = "You are about to empty the hash table. " + "Click Yes to Empty. Click No or Cancel to exit.";
  NextUserAction = JOptionPane.showConfirmDialog(null, RemovalPrompt);
  if (NextUserAction == JOptionPane.YES_OPTION) {PatronsList.clear(); KeyValues.clear();}
} // End of Empty Method

} // End of PatronsHashMonitor class
```

Figure A2.10c (Continued) Essential Java Code for the PatronsHashMonitor class.

A2.5 Summary and Concluding Remarks

Let us summarize what we have covered in this chapter:

- A hash function takes a key value and applies an algorithm to determine its location.
- The hash function maps key values from a relatively large domain to a relatively smaller range of address locations. Because of this, collision may occur and must be resolved.
- Among the commonly used hashing techniques are absolute addressing, direct table lookup, division-remainder, mid-square, folding, and truncation.
- Three commonly used collision resolution strategies are linear probing, synonym chaining, and rehashing.
- In linear probing, consecutive address locations are checked until a free location is found for insertion, or the sought item is found.
- In synonym chaining, each hash address is a reference to a linked list of items that have the hash address.
- In rehashing, successive hash functions are applied until the collision is resolved.
- Java facilitates management of hash tables via the **Hashtable**, **HashMap**, and **TreeMap** classes.

Hashing is widely used in the implementation of filing systems for compilers, operating systems, and database management systems.

References and Recommended Readings

Carrano, Frank. 2012. *Data Abstraction & Abstractions with Java* 3rd ed. Boston, MA: Pearson Education. See chapters 21& 22.

Drake, Peter. 2006. *Data Structures and Algorithms in Java.* Upper Saddle River, NJ: Prentice Hall. See chapter 11.

Ford, William H., & William R. Topp. 2005. *Data Structures with Java.* Upper Saddle River, NJ: Prentice Hall. See chapters 19–21.

Goodrich, Michael T., & Roberto Tamassia. 2010. *Data Structures & Algorithms* 5th ed. New York: Wiley. See chapter

Kruse, Robert L., & Alex Ryba. 1999. *Data Structures and Program Design in C++*. Eaglewood Cliffs, NJ: Prentice Hall. See chapter 9.

Langsam, Yedidya, Moshe Augenstein, & Aaron M. Tanenbaum. 2003. *Data Structures Using Java*. Upper Saddle River, NJ: Prentice Hall. See chapter 4.

Prichard, Janet J., & Frank Carrano. 2011. *Data Abstraction & Problem Solving with Java* 3rd ed. Boston, MA: Pearson Education. See chapters 12 & 13.

Venugopal, Sesh. 2007. *Data Structures Outside In with Java*. Upper Saddle River, NJ: Prentice Hall. See chapter 12.

Weiss, Mark Allen. 2012. *Data Structures and Algorithm Analysis in Java* 3rd ed. Boston, MA: Pearson Education. See chapter 5.

Appendix 3: Review of Information-Gathering Techniques

In order to accurately and comprehensively specify the system, the software engineer gathers and analyzes information via various methodologies. This chapter discusses these methodologies as outlined below:

- Rationale for Information Gathering
- Interviews
- Questionnaires and Surveys
- Sampling and Experimenting
- Observation and Document Review
- Prototyping
- Brainstorming and Mathematical Proof
- Object Identification
- End-User Involvement
- Summary and Concluding Remarks

A3.1 Rationale for Information Gathering

What kind of information is the software engineer looking for? The answer is simple, but profound: You are looking for information that will help to accurately and comprehensively define the requirements of the software to be constructed. The process is referred to as *requirements analysis* and involves a range of activities that eventually lead to the deliverable that we call the *requirements specification* (RS). In particular, the software engineer must determine the following:

- **Synergistic interrelationships** of the system components: This relates to the components and how they (should) fit together.
- **System information entities** (*object types*) and their interrelatedness: An entity refers to an object or concept about data is to be stored and managed.
- **System operations** and their interrelatedness: Operations are programmed instructions that enable the requirements of the system to be met. Some operations are system-based and may be oblivious to the end user; others facilitate user interaction with the software; others are internal and often operate in a manner that is transparent to the end user.
- **System business rules:** Business rules are guidelines that specify how the system should operate. These relate to data access, data flow, relationships among entities, and the behavior of system operations.
- **System security mechanism(s)** that must be in place: It will be necessary to allow authorized users to access the system while denying access to unauthorized users. Additionally, the privileges of authorized users may be further constrained to ensure that they have access only to resources that they need. These measures protect the integrity and reliability of the system.

As the software engineer embarks on the path toward preparation of the requirements specification, these objectives must be constantly borne in mind. In the early stages of the research, the following questions should yield useful pointers:

- WHAT are the (major) categories of information handled? Further probing will be necessary, but you should continue your pursuit until this question is satisfactorily answered.
- WHERE does this information come from? Does it come from an internal department or from an external organization?
- WHERE does this information go after leaving this office? Does it go to an internal department or to an external organization?
- HOW and in WHAT way is this information used? Obtaining answers to these questions will help you to identify business rules and operations.
- WHAT are the main activities of this unit? A unit may be a division or department or section of the organization. Obtaining the answer to this question will also help you to gain further insights into the operations.
- WHAT information is needed to carry out this activity? Again here, you are trying to refine the requirements of each operation by identifying its input(s).
- WHAT does this activity involve? Obtaining the answer to this question will help you to further refine the operation in question.
- WHEN is it normally done? Obtaining the answer to this question will help you to further refine the operation in question by determining whether there is a time constraint on an operation.
- WHY is this important? WHY is this done? WHY...? Obtaining answers to these probes will help you to gain a better understanding of the requirements of the software system.

Of course, your approach to obtaining answers to these probing questions will be influenced by whether the software system being researched is to be used for in-house purposes or marketed to the public. The next few sections will examine the commonly used information-gathering strategies.

A3.2 Interviewing

Interviewing is the most frequent method of information gathering. It can be very effective if carefully planned and well conducted. It is useful when the information needed must be elaborate or clarification on various issues is required. The interview also provides an opportunity for the software engineer to win the confidence and trust of clients. It should therefore not be squandered.

A3.2.1 Steps in Planning the Interview

In panning to conduct an interview, please observe the following steps:

1. Read background information
2. Establish objectives
3. Decide whom to interview
4. Prepare the interviewee(s)
5. Decide on structure and questions

A3.2.2 Basic Guidelines for Interviews

Figure A3.1 provides some guidelines for successfully planning and conducting an interview. These guidelines are listed in the form of a do-list, and a don't-list.

Do-List for Interviews:

1. Make an appointment.
2. Plan. Consider topics to be covered.
3. Make sure information requested is impersonal and objective.
4. Prepare for the interview (theme and questions).
5. Ask questions at the right level.
6. State the purpose clearly, up front.
7. Communicate in the interviewee's language.
8. If compliments become necessary, be sincere.
9. Be relaxed and help the respondent to be relaxed.
10. Listen attentively.
11. Identify facts as opposed to opinions. Both are important.
12. Accept ideas and hints.
13. Check the facts.
14. Collect source documents and forms.
15. Make effective use of open-ended and closed questions.
16. Part pleasantly.

Don't-List for Interviews:

1. Don't be late.
2. " be too formal of too casual.
3. " interrupt the speaker.
4. " use technical jargons.
5. " jump to conclusions.
6. " argue or criticize (constructive or destructive).
7. " make suggestions (not as yet; you will get your opportunity to do so later).

Figure A3.1 Basic guidelines for interview.

A3.3 Questionnaires and Surveys

A questionnaire is applicable when any of the following situations hold:

- A small amount of information is required of a large population.
- The time frame is short but a vast area (and/or dispersed population) must be covered.
- Simple answers are required to a number of standard questions.

A3.3.1 Guidelines for Questionnaires

Figure A3.2 provides a set of basic guidelines for preparing a questionnaire. The questionnaire must begin with an appropriate heading and a clearly stated purpose.

1. State purpose clearly } Usually in the form of a cover note or letter
2. Thank the participants
3. Must have a topic or heading which reflects an apt summary of the information sought.
4. Should adhere to the principles of forms design.
5. Avoid ambiguity.
6. Decide when to use open-ended questions, closed questions or scalar questions.
7. Order questions appropriately.
8. State questions in a language the respondent will readily understand.
9. Be consistent in style.
10. Ask questions of importance to the respondents first.
11. Bring up less controversial questions first.
12. Cluster related questions.

Figure A3.2 Guidelines for questionnaires.

A3.3.2 Using Scales in Questionnaires

Scales may be used to measure the attitudes and characteristics of respondents, or have respondents judge the subject matter in question. There are five types of scales as outlined below:

1. **Nominal Scale:** Used to classify things. A number represents a choice. One can obtain a total for each classification.
2. **Ordinal Scale:** Similar to nominal, but here the number implies ordering or ranking.
3. **Interval Scale:** Ordinal with equal intervals.
4. **Ratio Scale:** Interval scale with absolute zero.
5. **Likert Scale:** The Likert scale has emerged as the default standard for measuring, quantifying, and analyzing user responses in surveys and research initiatives that involve the use of questionnaires. Likert scales are typically designed to be symmetrical — involving an equal number of possible positive or negative responses and a neutral point. Likert scale implementations involving five points or seven points appear to be very common. The five-point implementation often appears with responses such as:
   1. Strongly disagree
   2. Disagree
   3. Neutral
   4. Agree
   5. Strongly agree

EXAMPLE A3.1: ILLUSTRATING AN ORDINAL SCALE

The financial status of an individual may be represented as follows:

1 = Extremely Rich; 2 = Very Rich; 3 = Rich; 4 = Not Rich; 5 = poor; 6 = very poor; 7 = Pauper

EXAMPLE A3.2: ILLUSTRATING INTERVAL SCALE

Usage of a particular software product by number of modules used (10 means high):

1 2 3 10

EXAMPLE A3.3: ILLUSTRATING RATIO SCALE

Distance traveled to obtain a system report:

0 1 2 3 .... 10 [Meters]

EXAMPLE A3.4: ILLUSTRATING RATIO SCALE

Average response time of the system:

0 1 2 3 .... 10 [Minutes]

A3.3.3 Administering the Questionnaire

Options for administering the questionnaire include the following:

- Convening respondents together at one time
- Handing out blank questionnaires and collecting completed ones
- Allowing respondents to self-administer the questionnaire at work and leave it at a centrally located place
- Mailing questionnaires with instructions, deadlines, and return postage
- Using the facilities of the World Wide Web (WWW), for example, email, user forums, and/or applications such as Survey Monkey

A3.4 Sampling and Experimenting

Sampling is useful when the information required is of a quantitative nature or can be quantified, no precise detail is available, and it is not likely that such details will be obtained via other methods. The findings are then used to influence decisions about software systems of interest. Figure A3.3 provides an example of a situation in which sampling is applicable.

Sampling is not confined to software engineering alone; rather, the technique is practiced in most (if not all) professional disciplines. Whenever one desires to draw credible inferences about a large population, analysis is conducted on a representative sample of that population. Sampling theory describes two broad categories of samples:

- *Probability sampling* involving random selection of elements
- *Non-probability sampling* where judgment is applied in selection of elements

Shipping	Number of Orders	% of Total
As Promised	186	37.2
1 day late	71	14.2
2 days late	49	9.8
3 days late	35	7.0
4 days late	38	7.6
5 days late	28	5.6
6 days late	93	18.6
	500	**100.0**

Figure A3.3 Examining the delivery of orders after customer complaints.

A3.4.1 Probability Sampling Techniques

There are four types of probability sampling techniques:

- **Simple Random Sampling** uses a random method of selection of elements.
- **Systematic Random Sampling** involves selection of elements at constant intervals. Interval = **N/n** where **N** is the population size and **n** is the sample size.
- **Stratified Sampling** involves grouping of the data in strata. Random sampling is employed within each stratum.
- **Cluster Sampling:** The population is divided into (geographic) clusters. A random sample is taken from each cluster.

The latter three techniques constitute *quasi-random sampling.* The reason for this is that they are not regarded as perfectly random sampling.

A3.4.2 Non-probability Sampling Techniques

There are four types of non-probability sampling techniques:

- **Convenience Sampling:** Items are selected in the most convenient manner available.
- **Judgment Sampling:** An experienced individual selects a sample (e.g. a market research).

- **Quota Sampling:** A subgroup is selected until a limit is reached (e.g., every other employee up to 500).
- **Snowball Sampling:** An initial set of respondents is selected. They in turn select other respondents; this continues until an acceptable sample size is reached.

A3.4.3 Sample Calculations

Figure A3.4 provides a summary of the formulas that are often used in performing calculations about samples: mean, standard deviation, variance, standard unit, unit error, and sample size. It is

Item	Clarification
Mean	$X' = \sum(X_i)/n$ OR $\sum(F_iX_i) / \sum(F_i)$ where n is the number of items (elements); F_i is the frequency of X_i; and X_i represents the data values.
Standard Deviation	$S = \sqrt{((\sum F_i (X_i - X')^2) / n)}$ where n is the sample size
Variance	Variance = S^2
Standard Unit	$Z = (X_i - X') / S$
Standard Error	$S_E = S / \sqrt{(n)}$
Unit Error	$r = ZS_E = ZS / \sqrt{(n)}$
Sample Size	From the equation for unit error above, $n = (ZS / r)^2$

Figure A3.4 Formulas for sample calculations.

assumed that you are familiar (even if minimally) with these basic concepts.

These formulas are best explained by examining the normal distribution curve (Figure A3.5). From the curve, observe that:

- Prob (–1 <= Z <= 1) = 68.27%
- Prob (–2 <= Z <= 2) = 95.25%
- Prob (–3 <= Z <= 3) = 99.73%

The confidence limit of a population mean is normally given by $X' \pm ZS_E$ where Z is determined by the normal distribution of the curve, considering the percentage confidence limit required. The following Z values should be memorized:

- 68% confidence => Z = 1.64
- 95% confidence => Z = 1.96
- 99% confidence => Z = 2.58

The confidence limit defines where an occurrence X may lie in the range $(X' - ZS_E) \leq X \leq (X' + ZS_E)$, given a certain confidence. As you practice solving sampling problems, your confidence in using the formulas will improve.

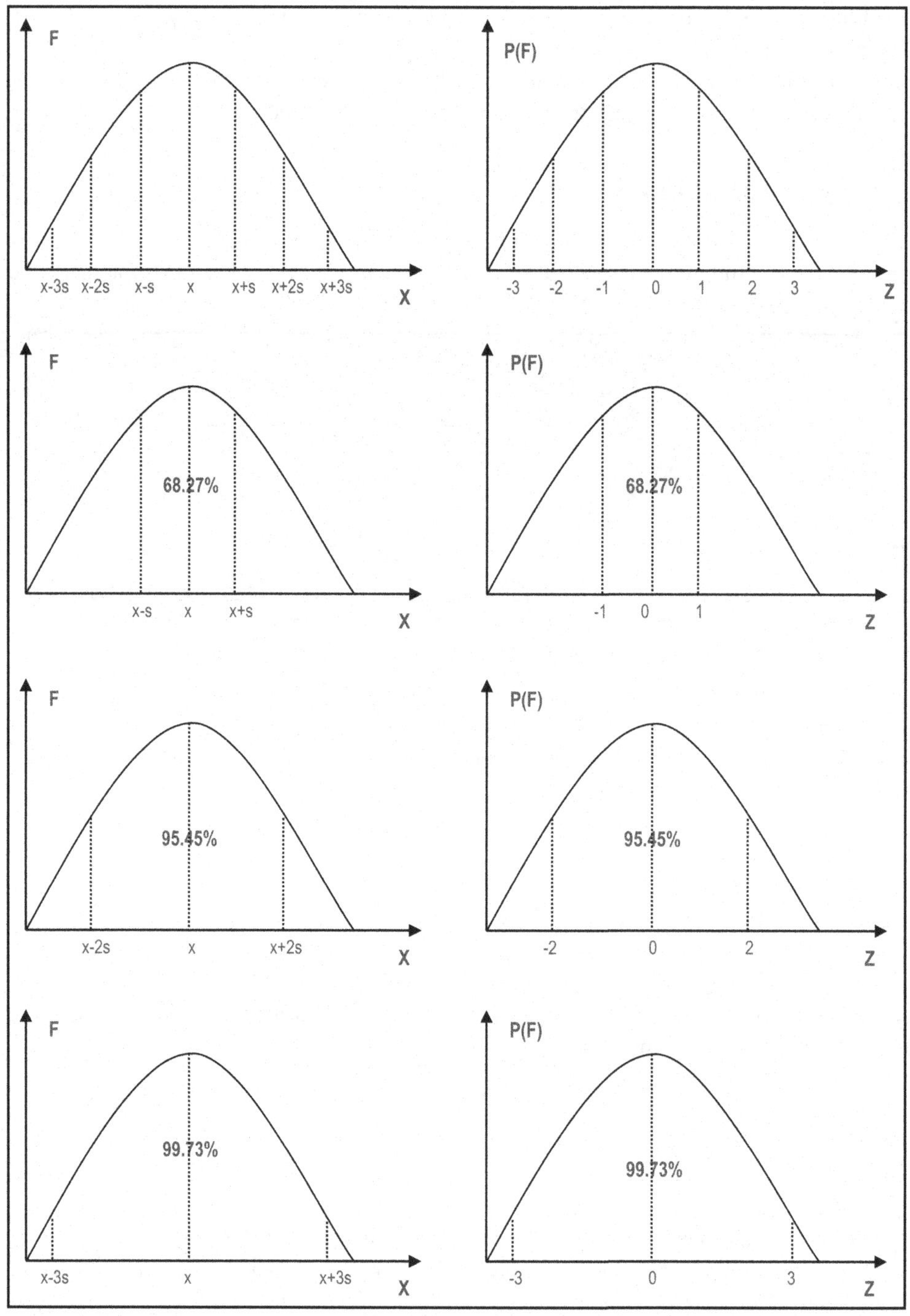

Figure A3.5 The normal distribution table.

A3.5 Observation and Document Review

Review of source documents will provide useful information about the input, data storage, and output requirements of the software system. This activity could also provide useful information on the processing requirements of the software system. To illustrate, get a hold of an application form at your organization and attempt to identify information entities (object types) represented on the form. With a little thought, you should be able to identify some or all of the following entities:

- Personal Information
- Family/Kin Contact Information
- Education History
- Employment History
- Professional References
- Extra-curricular Activities

Internal documents include forms, reports, and other internal publications; external documents are mainly in the form of journals and other professional publications. These source documents are the raw materials for gaining insights into the requirements of the system, so you want to pay keen attention to them.

With respect to observation, it is always useful to check what you have been told against what you observe and obtain clarification where deviations exist. Through keen observation, the software engineer could gather useful information not obtained by other means.

A3.6 Prototyping

In prototyping, the user gets a "trial model" of the software and is allowed to critique it. User responses are used as feedback information to revise the system. This process continues until a software system meeting user satisfaction is obtained.

The following are some basic guidelines for developing a prototype:

- Work in manageable modules.
- Build prototype rapidly.
- Modify prototype in successive iterations.
- Emphasize the user interface — it should be friendly and meeting user requirements.

There are various types of prototypes that you will find in software engineering. Following is a brief description of five common categories of prototypes.

Patched-up Prototype or Production Model: This prototype is functional for the first time, albeit inefficiently constructed. Further enhancements can be made with time.

Non-operational or Interactive Prototype: This is a prototype that is intended to be tested in order to obtain user feedback about the requirements of the system represented. A good example is where screens of the proposed system are designed; the user is allowed to pass through these screens, but no actual processing is done. This approach is particularly useful in user interface design.

First of Series Prototype: This is an operational prototype. Subsequent releases are intended to have identical features, but without glitches that the users may identify. This prototype is typically used as a marketing experiment: it is distributed free of charge or for a nominal fee; users are encouraged to use it and submit their comments about the product. These comments are then used to refine the product before subsequent releases.

Selected Features Prototype or Working Model: In this prototype, not all intended features of the (represented) software are included. Subsequent releases are intended to be enhancements with additional features. For this reason, it is sometimes referred to as *evolutionary prototype*. The initial prototype is progressively refined until an acceptable system is obtained.

Throw-away Prototype: An initial model is proposed for the sole purpose of eliciting criticism. The criticisms are then used to develop a more acceptable model, and the initial prototype is abandoned.

A3.7 Brainstorming and Mathematical Proof

The methodologies discussed so far all assume that there is readily available information which when analyzed will lead to accurate capture of the requirements of the desired software. However, this is not always the case. There are many situations in which software systems are required, but there is no readily available information that would lead to the specification of the requirements of such systems. Examples include (but are not confined to) the following:

- Writing a new compiler
- Writing a new operating system
- Writing a new CASE tool, RAD tool, or DBMS
- Developing certain expert systems
- Developing a business in a problem domain for which there is no perfect frame of reference

For these kinds of scenarios, a non-standard approach to information gathering is required. Brainstorming is particularly useful here. A close to accurate coverage of the requirements of an original software product may be obtained through brainstorming: a group of software engineering experts and prospective users come together, and through several stages of discussion, hammer out the essential requirements of the proposed software. The requirements are then documented and, through various review processes, are further refined. A prototype of the system can then be developed and subjected to further scrutiny.

Even where more conventional approaches have been employed, brainstorming is still relevant, as it forces the software engineering team to really think about the requirements identified so far, and ask tough questions to ascertain whether the requirements have been comprehensively and accurately defined.

Mathematical proofs can also be used to provide useful revelations about the required computer software. This method is particularly useful in an environment where formal methods are used for software requirements specification. This approach is often used in the synthesis of integrated circuits and chips, where there is a high demand for precision and negligible room for error. As mentioned in Chapter 1 of Foster (2021), formal methods are not applicable to every problem domain.

A3.8 Object Identification

We have discussed six different information-gathering strategies. As mentioned in Section A3.1, these strategies are to be used to identify the core requirements of the software system. As mentioned then, one aspect that we are seeking to define is the set of information entities (object types). Notice that the term information entity is used as an alternative to object type. The two terms are not identical, but for most practical purposes, they are similar. An *information entity* is a concept, object, or thing about which data is to be stored. An *object type* is a concept, object, or thing about which data is to be stored, and upon which a set of operations is to be defined.

In object-oriented environments, the term *object type* is preferred to information entity. However, as you will more fully appreciate later, in most situations, the software system is likely to be implemented in a *hybrid environment* as an object-oriented (OO) *user interface* superimposed on a *relational database*. This is a loaded statement that will make more sense after learning more about databases. For now, just accept that we can use the terms *information entity* and *object type* interchangeably in the early stages of software planning.

Early identification of information entities is critical to successful software engineering in the OO paradigm. This is so because your software will be defined in terms of objects and their interactions. For each object type, you want to be able to describe the data that it will host and related operations that will act on that data. Approaching the software planning in this way yields a number of significant advantages; moreover, even if it turns out that the software development environment is not object oriented, the effort is not lost (in light of the previous paragraph).

Several approaches to object identification (more precisely object-type identification) have been proposed. The truth, however, is that these approaches do not offer any guarantee of successful identification of all object types required by a given system. Among the approaches that have been proposed are the following:

- Using the Descriptive Narrative Approach
- Using the Rule-of-Thumb Method
- Using Things to be Modeled
- Using Definitions of Objects, Categories and Types
- Using Decomposition
- Using Generalizations and Subclasses
- Using OO Domain Analysis or Application Framework
- Reusing Individual Hierarchies, Objects, and Classes
- Using Personal Experience
- Using Class–Responsibility–Collaboration Card

A3.8.1 The Descriptive Narrative Approach

To use the *descriptive narrative approach*, start with a descriptive overview of the software system. For larger systems consisting of multiple subsystems, prepare a descriptive narrative of each component subsystem. From each descriptive overview, identify nouns (objects) and verbs (operations). Repeatedly refine the process until all nouns and verbs are identified. Represent nouns as object types and verbs as operations, avoiding duplication of effort.

To illustrate, the *Purchase Order and Receipt Subsystem* (of an Inventory System) might have the following overview (Figure A3.6).

Purchase orders (PO) are sent to suppliers, requesting inventory items in specific quantities. If a PO is incorrectly generated, it is immediately removed and a new PO generated. The purchase invoice is the official document used to recognize receipt of goods from suppliers. All goods received are accompanied by invoices. Once received, the invoice is recorded. Items received are also recorded and appropriate inventory adjustments made to the inventory item master file. Receipt quantities can be adjusted, but if wrong items are recorded on receipt, or omissions are made, the whole invoice must be removed and re-recorded. When a receipt is correctly recorded, the associated PO status is adjusted.

Figure A3.6 Descriptive narrative of purchase order and invoice receipt subsystem.

From this narrative, an initial list of object types and associated operations can be constructed, as shown in Figure A3.7. Further refinement would be required; for instance, additional operations may be defined for each object type (left as an exercise); also, the data description can be further refined.

Object Type	Data Description	Operations
Purchase Order	Stores Order Number, Order Date, Supplier, Items Ordered and related Quantity Ordered, etc.	Generate, Remove, Adjust-Status
Supplier	Stores Supplier Code, Supplier Name, Supplier Address, Contact Person, Telephone, E-mail, etc.	Sent-Invoice
Purchase Invoice	Stores Invoice Number, Invoice Date, Related Supplier, Items Shipped and related Quantity Shipped, Invoice Amount, Discount, Tax, etc.	Record, Remove, Adjust-Quantity
Inventory Item	Stores Item Code, Item Name, Item Category, Quantity on Hand, Last Purchase Price, etc.	Adjust-Inventory

Figure A3.7 Object types and operations for purchase order and invoice receipt subsystem.

A3.8.2 The Rule-of-Thumb Approach

As an alternative to the descriptive narrative strategy, you may adopt an intuitive approach as follows: Using principles discussed earlier, identify the main information entities (object types) that will make up the system. Most information entities that make up a system will be subject to some combination of the following basic operations:

- **ADD:** Addition of data items
- **MODIFY:** Update of existing data items
- **DELETE:** Deletion of existing data items
- **INQUIRE/ANALYZE:** Inquiry and/or analysis on existing information
- **REPORT/ANALYZE:** Reporting and/or analysis of existing information
- **RETRIEVE:** Retrieval of existing data
- **FORECAST:** Predict future data based on analysis of existing data

Obviously, not all operations will apply for all object types (data entities); also, some object types (entities) may require additional operations. The software engineer makes intelligent decisions about these exceptions, depending on the situation. Additionally, the level of complexity of each operation will depend to some extent on the object type (data entity).

In a truly OO environment, the operations may be included as part of the object's services. In a hybrid environment, the information entities may be implemented as part of a relational database, and the operations would be implemented as user interface objects (windows, forms, etc.).

A3.8.3 Using Things to be Modeled

This is the preferred approach of experienced software engineers. It is summarized in two steps:

1. Identify tangible objects (more precisely, object types) in the application domain that are to be modeled.
2. Put these objects into logical categories. These categories will become super-classes.

In order for this method to be successful, the software engineer must make a paradigm shift to an object-oriented mindset. For this reason, software engineers who have not made that transition usually have difficulties employing it.

The main drawback of this approach is that on the surface, it does not help in the identification of abstract (intangible) object types. Of course, the counter argument here is that with experience, identifying abstract object types will not be a problem.

A3.8.4 Using the Definitions of Objects, Categories, and Interfaces

This technique assumes that the most effective way to identify object types is to do so directly based on the software engineer's knowledge of the application domain, as well as his/her knowledge of and experience in object abstraction and object categorization. Objects, categories (classes), and interfaces are identified intuitively.

This approach is very effective for the experienced software engineer who has made the paradigm shift of OO methodologies.

A3.8.5 Using Decomposition

This approach assumes that the system will consist of several component objects, and it works very well in situations where this assumption is true. The steps involved are:

1. Identify the aggregate objects or categories.
2. Repeatedly decompose aggregate objects into their components until a stable state is reached.

There are two drawbacks with this approach:

- Not all systems have an abundance of aggregate object types.
- If the approach is strictly followed, one might end up proposing component relationships where one-to-many (or many-to-many) relationships would better serve the situation.

A3.8.6 Using Generalizations and Subclasses

This technique contends that objects are identified before their categories (object types). The steps involved are:

1. Identify all objects.
2. Identify objects that share the same attributes and services (operations). Generalize these objects into categories (object types).
3. Identify categories (classes) that share common resources (attributes, relationships, operations, etc.).
4. Factor out the common resources to form super-classes, then use generalization or specialization for all categories that share these common resources to form subcategories (subclasses).
5. If the only common factors are service prototypes, use the *interface* to factor out these common factors.

The main advantage of this technique is reuse — it is likely to produce a design that is very compact, due to a high degree of code reuse.

The main disadvantage of the technique is that it could be easily misused to produce a design of countless splinter classes that reuse logically unrelated resources. This could result in a system that is difficult to maintain.

A3.8.7 Using OO Domain Analysis or Application Framework

This approach assumes that an OO domain analysis (OODA) and/or application framework of an application in the same problem domain was previously done, and can therefore be used. The steps involved are as follows:

1. Analyze the existing OODA or application framework for the problem.
2. Reuse (with modifications where necessary) objects and/or categories as required.

The main advantage of this approach is that if such reusable components can be identified, system development time can be significantly reduced.

The main drawback of the approach is that it is not always applicable. Most existing systems either have incomplete OODA or no OODA at all. This should not be surprising since software engineering is a fairly youthful discipline. With time, this approach should be quite useful.

A3.8.8 Reusing Hierarchies, Individual Objects, and Classes

This technique is relevant in a situation where there is a repository with reusable class hierarchies, which is available. The steps involved are as follows:

1. Use the repository to identify classes and class hierarchies that can be reused.
2. Where necessary, adopt and modify existing classes to be reused.
3. Introduce new classes (categories) where necessary.
4. If the classes are parameterized, supply the generic formal parameters (this is not currently supported in Java).

The advantages and disadvantages of this approach are identical to those specified in the previous subsection.

A3.8.9 Using Personal Experience

This approach is likely to become more popular as software engineers become more experienced in designing software via object technology. The software engineer draws on his/her experience in previously designed systems to design a new one. The required steps are as follows:

1. Identify objects, categories (classes), and interfaces that correspond to the ones used in previous models that are in the same application domain.
2. Utilize these resources (with modification where applicable) in the new system.
3. Introduce new categories (object types) and interfaces where necessary.

This approach could significantly reduce the development time of a new system, while building the experience repertoire of the software engineer(s) involved.

Where the relevant prior experience is lacking, this approach breaks down and becomes potentially dangerous. Also, the approach could facilitate the proliferation of shoddy design with poor documentation, or no documentation at all. This could result in a system that is very difficult to maintain.

A3.8.10 Using the Class–Responsibility–Collaboration Method

A popular concept in object identification is *responsibility-driven design*, a term used to mean, we identify classes and determine their responsibilities. This must be done well before the internals of the class can be tackled.

The *CRC* (*class–responsibility–collaborator*) methodology, proposed by Kent Beck and Ward Cunningham (see Beck & Cunningham 1989), defines for each class its responsibilities and collaborators that the class may use to achieve its objectives (hence the acronym). The responsibilities of a class are the requests it must correctly respond to; the collaborators of a class are other classes it must invoke in order to carry out its responsibilities. In carrying out its responsibilities, a class may use its own internal methods or it may solicit help via a collaborator's method(s).

The CRC methodology is summarized in the following steps:

1. In a brainstorming session, identify object types (categories) that may be required for the system.
2. For each object type (category), construct a list of (possible) services to be provided (or operations to be performed on an instance of that object type).
3. For each object type (category), identify possible collaborators.
4. Identify missing object types (categories) and interfaces that should be added. To identify new object types, look for attributes and services that cannot be allocated to the current set of object types. To identify new interfaces, look for common service protocols.
5. Develop a CRC card for each object type and place them on a whiteboard (or desk surface). Draw association arcs from object types (classes) to collaborators.
6. Test and refine the model by utilizing *use cases*.

Figure A3.8 provides an illustration of a CRC card. The CRC card must be stored electronically as well as physically to aid the design process. For instance, CRC cards can be easily classified, printed, and strung out on a table, during a brainstorming session, to assist designers with gaining a comprehensive overview of the system. Class names and responsibilities must be carefully worded to avoid ambiguities.

Class Name	
Responsibilities	Collaborators
•••	•••

Figure A3.8 The CRC card.

The main advantages of the technique are:

- It is easy to learn and follow.
- When used properly, it helps designers gain a comprehensive overview of the system.

The main drawback of the technique is that it requires experienced software engineer(s) if success is to be achieved. Moreover, as you will soon see (or probably have already observed), CRC cards are no longer used because the principles have been subsumed into the Unified Modeling Language (UML) standards.

A3.9 End-User Involvement

Object-oriented software engineering (OOSE) must thoroughly involve end users to ensure their satisfaction and acceptance. The software engineer should not operate in a manner that is oblivious of the end user. Rather, his/her role is to extract information needs from the end users and model it in a way that the users understand and are satisfied with. Workshops are usually effective means. Three types of workshops are:

- JEM — Joint Enterprise Modeling
- JRP — Joint Requirements Planning
- JAD — Joint Application Design

Each session is guided by a facilitator (usually a professional, skilled in object-oriented methodologies [OOM]). Key end users do most of the talking. The facilitator does the modeling for end users to see and accept or revise. The facilitator should be well experienced in system and software design, an excellent motivator and communicator; must have a good reputation that inspires confidence; and a skilled negotiator.

Martin makes a number of specific recommendations about workshops: Workshops typically last for no longer than a week at-a-time, and should not be allowed to spread out over an extended period of time. Nothing should be allowed to stall or slow down the progress of the workshop. Issues that cannot be resolved in a reasonable timeframe must be declared open (by the facilitator), and subject to subsequent analysis and/or discussion leading to satisfactory resolution (see Martin 1993).

In order for a user workshop to be successful, thorough preparation must take place. The software engineers on the project must do their homework in obtaining relevant information, analyzing it, and preparing working models and/or proposals that will be examined and/or used in the workshop. Additionally, training of participants may be required prior to the workshop in order to achieve optimum benefits. The training sessions should be carefully planned and administered. Here, the participants should be briefed on the expectations and activities of the workshops.

A3.10 Summary and Concluding Remarks

Here is a summary of what we have covered in this chapter:

- It is important to conduct a research on the requirements of a software system to be developed. By so doing, we determine the synergistic interrelationships, information entities, operations, business rules, and security mechanisms.
- In conducting the software requirements research, obtaining answers to questions commencing with the words WHAT, WHERE, HOW, WHEN, and WHY is very important.
- Information-gathering strategies include interviews, questionnaires and surveys, sampling and experimenting, observation and document review, prototyping, brainstorming, and mathematical proofs.
- The interview is useful when the information needed must be elaborate or clarification on various issues is required. The interview also provides an opportunity for the software engineer to win the confidence and trust of clients. In preparing to conduct an interview, the software engineer must be thoroughly prepared and must follow well-known interviewing norms.
- A questionnaire is viable when any of the following situations hold: A small amount of information is required of a large population; the time frame is short, but a vast area (and/or dispersed population) must be covered; simple answers are required to a number of standard questions. The software engineer must follow established norms in preparing and administering a questionnaire.
- Sampling is useful when the information required is of a quantitative nature or can be quantified, no precise detail is available, and it is not likely that such details will be obtained via other methods. The software engineer must be familiar with various sampling techniques and know when to use a particular technique.
- Review of source documents will provide useful information about the input, data storage, and output requirements of the software system. This activity could also provide useful information on the processing requirements of the software system.
- Prototyping involves providing a trial model of the software for user critique. User responses are used as feedback information to revise the system. This process continues until a software system meeting user satisfaction is obtained. The software engineer should be familiar with the different types of prototypes.
- Brainstorming is useful in situations in which software systems are required, but there is no readily available information that would lead to the specification of the requirements of such systems. Brainstorming involves a number of software engineers coming together to discuss and hammer out the requirements of a software system.
- Mathematical proof is particularly useful in an environment where formal methods are used for software requirements specification.
- One primary objective of these techniques is the identification and accurate specification of the information entities (or object types) comprising the software system. There are various techniques for identifying and refining object types. Among the various object identification techniques that have been proposed are the following: the descriptive narrative strategy; the rule-of-thumb strategy; using things to be modeled; using definitions of objects, categories, and types; using decomposition; using generalizations and subclasses; using OO domain analysis or application framework; reusing individual hierarchies, objects, and classes; using personal experience; using class–responsibility–collaboration cards.

- OOSE must thoroughly involve end users to ensure their satisfaction and acceptance. Three types of workshops that are common are Joint Enterprise Modeling (JEM), Joint Requirements Planning (JRP), and Joint Application Design (JAD).

Accurate and comprehensive information gathering is critical to the success of a software engineering venture. In fact, the success of the venture depends to a large extent on this. Your information-gathering skills will improve with practice and experience.

In applying these techniques, the software engineer will no doubt gather much information concerning the requirements of the software system to be constructed. How will you record all this information? If you start writing narratives, you will soon wind up with huge books that not many people will bother to read. In software engineering, rather than writing voluminous narratives to document the requirements, we use unambiguous notations and diagrams (of course, you still need to write but not as much as you would without the notations and diagrams).

References and Recommended Readings

Beck, Kent, & Ward Cunningham. 1989. "A Laboratory for Teaching Object-Oriented Thinking." In OOPSLA'89 Conference Proceedings, October 1–6, 1989. New Orleans, LA. http://c2.com/doc/oopsla89/paper.html

Bruegge, Bernd, & Allen H. Dutoit. 2010. *Object-Oriented Software Engineering* 3rd ed. Boston, MA: Pearson. See chapters 4 & 5.

DeGroot, Morris, & Mark Schervish. 2014. *Probability and Statistics* 4th ed. Boston, MA: Pearson.

Due', Richard T. 2002. *Mentoring Object Technology Projects*. Saddle River, NJ: Prentice Hall. See chapters 3–6.

Foster, Elvis C. 2021. *Software Engineering: A Methodical Approach* 2nd ed. New York: CRC Press.

Kendall, Kenneth E., & Julia E. Kendall. 2014. *Systems Analysis and Design* 9th ed. Boston, MA: Pearson. See chapters 4–6.

Lee, Richard C., & William M. Tepfenhart. 2002. *Practical Object-Oriented Development with UML and Java*. Upper Saddle River, NJ: Prentice Hall. See chapter 4.

Martin, James, & James Odell. 1993. *Principles of Object-Oriented Analysis and Design*. Eaglewood Cliffs, NJ: Prentice Hall. See chapters 4 & 5.

Pfleeger, Shari Lawrence. 2006. *Software Engineering Theory and Practice* 3rd ed. Upper Saddle River, NJ: Prentice Hall. See chapter 4.

Sommerville, Ian. 2016. *Software Engineering* 10th ed. Boston, MA: Pearson. See chapter 4.

Van Vliet, Hans. 2008. *Software Engineering* 3rd ed. New York: Wiley. See chapter 9.

Appendix 4: BNF Syntax for Selected SQL Statements

This appendix provides the Backus–Naur Form (BNF) syntax for selected Structured Query Language (SQL) statements referenced throughout the text. This is by no means a comprehensive coverage of SQL, and should therefore not take the place for an SQL reference or appropriate product documentation. The intent here is to prove a quick reference point for familiarization to selected SQL statements. Areas covered include:

- Database Management
- Tablespace Management
- Table Management
- Index Management
- Data Insertion, Update, and Deletion
- Transaction Management
- Building Queries
- Managing Logical Views
- Managing System Security

A4.1 Database Management

A4.1.1 Syntax for Create-Database Statement

Create-Database ::=
CREATE DATABASE <DatabaseName>
[USER SYS IDENTIFIED BY <Password>]
[USER SYSTEM IDENTIFIED BY <Password>]
[CONTROLFILE REUSE]
[LOGFILE [GROUP <n> (<FileSpec> {,<FileSpec>})
{ GROUP <n> (<FileSpec> {,<FileSpec>})}
[MAXLOGFILES <n>] [MAXLOGMEMBERS <n>] [MAXLOGHISTORY <n>]
[MAXDATAFILES <n>] [MAXINSTANCES <n>]
[ARCHIVELOG|NOARCHIVELOG] [FORCE LOGGING]
[CHARACTER SET <Charset>] [NATIONAL CHARACTER SET <Charset>]
[DATAFILE <FileSpec> [<Auto-extend-Clause>] {, <FileSpec> [<Auto-extend-Clause>]}
[EXTENT MANAGEMENT LOCAL]
[<Default-temporary-tablespace-Clause>]
[<Undo-tablespace-Clause>]
[Set-Time-Zone-Clause]

Auto_Extend-Clause ::=
AUTOEXTEND ON|OFF NEXT <n> K|M MAXSIZE <n>|UNLIMITED K|M

Default-temporary-tablespace-Clause ::=
DEFAULT TEMPORARY TABLESPACE <TablespaceName>
[TEMPFILE <FileSpec>]
[extent management local]
[UNIFORM SIZE <n> K|M]

Undo-tablespace-Clause ::=
Undo tablespace <TablespaceName> DATAFILE File-spec

Set-Time-Zone-Clause ::=
SET TIME_ZONE = '<Time-Zone-Spec>'

FileSpec ::= '<PathtoFile>' SIZE <Integer> K|M [REUSE]

Note:

1. FileSpec is a path to a physical operating system file that you specify, based on your environment and database standards.
2. If you specify locally managed TBS (EXTENT MANAGEMENT LOCAL), you must specify the default temporary TBS.
3. If you specify locally managed TBS and the **Datafile-Clause**, you must specify the default temporary TBS and a datafile for that TBS.
4. If you specify locally managed TBS but do not specify the **Datafile-Clause**, you can omit the default temporary TBS clause. Oracle will create a temporary TBS called TEMP with a 10 M datafile.
5. The default temporary TBS size is (uniform) 1 M.
6. The default log-file size (if you do not specify the **Log-File-Clause**, Oracle creates two log-files) is 100 M.
7. Time zone is specified as a number of hours and minutes ahead of (+) the standard GMT, or by using a predetermined time-zone name (as obtained from the view **v$TIMEZONE_NAMES**).

A4.1.2 Syntax for Alter-Database Statement

Alter-Database ::=
Alter Database [<DatabaseName>]
Startup-Clauses | Recovery-Clauses | Datafile-Clauses | Logfile-Clauses | Controlfile-Clauses |
Standby-Database-Clauses | Default-Settings-Clauses | Conversion-Clauses |
Redo-Thread-Clauses | Security-Clause;

Startup-Clauses::= Startup1 | Startup2 | Startup3
Startup1 ::= MOUNT [Standby | Clone Database]
Startup2 ::= Open Read Only
Startup3 ::= Open Read Write [Resetlogs | Noresetlogs] [Migrate]

Recovery-Clauses ::= General-Recovery-Clause | Managed-Standby-Recovery | End Backup

General-Recovery-Clause ::=
Recover [Automatic [From '<PathtoFile>']]
Full-Database-Recovery-Clause | Partial-Database-Recovery-Clause | LOGFILE '<PathtoFile >'
TestCorruption | RecoveryOption1 | RecoveryOption2

TestCorruption ::= [Test] | [Allow <n> Corruption] | [Noparallel] | [Parallel <n>]
RecoveryOption1 ::= Recover [Automatic [From '<PathtoFile >']] Cancel
RecoveryOption2 ::= Recover [Automatic [From '<PathtoFile >']] Continue [Default]
Full-Database-Recovery-Clause ::= Standby1 | Standby2 | Standby3 | Standby4
Standby1 ::= [Standby] Database Until Cancel
Standby2 ::= [Standby] Database UNTIL Time <Timestamp>
Standby3 ::= [Standby] Database UNTIL Change <n>]
Standby4 ::= [Standby] Database Using Backup Controlfile

Partial-Database-Recovery-Clause ::= PartialR1 | PartialR2 | PartialR3 | PartialR4
PartialR1 ::= TABLESPACE <Tablespace> {, <Tablespace>}
PartialR2 ::= DATAFILE '<PathtoFile>' | <File-Number> {, '<PathtoFile >' | <File-Number>}
PartialR3 ::= Standby Tablespace <Tablespace> {, <Tablespace>}
Until [Consistent With] Controlfile
PartialR4 ::= Standby Datafile '<PathtoFile >' | <File-Number> {, '<PathtoFile >' | <File-Number>}
Until [Consistent With] Controlfile

Managed-Standby-Recovery-Clause ::=
Recover Managed Standby Database Recover-Clause | Cancel-Clause | Finish-Clause

Cancel-Clause ::=
Cancel [Immediate] [Wait | Nowait]

Finish-Clause ::=
Disconnect [From Session] [Noparallel] | [Parallel <n>]
Finish [Skip [Standby Logfile]][Wait | Nowait]

Recover-Clause ::= RClause1 | RClause2 | RClause3 | RClause4 | RClause5 | RClause6 | RClause7 | RClause8 | RClause9 | RClause10 | RClause11 | RClause12 | RClause13 | RClause14
RClause1 ::= Disconnect [From Session]
RClause2 ::= Timeout <n>
RClause3 ::= Notimeout
RClause4 ::= Nodelay
RClause5 ::= Default Delay
RClause6::= Delay <n>
RClause7::= Next <n>
RClause8::= Expire <n>
RClause9 ::= No Expire
RClause10 ::= NOPARALLEL
RClause11::= Parallel <n>
RClause12 ::= Through [Thread <n>] Sequence <n>
RClause13 ::= Through All Archivelog
RClause14::= Through All | Last | Next Switchover

Datafile-Clause ::= DFSpec1 | DFSpec2 | DFSpec3 | DFSpec4 | DFSpec5 | DFSpec6 | DFSpec7 | DFSpec8 | DFSpec9 | DFSpec10
DFSpec1 :: = Create Datafile <File-Number> | '<File-Name>' {,<File-Number> | '<File-Name>'} [As New] | [AS <File-spec> {, <File-spec>}]
DFSpec2 :: = DATAFILE <File-Number> | '<File-Name>' {, <File-Number> | '<File Name>'} ONLINE
DFSpec3 :: = DATAFILE <File-Number> | '<File-Name>' {, <File-Number> | '<File Name>'} OFFILINE [Drop]
DFSpec4 :: = DATAFILE <File-Number> | '<File-Name>' {, <File-Number> | '<File Name>'} Resize <n> [K|M]
DFSpec5 :: = DATAFILE <File-Number> | '<File-Name>' {, <File-Number> | '<File Name>'} Autoextend-Clause
DFSpec6 :: = DATAFILE <File-Number> | '<File-Name>' {, <File-Number> | '<File Name>'} End Backup
DFSpec7 :: = TEMPFILE <File-Number> | '<File-Name>' {, <File-Number> | '<File Name>'} Online | Offline | Autoextend_Clause
DFSpec8 :: = TEMPFILE <File-Number> | '<File-Name>' {, <File-Number> | '<File Name>'} Drop [Including Datafiles] **DFSpec9 :: =** TEMPFILE <File-Number> | '<File-Name>' {, <File-Number> | '<File Name>'} Resize <nr> [K|M]
DFSpec10 :: = Rename File '<PathtoFile>' {, '<PathtoFile>' } TO '<PathtoFile>' {, '<PathtoFile>' }

Autoextend-Clause ::= AutoOnOption | AutoOffOption
AutoOnOption ::= Autoextend Off
AutoOffOption ::= Autoextend On [Next <nr> [K|M] [Maxsize-Clause]]

Maxsize-Clause ::= FixedSize | VariableSize
VariableSize ::= Maxsize Unlimited
FixedSize ::= MAXSIZE <nr> [K|M]

Logfile-Clauses ::= LogClause1 | LogClause2 | LogClause3 | LogClause4 | LogClause5 | LogClause6 | LogClause7 | LogClause8 | LogClausen9 | LogClause10
LogClause1 ::= Archivelog | Noarchivelog
LogClause2::= [No] Force Logging
LogClause3::= Add [Standby] Logfile [Thread <n>] [Group <n>] Logfile-Spec {, [Group <n>] Logfile-Spec}
LogClause4::= Drop [Standby] LOGFILE Logfile-Spec
LogClause5::= Add [Standby] Logfile Member '<PathtoFile>' [Reuse] {, '<PathtoFile>' [Reuse]} To Logfile-Spec {, Logfile-Spec}
LogClause6::= Drop [Standby] Logfile Member '<PathtoFile>' {, '<PathtoFile>'}
LogClause7::= Add Supplemental Log Data [(Primary Key | Unique Index {, Primary Key | Unique Index}) Columns]
LogClause8::= Drop Supplemental Log Data
LogOClause9::= Rename File '<PathtoFile>' {, '<PathtoFile>'} TO '<PathtoFile>' {, '<PathtoFile>'}
LogClause10::= Clear [Unarchived] Logfile Logfile-Spec {, Logfile-Spec} [Unrecoverable Datafile]

Logfile-Spec ::= LogOption1 | LogOption2 | LogOption3
LogOption1 ::= Group <n>
LogOption2 ::= ('<PathtoFile>' {, '<PathtoFile>'})
LogOption3 ::= '<PathtoFile>'

Controlfile-Clauses ::= CFOption1 | CFOption1 | CFOption3
CFOption1 ::= Create Standby Controlfile As '<PathtoFile>' [Reuse]
CFOption2 ::= Backup Controlfile To Tracefile-Clause
CFOption3 ::= Backup Controlfile To '<PathtoFile>' [Reuse]

Tracefile-Clause ::=
Trace [As '<PathtoFile> [Reuse] [Resetlogs | Noresetlogs]]

Standby-Database-Clauses ::= SClause1 | SClause2 | SClause3 | SClause4 | SClause5 | SClause6 | SClause7 |
SClause8 | SClause9
SClause1 ::= Activate [Physical | Logical] Standby Database [Skip [Standby Logfile]]
SClause2::= Set Standby Database To Maximize Protection | Availability | Performance
SClause3 ::= Register [Or Replace] [Physical | Logical] Logfile Logfile-Spec {, Logfile-Spec}
SClause4 ::= Start Logical Standby Apply [[New Primary <DBlink>] | [Initial <Scan-value>]]
SClause5 ::= Stop | Abort Logical Standby Apply
SClause6 ::= NOPARALLEL
SClause7 ::= PARALLEL <n>
SClause8 ::= Commit To Switchover To Physical | Logical Primary | Standby
SClause9::= [With | Without Session Shutdown] [Wait | Nowait]

Default-Settings-Clauses ::= DSClause1 | DSClause2 | DSClause3 | DSClause4
DSClause1 ::= [NATIONAL] Character Set <Charset>
DSClause2 ::= Set-Time-Zone-Clause
DSClause3 ::= Default Temporary Tablespace <TablespaceName>
DSClause4 ::= Rename Global_Name To <Database-Name>.<Domain>

Set-Time-Zone-Clause ::=
Set Time_Zone = '± <HH:MM>' | <Time-Zone-Region>

Conversion-Clauses ::=
CONVERT | ResetOption
ResetOption ::= Reset Compatibility

Redo-Thread-Clauses ::= EnableThread | DisableThread
EnableThread ::= Enable [Public] Thread <n>
DisableThread ::= Disable Thread <n>

Security-Clause ::=
Guard All | Standby | None

Examples:

ALTER DATABASE SampleDB MOUNT;

ALTER DATABASE SampleDB READ ONLY;

ALTER DATABASE SampleDB Backup Controlfile To TRACE AS 'C:\Oracle\EFDB06\SampleTrace';

A4.2 Tablespace Management

A4.2.1 Syntax for Create-Tablespace Statement

Create-Tablespace::=
CREATE [BIGFILE | SMALLFILE] <Permanent-Tablespace-Clause> | <Temporary-Tablespace-Clause> |
<Undo-Tablespace-Clause>;

Permanent-Tablespace-Clause ::=
TABLESPACE <TablespaceName>
[DATAFILE <FileSpec> {, <FileSpec>}]
[MINIMUM EXTENT <n> [K|M|G|T|P|E]] /*NA if locally managed; see Extent-Management-Clause */
[BLOCKSIZE <n> [K]]
[LOGGING | NOLOGGING] [FORCE LOGGING] [ONLINE | OFFLINE]
[DEFAULT] [COMPRESS | NOCOMPRESS] <Storage-Clause>]
[<Extent-Management-Clause>]
[<Segment-Management-Clause>]
[FLASHBACK ON|OFF]

Temporary-Tablespace-Clause ::=
TEMPORARY TABLESPACE <TablespaceName>
[TEMPFILE <FileSpec> {,<FileSpec>}]
[TABLESPACE GROUP <TablespaceGroupName> | ' ']
[<Extent-Management-Clause>]

Undo-Tablespace-Clause ::=
UNDO TABLESPACE <TablespaceName>
[DATAFILE <FileSpec> {,<FileSpec>}]
[<Extent-Management-Clause>]
[RETENTION GUARANTEE | NONGUARANTEE]

Extent-Management-Clause ::= DictionaryOption | LocalOption1 | LocalOption2
DictionaryOption ::= EXTENT MANAGEMENT DICTIONARY
LocalOption1 ::= EXTENT MANAGEMENT LOCAL AUTOALLOCATE
LocalOption2 ::= EXTENT MANAGAEMENT LOCAL [UNIFORM [SIZE <n> [K|M|G|T|P|E]]]

FileSpec ::= '<PathtoFile>' SIZE <n> K|M|G|T|P|E [REUSE] [<Autoextend-Clause>]

Storage-Clause ::=
STORAGE (
[INITIAL <n> [K|M|G|T|P|E]] [NEXT <n> [K|M|G|T|P|E]] [MINEXTENTS <n>] [MAXEXTENTS <In> | UNLIMITED]
[PCTINCREASE <n>] [FREELISTS <n>] [FREELIST GROUPS <n>] [OPTIMAL <n> [K|M|G|T|P|E]] [OPTIMAL NULL]
[BUFFER_POOL KEEP | RECYCLE | DEFAULT] **)**

Autoextend-Clause ::= AUTOEXTEND OFF /* Or */
AUTOEXTEND ON [NEXT <n> K|M|G|T|P|E] [<Maxsize-Clause>]

Maxsize-Clause ::= MAXSIZE <n> K|M|G|T|P|E /* Or */
MAXSIZE UNLIMITED

Segment-Management-Clause ::= SEGMENT SPACE MANAGEMENT MANUAL | AUTO /* AUTO recommended */

A4.2.2 Syntax for Alter-Tablespace Statement

```
Alter-Tablespace::=
ALTER TABLESPACE <TablespaceName>
[DEFAULT [COMPRESS | NOCOMPRESS] <Storage-Clause>]
[MINIMUM EXTENT <n> K|M|G|T|P|E]
[RESIZE <n> K|M|G|T|P|E] [COALESCE]
[SHRINK SPACE [KEEP <n> K|M|G|T|P|E]]
[RENAME TO <NewTablespaceName>]
[BEGIN | END BACKUP]
[<Datafile-Tempfile-Clauses>]
[<Tablespace-Logging-Clause>]
[TABLESPACE GROUP <TablespaceGroupName> | ' ']
[<Tablespace-State-Clause>]
[<Autoextend-Clause>]
[FLASHBACK ON | OFF]
[<Tablespace-Retention-Clause>];

Datafile-Tempfile-Clauses ::=
[ADD DATAFILE | TEMPFILE <FileSpec> {,<FileSpec>}]
[DROP DATAFILE | TEMPFILE <FileNumber> | <'FileName'>]
[SHRINK TEMPFILE <FileNumber> | <'FileName'> [KEEP <n> K|M|G|T|P|E]]
[RENAME DATAFILE <'FileName'> {, <'FileName'>} TO <'NewFileName'> {, <'NewFileName'>}]
[DATAFILE | TEMPFILE ONLINE | OFFLINE]

Tablespace-State-Clause ::=
[ONLINE | OfflineSpec]
[READ ONLY | WRITE]
[PERMANENT | TEMPORARY]

Offline-Spec ::= OFFLINE [NORMAL | TEMPORARY | IMMEDIATE]

Tablespace-Logging-Clause ::=
LOGGING | NOLOGGING /* Or*/
[NO] FORCE LOGGING

Tablespace-Retention-Clause ::= RETENTION GUARANTEE | NOGUARANTEE
FileSpec ::=              /* As defined in Create-Tablespace */
Storage-Clause ::=        /* As defined in Create-Tablespace */
Autoextend-Clause ::=     /* As defined in Create-Tablespace */
Maxsize-Clause ::=        /* As defined in Create-Tablespace */
```

A4.3 Table Management

A4.3.1 Syntax for Create-Table Statement

Create-Table::=
Create-Relational-Table | Create-Object-Table | Create-XML-Table

Create-Relational-Table ::=
CREATE [GLOBAL TEMPORARY] TABLE [<Schema>·] <TableName> [(Relational-Properties)]
[ON COMMIT DELETE | PRESERVE ROWS]
[Physical-Properties Table-Properties] [Storage-clause];

Relational-Properties ::=
Column-Definition-Clause {, Column-Definition-Clause }
[Constraint-Definition-Clause {, Constraint-Definition-Clause}]

Column-Definition-Clause ::=
<Column-Name> Data-Type-Spec [DEFAULT <Expression>]
[Inline-Constraint | Inline-Ref-Constraint]
// Data-Type-Spec refers to any valid type specification as outlined in figure 11.11

Constraint-Definition-Clause ::=
Inline-Constraint | Out-of-line-Constraint | Inline-Ref-Constraint | Out-of-Line-Ref-Constraint

Out-of-line-Constraint ::= Out-Constraint-Unique | Out-Constraint-Primary | Out-Constraint-Foreign | Out-Constraint-Check

Out-Constraint-Unique ::= [CONSTRAINT <Constraint-Name>] UNIQUE (<Column> {,<Column>}) [Constraint-State]

Out-Constraint-Primary ::=
[CONSTRAINT <Constraint-Name>] PRIMARY KEY (<Column>[ASC/DESC]{, <Column>[ASC/DESC]}) [Constraint-State]

Out-Constraint-Foreign ::=
[CONSTRAINT<Constraint-Name>] FOREIGN KEY (<Column> {,<Column>}) Reference-Clause
[Constraint-State]

Out-Constraint-Check ::= [CONSTRAINT<Constraint-Name>] Check-Spec [Constraint-State]

Check-Spec ::= CFormat1 | CFormat2 | CFormat3
CFormat1 ::= Check (<Column> BETWEEN <Value1> AND <Value2>)
CFormat2 :::= CHECK (<Column> operator <Value>)
CFormat3 :::= CHECK (<Column> IN <Value> {,<Value>})

Inline-Constraint ::= In-Constraint-Null | In-Constraint-Unique | In-Constraint-Primary | In-Constraint-Foreign | In-Constraint-Check
In-Constraint-Null ::= [CONSTRAINT <Constraint-Name> [NOT] NULL [Constraint-State]
In-Constraint-Unique ::= [CONSTRAINT <Constraint-Name> UNIQUE [Constraint-State]
In-Constraint-Primary ::= [CONSTRAINT <Constraint-Name> PRIMARY KEY [Constraint-State]
In-Constraint-Foreign ::= [CONSTRAINT <Constraint-Name> Reference-Clause [Constraint-State]
In-Constraint-Check ::= [CONSTRAINT <Constraint-Name> Check-Spec [Constraint-State]

Reference-Clause ::=
REFERENCES [<Schema>·] <ObjectName> [(<column>)]
[ON DELETE CASCADE] | [ON DELETE SET NULL]

Constraint-State ::= Exception-Clause | Using-Index-Clause | CState1 | CState2 | CState3 |
CState4 | CState5
CState1 ::= [NOT] DEFERRABLE
CState2 ::= INITIALLY IMMEDIATE | DEFERRED **CState3 ::=** ENABLE | DISABLE
CState4 ::= VALIDATE | NOVALIDATE
CState5 ::= REPLY | NO REPLY

Inline-Ref-Constraint ::= IROption1 | IROption2 | IROption3
IROption1 ::= SCOPE IS [<Schema>·] <Scope-Table>
IROption2 ::= WITH ROWID
IROption3 ::= [CONSTRAINT <Constraint-Name>] Reference-Clause [Constraint-State]

Out-of-Line-Ref-Constraint ::= OROption1 | OROption2 | OROption3
OROption1 ::= SCOPE FOR (<Ref-Column> | Ref-Attribute>) IS [<Schema·>] <Scope-Table>
OROption2 ::= REF (<Ref-Column> | <Ref-Attribute>) WITH ROWID
OROption3 ::= [CONSTRAINT <Constraint-Name>] FOREIGN KEY (<Ref-Column> |
<Ref-Attribute>)
Reference-Clause [Constraint-State]

Using-Index-Clause ::= UIOption1 | UIOption2 | UIOption3 | UIOption4 | UIOption5 | UIOption6 |
UIOption7 | UIOption8
UIOption1 ::= USING INDEX <Index Name>
UIOption2 ::= USING INDEX (Create-Index)
UIOption3 ::= USING INDEX PCTFREE | INITRANS | MAXTRANS <Integer>
UIOption4 ::= USING INDEX Storage-Clause
UIOption5 ::= USING INDEX TABLESPACE <Tablespace>
UIOption6 ::= USING INDEX SORT | NOSORT
UIOption7 ::= USING INDEX LOGGING | NOLOGGING
UIOption8 ::= USING INDEX LOCAL | Global-Partition-Index

Global-Partition-Index ::=
GLOBAL PARTITION BY RANGE (<Column> {,<Column>})
(Index-Partitioning-Clause)

Index-Partitioning-Clause ::=
PARTITION [<Partition>] VALUES LESS THAN (<Value> {,<Value>})
[Segment-Attributes- Clause]

Segment-Attributes-Clause ::=
[TABLESPACE <Tablespace>]
[Physical-Attributes-Clause] [LOGGING | NOLOGGING]

Physical–Attributes-Clause ::=
[PCTFREE <n>] [PCTUSED <n>] [INITRANS <n>] [MAXTRANS <n>]
[Storage–Clause]

Exception-Clause ::=
EXCEPTIONS INTO [<Schema•>] <Table>

Physical-Properties ::= PPOption1 | PPOption2 | PPOption3 | PPOption4 | PPOption5
PPOption1 ::= Segment- Attributes-Clause [COMPRESS | NOCOMPRESS]
PPOption2 ::= ORGANIZATION HEAP [Segment –Attributes-Clause] [COMPRESS | NOCOMPRESS}
PPOption3 ::= ORGANIZATION INDEX [Segment-Attributes-Clause] IOT-Clause
PPOption4 ::= ORGANIZATION EXTERNAL External-Table-Clause
PPOption5 ::= CLUSTER <Cluster-Name> (<Column> {,<Column>})

External-Table-Clause ::=
([TYPE Access-Driver-Type] External-Data-Properties)
[RESET LIMIT <Integer>|UNLIMITED]

External-Data-Properties ::=
DEFAULT DIRECTORY <Directory>
[ACCESS PARAMETERS (<Opaque-Format-Spec>)] | [ACCESS PARAMETERS USING CLOB <Sub-query>]
LOCATION ([<Directory>:} '<Path-to-File>' {,[<Directory>:] '<Path-to-File >'})

Table-Properties ::=
[Column-Properties] [Table-Partitioning-Clause] [CACHE | NOCACHE]
[Parallel-Clause] [ROWDEPENDENCIES | NOROWDEPENDENCIES]
[MONITORING | NOMONITORING]
[Enable-Disable-Clause {Enable-Disable-Clause}] [Row-Moment-Clause]
[AS <Sub-query>]

Column-Properties::= CPOption1 | CPOption2
CPOption1 ::= Object-Type-Col-Properties | Nested-Table-Col-Properties | XML-Type-Col-Properties
CPOption2 ::= Varrey-Col-Properties | LOB-Storage-Clause [LOB-Partition–Storage]

Paralled-Clause::= NOPARALLEL | Parallel-Option
Parallel-Option ::= PARALLEL [<Integer>]
// **Note:** Default is NOPARALLEL.

Row-Movement-Clause::=
ENABLE | DISABLE ROW MOVEMENT

Table-Partitioning-Clause ::=
Range-Partition | Hash-Partition | List-Partition | Composite-Partition

Enable-Disable-Clause::= EDOption1 | EDOption2 | EDOption3
EDOption1 ::= ENABLE | DISABLE [VALIDATE | NOVALIDATE] UNIQUE (<Column >{, <Column>)
[Using- Index-Clause] [Exception-Clause] [CASCADE] [KEEP | DROP INDEX]
EDOption2 ::= ENABLE | DISABLE [VOLATE | NOVOLATILE] PRIMARY KEY
[Using-Index-Clause] [Exception- Clause] [CASCADE] [KEEP | DROP INDEX]
EDOption3 ::= ENABLE | DISABLE [VOLATILE | NOVOLATILE] CONSTRAINT <Constraint-Name>
[Using-Index-Clause] [Exception-Clause] [CASCADE] [KEEP DROP | INDEX]
Using-Index-Clause :: = UIOption9 | UIOption10 | UIOption11
UIOption9 ::= USING INDEX [<Schema>.] <IndexName>
UIOption10 ::= USING INDEX (Create-Index)
UIOption11 ::= USING INDEX Index-Alternatives {Index-Alternatives}
// Note: Create Index is defined in the Create-Index statement.
Index-Alternatives ::= LOCAL | Global-Partitioned-Index | IAOption1 | IAOption2 | IAOption3 | IAOption4 |
IAOption5 | IAOption6 | IAOption7 |
IAOption1 ::= PCTFREE <Integer>
IAOption2 ::= INITRANS <Integer>
IAOption3 ::= MAXTRANS <Integer>
IAOption4 ::= TABLESPACE <Tablespace>
IAOption5 ::= Storage-Clause
IAOption6 ::= SORT | NSORT
IAOption7 ::= LOGGING | NOLOGGING /* Default is LOGGING */

Storage-Clause :: =
STORAGE (INITIAL <n> [K|M] NEXT <n> [K|M]
PCTINCREASE <n> MINEXTENTS <n>
MAXEXTENTS <In> | UNLIMITED
FREELISTS <n> FREELIST GROUPS <n>
OPTIMAL <n> [K|M]
[PCTUSED <n>] [PCTFREE <n>]
[INITRANS <n>] [MAXFRANS <n>]
[BUFFER_POOL KEEP | RECYCLE | DEFAULT])

Range-Partition::=
PARTITION BY RANGE (<Column>{, <Column>})
(PARTITION [<Partition-Name>] Range-Values-Clause Table-Partition-Desc
{, PARTITION[<Partition-Name>] Range-Values-Clause Table-Partition-Desc})

Range-Values-Clause::=
VALUES LESS THAN (<Value>|MAXVALUE {, <Value>|MAXVALUE})

Table-Partition-Desc ::=
[Segment-Attributes-Clause]
[[COMPRESS[<Integer>]] | NOCOMPRESS]
[OVERFLOW [Segment-Attributes-Clause]]
[LOB-Storage-Clause | Varray-Col-Properties]
[Partition-Level-Subpartition]

Lob-Storage-Clause::= LOB-Option1 | LOB-Option2
LOB-Option1 ::= LOB (<LOB-Item>{, <LOB-Item>}) STORE AS (LOB-Parameters) | <LOB-Segment-Name >
LOB-Option2 ::= LOB (<LOB-Item>{, <LOB-Item>}) STORE AS <LOB- Segment-Name> (LOB-Parameters)

LOB-Parameters ::=
TABLESPACE <Tablespace>
[ENABLE | DISABLE STORAGE IN ROW]
[Storage-Clause]
[CHUNK <Integer>] [PCTVERSION <Integer>] [RETENTION]
[FREEPOOLS <Integer>][CACHE [READS] [LOGGING | NOLOGGING]]
[NOCACHE]

LOB-Partition-Storage ::=
PARTITION <Partition-Name> LOB-Storage-Clause | Varray-Col-Properties
[(SUBPARTITION <Subpartition-Name> LOB- Storage -Clause | Varrey-Col- Properties
{LOB- Storage -Clause |Varray-Col-Properties })]

Partition-Level-Subpartition ::= PLOption1 | PLOption2
PLOption1 ::= SUBPARTITIONS <Hash-Subpartition-Quantity>
[STORE IN (<Tablespace> {, <Tablespace> })]
PLOption2 ::= (Subpartition-Spec {, Subpartition-Spec })

Subpartition- Spec ::=
SUBPARTITION [<Subpartition-Name>]
[List-Values-Clause] [Partitioning- Storage -Clause]

Partitioning- Storage-Clause ::= PSOption1 | PSOption2 | PSOption3 | PSOption4 | PSOption5
PSOption1 ::= [TABLESPACE<Tablespace>]
PSOption2 ::= [OVERFLOW [TABLESPACE <Tablespace>])
PSOption3 ::= [LOB (<LOB-Item> STORE AS (TABLESPACE <Tablespace>)]
PSOption4 ::= [VARRAY <Varray-Item> STORE AS LOB <LOB-Segment-Name>]
PSOption5 ::= [LOB (<LOB-Item>) STORE AS <LOB-Segment-Name> [(TABLESPACE<Tablespace>)]]

List-Values-Clause ::= LVOption1 | LVOption2
LVOption1 ::= VALUES (DEFAULT | NULL)
LVOption2 ::= VALUES (<Value> {, <Value>})

Hash-Partition ::=

PARTITION BY HASH (<Column> {, <Column>})
Individual- Hash-Partitions / Hash- Partitions-By- Quantity

Individual-Hash-Partitions::=
(PARTITION [<Partition-Name> Partitioning-Storage-Clause]
{,PARTITION [<Partition-Name> Partitioning-Storage-Clause}])

Hash-Partitions-by-Quantity ::=
PARTITIONS <Integer> [STORE IN (<Tablespace> {, <Tablespace>})]
[OVERFLOW STORE IN (<Tablespace>{, <Tablespace>})]
List-Partition ::=

PARTITION BY LIST (<Column>)
(PARTITION [<Partition-Name>] List-Values-Clause Table-Partition-Descr
{ , PARTITION [<Partition-Name>] List Values-Clause Table-Partition Descr})

Composite-Partition ::=
PARTITION BY RANGE (<Column>{,<Column>})
Subpartition-by-List | Subpartition-by-Hash
(PARTITION [<Partition-Name>] Range-Values-Clause Table-Partition-Descr
{, PARTITION [<Partition-Name>] Range Values Clause Table-Partition-Descr})

Subpartition-by-Hash::= SHOption1 | SHOption2
SHOption1 ::= SUBPARTITION BY HASH (<Column> {,<Column>}) Subpartition-Template
SHOption2 ::= SUBPARTITION BY HASH(<Column>{,<Column>})
SUBPARTITIONS <Integer> [STORE IN (<Tablespace> {, <Tablespace>})])

Subpartition-Template ::= STOption1 | STOption2 | STOption3 | STOption4
STOption1 ::= SUBPARTITION TEMPLATE Hash-Subpartition-by-Quantity
STOption2 ::= SUBPARTITION TEMPLATE Individual-Hash-Subparts {, Individual-Hash-Subparts}
STOption3 ::= SUBPARTITION TEMPLATE (List-Subpartition-Desc {, List-Subpartition-Desc})
STOption4 ::= SUBPARTITION TEMPLATE Range-Subpartition-Desc {, Range-Subpartition-Desc}

Hash-Subpartition-by-Quantity ::= PARTITIONS <Integer> [STORE IN (<Tablespace> {, <Tablespace>})]
Individual-Hash-Subparts ::= SUBPARTITION [Subpartition] Partitioning-Storage-Clause
List-Subpartition-Desc ::= SUBPARTITION [Subpartition] Partitioning-Storage-Clause
Range-Subpartition-Desc ::= SUBPARTITION [Subpartition] List-Values-Clause [Partitioning-Storage-Clause]

Subpartition-by-List ::=
SUBPARTITION BY LIST (<Column>)[Subpartition-Template]

Range-Values-Clause ::=
VALUES LESS THAN (<Value>|MAXVALUE {,<Value>|MAXVALUE})

IOT-Clause ::= IOTOption1 | IOTOption2 | IOTOption3 | IOTOption4
IOTOption1 ::= Mapping-Table-Clause [Index-Org-Overflow-Clause]
IOTOption2 ::= PCTHRESHOLD <Integer> [Index-Org-Overflow- Clause]
IOTOption3 ::= COMPRESS [<Integer>] [Index Org-Overflow- Clause]
IOTOption4 ::= NOCOMPRESS [Index-Org-Overflow- Clause]

Mapping-Table-Clause ::= NOMAPPING | MappingOption
MappingOption ::= MAPPING TABLE

Index-Org-Overflow-Clause ::=
[INCLUDING <Column>] OVERFLOW [Segment-Attributes-Clause]

Object-Type-Col-Properties ::=
COLUMN <Column> Substitutable-Column-Clause

Substitutable-Column-Clause ::= SCOption1 | SCOption2
SCOption1 ::= [ELEMENT] IS OF [TYPE] (ONLY <Type>)
SCOption2 ::= [NOT] SUBSTITUTABLE AT ALL LEVELS

Nested-Table-Col-Properties ::=
NESTED TABLE <Nested-Item >|COLUMN-VALUE [Substitutable-Column-Clause]
STORE AS <Storage-Table> [((Object-Properties) [Physical-Properties] [Column-Properties])]
[RETURN AS LOCATOR | VALUE]

/* Note:
1. Object-properties are the same as table-properties for relational tables. However, instead of specifying column, you specify attributes of the object.
2. <Nested-item> represents the users supplied column-name (or top level attribute of the table's object type) whose type is nested table.
3. <Storage-Table>represents the user-supplied table where rows of the nested-item reside. */

Varray-Col-Properties ::= VPOption1 | VPOption2 | VPOption3
VPOption1 ::= VARRAY <Varray-Item> Subtitutable–Column-clause
VPOption2 ::= VARRAY < Varray-Item> STORE AS LOB <LOB-Segment-Name>
VPOption3 ::= VARRAY < Varray-Item> STORE AS [<LOB-Segment-Name>] (LOB-Parameters)

/* Note:
1. <LOB-Segment-Name> represents the user-supplied name for the LOB data segment.
2. It assumes one LOB-item only.
3. <Varray-Item> represents the user-supplied name of variable array. */

XML-Type-Col-Properties ::=
XMLTYPE [COLUMN] <Column>[XML-Type-Storage] [XML-Schema-Spec]

XML-Type-Storage ::= XTOption1 | XTOption2 | XTOption3
XTOption1 ::= STORE AS OBLJECT RELATIONAL
XTOption2 ::= STORE AS CLOB [LOB-Parameters]
XTOption3 ::= STORE AS CLOB [<LOB-Segment-Name> [(LOB-Parameters)]
XML-Schema-Spec ::= XSOption1 | XSOption2
XSOption1 ::= [XMLSCHEMA <XML-Schema-URL>] ELEMENT <Element>
XSOption2 ::= [XMLSCHEMA <XML-Schema-URL>] ELEMENT <XML-Schema-URL> # <Element>

/* **Note:**
<XML-Schema-URL>represents the user-supplied URL of a registered XML schema (URL schema must have been registered via package DBMS-XMLSCHEMA)
<Element> represents the user-supplied element name. */

Create-Object-Table ::=
CREATE [GLOBAL TEMPORARY] TABLE [<SCHEMA>·]
<TableName> OF [<Schema>·] <Object-Type>
[Object-Table-Substitution]
[(Object-Properties)] [ON COMMIT DELETE | PRESERVE ROWS]
[OID-Clause] [OID-Index-Clause]
[Physical-Properties] [Table-Properties];

Object-Table-Substitution ::= [NOT] SUBSTITUTABLE AT ALL LEVELS

Object-Properties ::= OPOption1 | OPOption2
OPOption1 ::= <Column> | <Attribute> [DEFAULT<Expression>]
[Inline-Constraint {Inline-Constraint}] | [Inline-Ref-Constraint]
OPOption2 ::= Out-of-line Constraint | Out-of-line-Ref-Constraint | Supplemental-Logging-Props

Supplemental-Logging-Pops ::=
SUPPLEMENTAL LOG GROUP<Log-Group-Number>(<Column>{,<Column>}) [ALWAYS]

OID-Clause ::= OCOption1 | OCOption2
OCOption1 ::= OBJECT IDENTIFIER IS SYSTEM GENERATED
OCOption2 ::= OBJECT IDENTIFIER IS PRIMARY KEY

OID-Index-Clause ::= OIOption1 | OIOption2
OIOption1 ::= OIDINDEX [<IndexName>] (Physical-Attributes-Clause)
OIOption2 ::= OIDINDEX [<IndexName>] (TABLESPACE <Tablespace>)

Create- XML-Table ::=
CREATE TABLE [<Schema>·] <TableName> OF XMLTYPE
[XMLTYPE XML-Type-Storage] [XML-Schema-Spec]

XML-Schema-Spec ::=
[XMLSCHEMA <XML-Schema-URL>]
ELEMENT [<XML-Schema-URL> #] <XML-Element>
/* **Note:** The XML-Element is specified as a string in double quotes. */

A4.3.2 Syntax for Alter-Table Statement

Alter-Table ::=
ALTER TABLE [<Schema>.] <Table-Name>
[Alter-Table-Props | Column-Clauses | Constraint-Clauses | Alter-Table-Partitioning | Alter-External-Table-Clauses |
Move-Table-Clause]
[Enable-Disable-Clause] | [ENABLE | DISABLE TABLE LOCK] | [ENABLE | DISABLE ALL TRIGGERS];

Alter-Table-Props::=
Physical-Attributes-Clause | LOGGING | NOLOGGING | COMPRESS | NOCOMPRESS | Supplemental-log-Group-Clauses | Allocate-Extent-Clause | Deallocate-Unused-Clause | CACHE | NOCACHE | MONITORING | NOMONITORING |
Upgrade-Table-Clause | Record-Per-Block clause | Row-Movement-Clause | Parallel-Clause | Renew-Table-Spec
[Alter-IOT-Clauses]

Supplemental-Log-Group-Clauses ::= SLClause1 | SLClause2
SLClause1 ::= ADD SUPPLEMENTAL LOG GROUP <LogGroup> (<Column> {,<Column>}) [ALWAYS]
SLClause2 ::= DROP SUPPLEMTAL LOG GROUP <LogGroup>

Allocate-Extent-Clauses::=
ALLOCATE EXTENT [(Extent-Spec {Extent-Spec})]

Extent-Spec ::= ESOption1 | ESOption2 | ESOption3
ESOption1 ::= Size [K|M]
ESOption2 ::= DATAFILE '<PathtoFile>'
ESOption3 ::= INSTANCE <n>

Deallocate-Unused-Clause::= DEALLOCATE UNUSED [KEEP <n> [K|M]

Upgrade-Table-Clause::= UPGRADE [[NOT] INCLUDING DATA] [Column-Properties]

Record Per-Block-Clause::= MINIMIZE | NOMINIMIZE RECORD PER BLOCK

Rename-Table-Spec::= Rename to <New-TableName>

Alter-IOT-Clauses ::= IOT-Clause | Alter-Overflow-Clause | Alter-Mapping-Table-Clause | Coalesce

// **IOT-Clause** is defined in Create-Table

Alter-Overflow-Clauses ::= AOClause1 | AOClause2
AOClause1 ::= Add-Overflow-Clause
AOClause2 ::= OVERFLOW {Allocate-Extent-Clause} | Deallocate-Unused-Clause

Add-Overflow-Clause ::=
ADD OVERFLOW [Segment-Attributes-Clause]
[(PARTITION [Segment-Attributes-Clause] {, PARTITION [Segment-Attributes-Clause]})]

// Segment-Attributes-Clause is defined in **Create-Table**

Alter-Mapping-Table-Clauses::= AMClause1 | AMClause2
MClause1 ::= MAPPING TABLE UPDATE BLOCK REFERENCES
MClause2 ::= MAPPING TABLE Allocate-Extent-Clause | Deallocate-unused-Clause

Column-Clauses ::=
{Add-Column-Clause}| {Modify-Column-Clause} | {Drop-Column-Clause}|
Rename-Column-Clause | {Modify-Collection-Retrieval} |
{Modify-LOB-Storage-Clauses} | {Alter-Varray-Col_Props}

Add-Column-Clause ::=
ADD (Column-Definition-Clause -{,Column-Definition-Clause})

// Column-Definition-Clause is defined in Create-Table

Modify-Column-Clause ::=
MODIFY Modify-Col-Props | Modify-Col-Substitutable

Modify-Col-Props ::= /*Similar but not identical to Column Definition*/
(<Column> [Data-type-Spec] [DEFAULT <Expression>] {Inline-Constraint}
{, <Column> [Data-type-Spec] [DEFAULT <Expression>] {Inline-Constraint})

// Inline-Constraints defined in **Create-Table**

Modify-Col-Substitutable ::=
COLUMN <Column> [NOT] SUBSTITUTABLE AT ALL LEVELS [FORCE]

Drop-Column-Clause ::= DCOption1 | DCOption2 | DCOption3 | DCOption4 | DCOption5 | DCOption6
DCOption1 ::= SET UNUSED COLUMN <Column> [[CASCADE CONSTRAINTS] | INVALIDATE]
DCOption2 ::= SET UNUSED (<Column> {,<Column>}) [[CASCADE CONSTRAINTS] | INVALIDATE]
{[[CASCADE CONSTRAINTS] | INVALIDATE]}
DCOption3 ::= DROP COLUMN <Column> [[CASCADE CONSTRAINTS] | INVALIDATE] [CHECKPOINT <n>]
DCOption4 ::= DROP COLUMN (<Column> {,<Column>} [[CASCADE CONSTRAINTS] | INVALIDATE]
{[[CASCADE CONSTRAINTS] | INVALIDATE]} [CHECKPOINT <n>]
DCOption5 ::= DROP UNUSED COLUMNS [CHECKPOINT <n>]
DCOption6 ::= DROP COLUMNS CONTINUE [CHECKPOINT <n>]

Rename-Column-Clause ::=
RENAME COLUMN <Old-Column> TO <New-Column>

Modify-Collection-Retrieval ::=
MODIFY NESTED TABLE <Collection-Item> RETURN AS LOCATOR | VALUE

Modify-LOB-Storage-Clause ::=
MODIFY LOB (<LOB-Item>) (Modify-LOB-Parms)

Modify-LOB-Parms ::= MLParm1 | MLParm2 | MLParm3 | MLParm4 | MLParm5 | MLParm6
MLParm1 ::= Storage-Clause | Allocate-Extent-Clause | Deallocate-Unused-Clause | RETENTION | CACHE
MLParm2 ::= PCTVERSION <n>
MLParm3 ::= FREEPOOLS <n>
MLParm4 ::= REBUILD FREEPOOLS
MLParm5 ::= NOCACHE [LOGGING | NOLOGGING]
MLParm6 ::= CACHE READS [LOGGING / NOLOGGING]

Alter-Varray-Col-Props ::= MODIFY-VARRAY <Varray-Item> (Modify-LOB-Parms)
Constraints-Clauses ::= Drop-Constraint-Clause | ACClause1 | ACClause2 | ACClause3 | ACClause4 | ACClause5 | ACClause6
ACClause1 ::= ADD Out-of-Line-Constraint <Out-of-Line-Constraint >
ACClause2 ::= ADD Out-of-Line-Ref-Constraint
ACClause3 ::= MODIFY CONSTAINT <Constraint-Name> <Constraint-State>
ACClause4 ::= MODIFY PRIMARY KEY <Constraint-Name>
ACClause5 ::= MODIFY UNIQUE (<Column> {,<Column>}) <Constraint-State>
ACClause6 ::= RENAME CONSTRAINT <Old-Constraint> To <New-Constraint>

/***Constraint-State**, **Out-of-Line-Constraint**, **Out-of-Line-Ref-Constraints** are as defined in Create-Table*/

Drop-Constraint-Clause::= DCOption1 | DCOption2 | DCOption3
DCOption1 ::= DROP PRIMARY KEY [CASCADE] [KEEP | DROP INDEX]
DCOption2 ::= DROP UNIQUE (<Column> {,<column>}) [CASCADE] [KEEP | DROP INDEX]
DCOption3 ::= DROP CONSTRAINT <Constraint> [CASCADE]

Alter-Table-Partitioning::=
Modify-Table-Default-Attributes | Set-Subpartition-Template | Modify-Table-Partition | Modify-Table-Subpartition | Move-Table-Partition | Move-Table-Subpartition | Add-Table-Partition | Coalesce-Table-Partition | Drop-Table-Partition | Drop-Table-Subpartition | Rename-Partition-Subpart | Truncate-Partition-Subpart |Split-Table-Partition | Split-Table-Subpartition | Merge-Table-Partitions | Exchange-Partition-Subpart | Merge-Table-Subpartition.

Modify-Table-Defaults-Attributes::=
MODIFY DEFAULT ATTRIBUTES [FOR PARTITION <Partition>]
[Segment-Attributes-Clause] [COMPRESS | NOCOMPRESS] [PTCTHRESHOLD <n>] [[COMPRESS <n>] |
NOCOMPRESS] [Alter-Overflow-Clause] [LOB {(<LOB-Item>) (<LOB-Parameters>)} {VARRAY <Varray> (<LOB-Parameters>)}]

// **Segment-Attributes-Clause** and **LOB-Parameters** are defined in **Create-Table**

Set-Subpartition-Template::= SSTOption1 | SSTOption2 | SSTOption3
SSTOption1 ::= SET SUBPARTITION TEMPLATE Hash-Subpartition-by-Quantity
SSTOption2 ::= SET SUBPARTITION TEMPLATE (Range-Subpartition-Desc {, Range-Subpartition-Desc }}
SSTOption3 ::= SET SUBPARTITION TEMPLATE (List-Subpartition-Desc {, List-Subpartition-Desc }}

/***Hash-Subpartition-by-Quantity**, **Range-Subpartition-Desc**, and **List-Subpartition-Desc** are defined in Create-Table */

Modify-Table-Partition ::=
Modify-Range-Partition | Modify-Hash-Partition | Modify-List-Partition

Modify-Range-Partition ::=
MODIFY PARTITION <Partition> Partition-Attributes | Add-Hash-Subpartition | Add-List-Subpartition | Alter-Mapping-Table-Clause | Coalesce-Subpartition | Rebuild-Unusable

Coalesce-Subpartition ::= COALESCE SUBPARTION [Update-Global-Index-Clause] [Parallel-Clause]

// **Parallel-Clause** is defined in Create-Table

Update-Global-Index-Clause ::= UPDATE | INVALIDATE GLOBAL INDEXES

Rebuild-Unusable ::= [REBUILD] UNUSABLE LOCAL INDEXES

Partition-Attributes ::=
[{Physical-Attributes-Clause | LOGGING | NOLOGGING | Allocate-Extent-Clause |
Deallocate-Unused-Clause}]
[OVERFLOW {Physical-Attributes-Clause | LOGGIN |NOLOGGING | Allocate-Extent-Clause |
Deallocate-Unused-Clause}]
[COMPRESS | NOCOMPRESS] [LOB-Varray-Option]

LOB-Varray-Option ::= LOBOption1 | LOBOption2
LOBOption1 ::= {LOB <LOB-Item> Modify-LOB-Parms}
LOBOption2 ::= VARRAY <Varray> Modify-LOB-Parms

Add-Hash-Subpartition ::=
ADD Subpartition-Specs [Update-Global-Index-Clause] [Parallel-Clause]

// **Parallel-Clause** and **Subpartition**-Specs are defined in Create-Table

Add-List-Subpartition ::= ADD Subpartition-Specs

Alter-Mapping-Table-Clause::=
MAPPING TABLE Allocate-Extent-Clause | Deallocate-Unused-Clause | Update-Block-Refs

Update-Block-Refs ::= UPDATE BLOCK REFERENCES
Modify-Hash-Partition ::=
MODIFY PARTITION <Partition> Partition-Attributes | Rebuild-Unusable |
Alter-Mapping-Table-Clause

Modify-List-Partition ::=
MODIFY PARTITION <Partition> Partition-Attributes | Rebuild-Unusable | Add-Drop-Values

Add-Drop-Values ::= ADD | DROP VALUES (<Partition-Value> {,<Partition-Value})

Modify-Table-Subpartition ::=
MODIFY SUBPARTITION <Subpartition> Modify-Hash-Subpartition | Modify-List-Subpartition

Modify-Hash-Subpartition ::=
Rebuild-Unusable | {LOB-Varray-Option} | Allocate-Extent-Clause | Deallocate-Unused-Clause

Modify-List-Subpartition ::=
Rebuild-Unusable | {LOB-Varray-Option} | Allocate-Extent-Clause | Deallocate-Unused-Clause |
Add-Drop-Subvalues

Add-Drop-Subvalues ::= ADD | DROP VALUES (<Value> {<Value>})

Move-Table-Partition::=
MOVE PARTITION <Partition> [MAPPING TABLE] [Table-Partition-Description]
[Update-Global-Index-Clause] [Parallel-Clause]

Table-Partition-Description ::=
[Segment-Attributes-Clause] | [[COMPRESS [<Integer>]] | NOCOMPRESS]
[Overflow [Segment-Attribute-Clause]]
[{LOB-Storage-Clause | Varray-Col-Properties}] [Partition-Level-Subpartition]
[Partition-Level-Subpartition]

Move-Table-Subpartition ::=
MOVE Subpartition-Spec [Update-Global-Index-Clause] [Parallel-Clause]

Add-Table-Partition ::=
ADD-Range-Partition-Clause | Add-Hash-Partition-Clause | Add-List-Partition-Clause

Add-Range-Partition-Clause ::=
ADD PARTITION [<Partition>] Range-Value-Clause [Table-Partition-Description]

// **Range-Value-Clause** defined in Create-Table
Add-Hash-Partition-Clause ::=
ADD PARTITION [<Partition>] Partitioning-Storage-Clause [Update-Global-Index-Clause] [Parallel-Clause]

// **Partitioning-Storage-Clause** and **Parallel-Clause** defined in **Create-Table**

Add-List-Partition-Clause ::=
ADD PARTITION [<Partition>] List-Value-Clause [Table-Partition-Description]

Coalesce-Table-Partition ::=
COALESCE PARTITION [Update-Global-Index-Clause] [Parallel-Clause]

Drop-Table-Partition ::=
DROP PARTITION <Partition> Update-Global-Index-Clause [Parallel-Clause]

Drop-Table-Subpartition ::=
DROP SUBPARTITION <Subpartition> [Update-Global-Index-Clause [Parallel-Clause]

Rename-Partition-Subpartition ::=
RENAME PARTITION | SUBPARTITION <Oldname> TO <Newname>
Truncate-Partition-Subpartition ::=
TRUNCATE PARTITION | SUBPARTITION <Partition | Subpartition>
[DROP | REUSE STORAGE] [Update-Global-Index-Clause [Parallel-Clause]]

Split-Table-Partition ::=
SPLIT PARTITION <Partition> AT | VALUES (<Value> {,<Value>}) [INTO (Partition-spec, Partition-Spec)]
[Update-Global-Index-Clause] [Parallel-Clause]

Partition-Spec ::=
PARTITION [<Partition>] [Table-Partition-Description]

Split-Table-Subpartition ::=
SPLIT SUBPARTITION <Subpartition> VALUES (<Value> | NULL {,<Value> | NULL}] INTO (Subpartition-Spec, Subpartition-Spec) [Update-Global-Index-Clause] [Parallel-Clause]

Merge-Table-Partitions ::=
MERGE PARTITIONS <Partition-1> , < Partition-2> INTO Partition-Spec
[Update-Global-Index-Clause] [Parallel-Clause]

Subpartition_Spec ::=
SUBPARTITION [<Subpartition>] [List-values-Clause] [Partitioning-Storage-Clause]

// **List-Values-Clause** and **Partitioning-Storage-Clause** are defined in **Create-Table**

Merge-Table-Subpartition ::=
MERGE SUBPARTITIONS <Subpartition-1> , <Subpartition-2> [INTO Subpartition-Spec]
[Update-Global-Index-Clause] [Parallel-Clause]

Exchange-Partition-Subpartition ::=
EXCHANGE Partition-Extended-Name | Subpartition-Extended-Name WITH TABLE <TableName>
[INCLUDING | EXCLUDING INDEXES] [WITH | WITHOUT VALIDATION]
[Exceptions-Clause] [Update-Index-Clauses] [Parallel-Clause]

Partition-Extended-Name ::= PNOption1 | PNOption2
PNOption1 ::= PARTITION <PartitionName>
PNOption2 ::= PARTITION FOR (<Partition-Key-Value> {, Partition-Key-Value>})

Subpartition-Extended-Name ::= SPNOption1 | SPNOption2
SPNOption1 ::= SUBPARTITION <SubpartitionName>
SPNOption2 ::= SUBPARTITION FOR (<Subpartition-Key-Value> {, Subpartition-Key-Value>})

Update-Index-Clauses ::= Update-Global-Index-Clause | Update-All-Indexes-Clause
Update-Global-Index-Clause ::= UPDATE | INVALIDATE GLOBAL INDEXES
Update-All-Index-Clause ::= UPDATE INDEXES [(<IndexName> {, <IndexName>})]

Exception-Clause ::=
EXCEPTIONS INTO [<Scheme>.] <TableName>
Alter-External-Table-Clause ::=
Add-Column-Clause | Modify-Column-Clause | Drop-Column-Clause | Parallel-Clause | External-Data-Properties |
Reject-Spec {Alter-External-Table-Clause}
Reject-Spec ::= REJECT LIMIT <Integer> | UNLIMITED

// **Parallel-Clause** and **External-Data-Properties** are defined in **Create-Table**

Move-Table-Clause ::=
MOVE [ONLINE] [Segment-Attributes-Clause] [COMPRESS | NOCOMPRESS]
[IOT-Clause] {LOB-Storage-Clause | Varray-Col-Properties} [Parallel-Clause]

Move-Table-Clause ::= MTOption1 | MTOption2 | MTOption3
MTOption1 ::= ENABLE | DISABLE [VALIDATE | NOVALIDATE] UNIQUE (<Column> {,<Column>}
[Unique-Index-Clause] [Exceptions-Clause] [CASCADE] [KEEP | DROP] INDEX
MTOption2 ::= ENABLE | DISABLE [VALIDATE | NOVALIDATE]
PRIMARY KEY [Unique-Index-Clause] [Exceptions-Clause]
[CASCADE] [KEEP | DROP] INDEX
MTOption3 ::= ENABLE | DISABLE [VALIDATE | NOVALIDATE] CONSTRAINT (<Constraint>
[Unique-Index-Clause] [Exceptions-Clause] [CASCADE] [KEEP | DROP] INDEX

// Components not defined here have been defined earlier, or in **Create-Table**

A4.4 Index Management

A4.4.1 Syntax for Create-Index Statement

Create-Index ::=
CREATE [UNIQUE | BITMAP] INDEX [<Schema>.] <IndexName>
ON Cluster-Index-Clause | Table-Index-Clause | Bitmap-Join-Index-Clause;

Cluster-Index-Clause ::=
CLUSTER [<Schema>.] <Cluster> Index-Attributes

Index-Attributes ::=
[Physical-Attributes-Clause] [LOGGING | NOLOGGING] [ONLINE] [REVERSE]
[SORT | NOSORT] [Parallel-Clause] [Key-Compression] [Compute-Stat] [Tablespace-Spec]
{ Index-Attributes }

Key-Compression ::= NOCOMPRESS | Compress-Option
Compress-Option ::= COMPRESS <Integer>

Compute-Stat ::=
COMPUTE STATISTICS
Tablespace-Spec ::=
TABLESPACE <Tablespace> | DEFAULT

/* **Physical-Attributes-Clause** and **Parallel-Clause** are defined in **Create-Table** */

Table-Index-Clause ::=
[<Schema>.] <Table> [<Table-Alias>]
(Index-Expr [ASC/DESC] {,Index-Expr [ASC/DESC] })
[Global-Partitioned-Index | Local-Partitioned-Index] [Index-Attributes] [Domain-Index-Clause]

Index-Expr ::=
<Column> | <Column-Expression>

// Column expressions occur in function-based indexes

Global-Partitioned-Index ::=
GLOBAL PARTITION BY RANGE (<Column>{,<Column>})
(Index-Partitioning-Clause>

Index-Partitioning-Clause ::=
PARTITION [<Partition>] VALUES LESS THAN (<Value> {,<Value>})
[Segment-Attributes-Clause]

// **Segment-Attributes-Clause** defined in **Create-Table**

Local-Partitioned-Index ::=
LOCAL [On-Range-Partitioned-Table | On-List-Partitioned-Table | On-Hash-Partitioned-Table |
On-Comp-Partitioned-Table]

On-Range-Partitioned-Table ::=
(PARTITION [<Partition> [Segment-Attributes-Clause]]
{,PARTITION [<Partition> [Segment-Attributes-Clause]]})

// **On-List-Partition-Table** is as for **On-Range-Partition-Table**

On-Hash-Partitioned-Table ::= OHPOption1 | OHPOption2
OHPOption1 ::= STORE IN (<TableSpace> {,<TableSpace>})
OHPOption2 ::= On-Range-Partitioned-Table

On-Comp-Partitioned-Table ::=
STORE IN (<TableSpace> {,<TableSpace>})
(PARTITION [<Partition> {Segment-Attribute-Clause} [Index-Subpartition-Clause]]
{,PARTITION [<Partition> {Segment-Attributes-Clause}[Index-Subpartition-Cause]]})

Index-Subpartition-Clause ::= ISCOption1 | ISCOption2
ISCOption1 ::= STORE IN (<TableSpace>{,<TableSpace>})
ISCOption2 ::= (SUBPARTITION[<Subpartition>[TABLESPACE<Tablespace>]]
{, SUBPARTITION[<Subpartition>[TABLESPACE<TableSpace>
]]})

// **Segment-Attributes-Clause** is defined in **Create-Table**

Domain-Index-Clause ::=
INDEXTYPE IS <IndexType> [Parallel-Clause]
[PARAMETERS ('<ODCI-Parameters>')]

Bitmap-Join-Index-Clause ::=
[<Schema>·] <Table> ([[<Schema>·]<Table>·] | [<Table-Alias>·] <Column>[ASC/DESC]
{,[[<Schema>·]<Table>·] | [<Table-Alias>·]<Column>[ASC/DESC]]})
FROM [<Schema>·]<Table> [<Table-Alias>]
{,[<Schema>·]<Table> [<Table-Alias>]}
WHERE <Condition>[Local-Partitioned-Index] Index-Attributes

A4.4.2 Syntax for Alter-Index Statement

Alter-Index ::=
ALTER INDEX [<Schema>·]<IndexName>
{Deallocate-Unused-Clause | Allocate-Extent-Clause | Parallel-Clause |
Physical-Attributes-Clause | LOGGING | NOLOGGING} | Rebuild-Clause | ENABLE | DISABLE |
UNUSABLE |
COALESCE | Rename-Index | Index-Parms1 | Monitor-Usage | Updated-Block-Ref |
Alter-Index-Partitioning};

Index-Parms1 ::=
PARAMETERS ('<ODCI-Parameters>')

Rename-Index ::=
RENAME TO <NewIndexName>

Monitor-Usage ::=
MONITORING | NONMONITORING USAGE

Update-Bloc-Ref ::=
UPDATE BLOCK REFERENCES

Alter-Index-Partitioning ::=
Modify-Index-Default-Attributes | Modify-Index-Partition | Rename-Index-Partition |
Drop-Index-Partition | Split-Index-Partition | Modify-Index-Subpartition

Rebuild-Clause ::=
REBUILD
[PARTITION <PartitionName>] |
[SUBPARTITION <SubpartitionName>] |
[REVERSE | NOREVERSE]
[<Parallel-Clause>]
[TABLESPACE <TablespaceName>]
[PARAMETERS ('<ODCI-Parameters>')]
[ONLINE]
[COMPUTE STATISTICS]
[<Physical-Attributes-Clause>]
[NOCOMPRESS | COMPRESS [<n>]]
[LOGGING | NOLOGGING]

Modify-Index-Default-Attributes ::=
MODIFY DEFAULT ATTRIBUTES [FOR PARTITION <Partition>]
{Physical-Attributes-Clause | LOGGING | NOLOGGING | TableSpace-Spec}
// **Physical-Attributes-Clause** and **Parallel-Clause** are defined in **Create-Table**

Modify-Index-Partition ::=
MODIFY PARTITION <Partition>
{Physical-Attributes-Clause | LOGGING | NOLOGGING | Allocate-Extent-Clause |
Deallocate-Unused-Clause}
Index-Parms2 | COALESCE | UNUSEABLE | Update-Block-Ref

Index-Parms2 ::=
PARAMETERS ('<Alter-Partition-Parms>')

// **Physical-Attributes-Clause** is defined in **Create-Table**

// **Allocate-Extent-Clause** and **Deallocate-Unused-Clause** defined in **Alter-Table**

Rename-Index-Partition ::= RIPOption1 | RIPOption2
RIPOption1 ::= RENAME PARTITION <partition> TO <NewName>
RIPOption2 ::= RENAME SUBPARTITION <Subpartition> TO <NewName>

Drop-Index-Partition ::=
DROP PARTITION <Partition>

Split-Index-Partition ::=
SPLIT PARTITION <Partition> AT (<Value> {,<Value>})
[INTO (Index-Partition-Description, Index-Partition-Description)] [Parallel-Clause]

Index-Partition-Description ::=
PARTITION [<Partition> {[Segment-Attributes-Clause | Key-Compression]}]

// **Segment-Attributes-Clause** is defined in **Create-Table**

Modify-Index-Subpartition ::=
MODIFY SUBPARTITION <Subpartition> UNUSABLE | Allocate-Extent-Clause |
Deallocate-Unused-Clause

// **Allocate-Extent-Clause** and **Deallocate-Unused-Clause** are defined in **Alter-Table**

A4.5 Data Insertion, Update, and Deletion

A4.5.1 Abridged Syntax for the Insert Statement

```
Insert ::= Insert1 | Insert2 | Insert3

Insert1 ::=
INSERT INTO <TableName> [(<Column> {, <Column>})]
VALUES (<Literal> {, <Literal>});

Insert2 ::=
INSERT INTO <TableName> [(<Column> {, <Column>})]
VALUES ([&|:] <PgmVariable> {, [&|:] <PgmVariable>});

Insert3 ::=
INSERT INTO <TableName> [(<Column> {, <Column>})]
<Subquery>;

// Subquery represents a Select statement — to be defined later
```

A4.5.2 Abridged Syntax for the Update Statement

```
Update1 ::= Update1 | Update2 | Update3

Update1 ::=
UPDATE <TableName>
SET <Column> = <Scalar-Expression> | <Literal> {, <Column> = <Scalar-Expression> |
<Literal>}
[WHERE <Condition>];

Update2 ::=
UPDATE <TableName>
SET <Column> = [&|:]<PgmVariable> {, <Column> = [&|:]<PgmVariable>}
[WHERE <Condition>];

pdate3 ::=
UPDATE <TableName>
SET (<Column> {,<Column>}) = (<Sub-query>)
[WHERE <Condition>];
```

A4.5.3 The Delete Statement and the Truncate Statement

```
Delete ::= DELETE FROM <TableName> [WHERE <Condition>];

Truncate ::= TRUNCATE TABLE <TableName>;
```

A4.6 Transaction Management

Commit ::= COMMIT;

Rollback ::= ROLLBACK [TO SAVEPOINT <SavePointName>];

Set-Transaction ::=
SET TRANSACTION <Mode>;
Mode ::= READ ONLY | READ WRITE

Save-Point ::=
SAVEPOINT <Save-point-name>;

A4.7 Building Queries

A4.7.1 Abridged Syntax for the Select Statement

Select-Statement ::=
SELECT [<Function-Spec-Clause>] [DISTINCT] [,] <Item-List>
[INTO <Retrieval-List]
FROM <Relation-List>
[WHERE <Condition>]
[GROUP BY <Group-List>]
[HAVING <Condition>]
[ORDER BY <Order-List>]
[Start-With-Clause]
[Connect-By-Clause];

Item-List ::=
<Attribute> | <Scalar-Expression> [[AS] <Alias>] {,<Attribute> | <Scalar-Expression> [[AS] <Alias>] }

Retrieval-List ::=
:<ProgramVariable> {,:<ProgramVariable>}
// The number of variables must match the number items in the items list.

Relation-List ::=
<Table> | <Select-Statement> [<TupleVariable>] {, <Table> | <Select-Statement> [<TupleVariable>]}

Condition ::=
Comparison | NotCondition | AndCondition | OrCondition | ExistsItem

NotCondition ::= NOT <Condition>
AndCondition ::= <Condition> AND < Condition >
OrCondition ::= <Condition> OR <Condition>
ExistsItem ::= EXISTS (<Subquery>)

omparison::= Comparison1 | Comparison2 | Comparison3 | Comparison4 | Comparison5
Comparison1 ::= <Variable> <Operator> <Variable>
Comparison2 ::= <Attribute> <Operator> <Attribute>
Comparison2 ::= <Attribute> <Operator> <Literal> | <Scalar-Expression>
Comparison4 ::= <Attribute> IN (Values-List | Subquery)
Comparisin5 ::= <Attribute> BETWEEN (<Scalar-Expression1> AND <Scalar-Expression2>)
Values-List ::= <Literal> | <Scalar-Expression> {,<Literal> | <Scalar-Expression>}

Attribute ::=
<TupleVariable.Column> | <Column>

Operator ::= = | < | <= | > | >= | <> | LIKE | IN | BETWEEN
Order-List::= <Column> [ASC/DESC] {,<Column> [ASC/DESC]}
Group-List::= <Column> {,<Column>}

Start-With-Clause ::=
START WITH <Condition> CONNECT BY [NOCYCLE} <Condition>
Connect-By-Clause ::=
CONNECT BY [NOCYCLE] <Condition> [START WITH <Condition>]

A4.7.2 Modified From-Clause for ANSI Join

From-Clause ::= From1 | From2 | From3 | From4 | From5

From1 ::= FROM <Table> NATURAL [INNER] JOIN <Table>
{NATURAL [INNER] JOIN <Table>}

From2 ::= FROM <Table> [INNER] JOIN <Table> USING <Column-list>
{[INNER] JOIN <Table> USING <Column-list>}

From3 ::= FROM <Table> [Alias] [INNER] JOIN <Table> [Alias] ON <Condition>
{[INNER] JOIN <Table> [Alias] ON <Condition>}

From4 ::= FROM <Table> CROSS JOIN <Table>

From5 ::= FROM <Table> [Alias] [NATURAL] LEFT|RIGHT|FULL [OUTER] JOIN <Table> [Alias] [ON <Condition>]
{[NATURAL] LEFT|RIGHT|FULL [OUTER] JOIN <Table> [Alias] [ON <Condition>]}

A4.7.3 Some Commonly Used SQL Row Functions

Function	*Explanation*
NVL(<Scalar-Expr1>, <Scalar-Expr2>)	Replaces null values in the specified expression 1 with the value of Expression 2
NVL2(<Scalar-Expr1>, <Scalar-Expr2>, <Scalar-Expr2>)	If Expression 1 is non-null, Expression 2 is returned; otherwise, Expression 3 is returned.
COALESCE(<Scalar-Expr> {,<Scalar-Expr>})	Returns the first non-null expression from the list; returns null if each expression in the list is null.
CONCAT(<Scalar-Expr1>, <Scaler-Expr2>)	Concatenates the two expressions specified. The expressions must be alphanumeric — of data-type CHAR, NCHAR, VARCHAR2, NVARCHAR2, CLOB, or NCLOB.
LOWER(<Scalar-Expr>)	Converts the alphanumeric argument to lower case
UPPER(<Scalar-Expr>)	Converts the alphanumeric argument to upper case
INITCAP(<Scalar-Expr>)	Converts the first character of each word in the alphabetic (CHAR, NCHAR, VARCHAR2, or NVARCHAR2) argument to upper case
LPAD(<Scalar-Expr1>, <n> [, <Scaler-Expr2>])	Returns Expression 1 left-padded up to length n with characters from Expression 2 (repeated as often as required); if Expression 2 is not specified, the default character is a single space. The expressions must be alphanumeric.
RPAD(<Scalar-Expr1>, <n> [, <Scaler-Expr2>])	Returns Expression 1 right-padded up to length n with characters from Expression 2 (repeated as often as required); if Expression 2 is not specified, the default character is a single space. The expressions must be alphanumeric.
LTRIM(<Scalar-Expr1> [, <Scaler-Expr2>])	Removes from the left of Expression 1 all characters of Expression 2; if Expression 2 is not specified, the default character is a single space. Both expressions must be alphanumeric.
LTRIM(<Scalar-Expr1> [, <Scaler-Expr2>])	Similar to LTRIM except that trimming occurs on the right.
TRIM([[LEADING|TRAILING| BOTH] [<Scalar-Expr2>] FROM] <Scalar-Expr1>)	Removes from the left, right, or both ends of Expression 1 all characters of Expression 2; if Expression 2 is not specified, the default character is a single space. Both expressions must be alphanumeric. The default trim position is BOTH.

(Continued)

Function	*Explanation*
SUBSTR(<Scalar-Expr>, <Start>,[<Length>])	Returns a substring of the specified length, starting at the start position. If the length is omitted, all character from **Start** to the end of the string will be returned. **Scalar-Expr** must be alphanumeric.
LENGTH(<Scalar-Expr>)	Returns the length of the alphanumeric string specified.
ABS(<Scalar-Expr>)	Returns the absolute value of the numeric argument or non-numeric argument that can be implicitly converted to numeric.
MOD(<Scalar-Expr1 >, <Scalar-Expr2>)	Returns the remainder of expression 1 divided by Expression 2; if Expression 2 is 0, then Expression 1 is returned.
CEIL(<Scalar-Expr>)	Rounds the numeric argument up to the nearest integer.
FLOOR(<Scalar-Expr>)	Rounds the numeric argument down to the nearest integer.
ROUND(<Scalar-Expr>, [<Precision>])	Rounds the numeric argument to the specified precision. If the precision is negative, round to the left of the decimal point. The default precision is 0 decimal places.
TRUNC(<Scalar-Expr>, [<Precision>])	Truncates to the specified precision. If the precision is negative, truncate to the left of the decimal point. The default precision is 0 decimal places.
GREATEST(<Scalar-Expr>{,<Scalar-Expr>})	Returns the highest value from a list of arguments; arguments are assumed to be of the same data-type; the first item is used to determine the data-type of the others and the return type.
LEAST(<Scalar-Expr>{,<Scalar-Expr>})	Similar to GREATEST but returns the lowest value of the list.
SQRT(<Scalar-Expr>)	Returns the square root of the numeric argument (or argument that can be converted to numeric form).
VSIZE(<Scalar-Expr>)	Returns the size in bytes of the argument.
DECODE(<Scalar-Expr>, <Search>, <Result> {,<Search> , <Result>})	Replaces the search argument with the result for the specified column or expression.
POWER(<Scalar-Expr1>, <Scalar-Expr2>)	Returns **Scalar-Expr1** raised to the power of **Scalar-Expr2**. The arguments must be numeric; moreover, if **Scalar-Expr1** is negative then **Scalar-Expr2** must be an integer.

A4.7.4 Some SQL Date Manipulation Functions

Function	*Explanation*
MONTHS_BETWEEN(<Date1>, <Date2>)	Returns the number of months between two dates.
ADD_MONTHS(<Date>, <Months>)	Returns a new date after adding a specified number of months.
NEXT_DAY(<Date>)	Returns the new date after a specified date.
LAST_DAY(<Date>)	Returns the last day of the month specified.
ROUND(<Date>, [,<Format>])	Truncates the date given to the nearest day, month or year, depending on the format (which is 'DAY' or 'MONTH' or 'YEAR').
TRUNC(<Date>, [,<Format>])	Truncates the date given to the nearest day, month or year, depending on the format (which is 'DAY' or 'MONTH' or 'YEAR').
EXTRACT(<DateField)> FROM <DateTime-Expr> | <Internal-Expr>)	Extracts and returns the value for the specified date-field from the stated date-time expression or internal expression — expression based on the DATE orTIMESTAMP data-type (review Figure 11.11 of the previous chapter). #
NEW_TIME(<Date>, <Zone1>, <Zone2>)	Converts a date and time in time zone **Zone1** to date and time in time zone **Zone2**. Time formats are in the form XST or XDT (S for Standard, D for Daylight saving) with two exceptions: GMT and there is no NDT for Newfoundland.
CURRENT_DATE	Returns the current system date; typically treated as a pseudo-column.
SYSDATE	An alternative to CURRENT_DATE; similar behavior.
# Date-Field ::=[YEAR | MONTH | DAY | HOUR | MINUTE | SECOND][TIMEZONE_HOUR | TIMEZONE_MINUTE][TIMEZONE_REGION | TIMEZONE_ABBR]	

A4.7.5 SQL Commonly Used Data Conversion Functions

TO_CHAR(<Scalar-Expr>[,<Format>])
/* Converts a number or date to a VARCHAR2 value, based on the format (specified in single quotes) */

TO_NUMBER(<Scalar-Expr> [,<Format>])
/* Converts a string with valid digits to number, based on the format specified */

TO_DATE(<Scalar-Expr>[,<Format>])
/* Converts a string to date, based on the format specified (default DD-MON-YY) */

CAST (<Scalar-Expr> | <Multiset-Spec>) AS <DataType>

Multiset-Spec ::=
MULTISET (<Subquery>)
/* Cast facilitates data belonging to one data-type to be construed as data of another data-type. */
/* The effect is typically on the result-set, and not the source. */

A4.7.6 Some Valid Date and Numeric Formats

Valid Date Formats	
Date Format	**Clarification**
YYYY	Four digit year
Y, YY, or YYY	Last 1, 2, or 3 digits of year
YEAR	Year spelled out
Q	Quarter of the year
MM	Two digits for month
MON	First three characters of month
MONTH	Month spelled out
WW or W	Week number of year or month
DDD, DD, or D	Day of year, month, or week
DAY	Day spelled out
DY	Three-letter abbreviation of day
DDTH	Ordinal number of day, e.g., 7th
HH, HH12, HH24	Hour of day, hour (0–12) or hour (0–23)
MI	Minute (0–59)
SS	Second (0–59)
Valid Numeric Formats	
Numeric Format	**Clarification**
9	Number of 9s determine width
0	Displays leading zeros
$	Displays floating dollar sign
L	Displays floating local currency
.	Displays decimal point
,	Displays thousand indicator as specified
PR	Displays negative numbers in parentheses

A4.7.7 Commonly Used SQL Aggregate Functions

Function	*Explanation*
AVG([DISTINCT | ALL] <Scalar-Expr>) [OVER (<Analytic-Clause)]	Returns average of tuple values for the expression specified. The column-related expression must be numeric or convertible to numeric form.
COUNT([DISTINCT | ALL |*] <Scalar-Expr>)[OVER (<Analytic-Clause)]	Returns number of tuples spanned by the query.
SUM([DISTINCT | ALL] <Scalar-Expr>) [OVER (<Analytic-Clause)]	Returns total of tuple values for the expression specified. The column-related expression must be numeric or convertible to numeric form.
MAX([DISTINCT | ALL] <Scalar-Expr>) [OVER (<Analytic-Clause)]	Returns largest tuple value for the expression specified. The expression is typically column related.
MIN([DISTINCT | ALL] <Scalar-Expr>) [OVER (<Analytic-Clause)]	Returns smallest tuple value for the expression specified. The expression is typically column related.
STDDEV([DISTINCT | ALL] <Scalar-Expr>) [OVER (<Analytic-Clause)]	Returns standard deviation of tuple values for the expression specified. The column-related expression must be numeric.
VARIANCE([DISTINCT | ALL] <Scalar-Expr>)[OVER (<Analytic-Clause)]	Returns variance (which is the square of the standard deviation) of tuple values for the expression specified. The column-related expression must be numeric.
Note: Each of the functions in this list may be used as an aggregation function or an analytics function; hence, the **Analytic-Clause** which will be discussed in the upcoming subsection.	

A4.7.8 Syntax for Using Analytic Function

Analytic-Function-Spec ::=
<AnalyticFunction> ([<Argument(s)>]) OVER (<Analytic-Clause>)

Analytic-Clause ::=
[<Query-Partitioning-Clause>]
[<Order-By-Clause> [<Windowing-Clause>]]

Query-Partitioning-Clause ::=
PARTITION BY <Scalar-Expr> {,<Scalar-Expr>} /* Or */
PARTITION BY (<Scalar-Expr> {,<Scalar-Expr>})

// The **Order-By-Clause** is as previously defined in figure 12-5
// The scalar expression specified is typically a column-based expression
// The **Windowing-Clause** has been omitted; see the Oracle Product Documentation

A4.7.9 Syntax for Nested Queries

SELECT <Item-List> FROM <Relation-List> WHERE <Column> | <Scalar-Expression>
<Boolean-Operator> (<Subquery>);
/* Or */
SELECT <Item-List> FROM <Relation-List> [WHERE <Condition>];

A4.8 Managing Logical Views

A4.8.1 Creating the Logical View

Create-View ::=
CREATE [OR REPLACE] [FORCE/NOFORCE] VIEW [<Schema>.] <ViewName>
[Column-Spec | Object-View-Clause | XML-Type-View-Clause | View-Constraint-Spec]
AS <Sub-query>
[WITH CHECK OPTION [CONSTRAINT <Constraint-name>]] | [WITH READ ONLY];

View-Constraint-Spec ::= VCOption1 | VCOption2
VCOption1 ::= (Out-of-line-Constraint)
VCOption2 ::= (<Alias> Inline-Constraint {, Inline-Constraint} {, <Alias> Inline-Constraint {, Inline-Constraint})

// **Inline-Constraint** and **Out-of-line-Constraint** are defined in **Create-Table** (review chapter11)

Object-View-Clause ::= OVOption1 | OVOption2
OVOption1 ::= OF [<Schema>.] <TypeName> UNDER [<Schema>.] <Super-View>
(Out-of-line-Constraint | [<Attribute> Inline-Constraint]
{, Out-of-line-Constraint | [<Attribute> Inline-Constraint]})
OVOption2 ::= OF [<Schema>.] <TypeName> WITH OBJECT IDENTIFIER DEFAULT |
[<Attribute> {, <Attribute>}]
(Out-of-line-Constraint | [<Attribute> Inline-Constraint]
{, Out-of-line-Constraint | [<Attribute> Inline-Constraint]})

XML-Type-View-Clause ::=
OF XMLTYPE XML-Schema-Spec
WITH OBJECT IDENTIFIER DEFAULT |[(<Expression> {, <Expression>})]

XML-Schema-Spec ::=
[XMLSCHEMA <XML-Schema-URL>] ELEMENT [<XML-Schema-URL> #] <Element>

Column-Spec ::=
(<Column> {, <Column>}

A4.8.2 Altering or Dropping the Logical View

Alter-View ::=
ALTER VIEW [<Schema>.] <ViewName>
[ADD Constraint-Clause] |
[MODIFY CONSTRAINT <Constraint> RELY | NORELY] |
[DROP CONSTRAINT <Constraint>] |
[DROP PRIMARY KEY] |
[DROP UNIQUE (<Column> {, <Column}] | COMPILE;

// **Constraints-Clause** is as defined in **Create-Table** (review chapter 11)

Drop-View ::=
DROP VIEW [<Schema>.] <ViewName> [CASCADE CONSTRAINTS];

A4.9 Managing System Security

A4.9.1 Syntax for the Create-Profile Statement

Create-Profile ::=
CREATE PROFILE <ProfileName> LIMIT Resource-Parms | Password-Parms {Resource-Parms | Password-Parms};

Resource-Parms :=
[SESSIONS_PER_USER <n> | UNLIMITED | DEFAULT] /* No. of concurrent sessions per user */
[CPU_PER_SESSION <n> | UNLIMITED | DEFAULT] /* CPU time limit for a session, expressed in hundredth of seconds*/
[CPU_PER_CALL <n> | UNLIMITED | DEFAULT] /* CPU time limit for a call (parse, execute or fetch), expressed in hundredth of seconds*/
[CONNECT_TIME <n> | UNLIMITED | DEFAULT] /* The total elapsed time limit for a session, expressed in minutes */
[IDLE_TIME <n> | UNLIMITED | DEFAULT] /* Inactive time during a session, expressed in minutes */
[LOGICAL_READS_PER_SESSION <n> | UNLIMITED | DEFAULT]
[LOGICAL_READS_PER_CALL <n> | UNLIMITED | DEFAULT]
[COMPOSITE_LIMIT <n> | UNLIMITED | DEFAULT] /* The total resource cost for a session, expressed in **service units**. Oracle calculates the total service units as a weighted sum of CPU_PER_SESSION, CONNECT_TIME, LOGICAL_READS_PER_SESSION, and PRIVATE_SGA. */
[PRIVATE_SGA [<n> [K|M]] | UNLIMITED | DEFAULT]

Password-Parms :=
[FAILED_LOGIN_ATTEMPTS <Expression> | UNLIMITED | DEFAULT] /* the number of failed attempts to log in to the user account before the account is locked */
[PASSWORD_LIFE_TIME <Expression> | UNLIMITED | DEFAULT] /* The number of days the same password can be used for authentication. The password expires if it is not changed within this period */
[PASSWORD_REUSE_TIME <Expression> | UNLIMITED | DEFAULT] /* The number of days before which a password cannot be reused */
[PASSWORD_REUSE_MAX <Expression> | UNLIMITED | DEFAULT] /* The number of password changes required before the current password can be reused*/
[PASSWORD_LOCK_TIME <Expression> | UNLIMITED | DEFAULT] /* The number of days an account will be locked after the specified number of consecutive failed login at */
[PASSWORD_GRACE_TIME <Expression> | UNLIMITED | DEFAULT] /* The number of days after the grace period beginsduring which a warning is issued and login is allowed. The password expires after the grace period. */
[PASSWORD_VERIFY_FUNCTION <Function> | NULL | DEFAULT] /* Function to verify password */

A4.9.2 Syntax for Altering or Dropping a Profile

Alter-Profile ::=
ALTER PROFILE <ProfileName> LIMIT Resource-Parms | Password-Parms {Resource-Parms | Password-Parms}

Drop-Profile ::=
DROP PROFILE <ProfileName> [CASCADE];

A4.9.3 Syntax for Creating User Account(s)

```
Create-User ::= CUOption1 | CUOption2 | CUOption3
CUOption1 ::= CREATE USER <UserName> IDENTIFIED BY <Password> User-Spec;
CUOption2 ::= CREATE USER <UserName> IDENTIFIED EXTERNALLY User-Spec;
CUOption3 ::= CREATE USER <UserName> IDENTIFIED GLOBALLY AS '<ExternalName>'
              User-Spec;

User-Spec ::=
[DEFAULT TABLESPACE <Tablespace>]
[TEMPORARY TABLESPACE <Tablespace>]
[QUOTA <n> [K|M] ON <Tablespace>]
[QUOTA UNLIMITED ON <Tablespace>]
[PROFILE <Profile>]
[PASSWORD EXPIRE]
[ACCOUNT LOCK | UNLOCK]
```

A4.9.4 Syntax for Altering or Dropping User Account(s)

```
Alter-User ::= AUOption1 | AUOption2 | AUOption3
AUOption1 ::= ALTER USER <UserName> IDENTIFIED BY <Password>
                [REPLACE <OldPassword>] Alter-User-Spec;
AUOption2 ::= ALTER USER <UserName> IDENTIFIED EXTERNALLY Alter-User-Spec;
AUOption3 ::= ALTER USER <UserName>
                IDENTIFIED GLOBALLY AS '<ExternalName>' Alter-User-Spec;

Alter-User-Spec ::=
[DEFAULT TABLESPACE <Tablespace>]
[TEMPORARY TABLESPACE <Tablespace>]
[QUOTA <n> [K|M] ON <Tablespace>]
[QUOTA UNLIMITED ON <Tablespace>]
[DEFAULT ROLE [ALL EXCEPT] <Role> {,<Role>}]
[DEFAULT ROLE NONE]
[PROFILE <Profile>]
[PASSWORD EXPIRE]
[ACCOUNT LOCK | UNLOCK]
[Proxy-Clause] /* See Oracle Documentation */

Drop-User ::=
DROP USER <User> [CASCADE]
```

A4.9.5 Syntax for Creating, Altering, or Dropping Role(s)

Create-Role ::=
CREATE ROLE [<Schema>.] <RoleName>
[NOT IDENTIFIED] |
[IDENTIFIED BY <Password>] |
[IDENTIFIED USING [<Schema>.] <Package>] |
[IDENTIFIED EXTERNALLY | GLOBALLY];

Alter-Role ::=
ALTR ROLE [<Schema>.] <RoleName>
[NOT IDENTIFIED] |
[IDENTIFIED BY <Password>] |
[IDENTIFIED USING [<Schema>.] <Package>] |
[IDENTIFIED EXTERNALLY | GLOBALLY];

Drop-Role ::=
DROP ROLE [<Schema>.] <RoleName>;

A4.9.6 Syntax for Granting or Revoking Privilege(s)

Grant ::=
GRANT [ALL PRIVILEGES] | [<Privilege> | <Role> {,<Privilege> | <Role>}]
[(<Column> {,<Column>}] [ON <Object Name>]
To <User> | <Role> | PUBLIC {,<User> | <Role>}
[WITH ADMIN OPTION] [WITH GRANT OPTION];

Revoke ::=
REVOKE ALL|<Privilege>|<Role> {,<Privilege>|<Role>} [On <Object Name>]
FROM <User>|<Role>|PUBLIC {, <User>|<Role>} [CASCADE CONSTRAINTS];

Appendix 5: Sample Exercises and Examination Questions

This final chapter of sample exercises, examinations, and case studies will proceed under the following captions:

- Introduction
- Sample Assignment 2A
- Sample Assignment 3B
- Sample Assignment 4A
- Sample Assignment 5A
- Sample Assignment 6A
- Sample Assignment 7A
- Sample Assignment 8
- Sample Assignment 9

A5.1 Introduction

This chapter provides you with some sample questions and case studies, designed to help you solidify in your mind, the concepts and principles covered in the course. The problems are arranged in assignments. The intent is to enlighten you on the type of problems that are likely to be asked to solve. The suggested weight (in points) for each question is indicated immediately following the question in curly brackets.

No solution is provided for the problems posed in this chapter, for the following reasons:

- The problems are intended to test your understanding of the materials covered. If you find that you are struggling with the solution to a problem, then you need to review the relevant sections before continuing.
- In some cases (particularly where you are asked to design a database), there may be more than one solution to the problem(s) posed. This is typical in software engineering as well as in database systems, where you analyze alternate solutions to a problem, and choose the most prudent one.
- If you are using this book as a prescribed text for a course in database systems, your professor will want to have a say in what questions you ought to focus on.

Solutions to the questions will be available on the resource website provided by the publisher. For best results, try developing a solution on your own before consulting the posted solutions.

A5.2 Sample Assignment 2A

Zealot Industries Inc. is a manufacturing firm with over 2000 employees. The firm has been expanding well beyond the founders' expectations. However, it faces a severe hindrance to additional expansion – the absence of a computerized Human Resource Management System (HRMS) that could facilitate management of human resource-related issues.

Precision Software Inc. is a software engineering firm contracted to develop and implement the HRMS project. After a preliminary meeting with Zealot's senior management, the lead software engineer on the Precision Software team documents his findings as follows:

- A detailed employee profile consists of the employee's personal information, employment history, education history, and beneficiary information.
- Every employee belongs to a department and is assigned a specific job description (there may be several employees with the same job description).
- Employees are classified according to the salary range that they fall in.
- Each employee has a compensation package which outlines basic salary and other benefits. Benefits include health insurance, retirement plan, life insurance, professional development allowance, housing allowance, traveling allowance, entertainment allowance, vacation allowance, and education allowance.
- A payroll log keeps track of remuneration paid out to employees.
- Each employee is assigned to at least one project; the number of employees assigned to a project depends on the size and complexity of the project.

1. Identify all information entities (object types) mentioned or implied in the case. {10}
2. Identify all relationships among the entities (object types) and represent them on an ERD or ORD. {20}
3. Use the information in the case combined with the information conveyed by the E–R diagram and your own intuitive design skills to propose relations and their associated attributes for each of the entities represented on the E–R diagram. You may also introduce additional relation(s) to take care of M:M relationship(s) that may exist. {60}

Total suggested points for this assignment: 90

A5.3 Sample Assignment 3B

1. What are the possible advantages of using distributed processing in a database system environment? State at least six advantages. {06}
2. What are the problems that can be experienced from having non-normalized tables in a database system? {06}
3. Relation R(A B C . . . K) is in 1NF only. The following FDs hold:
   FD1: Primary key is (ABC)
   FD2: A D
   FD3: B E
   FD4: F G,H,I
   FD5: J B
   FD6: K C
   FD7: (AF) > J/K

   By repeatedly using Heath's or Fagin's theorems, decompose R into a set of 5NF relations. Show how the new relations will be keyed. {15}
4. The following atomic data elements were taken from the data dictionary of an inventory system:

Item Number [ITM#]	Item Name [ITMNAME]
Item Last Price [ITMLP]	Item On Hand Quantity [ITMQOH]
Item Average Price [ITMAP]	Item Category Code [ITMCTG]
Category Description [CTGDES]	Item Supplier [ITMSPLR]
Order# [ORD#]	Order Date [ORDDATE]
Item Ordered [ORDITM]	Quantity Ordered [ORDQTY]
Order Status [ORDSTS]	Invoice Number [INV#]
Invoice Received Date [RCVDATE]	Item Received [RCVITM#]
Quantity Received [RCVQTY]	Invoice's Related Order [INVORD#]
Requisition Date [RQSDATE]	Requisition Number [RQS#]
Department Requesting [RQSDEPT]	Issue Date [ISSDATE]
Requisition Honor Date [RQSHDATE]	Quantity Issued [ISSQTY]
Issue Number [ISS#]	Item Issued [ISSITM]
Department Receiving the Issue [ISSDPT]	Item Issue Price [ISSPRC]

   Group the attributes into related entities. Then for each entity, identify the FDs, and then normalize until you obtain a set of normalized relations. Grade will be assigned as follows:

- Identification of all possible FDs. {11}
- Putting the elements into normalized relations. You may introduce elements as required with appropriate explanation. {44}

5. Construct an ERD representing the case in Question 4 with its normalized relations. {30}

Total suggested points for this assignment: 112

A5.4 Sample Assignment 4A

1. Define the following terms:
   a. Candidate key {02}
   b. Foreign key {02}
   c. Functional dependence {03}
   d. Transitive dependence {03}

2. Give the primary key and the highest normal form of each of the following relations. State any assumptions made and give reasons for your answers.
   a. **Student** {Student#, Name, Address, Gender, DoB} {03}
   b. **Flight** {Flight#, Date, Seats_Available, Flight_Time, From_City, To_City} {03}
   c. **Account** {Account#, Customer#, Customer_Name, Balance} {03}
   d. **Supply** {Part#, Quantity, Supplier#, Supplier_Address} {03}

3. In the relation **Participation**, attributes are Manager, Project, Hours (spent on project per month), Salary. A manager's salary is fixed, and he/she can work on many projects.
   a. Draw an FD diagram of the relation. {03}
   b. Would you store **Participation** as defined? Justify your answer. {04}
   c. Provide some sample data for the relation. {04}
   d. What is the highest normal form of **Participation**? {03}
   e. Show how you would design a conceptual schema to store the data mentioned for relation **Participation**? {11}

4. A scientific research establishment organizes its work in projects, each project consisting of several experiments. For each experiment, the following is recorded:
   - Project number
   - Project manager
   - Project name
   - Experiment number
   - Experiment name
   - Lab in which experiment is to be conducted
   - List of scientists to work on the project

   Note: A project manager may work on an experiment. A scientist may work on more than one experiment, provided that they are done in the same lab. A manager may manage more than one project.

   a. From the information given, develop an ERD. {10}
   b. Using the XR model of Chapter 5 (Section 5.5), derive a set of BCNF relations for the system. Use self-explanatory relation and attribute names (you may introduce new attributes). {20}
5. Figure A5.1 provides a snapshot of a live database. Primary keys are highlighted. Based on the data, derive an E–R diagram (state any assumptions made). {10}

Warehouse		
WhNo	**City**	**Size (Sq.ft.)**
WH1	Seattle	37,000
WH2	New York	50,000
WH3	Miami	20,000
WH4	Boston	13,000
...	...	...

Employee		
WhNo	**EmpNo**	**Salary ($US)**
WH2	E1	42,000
WH1	E3	61,000
WH2	E4	65,000
WH3	E6	63,000
WH1	E7	55,000
...	...	...

Supplier		
SNo	**Sname**	**Location**
S3	Wilson	Jamaica
S4	Barnes	USA
S6	Jones	UK
S7	Lewis	Singapore
...	...	...

PurchaseOrder			
EmpNo	**SNo**	**OrderNo**	**OrderDate**
E3	S7	OR67	840623
E1	S4	OR73	840728
E7	S4	OR76	840525
E6	S6	OR77	840619
E3	S4	OR79	840623
E1	S6	OR80	840729
E3	S6	OR90	850622
E3	S3	0R91	850713
...	...		...

Figure A5.1 Snapshot of a database.

Total suggested points for this assignment: 85

A5.5 Sample Assignment 5A

Figure A5.2 provides basic specifications for a music database. Carefully analyze the figure and answer the questions that follow.

Music Database Relations
Musicians {Mno, MName, MDoB, MCountry} PK [Mno]
Compositions {Cno, CTitle, CMno, CDate} PK [Cno]; FK [CMno references **Musicians**]
Ensembles {Eno, EName, ECountry, EMnoMgr} PK [Eno]; FK [EMnoMgr references **Musicians**]
Performances {Pno, PDate, PCno, PCity, PCountry, PEno} CKs [PDate, PEno], [Pno] ; FK [PCno references **Compositions**]; FK [PEno references **Ensembles**]
Ensemble_Members {EmEno, EmMno, EmInstrument} PK [EmEno, EmMno]; FK [EmEno references **Ensembles**]; FK EmMno references **Musicians**]

Figure A5.2 Basic specifications music database.

1. Prepare a database specification — either a RAL (of Section 3.9) or an ESG (of Section 5.8) showing database-related details only — for the database. This database specification may be refined by introducing a sixth relational table and adjusting three other tables to each having a foreign key that references this additional table.
   a. Identify the additional table that is required and clearly describe the adjustments that need to be madeto three tables in order to have a normalized database. {08}
   b. Propose an ESG or RAL that provides specs for the six relational tables of the database. {36}
   c. Provide some sample data for the music database. Your data should demonstrate that you understand the important role of foreign keys. {10}
2. Write relational calculus statements to realize the following:
   a. List the registered musicians from USA or JAM (where "USA" and "JAM" are abbreviated codes for United States and Jamaica, respectively). {03}
   b. Give the Eno & EName of every ensemble that includes a SAXAPHONE or CLARINET player. {03}
   c. Give the Eno & EName of ensembles that include a SAXAPHONE but not a CLARINET player. {04}
   d. List all compositions (Cno and CTitle) by MOZART {04}
   e. List all performances (Pno, Cno, Mno, & PCountry) of compositions in the country of origin. {06}
   f. Give the Eno & EName of every ensemble that includes a SAXAPHONE or CLARINET player, but not both. {06}
   g. Find Cno & CTitle for compositions all of which have been performed in USA. {03}
   h. List countries in which MOZART's compositions have been performed. {03}
   i. Give EName of ensembles whose manager is JAMAICAN. {03}
3. Write relational algebra statements corresponding to each of the relational calculus statements of Question #2. {35}

Total suggested points for this assignment: 124

A5.6 Sample Assignment 6A

The (cross section of the) college database introduced in Chapter 7 (and discussed in subsequent chapters) is repeated here (Figure A5.3).

Relation	Attributes	Primary Key	Foreign Key
Student	{Stud#, StudFName, StudLName, StudSex, StudAddr, StudPgm#, StudHall#, StudDoB ...}	[Stud#]	StudPgm# references **AcademicProgram.Pgm#** StudHall# references **Hall.Hall#**
AcademicProgram	{Pgm#, PgmName ...}	[Pgm#]	None
Hall	{Hall#, HallName ...}	[Hall#]	None
Department	{Dept#, DeptName, DeptHead#, DeptDiv#}	[Dept#]	DeptHead# references **Employee.Emp#** DeptDiv# references **Division.Div#**
Employee	{Emp#, EmpName, EmpDept# ...}	[Emp#]	EmpDept# references **Department.Dept#**
Course	{Crs#, CrsName, CrsDept# ...}	[Crs#]	None
Pgm_Struct	{PSPgm#, PSCrs#, PSCrsSeqn}	[PSPgm#, PSCrs#]	PSPgm# references **AcademicProgram.Pgm#** PSCrs# references **Course.Crs#**
Division	{Div#, DivName, DivHead# ...}	[Div#]	DivHead# references **Employee.Emp#**

Each relation and each attribute would need additional clarification prior to database construction and table creation.

Figure A5.3 Relation–attributes list for cross section of a college database.

1. Based on the principles of database design discussed in Lectures 4 and 5, construct an ESG of the college database for the entities specified above. {50}
2. Write SQL statements to create these tables in your schema of the class database. You may add additional attributes to the structure of each database table, as you deem appropriate. Store these statements in an SQL script file. {40}
3. Use your SQL script file to create the tables in the class database. {16}
4. Populate your tables with sample data (at least six records per table). {16}

Total suggested points for this assignment: 122

A5.7 Sample Assignment 7A

Figure A5.2 is repeated as Figure A5.4 for ease of reference. Carefully analyze the figure and answer the questions that follow.

Music Database Relations
Musicians {Mno, MName, MDoB, MCountry} PK [Mno]
Compositions {Cno, CTitle, CMno, CDate} PK [Cno]; FK [CMno references **Musicians**]
Ensembles {Eno, EName, ECountry, EMnoMgr} PK [Eno]; FK [EMnoMgr references **Musicians**]
Performances {Pno, PDate, PCno, PCity, PCountry, PEno} CKs [PDate, PEno], [Pno] ; FK [PCno references **Compositions**]; FK [PEno references **Ensembles**]
Ensemble_Members {EmEno, EmMno, EmInstrument} PK [EmEno, EmMno]; FK [EmEno references **Ensembles**]; FK EmMno references **Musicians**]

Figure A5.4 Basic specifications for music database.

1. As mentioned in Section A5.5, this database specification may be refined by introducing a sixth relational table and adjusting three other tables to each having a foreign key that references this additional table. Reiterate your database specification from Assignment 5A (Section A5.5) — either a RAL (of Section 3.9) or an ESG (of Section 5.8) showing database-related details only — for the database.
   a. Identify the additional table that is required and clearly describe the adjustments that need to be made to three tables in order to have a normalized database. {08}
   b. Propose an ESG or RAL that provides specs for the six relational tables of the database. Be sure to clearly indicate an appropriate data type and length for each attribute. {36}
2. Create the tables and insert some sample data for the music database:
   a. Write appropriate SQL statements to create the tables specified in ESG or RAL; be sure to include important integrity constraints where relevant. Use your SQL statements to create the tables. {36}
   b. Insert sample data into your database tables (at least six records per table). Your data must illustrate the role of foreign keys and referential integrity. {18}
3. Write SQL statements to realize the following:
   a. List all registered musicians from USA or JAM (where "USA" and "JAM" are abbreviations for United States and Jamaica). {03}
   b. Give the Eno & EName of every ensemble that includes a SAXAPHONE or CLARINET player. {03}
   c. Give the Eno & EName of every ensemble that includes a SAXAPHONE but not a CLARINET player. {04}
   d. List all compositions (Cno and CTitle) by MOZART {04}
   e. List all performances (Pno, Cno, Mno, & PCountry) of compositions in the country of origin. {06}
   f. Give the Eno & EName of every ensemble that includes a SAXAPHONE or CLARINET player, but not both. {06}
   g. Find Cno & CTitle for compositions that have been performed in USA. {03}
   h. List countries in which MOZART's compositions have been performed. {03}
   i. Give EName of ensembles whose manager is AMERICAN, JAMAICAN, or RUSSIAN. {03}
4. Define and create SQL views to realize the above requirements (of Question 3). {35}

Total suggested points for this assignment: 168

A5.8 Sample Assignment 8

This assignment requires you to work with your Oracle user account (which typically would be created by your course instructor or someone with DBA privilege). For each question, show the SQL statement(s) that you have used in order to address the question.

1. Create a test table called **JonesB_TestTable** (where "JonesB" represents your username) and populate it with some sample data. {10}
2. The next four instructions guide you through manipulation of selected object privileges on your test table as well as a test role that you will create.
   a. Create a role and for its name, use the concatenation of your user account and the word "Role." For instance, the role may be called **JonesB_Role** (where "JonesB" represents your username). {02}
   b. Grant to your role the system privileges that facilitate application development (see Section 13.2.2 for guidelines on this). {11}
   c. Review Section 13.2.3 on object privileges and grant SUDI privileges on **JonesB_TestTable** to your test role. {04}
   d. There is a catalog view that you can access to see all the system privileges granted to all roles in the system. Its name is **Role_Sys_Privs**. Study its structure and then issue an appropriate SQL statement to display the system privileges associated with your role. Show the SQL statement used. {06}
3. The next seven instructions guide you through a series of tests designed to help you develop a better understanding or how security works in the Oracle environment.
   a. Create a test user account called **JonesB_User** (where "JonesB" represents your username). Assign the user to your tablespace; also use a password that you will remember. Your current tablespace name can be determined by either asking your course instructor or examining the catalog view **User_Users**. {04}
   b. Try logging on to the database as **JonesB_User**. Record the result of your attempt and provide an explanation for this observation. {04}
   c. Log on with your normal user account and then grant your test role (**JonesB_Role** created in 2a) to test user **JonesB_User**. {02}
   d. Try logging on as **JonesB_User** once more and record the result of your attempt. Provide an explanation for this observation. {04}
   e. Log on as **JonesB_User** and try running a query on your test table. To do this, you must either create synonym for **JonesB_TestTable** in the schema of **JonesB_User**, and then run the query on the synonym, or you must qualify the table name when you run the query on **JonesB_TestTable.** Record the result of your attempt and explain why you obtained that result. {10}
   f. Switch to your normal user account and revoke the SUDI privileges on **JonesB_TestTable** from your role. {04}
   g. Log on as **JonesB_User** again and try running a query on your **JonesB_TestTable**. Record the result of your attempt. Provide an explanation for this result. {06}
4. Study your music database that you created for the previous assignment and reflect on what you have learned about the system catalog. Write SQL statements to show what the contents of the system catalog views **User_Tables** (attributes Table_Name and Tablespace_Name) and **User_Tab_Columns** (attributes Column_Name and Table_Name) would be. Include a screenshot of the query results. {10}

Total suggested points for this assignment: 77

A5.9 Sample Assignment 9

1. Identify and summarize five issues to be addressed in database administration. {12}

2. Review Chapter 17 and answer the following questions:
   a. Give the definition of a distributed database system. {03}
   b. What is the difference between database fragmentation and database replication? Describe a situation that would warrant each strategy. {04}
   c. State and clarify 4 of Christopher Date's 12 rules for distributed databases. {12}
   d. Discuss two challenges to distributed databases and alternate approaches for addressing each challenge. {08}

3. Review Chapter 18 and answer the following questions:
   a. Describe two challenges to OO database management systems. {06}
   b. Describe one hybrid approach that facilitates the peaceful coexistence of relational databases and object technology. {04}

4. Review Chapter 19 and answer the questions related to the following case: Boch Toyota is a leading automobile distribution company with branches scattered across the US mainland. The company is desirous of keeping track of its sale of multiple product lines and products at various locations on a monthly and yearly basis. Sale is tracked in terms of the number of units sold as well as the equivalent sale amount.
   a. Construct a database conceptual schema that best represents this scenario and illustrate it in an ERD. Your model may include additional attributes not mentioned in the case providing that you consider them essential. However, no additional entity should be introduced other than those implied or mentioned in the case. {20}
   b. This database schema appears in multiple operational databases operated by the company. However, it is desirable to preserve this schema for the entire organization over an extended time horizon, where information aggregated from the various operational databases will be stored. Based on what you have learned in this course, propose a relevant solution for this problem. {02}
   c. Using an appropriate diagram, propose a suitable architecture for your solution. Provide a brief clarification of the diagram. {12}

Total suggested points for this assignment: 83

Index

D

E

F

G

H

S

T

U

Printed in the United States
by Baker & Taylor Publisher Services